EXPLAINABLE AI IN HEALTHCARE IMAGING FOR MEDICAL DIAGNOSES

EXPLAINABLE AI IN HEALTHCARE IMAGING FOR MEDICAL DIAGNOSES

Digital Revolution of Artificial Intelligence

Edited by

TANZILA SABA

Research Professor, Associate Director of Research and Initiative Center, Leader, Artificial Intelligence and Data Analytics Lab, College of Computer and Information Sciences, Prince Sultan University, Riyadh, Saudi Arabia

AHMAD TAHER AZAR

College of Computer and Information Sciences, Prince Sultan University, Riyadh, Saudi Arabia; Automated Systems and Soft Computing Lab (ASSCL), Prince Sultan University, Saudi Arabia

SEIFEDINE KADRY

Department of Computer Science and Mathematics, Lebanese American University, Beirut, Lebanon; Department of Applied Data Science, Noroff University College, Kristiansand, Norway

ELSEVIER

ACADEMIC PRESS

An imprint of Elsevier

Academic Press is an imprint of Elsevier
125 London Wall, London EC2Y 5AS, United Kingdom
525 B Street, Suite 1650, San Diego, CA 92101, United States
50 Hampshire Street, 5th Floor, Cambridge, MA 02139, United States

Notices
Knowledge and best practice in this field are constantly changing. As new research and experience broaden our understanding, changes in research methods, professional practices, or medical treatment may become necessary.

Practitioners and researchers must always rely on their own experience and knowledge in evaluating and using any information, methods, compounds, or experiments described herein. In using such information or methods they should be mindful of their own safety and the safety of others, including parties for whom they have a professional responsibility.

To the fullest extent of the law, neither the Publisher nor the authors, contributors, or editors, assume any liability for any injury and/or damage to persons or property as a matter of products liability, negligence or otherwise, or from any use or operation of any methods, products, instructions, or ideas contained in the material herein.

ISBN 978-0-443-23979-3

For information on all Academic Press publications
visit our website at https://www.elsevier.com/books-and-journals

Publisher: Mara Conner
Acquisitions Editor: Sonnini Yura
Editorial Project Manager: Vaishali Panwar
Production Project Manager: Jayadivya Saiprasad
Cover Designer: Mark Rogers

Typeset by STRAIVE, India

Working together
to grow libraries in
developing countries

www.elsevier.com • www.bookaid.org

Contents

Contributors

Shahzad Akbar
Riphah College of Computing, Riphah International University, Faisalabad, Pakistan

Arefeh Amiri
Shahid Madani Hospital, Lorestan University of Medical Sciences, Khorramabad, Lorestan, Iran

Mayukha Mandya Ammangatambu
The University of Utah, Salt Lake City, UT, United States

Shirley Chellathurai Pon Anna Bai
Department of Computer Science and Engineering, Karunya Institute of Technology and Sciences, Coimbatore, Tamil Nadu, India

Nafiz Arica
Information Systems Engineering, Faculty of Engineering, Piri Reis University, Istanbul, Turkey

Noor Ayesha
Center of Excellenec in Cyber Security (CYBEX), Prince Sultan University Riyadh, Saudi Arabia; School of Clinical Medicine, Zhengzhou University, Zhengzhou, Henan, China

Ahmad Taher Azar
College of Computer and Information Sciences; Automated Systems and Soft Computing Lab (ASSCL), Prince Sultan University, Riyadh, Saudi Arabia; Faculty of Computers and Artificial Intelligence, Benha University, Benha, Egypt

Sundaravadivazhagan Balasubramanian
Department of Information Technology, University of Technology and Applied Sciences, Al Mussanah, Oman

Duygu Cakir
Software Engineering, Faculty of Engineering and Natural Sciences, Bahcesehir University, Istanbul, Turkey

Aayushi Chaudhari
U & P U Patel Department of Computer Engineering, Faculty of Technology (FTE), Chandubhai S. Patel Institute of Technology (CSPIT), Charotar University of Science and Technology (CHARUSAT), Changa, India

D.N. Disha
Department of Artificial Intelligence and Data Science, NMAMIT, Udupi, India

Mohammad Bagher Dowlatshahi
Department of Computer Engineering, Faculty of Engineering, Lorestan University, Khorramabad, Lorestan, Iran

Prisha Faldu
U & P U Patel Department of Computer Engineering, Faculty of Technology (FTE), Chandubhai S. Patel Institute of Technology (CSPIT), Charotar University of Science and Technology (CHARUSAT), Changa, India

Srđan Filipović
Department of Applied Data Science, Noroff University College, Kristiansand, Norway

Dulari Gajjar
U & P U Patel Department of Computer Engineering, Faculty of Technology (FTE), Chandubhai S. Patel Institute of Technology (CSPIT), Charotar University of Science and Technology (CHARUSAT), Changa, India

Rakesh Gnanasekaran
Department of Computer Science, Thiagarajar College, Madurai, Tamil Nadu, India

Sahar Gull
Riphah College of Computing, Riphah International University, Faisalabad, Pakistan

Weilin He
School of Automation Science and Engineering, South China University of Technology, Guangzhou, China

Aiesha Mahmoud Ibrahim
Department of Computer Science, College of Computer Science and Information Technology, University of Anbar, Anbar, Iraq

Sajid Iqbal
Department of Information Systems, College of Computer Sciences and Information Technology, King Faisal University, Al Ahsa, Saudi Arabia

Vishal Jain
KIIT School of Liberal Studies; KIIT School of Management, KIIT DU, Bhubaneswar; Department of Computer Science and Engineering, School of Engineering and Technology, Sharda University, Greater Noida, India

Seifedine Kadry
Department of Applied Data Science, Noroff University College, Kristiansand, Norway

Nashwa Ahmad Kamal
Faculty of Engineering, Cairo University, Giza, Egypt

Seema Kashyap
School of Computer Science and Engineering, IFTM University, Moradabad, Uttar Pradesh, India

Amjad R. Khan
Artificial Intelligence and Data Analytics (AIDA) Lab, CCIS Prince Sultan University, Riyadh, Kingdom of Saudi Arabia

Deep Kothadiya
U & P U Patel Department of Computer Engineering, Faculty of Technology (FTE), Chandubhai S. Patel Institute of Technology (CSPIT), Charotar University of Science and Technology (CHARUSAT), Changa, India

Bin Li
School of Automation Science and Engineering, South China University of Technology, Guangzhou, China

Tariq Mahmood
Artificial Intelligence and Data Analytics (AIDA) Lab, CCIS Prince Sultan University, Riyadh, Kingdom of Saudi Arabia

Elham Moeini
Computer Engineering, Graduate School of Engineering, Bahcesehir University, Istanbul, Turkey

Mazin Abed Mohammed
Department of Artificial Intelligence, College of Computer Science and Information Technology, University of Anbar, Anbar, Iraq

Nabaa Abd Mohammed
Department of Computer Science, College of Computer Science and Information Technology, University of Anbar, Anbar, Iraq

Iram Naim
Faculty of Engineering and Technology, MJP Rohilkhand University, Bareilly, India

Najmusseher
Department of Computer Science, CHRIST (Deemed to be University), Central Campus, Bangalore, Karnataka, India

Gnanasankaran Natarajan
Department of Computer Science, Thiagarajar College, Madurai, Tamil Nadu, India

Arpita Nayak
KIIT School of Liberal Studies; KIIT School of Management, KIIT DU, Bhubaneswar; Department of Computer Science and Engineering, School of Engineering and Technology, Sharda University, Greater Noida, India

P.K. Nizar Banu
Department of Computer Science, CHRIST (Deemed to be University), Central Campus, Bangalore, Karnataka, India

Harshini Reddy Penthala
The University of Texas at Arlington, Arlington, TX, United States

Abdulvahap Pinar
Rectorate Unit, Adıyaman University, Adıyaman, Turkey

Rizwan Qureshi
Fast School of Computing, National University of Computer and Emerging Sciences, Karachi, Pakistan; Center for Research in Computer Vision, The University of Central Florida, Orlando, FL, United States

Ferhat Sadak
Department of Mechanical Engineering, Bartin University, Bartin, Türkiye

Khadija Safdar
Riphah College of Computing, Riphah International University, Faisalabad, Pakistan

Ipseeta Satpathy
KIIT School of Liberal Studies; KIIT School of Management, KIIT DU, Bhubaneswar; Department of Computer Science and Engineering, School of Engineering and Technology, Sharda University, Greater Noida, India

Usama Shahzore
Riphah College of Computing, Riphah International University, Faisalabad, Pakistan

M. Shanmuga Sundari
Computer Science and Engineering, BVRIT HYDERABAD College of Engineering for Women, Hyderabad, India

K. Aditya Shastry
Department of Information Science and Engineering, Nitte Meenakshi Institute of Technology, Bengaluru, India

Armin Shoughi
Department of Computer Engineering, Faculty of Engineering, Lorestan University, Khorramabad, Lorestan, Iran

Arvind Kumar Shukla
School of Computer Science and Engineering, IFTM University, Moradabad, Uttar Pradesh, India

Yu Wang
Shandong Research Institute of Industrial Technology, Jinan, China

Fatma Hilal Yagin
Department of Biostatistics and Medical Informatics, Faculty of Medicine, Inonu University, Malatya, Turkey

About the editors

Tanzila Saba

Prof. Tanzila Saba is Associate Director of the Research and Initiative Center and Full Professor at the College of Computer and Information Sciences, Prince Sultan University (PSU), Riyadh, Saudi Arabia. She leads the Artificial Intelligence and Data Analytics (AIDA) Lab and is a senior member of IEEE, as well as an active member of ACM, AIS, and IAENG. Prof. Saba has been recognized in Marquis Who's Who in Science and Technology (2012) and has been consistently ranked among the world's top 2% of scientists (2020–2024) in a study by Stanford University and Elsevier. With more than 400 research publications, she has amassed nearly 21,534 citations and holds an h-index of 85. Her research spans medical imaging, machine learning, cybersecurity, and big data analytics, with a strong focus on AI-powered decision support systems, predictive analytics, and ethical AI development. Her outstanding research contributions have earned her multiple accolades at PSU, including the Research Excellence Award and the Highest Impact Researcher's Award for the academic years 2020–2024. Since December 2019, she has been recognized as one of Saudi Arabia's leading computer scientists. Additionally, as the PSU Women in Data Science (WiDS) ambassador at Stanford University, she actively promotes women's participation in STEM fields. Beyond academia, Prof. Saba is a passionate advocate for AI for social good, emphasizing ethical AI development, responsible data governance, and sustainable technology solutions to address real-world challenges. A highly sought-after speaker, she has delivered keynote addresses at numerous international conferences, sharing her expertise in AI, data science, and their applications in healthcare and beyond. She also serves as an editor and coeditor for several renowned peer-reviewed journals and plays a leading role in national and international conferences. In her administrative capacity, Prof. Saba has made significant contributions to curriculum development and quality assurance at PSU, with expertise in ABET and NCAAA accreditation processes. She is also actively involved in institutional strategic planning and research committees, further advancing the university's research and academic excellence.

Ahmad Taher Azar
College of Computer and Information Sciences, Prince Sultan University Riyadh, Saudi Arabia.
Automated Systems and Soft Computing Lab (ASSCL), Prince Sultan University, Saudi Arabia.

Prof. Ahmad Taher Azar is Full Professor at Prince Sultan University in Riyadh, Saudi Arabia, where he leads the Automated Systems and Soft Computing Lab (ASSCL). He serves as the Editor in Chief of the International Journal of Intelligent Engineering Informatics (IJIEI), published by Inderscience, United Kingdom. Additionally, he is an editor for several prestigious journals, including IEEE Transactions on Fuzzy Systems, IEEE Systems Journal, and IEEE Transactions on Neural Networks and Learning Systems. He also contributes to Springer's Human-centric Computing and Information Sciences and Elsevier's Engineering Applications of Artificial Intelligence. Prof. Azar's expertise encompasses artificial intelligence, machine learning, control theory, robotics, and computational intelligence. He has authored or coauthored over 550 research papers, book chapters, and conference proceedings in renowned peer-reviewed journals. His academic excellence has been recognized with numerous awards, such as the Benha University Prize for Scientific Excellence and the Egyptian State Encouragement Award in Engineering Sciences. In addition, he was elected as a senior member of the International Rough Set Society (IRSS) in 2018. Since December 2019, Prof. Azar has been acknowledged as one of the top computer scientists in Saudi Arabia. He received the Egyptian President's Distinguished Egyptian Order of the First Class in 2020. Furthermore, he was honored with the Abdul Hameed Shoman Arab Researchers Award in Machine Learning and Big Data Analytics in October 2020. Prof. Azar has been consistently recognized as one of the top 2% of scientists worldwide in Artificial Intelligence by Stanford University from 2020 to 2024. His outstanding research contributions have also earned him multiple awards at Prince Sultan University, including the Research Excellence Award and Highest Impact Researcher's Award for the academic years 2020–2024.

Seifedine Kadry

Department of Computer Science and Mathematics, Lebanese American University, Beirut, Lebanon; Department of Applied Data Science, Noroff University College, Kristiansand, Norway. He has a bachelor's degree in 1999 from Lebanese University, MS degree in 2002 from Reims University (France) and EPFL (Lausanne), PhD in 2007 from Blaise Pascal University (France), and HDR degree in 2017 from Rouen University. At present, his research focuses on data Science, education using technology, ystem prognostics, stochastic systems, and applied mathematics. He is an ABET program evaluator for Computing and Engineering Tech. He is Fellow of IET, IETE, and ACSIT. He is a distinguished speaker of IEEE Computer Society.

Preface

In recent years, the convergence of Artificial Intelligence (AI) and healthcare has begun to reshape the landscape of modern medicine. Among the many advances in AI, Explainable AI (XAI) stands out as a critical field of study, addressing an essential need for transparency, trust, and accountability. As AI models become increasingly integrated into medical imaging and diagnostics, the "black box" nature of many sophisticated algorithms has drawn attention to the fundamental question: How do we ensure that AI's decisions are understandable, traceable, and reliable? This is particularly crucial in healthcare, where the decisions made by AI can have life-altering implications for patients.

In healthcare, trust in AI systems cannot be built solely on the accuracy of predictions or classifications; it also depends on the ability of these systems to explain how they arrive at specific conclusions. XAI, with its focus on making AI decisions more interpretable, offers a promising solution by bringing transparency to the opaque mechanisms behind AI models. As AI continues to advance and find applications in precision medicine—where treatment plans and diagnoses are tailored to individual patients—the need for explainable, trustworthy AI is more pressing than ever. Clinicians, researchers, and policymakers are increasingly recognizing that the future of AI in healthcare hinges on the ability to provide interpretable insights, rather than mere results.

This book, *Explainable AI in Healthcare Imaging for Precision Medicine*, delves into the intersection of XAI and precision medicine, highlighting cutting-edge research, emerging challenges, and opportunities in this rapidly evolving field. With a focus on applications in healthcare imaging, the book explores how explainable AI can enhance patient outcomes, improve clinical decision-making, and foster greater collaboration between AI technologies and healthcare professionals. By illuminating the inner workings of AI models, XAI helps bridge the gap between complex algorithms and the clinicians who rely on them, ensuring that AI-assisted decisions are not only accurate but also understandable and actionable.

We have gathered insights from leading researchers and practitioners who are driving the development of XAI in healthcare, as well as reflections on future trends and open questions in this field. Whether you are a researcher, student, healthcare professional, or policymaker, this book aims to provide a comprehensive overview of XAI's pivotal role in shaping the future of precision medicine. It is our hope that the content herein will inspire further innovation, foster cross-disciplinary collaboration, and contribute to building more transparent, ethical, and effective AI systems for the benefit of patient care.

1. Objectives of the book

The primary objective of this book is to explore how Explainable AI (XAI) can enhance precision medicine by providing transparency, interpretability, and trust in AI-driven healthcare systems, particularly in medical imaging. It aims to present the latest research and applications of XAI in healthcare, address the challenges and opportunities in integrating explainability into AI models, and foster collaboration between AI researchers, clinicians, and healthcare professionals. The book also seeks to promote ethical and responsible use of AI technologies, ensuring that AI systems are accountable and interpretable in critical decision-making processes. Additionally, it provides a comprehensive resource for students, researchers, and professionals interested in the growing intersection of AI, explainability, and personalized medicine.

2. Organization of the book

This book, Explainable AI in Healthcare Imaging for Precision Medicine, is structured into 21 chapters, each addressing a crucial aspect of the rapidly evolving field of Explainable AI (XAI) in healthcare. The chapters are organized to provide readers with a progressive understanding, from foundational concepts to advanced applications and future perspectives in XAI for precision medicine.

3. Features of the book

This book, *Explainable AI in Healthcare Imaging for Precision Medicine*, offers several key features that make it a comprehensive and valuable resource for both experts and newcomers to the field:

- **Comprehensive coverage of XAI and precision medicine**: The book provides an in-depth exploration of Explainable AI (XAI) techniques, methodologies, and applications specific to healthcare, particularly in the context of precision medicine and medical imaging.
- **Latest research and innovations**: It includes the most recent research findings, technological advances, and innovations in XAI, helping readers stay up-to-date with cutting-edge developments in AI and healthcare.
- **Interdisciplinary approach**: The content bridges multiple disciplines, including AI, healthcare, medicine, ethics, and policy, fostering collaboration and understanding across fields crucial to the future of AI in healthcare.
- **Real-world case studies**: The book features numerous case studies and examples of XAI applied to real-world healthcare scenarios, helping readers understand practical implementations and challenges in clinical settings.

- **Ethical and regulatory insights**: It addresses the ethical, legal, and regulatory aspects of AI in healthcare, focusing on transparency, fairness, and accountability in AI assisted medical decisions.
- **Expert contributions**: Contributions from leading researchers, clinicians, and industry professionals provide diverse perspectives on how XAI can shape the future of healthcare and precision medicine.
- **Focus on patient-centered AI**: The book emphasizes how XAI can improve patient trust, communication, and engagement by making AI-driven decisions more understandable and transparent to both healthcare providers and patients.
- **Clear and accessible writing**: While technical in nature, the book is written in an accessible style, making complex AI concepts understandable for readers with varying levels of expertise, from students to professionals.
- **Future directions and trends**: Each chapter highlights emerging trends, future research directions, and unresolved questions in XAI and precision medicine, inspiring further innovation and exploration in the field.

These features make this book an essential guide for anyone seeking to understand the transformative impact of Explainable AI in healthcare, from academics and researchers to healthcare practitioners and policymakers.

Acknowledgments

The editors would like to thank Prince Sultan University, Riyadh, Saudi Arabia, for supporting this publication. The editors specially acknowledge the Automated Systems and Soft Computing Lab (ASSCL) at Prince Sultan University, Riyadh, Saudi Arabia, Artificial Intelligence and Data Analytics (AIDA) Lab, Prince Sultan University Riyadh, 11586, Saudi Arabia. Special acknowledgment to Noroff University College, Norway.

CHAPTER 1

Ensuring trust in healthcare robotics: The essential role of explainable AI

Ferhat Sadak[a] and Rizwan Qureshi[b,c]
[a]Department of Mechanical Engineering, Bartin University, Bartin, Türkiye
[b]Fast School of Computing, National University of Computer and Emerging Sciences, Karachi, Pakistan
[c]Center for Research in Computer Vision, The University of Central Florida, Orlando, FL, United States

1. Introduction

Developing medical robotics technologies, such as rehabilitation and social robots, has been a challenging but rewarding research endeavor, due to their significant potential benefits for various human-oriented applications [1]. For instance, rehabilitation robots are useful in healthcare for a variety of reasons. They provide precise help for recovering from physical abilities in elderly patients from degenerative diseases [2], or knee or hip replacement surgeries [3]. These robots may perform tasks that would be challenging for humans due to their medical conditions. For example, robotic exoskeletons can help people with lower limb disabilities walk and regain mobility [4]. They provide steady support and can be precisely adjusted to match the individual demands of each patient. Rehabilitation robots are also highly effective at giving repetitive and consistent therapy sessions, which are essential for rehabilitation.

Robotic physical therapy devices, for example, can assist patients restore strength and dexterity in a controlled and regular manner. This is especially helpful for stroke survivors since these robots can deliver constant and repetitive activities that are necessary for recovery [5].

Social robots in healthcare, on the other hand, offer a unique combination of interaction and assistance. These robots are intended to engage and empathetically communicate with patients. Social robots can provide emotional support and ease symptoms of social isolation for elderly people, who may feel isolated or lonely [6]. Furthermore, social robots can help patients for dementia rehabilitation in hospitals [7] as well as cognitive impairment [8]. Overall, rehabilitation and social robots play an important role in healthcare by improving patients' quality of life. These robots exemplify a synthesis of technological accuracy and human-like interaction, and their significance in healthcare, particularly in improving the lives of people in need.

Explainable AI in Healthcare Imaging for Medical Diagnoses
https://doi.org/10.1016/B978-0-443-23979-3.00001-4

The integration of artificial intelligence (AI) has enormous promise in the field of healthcare robots for enhancing the capacities of healthcare workers [9–11]. The National Academies of Science, Engineering, and Medicine, in collaboration with the Royal Society, published a paper that identified trust, transparency, and interpretability as major problems in developing, testing, and deploying AI systems [12]. According to studies, making the decision-making process of AI intelligible and explicit can help users trust these intelligent systems considerably [13]. The concept of "Trust in Robots," which can be defined as the level of trust and reliability that individuals, particularly in healthcare settings, place in robotic systems, where they have faith in the robots' ability to perform tasks accurately, safely, and in alignment with their intended purposes, while also understanding the robots' decision-making processes using explainable artificial intelligence (XAI) methods [14].

Before diving into the details of the concept of "Trust in Robots," it is critical to distinguish between trust and trustworthiness. Trust is defined as a person's feelings for an agent [14]. In contrast, trustworthiness is about the agent and is not influenced by the user's emotions [15]. This means that a person may not trust a trustworthy robot, and vice versa. In simpler terms, trust is about how you feel, and trustworthiness is about the characteristics of the agent. Even if a robot is trustworthy, you may not trust it at times, as illustrated in Fig. 1.

The inclusion of socially assistive robots (SAR) in rehabilitation is dependent on physicians and patients establishing trust. Clinicians must assess the dependability of these robots to ensure seamless inclusion into rehabilitation protocols since they are experts and end-users. Patient trust is based on the robots' ability to execute accurate activities and adjust to therapeutic demands. The clinician's confidence in SAR extends to entrusting the robot with the patient. As an example, it should follow the interdependency to

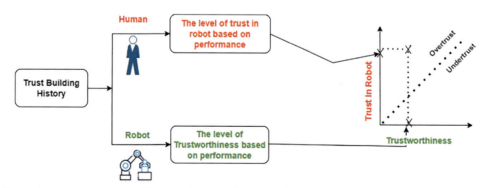

Fig. 1 A conceptual trust calibration diagram illustrates the outcomes of a divergence between a user's trust in the robot and the robot's true trustworthiness, potentially resulting in overtrust or undertrust.

the neurorehabilitation robot regulations defined based on three laws, as in [16]. This includes the robot's commitment to keeping the patient safe and rigorously following therapists' instructions. This is actually inspired by Isaac Asimov's famous "Three Laws of Robotics" [17]:

- A robot may not injure a human being or, through inaction, allow a human being to come to harm.
- A robot must obey orders given it by human beings except where such orders would conflict with the First Law.
- A robot must protect its own existence as long as such protection does not conflict with the First or Second Law.

Previous research has shown that using SARs for pediatric rehabilitation across numerous sessions increases physiotherapist confidence [18]. According to a study in [19], the level of trust in a specific technology is mainly reliant on recommendations from recognized individuals or colleagues. In conclusion, these data show that creating trust among practitioners in SARs entails ensuring the dependable protection of patients' privacy and transparently conveying the measures used to protect privacy. Furthermore, regular SAR use and good evaluations of SAR use from colleagues and therapists in the field may increase clinician trust. On the other hand, integrating SARs into care demands a thorough understanding of the factors that influence patient trust, particularly in the setting of rehabilitation. Patients undergoing rehabilitation, who are frequently dealing with a variety of impairments, demonstrate heightened vulnerability, emphasizing the importance of perceived safety and past technological experience in shaping patient-SAR confidence during rehabilitation. In the rehabilitation setting, where familiarity with robots may be lacking, studies demonstrate that prolonged interactions contribute to heightened trust in social robots [20, 21].

Hence, it is shown that the type of embodiment had no effect on the amount of trust. However, time proved to be an important component, with a considerable increase in user trust observed throughout the trial. AI integration in the context of SARs holds significant promise for advancing rehabilitation procedures. SARs powered by AI can improve flexibility and responsiveness by adapting treatment approaches to individual patient demands in real time. This increased level of customization, along with enhanced learning capabilities, has the potential to have a substantial influence on both patient and physician trust. The intuitive and adaptive nature of AI-enhanced SARs may provide patients with confidence, as these systems can continuously change to meet changing rehabilitation needs. Clinicians may be more trusting of AI-assisted SARs as a result of their capacity to deliver precise, data-driven insights regarding patient progress and therapy success. The incorporation of AI thus constitutes a transformative step, magnifying the potential for SARs in rehabilitation not only by maximizing patient outcomes but also by strengthening patient and clinician trust in the efficacy of these advanced technical solutions.

However, the most successful AI models have a significant challenge because of their high dimensionality, nonlinearity, and inherent complexity making them difficult to understand by human specialists [22]. These limitations become a crucial matter where transparency and explainability are not only important but are also required for legal and ethical reasons. In robotic healthcare applications, the rationale behind AI-driven decisions often holds greater importance than the outcomes themselves. To solve this difficulty, the field of XAI has grown in prominence, notably within the framework of healthcare robotics [23]. XAI aims to ensure that human experts can understand and provide a solid rationale for machine decisions, going beyond traditional evaluation metrics such as accuracy, precision, recall, etc. [24].

Building trust to use automated systems in healthcare requires fulfillment of several key parameters, including stability, responsibility, government compliance, and explainability at each step of the project. Notably, XAI, a type of AI that explains the reasoning behind its actions, has the potential to have a substantial impact on healthcare robots. Its potential is to provide clear and understandable explanations for medical diagnoses and suggestions [25], hence improving the model's decision-making mechanism. XAI methods increase robustness and interpretability, which leads to stability and contribute to trust building, as shown in Fig. 2.

This transparency, in turn, helps patients to have a more nuanced awareness of their health situations, allowing them to make informed healthcare decisions. Furthermore, the incorporation of XAI has the potential to improve the interaction dynamics between

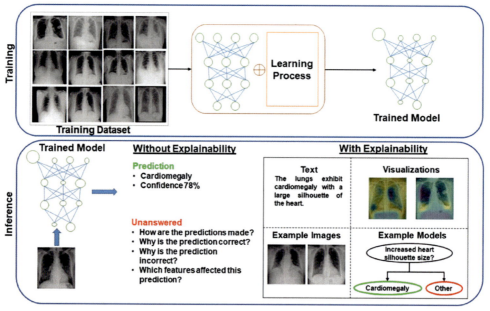

Fig. 2 XAI assists stakeholders in understanding the model's decision-making process [26].

healthcare robots and patients. By allowing these robots to explain their decisions transparently, a foundation of confidence and credibility is formed, along with precise information on recommended therapies [27]. In this regard, a major aspect of XAI is strengthening the safety procedure of healthcare robots. The ability of these entities to describe the reasoning behind their decisions and actions not only mitigates the potential harm caused by erroneous decisions but also allows for more effective interactions with users [23]. In conclusion, incorporating XAI delivers numerous benefits for healthcare robots. These robots can encourage trust, improve healthcare outcomes, and contribute to safety in the healthcare technology area by providing transparent explanations for decisions and actions. In Fig. 3, we show the connections between various components of XAI to build trust in healthcare robotics.

Fig. 3 Relations among XAI concepts. The explainability concepts usually seek to accomplish one or more goals with the explanations they produce [28].

In this chapter, we also comprehensively examine the challenges in AI-integrated healthcare robotics, such as encompassing ethical concerns, safety considerations, and the inherent complexity of AI systems, which are thoroughly discussed. Following this analysis, the exploration will shift toward solutions for fostering trust in healthcare robotics through the lens of XAI. This section will delve into the potency of XAI and its real-world benefits, shedding light on how it addresses ethical, safety, and complexity issues. Subsequently, the focus will turn to the exploration of emerging trends and the future landscape of trust in healthcare robotics. This will include discussions on advancements in XAI and hardware, evolving ethical and regulatory frameworks, and a visionary perspective on the future shaped by trust-based healthcare robotics. The concluding section will provide a synthesis of key insights derived from the exploration of challenges, solutions, and future trajectories in the integration of XAI within healthcare robotics.

2. Challenges in AI-integrated healthcare robotics

The integration of AI with healthcare robotics opens up a world of exciting possibilities while also posing numerous problems. This section addresses the specific challenges encountered in the field of AI-integrated healthcare robotics, examining ethical and safety concerns, and the inherent complexities of AI-based systems.

2.1 Ethical concerns

It is worth noting that the integration of AI and robots has the potential to assist the healthcare industry in a variety of ways, including well-being, diagnosis, and individualized care. Nonetheless, the deployment of these technologies must not violate ethical principles or threaten user privacy and safety. As a result, resolving the ethical challenges raised by the integration of AI and robotics, particularly in the healthcare industry, demands new authority initiatives. Existing frameworks for assessing and regulating medical devices are not appropriately suited to include AI-based systems, such as robots. As a result, the European Commission advocates for the monitoring of these technologies and the establishment of corresponding frameworks to determine the potential liabilities of various stakeholders such as medical experts, patients, or even industrial companies involved in the use of such technology. For this purpose, specific regulation at the intersection of AI and robotics from an ethical point of view was published in 2020 [29], followed by further amendments, especially in the use of AI in 2023 [30].

Robots frequently interact with vulnerable individuals [31]. It is critical to understand the concept of responsible robotics to manage this appropriately. According to a study in [31], responsible robotics includes legal and ethical considerations in addition to developing and using robots. This larger perspective resulted in the establishment of the

Foundation for Responsible Robotics (FRR), established in the Netherlands in 2016, which provides assessment guidelines. The FRR prioritizes fundamental concepts such as security, safety, privacy, and transparency. These concepts serve as the foundation for assessing responsible robotics activities. Essentially, the FRR stresses not only the technical components of robotics initiatives but also the legal and ethical dimensions, ensuring a well-rounded evaluation of robotics tasks. As technology progresses, the necessity to balance innovation and ethics becomes clearer.

In another study, researchers looked into the integration of autonomy in healthcare robots [32]. In this context in [32], autonomy refers to a robot's ability to do tasks independently, without constant human guidance. Notably, not all healthcare robots operate autonomously; for example, surgical robots are often controlled by surgeons. Nonetheless, there is a determined effort in research to improve robot self-sufficiency. The term "autonomous systems" is significant here, emphasizing the desire for robots capable of doing tasks without constant human intervention. This endeavor, however, raises concerns about the future replacement of human workers by robots, posing issues of ethics and job placements.

Even if robots do not completely replace human workers, a fundamental issue arises: establishing the degree of autonomy, or the amount to which a robot should operate without continual human supervision. This becomes a critical concern, especially in sensitive areas such as pediatric therapy. It is critical to determine the level of supervision and the extent to which a robot should perform without direct human assistance. These ethical considerations highlight the significance of finding a delicate balance between giving healthcare robots' autonomy and preserving appropriate human interaction [33].

Considering the AI-based automation demand in healthcare [34, 35], this becomes a very significant issue. Based on this, the question arises "Can we trust robots" as explored in [36]. One of their paper's key points in [36] is that trusting technology, particularly robots, is presented as more than a need for control. The traditional emphasis on instrumental utility and individualistic trust frameworks may not adequately convey the complexities of human-robot relationships. This point of view promotes a paradigm change, pushing a more sophisticated understanding that goes beyond standard concepts of control and power relations. As these robots become more involved in medical decision-making, the ethical imperative increases to include the transparency and interpretability of the algorithms that control their activities. Making AI-powered healthcare robots explainable is critical not just for developing trust among patients and healthcare practitioners, but also for maintaining accountability and protecting against any biases in decision-making processes. The intersection of trust, ethics, and XAI highlights the importance of taking a comprehensive approach to designing the future landscape of healthcare technology [37].

2.2 Safety concerns

The incorporation of AI and robotics into healthcare services is projected to introduce new safety and risk management challenges. These issues may emerge during routine operations and can be linked to a variety of sources such as design flaws, programming errors, configuration disparities, or data preparation deficiencies. To ensure the safe and effective use of AI and robotics in healthcare delivery, it is critical to address these concerns in advance [38]. As another example concerning safety in AI-based robots, robots interacting directly with patients in healthcare settings might expose problems with security, potentially leading to failures that pose risk to both patients and professionals. The injury could be caused directly by the robot's activities or indirectly by providing inaccurate feedback to a surgeon, for example. Furthermore, in cases of unexpected robot activity, users may have difficulty determining if the robot is operating as planned or is susceptible to a security breach. This uncertainty highlights the significance of resolving security concerns to protect both the dependability of robotic systems and the well-being of humans getting involved in healthcare tasks [39].

The European Commission developed a regulatory framework concentrating on safety concerns associated with AI, which goes beyond the realm of healthcare, in 2021 [40]. The major goal is to increase the trust and safety of AI systems. Risks are classified into four levels in this framework: unacceptable risk, high risk, limited risk, and minimal risk. AI systems found to pose unacceptable potential risks will be prohibited. Before they are released, high-risk AI systems are subjected to severe standards, including detailed risk assessments and mitigation procedures, as well as the traceability of outcomes. Chatbots used in telemedicine, for example, will require "labeling" to alert consumers that they are interacting with an AI-powered system. This approach strives to regulate and assure responsible AI development and deployment across multiple sectors, while also fostering user safety and confidence. In 2023, a new EU regulation was developed, which is more restrictive [41] than the published one in 2021. The AI Act uses a risk-based approach to categorize AI systems using a four-tier paradigm. This classification is based on the risks that these systems may pose to users and potential third parties. The severity of risks associated with the AI system increases the level of regulatory requirements. Essentially, the framework tailors its restrictions to each AI system's individual risk profile, ensuring that more stringent controls are applied to those offering greater risks. This approach shows a complex and flexible regulatory strategy for dealing with the numerous difficulties offered by AI applications. The demonstration of EU AI Act based on the level of risk is depicted in Fig. 4.

2.3 Complexity of AI systems

The incorporation of AI into healthcare robotics adds another layer of complexity to AI-based healthcare robotic systems. As AI-based technologies become more integrated into

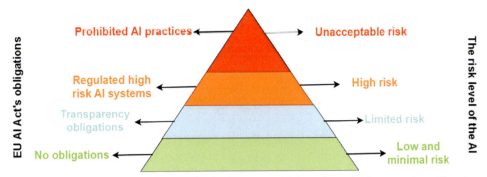

Fig. 4 The EU AI Act is demonstrated, which uses a risk-based method to regulate AI systems depending on their level of risk.

healthcare procedures, their inherent complexity raises a new set of challenges that must be carefully considered. AI relies on complex algorithms and models to manage and evaluate data, necessitating significant processing capacity and specialized programming techniques. Robotics, on the other hand, comprises the development and integration of physical components such as sensors, actuators, and control systems in order to create operational robots. This task necessitates a thorough understanding of both AI and robotic systems. The goal is to ensure that the robots are not only automated but also intelligent and capable of learning and adapting things in the complex clinical environment.

To begin with, visual servoing is a control approach that uses visual information to guide robotic system movement and actions. It entails employing camera or other visual sensor feedback to control the robot's position, orientation, or both. The main task is to increase the robot's autonomy by allowing it to react in real time, and adapt to changes in its environment. AI-based visual servoing is important in the field of healthcare robotics because it provides precision, versatility, and adaptability. In a recent study, an accelerated recurrent neural network (RNN) is employed [42] for visual servo control of a limited robotic flexible endoscope. Their robotic system [42], which consists of a patient-side manipulator and a flexible endoscope, is subjected to kinematic modeling in order to simplify visual servoing. The suggested method uses a quadratic programming control framework that takes into account both kinematic and physical restrictions, with the RNN accelerated for finite-time convergence using a sign-bipower activation function [43].

A portable robotic system designed for autonomously placing needles and catheters into deformable tissues, such as blood vessels, was introduced in [44]. The robotic cannulation is guided by predictions from deep convolutional neural networks, processing multimodal image sequences for real-time servoing. They demonstrate the device's ability to segment, identify, locate, and track peripheral vessels despite anatomical heterogeneity and motion in volunteers through imaging and tracking tests. Their device outperformed manual cannulations in success rates and procedure times, even under challenging conditions. Another study describes a real-time photoacoustic visual servoing system for robotic surgical and interventional operations. This imaging technology is also currently widely adapted within the microrobotic community to guide microrobots in an automated manner in various life science applications [45]. Unlike traditional approaches that rely on image generation and segmentation, this technology directly uses raw acoustic sensor data for robot path planning, allowing for effective tool tip monitoring and maintenance [46]. Their findings highlight the benefits of incorporating deep learning into robotic systems for better automation and visual tool tip servoing.

The incorporation of AI into robotic systems, particularly in healthcare applications, adds a layer of complexity due to the advanced algorithms and models used. Simultaneously, robotics requires a thorough understanding of both AI and robotic systems due to the deep integration of physical components such as sensors, actuators, and control systems. The goal of this fusion is to produce intelligent and adaptable robots that can learn and respond to their surroundings. As demonstrated from examples in the literature, visual servoing, an important feature of healthcare robotics automation, improves precision and adaptability. However, difficulties arise when decision-making processes are transparent, particularly in the absence of XAI. Visual servoing, which uses AI algorithms to make real-time decisions, adds another layer of complexity. The essence of the issue is the potential lack of transparency, which prevents a clear understanding of why a robotic system makes specific judgments. To solve this, XAI integration comes as a critical solution. XAI approaches shine light on the underlying workings of AI models, making the complex algorithms and the need for transparency in decision-making processes more understandable. This harmonic integration attempts to achieve a careful balance between the inherent complexities of AI-based robotic systems and the requirement for transparency.

3. XAI solutions for trust healthcare robotics

XAI helps build trust by opening the "black box" nature of AI algorithms. Recently, due to computational advancements, most of the state-of-the-art AI solutions are based on deep learning, which are even more complex, and difficult to interpret. In this context, the role of XAI becomes even more critical [47]. XAI allows users to comprehend not only the final output but also the underlying factors and patterns considered by the AI system [48]. This transparency helps healthcare professionals validate and verify the decisions made by AI systems, leading to increased confidence and trust in their capabilities. Here are some XAI methods:

- **Exploratory data analysis**: In most of the health data-science projects, the fundamental step is exploratory data analysis, such as its dimensionality, mean, standard deviation, range, distribution, correlation, and missing samples for input data and extracted features. Moreover, data visualization tools also not only help us design a transparent model for a downstream task, but also aids in building a new hypothesis [49].

Real-world healthcare robotics datasets are challenging and high dimensional. Visualizing data with dimensions beyond human perception can pose a difficulty in visualization. Overcoming this challenge involves utilizing specialized charts, such as parallel coordinate plots [50], which allow individuals to comprehend data with multiple dimensions. To simplify a high-dimensional dataset without losing its inherent structure, it can be projected onto a lower-dimensional representation. Principal component analysis [51] and t-distributed stochastic neighbor [52] embedding are two widely recognized methods in this field. These techniques aim to retain as much of the underlying structure as possible while reducing the dimensionality of the dataset.

- **Model explainability**: The traditional approach to constructing explainable models involves choosing interpretable modeling techniques, such as logistic regression [53], decision tress [54], rule sets [55], case-based reasoning [56], interpretable fuzzy systems [57], and generalized additive models [58]. Ref. [59] proposed three phases, algorithmic transparency, simulatability, and decomposability, to ensure interpretability. Algorithmic transparency aims to make the model's inner workings transparent and understandable. Simulatability emphasizes the ability to simulate and comprehend the model's behavior. Decomposability involves breaking down complex models into simpler components to enhance interpretability, such as ablation studies.
  - **Hybrid models**: To create a high-performance and explainable model, it is possible to combine an interpretable modeling technique with a sophisticated black-box method [60], such as DKNN [61].
  - **Self-explaining neural networks (SENN)**: The core idea of SENN is to extend a linear classifier by leveraging neural networks to learn the features, their corresponding coefficients, and the aggregation of networks for making predictions [62].
  - **Physics informed neural networks (PINN)**: PINNs [63] merge physical rules and constraints with the network's architecture. This integration of deep learning with physical modeling, process understanding, and domain knowledge enhances model interpretability and generalization.
  - **Knowledge distillation**: Knowledge distillation involves transferring information, often referred to as "dark knowledge," from a teacher network (e.g., a DNN) to a student network (e.g., a shallow NN) through model compression [64]. Knowledge distillation was first suggested to decrease a model's runtime computing cost, but it has subsequently been used to improve explainability [65].

- **Gradient-based methods**: Gradient-based methods utilize the gradients of the model's output with respect to its input features to quantify their importance [66]. Integrated gradients assign importance scores to input features based on the magnitude of their gradients. These scores can be used to generate explanations, highlighting the features that had the most influence on the model's prediction.
- **Saliency maps**: Saliency maps visualize the regions of an input image that are most relevant to the model's decision. They are derived by computing the gradient of the model's output with respect to the input image pixels [67]. The higher the gradient magnitude, the more important the corresponding pixel is for the model's prediction.
- **Layer-wise relevance propagation (LRP)**: LRP is a technique that propagates the relevance or importance scores backward through the layers of a deep neural network. It assigns relevance values to each neuron in the network, indicating their contribution to the final prediction. LRP can provide detailed explanations at the neuron level, helping to understand the model's decision-making process [68].
- **SHapley Additive exPlanations (SHAP)**: SHAP is a unified framework based on cooperative game theory and assigns importance values to each input feature [69]. It quantifies the contribution of each feature by considering all possible combinations of features and their impact on the model's predictions. SHAP values provide a comprehensive and theoretically grounded explanation of feature importance.
- **Sensitivity analysis (SA)**: SA is an alternative method to identify crucial input features, crucially influencing output across diverse domains like mutagenicity predictions, medical diagnosis, and ecological modeling [70, 71].
- **Local interpretable model-agnostic explanations (LIME)**: This approach involves creating local surrogate models to explain individual predictions made by black-box machine learning techniques. LIME specifically examines the impact on predictions when various types of data are inputted into the ML model [72].

XAI systems in healthcare robotics aim to enhance transparency and trust in AI-driven models and robotics. While there may be a trade-off between accuracy and interpretability, XAI techniques prioritize providing meaningful explanations without significantly compromising the overall performance of the AI system. The level of explanation and its impact on accuracy can vary depending on the chosen techniques and the complexity of the underlying AI model [73]. In Table 1, we present a nonexhaustive list of XAI methods with advantages and disadvantages. Table 2 presents software packages for XAI applications.

Table 1 XAI methods for ensuring trust in healthcare robotics.

XAI method	Advantages	Disadvantages	Applications
Model explainability	Insights into the inner workings of the model	May not capture complex interactions and limited to explaining the model, and not the underlying data	Reasoning behind diagnosis or treatments; identifying potential biases in the model's predictions
Local interpretable model-agnostic explanations (LIME)	Offers local explanations for individual predictions	Approximations may not capture the entire model behavior	Explaining individual predictions in robotic-assisted surgeries, trustworthiness for individual clinical decisions
Shapley values	Provides feature-level importance scores. Fairly attributes contributions to different features	Computationally expensive for large feature spaces. Requires considering different feature combinations	Assessing the impact of different sensor inputs on robotic diagnostics, identifying critical features affecting robotic treatment plans
Anchors	Generates concise if-then rules for explaining predictions. Easy to understand and interpret	May not capture complex decision boundaries. Limited to explaining specific aspects of the model's behavior	Explaining the rationale behind robotic assistance in patient monitoring, enhancing transparency in autonomous robotics
Counterfactual explanations	Generates alternative instances for desired outcomes	Requires defining appropriate distance metrics for counterfactual generation	Assessing the impact of changing input parameters on robotic treatment outcomes

Table 2 Software tools for XAI.

Software tool	Methods	Evaluation metrics
Shapash	Interpretable and understandable machine learning	Stability, consistency
OmniXAI	Visualization and grad-cam plots	NA
Alibi	Accumulated local effects	Trust score, linearity measure
AIX360	Boolean decision trees and others	NA
Skater	PDP plots, lime, and others	Interpretability and transparency

3.1 Accuracy trade-off

While XAI techniques aim to enhance interpretability and transparency, there can be a trade-off with the accuracy of the underlying AI model. This gives rise to the interpretability-performance trade-off, wherein a model's interpretability decreases as its performance increases, and vice versa [74]. Fig. 5 shows the trade-off between accuracy vs. transparency for various algorithms. Some explanation methods, such as simplifying complex models or using approximations, may result in a slight decrease in accuracy compared to the original model [75]. However, it is important to note that XAI techniques often prioritize interpretability without significantly compromising the overall performance of the AI system [76].

3.2 Explanation depth

The explanation level depends on the complexity of the underlying AI model and the chosen explanation technique. Some methods offer high-level explanations, such as feature importance rankings or global model summaries, which provide a broad

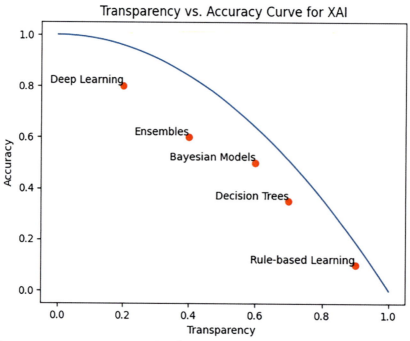

Fig. 5 The accuracy vs. transparency trade-off. As AI continues to reshape our world, the demand for XAI solutions with good accuracy is further expected to grow. Deep learning models (containing hundreds of millions of parameters) are less transparent, but more accurate. Rule-based or logistic regression are the most transparent, but there is a significant decrease in the performance (values in the curves are approximated).

understanding of the model's behavior. Other techniques, such as local explanations or visualizations, can provide more detailed insights into individual predictions or model components [77].

3.3 Contextual understanding

XAI systems provide explanations that are meaningful and understandable to humans [78]. However, it is important to recognize that AI models, especially deep learning models, often operate in a highly complex and abstract space. While XAI techniques can provide insights into the factors that contribute to the model's predictions, the explanations may not fully capture the nuanced reasoning and decision-making processes of the AI system. Therefore, it is crucial to interpret and utilize XAI explanations in the appropriate context, considering them as supportive tools rather than complete representations of the AI model's inner workings.

3.4 Post hoc explainability techniques

Post hoc approaches refer to methods or techniques applied after a machine-learning model has made predictions. These methods aim to provide interpretability and understanding of the model's decision-making process of a black-box model, such as deep neural networks. The post hoc models are explained by resorting to diverse means to enhance their interpretability, such as text explanations, visual explanations, and local explanations. Various techniques, such as text and visual explanations, local interpretations, explanations by example, explanations by simplification, and feature relevance analysis, are used to enhance the interpretability of post hoc models. These approaches provide insights into the models' decision-making processes, promoting transparency and trust in their predictions [79].

3.5 Evaluation metrics

To accurately assess the performance of XAI systems for healthcare robotics, it is crucial to employ appropriate metrics. A common challenge in measuring XAI systems arises from using the same evaluation metrics across different design objectives. Here, we discuss some of the common evaluation metrics.

- **Stability**: A stable XAI system should produce consistent explanations for similar inputs and maintain its performance despite variations in the data or the underlying model [80]. Stability is crucial in healthcare to instill trust and confidence in the XAI system's outputs and explanations.
- **Consistency**: Consistency involves ensuring that the explanations generated by the XAI system align with the underlying model's behavior and decision-making process [81]. Consistency is particularly important in healthcare to ensure that the XAI system's explanations are clinically meaningful and provide accurate justifications for its decisions.

- **Trust score**: A trust score measures the level of trustworthiness or reliability of an AI system and its explanations [14]. It provides a numerical value or score that indicates the degree of confidence users can place in the system's outputs and explanations. It is composed of several variables, including, accuracy, stability, consistency, transparency, and robustness.

3.6 The power of XAI

The power of XAI lies in its ability to bridge the gap between complex AI models and human understanding [82]. By providing explanations for AI-driven decisions and actions, XAI empowers users to trust, validate, and effectively collaborate with AI systems. XAI enables healthcare professionals to interpret the reasoning behind AI predictions, understand the factors influencing outcomes, and make informed decisions. It promotes transparency, accountability, and fairness in AI applications, fostering trust between humans and machines [83]. As the use of AI technology becomes more prevalent in practical applications, especially in situations where automated decisions could potentially impact human lives negatively, the demand for trustworthy AI is expected to grow.

3.7 Examples of XAI benefits in healthcare robots

Explainability in AI and robotics is becoming increasingly important in interdisciplinary discussions including machine learning, computer science, and ethics. This discussion provides a foundational look at the diverse terrain of real-world XAI advantages. Transparency is recognized as an important component for AI systems, and it is critical in unlocking the potential benefits of XAI in actual applications. This section looks into the larger context of real-world scenarios, demonstrating how including explainability improves the efficacy and trustworthiness of AI systems across multiple domains. Given our emphasis on XAI in the field of healthcare robots, we will use the term "Explainable robotics" to clarify our investigation of real-world XAI benefits in the healthcare robot domain. This word refers to the field of research and development that focuses on understanding and interpreting the actions and behaviors of robotic systems in a way that humans can comprehend and trust.

The primary goal is to bridge the gap between robots' complicated internal mechanics and humans' capacity to understand and effectively interact with them in healthcare applications. To demonstrate clearly what XAI could benefit in robotic research in the healthcare domain, we are going to first provide a hypothetical example that is discussed in [23]. In this scenario, a robot helps an elderly person with dementia by learning from a caregiver, gaining declarative and procedural information about the individual's routine, as shown in Fig. 6. Interactions with individuals, friends, and family expand the robot's understanding of real-world situations.

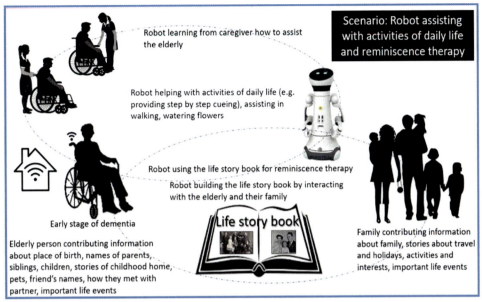

Fig. 6 Explainable robotics scenario of robot assisting with activities of daily life and reminiscence therapy for dementia sufferers. *(Reprinted with permission from R. Setchi, M.B. Dehkordi, J.S. Khan, Explainable robotics in human-robot interactions, Procedia Comput. Sci. 176 (2020) 3057–3066, Elsevier, Copyright (2024), 5722360979180.)*

For example, during reminiscence therapy with a life story book, the robot draws on external knowledge, displaying sophisticated explainable robotics abilities. An example is the robot addressing an elderly person's speaking difficulties. By looking at an unusual thing (dentures) and reasoning through continuous observation, the robot builds a causal relationship between speaking difficulties and the lack of dentures, motivating it to give aid. Explanations in robotics are critical for understanding the reasoning behind robot actions. The debate emphasizes the importance of providing explanations that are both human-friendly and trustworthy and robust, including previous experiences, and real-time information. While real-time trust assessment is difficult, particularly in a high-dimensional healthcare space, features such as emotional feedback and gestures can be utilized to enhance the robot's understanding. Mechanisms for assessing trust levels, detecting manipulations, and dealing with unexpected behaviors are currently being developed.

Another example is that as robots grow more integrated into our daily lives, the complexity of human surroundings becomes more difficult to manage, resulting in occasional failures in robotic systems. Existing solutions mostly focus on domain experts, leaving numerous instances of robotic failures that require aid from nonexpert users for recovery. To demonstrate the advantages of XAI in the context of robotics, consider a scenario where a household robot suffers an unexpected failure while executing a pick-and-place manipulation task [84]. This example is specially important since pick-and-place and manipulation task are widely performed in healthcare robots such as for pick and place [85] and manipulation of zebrafish embryo [34]. In response to this, their study introduces a novel explanation, denoted as ϵ_{err}, specifically tailored to elucidate the causes of such failures to nonexperts. To make ϵ_{err} meaningful, they delve into hand-scripted explanations, identifying the most informative sorts of information that can assist nonexperts in recognizing both the failure and potential solutions. Furthermore, they look into the autonomous development of explanations by improving an existing encoder-decoder model. This guarantees that the explanations are adaptable across a variety of situations, making them usable beyond their initial context of the household, which may be beneficial to be used in healthcare settings. Through their research focusing on the household robot, they find that explanations incorporating the context of failure and a history of past actions are particularly effective for nonexperts in identifying both the failure and potential solutions. Additionally, a second user evaluation affirms that the explanations generated by their model can extend to an unseen office environment, demonstrating efficacy comparable to hand-scripted explanations. One of the study's limitations is that context-based history explanations promote failure recovery, but their universal usefulness for nonexperts is unknown, necessitating explanations for individual users. Reinforcement learning could be a beneficial and promising method to overcome the limitations.

For instance, Beyret et al. [86] introduce a novel approach, called Dot-to-Dot, for robotic manipulation. Consider a scenario in which a robot must learn complex tasks by collaborating with humans. The problem is making the robot's decision-making process understandable to humans. Dot-to-Dot addresses this by employing a hierarchical reinforcement learning system in which a high-level agent generates explicit and interpretable subgoals, while a low-level agent effectively navigates the robot's action and state space to achieve these subgoals. Consider a practical case in which a robot, such as the Fetch Robotics Manipulator used in their study, needs to move a cube to a certain location. The high-level agent strategically creates subgoals, breaking down the larger objective into achievable pieces. These subgoals serve as reference points for the low-level agent, allowing it to learn and execute complex tasks more effectively. The fundamental innovation is that the high-level agent generates a representation of the task that is understandable to a human operator, promoting transparency in robot decision-making. In this context, Dot-to-Dot is useful in human-robot interactions. As the robot learns to interpret complex tasks and provide clear justifications for its behaviors, it fosters trust and understanding between humans and robots. This approach ensures that the robot's decision making is not a "black box," representing a significant step forward in the field of XAI for robotics.

In the field of urinary bladder diagnostics, cystoscopy has long been recognized as the gold standard for detecting bladder cancer (BCa); nonetheless, the difficulties associated with its various and sometimes misleading findings demand a paradigm shift. A paper published in 2022 [87] suggests and outlines the incorporation of XAI into robotic-assisted cystoscopy, which aligns with the present diagnostic constraints. The use of XAI robot-assisted cystoscopes is critical in addressing the current concerns about false negatives and positives, ultimately enhancing the precision and reliability of BCa diagnosis. This is significant because it has the potential to provide a standardized, fast, and patient-friendly diagnostic technique. By entrusting trained nurses with the semiautonomous operation of these XAI-enabled cystoscopes, the proposed method not only provides faster and more comfortable diagnostic scans, but also lessens the workload on urologists. The use of XAI in this robotic application is deemed critical because of its ability to deliver interpretable explanations, so overcoming the limitations associated with traditional cystoscopic methods.

This explainability aspect not only increases trust among medical experts, such as urologists and nurses, but also gives patients confidence in the reliability of diagnostic results. Furthermore, this is in line with regulatory considerations, addressing the issues faced by AI medical devices, particularly in the European Union. The emphasis on ethical principles, human oversight, and responsibility in the context of XAI demonstrates a commitment to the responsible and transparent integration of advanced technology into medical practice. Incorporating XAI into robotic-assisted cystoscopy, as proposed in [87], is a critical and progressive step forward in the aim of improving diagnostic accuracy, patient safety, and overall efficiency in urinary BCa detection.

In conclusion, the study of XAI in robotics, particularly its use in healthcare, demonstrates the transformational power of transparent and interpretable AI systems. As we look at real-world scenarios ranging from supporting dementia patients to addressing faults in household robots, the importance of explainability becomes clear. The examples provided, such as the explication of unexpected robot failures and the incorporation of XAI into urinary bladder diagnostics, effectively demonstrate the real benefits of boosting efficacy, trustworthiness, and overall results. Embracing the concept of "explainable robotics," we investigate the complex interaction between robot internal mechanics and human understanding, particularly in healthcare applications. The innovations described here not only improve human-robot interaction, but they also have the potential to revolutionize medical diagnostics. In summary, XAI integration is a critical step in advancing the responsible and transparent integration of AI technologies, paving the way for a future of human-robot collaboration marked by trust, understanding, and increased efficiency.

4. Emerging trends and the future of trust-based healthcare robotics

In the field of healthcare robotics, there are several questions, which need to be addressed, to fully exploit the potential of XAI and healthcare robotics in real-world problems. For example, XAI-based methods can help in model explanation and transparency, trust measurement, and bridging the gap between healthcare robotics and patient–doctor trust. Here, we discuss some of the emerging trends to ensure trust in healthcare robotics.

- **Licensed XAI robotics systems**: There is a need to develop certain benchmarks or license examinations for XAI robotics systems, to be eligible in clinical practice [88]. For example, a doctor or a surgeon passes several examinations, before entering into the clinics. We need to develop similar performance metrics and examinations for an XAI healthcare robotics systems.
- **Privacy and data governance protocols**: Healthcare robots equipped with XAI capabilities collect and process a vast amount of sensitive patient data, including medical records, images, and real-time health monitoring data. A robust mechanism to ensure data privacy and governance is required for the safe deployment of robots in healthcare settings [89, 90].
- **Responsible AI**: Responsible AI [91] methods advocate for explainability and transparency; these methods can help ensure privacy and data governance. This allows users and healthcare professionals to understand how the AI system makes decisions and uses personal data, thereby ensuring transparency and accountability in privacy-sensitive contexts.
- **Personalized and patient-centric care**: Trust-based healthcare robotics can contribute to personalized and patient-centric care [70]. Robots can offer personalized recommendations for self-care. There is a need to bridge the gap between robot, patient, and doctor, which will lead to building trust.
- **Enhanced surgical robotics**: Surgical robotics has already revolutionized many procedures, offering precision, dexterity, and minimally invasive techniques. The future will see further advancements in trust-based surgical robotics, including improved haptic feedback, enhanced visualization, and increased autonomy [92].
- **Data analytics and predictive insights**: By analyzing patient records, medical imaging, and other data sources, robots can assist healthcare professionals in making more accurate diagnoses, predicting disease progression, and recommending personalized treatment options [93].
- **System performance and ground truth evaluations**: Monitoring performance in the clinics and the interactions of XAI robots with the doctors and patients is a crucial aspect, for ensuring trust and safety. Moreover, the choice of the ground truth is a crucial aspect in performance assessment of XAI outcomes. Controlled experiments are needed to understand the impact of model explanations on a control group compared to a baseline.
- **Vision language models for healthcare robotics**: Large vision language models that combine the modalities of vision and language to perform various tasks in healthcare robotics may be developed for the ease of interaction between robots and patients. This will increase the interpretability and help us in building trust [94].

4.1 Advancing XAI hardware

XAI can be a computationally intensive task, especially when dealing with complex models and large datasets. The process of generating explanations often involves analyzing the internal workings of machine learning models, attributing feature importance, conducting sensitivity analysis, or creating visualizations [95]. In [95], a tensor processing unit for XAI applications was proposed. The proposed method explores several XAI methods for effective matrix multiplication. The method exploits the synergy between convolution and Fourier transform, which takes full advantage of TPU's (tensor processing units) inherent ability in accelerating matrix computations, and has the potential to become a real-time interpretation method. Edge-level XAI is also a promising solution for providing real-time interpretation, locally on edge devices in healthcare robotics [96]. It also allows for efficient resource utilization and lower latency, making it suitable for time-sensitive applications and scenarios where cloud-based solutions may not be feasible or practical [97].

4.2 Challenges in current methodologies

Optimizing the trade-off between selecting the best-performing model and a trustworthy model with a high level of explainability poses a challenging task. In reality, we still have a considerable way to go in terms of regulated top-performance models. For example, at the premium AI conference **NIPS** in 2018, two pictures were shown: a human surgeon with a 15% risk of patient death and a robotic arm with a 2% failure rate. Everyone voted for the human surgeon, except one person [98]. Transparent models, while interpretable, often struggle to handle complex real-world applications effectively, and we need to build a better bridge between (i) the model performance and XAI, (ii) the virtual and the real world, and (iii) effective regulations.

Another challenge is the measurement of trust in the wild [14]. It is difficult to measure trust in a robot, using a quantitative metric; however, there are efforts to use instruments that can measure the trust in the form of a self-report [99]. Recently, there are trends to measure the trust in terms of psychological behavior of robots such as eye tracking [100], skin response [101], and neural measures [102]. Moreover, with the advancement in large language models [103] and tools such as ChatGPT, it may be an appropriate solution to design a foundation model for XAI applications in healthcare robotics.

4.2.1 Regulatory standards and guidelines

Regulatory bodies collaborate with researchers and industry experts to establish requirements related to safety, data protection, interoperability, and ethical considerations. The demand for XAI has arisen as a response to new regulations, including the GDPR legislation, with the expectation that it provides explanations not only to individual users but also to society as a whole, particularly in areas such as healthcare robotics [82].

Scientific studies in bioethics and healthcare law examine the ethical and legal implications of trust-based healthcare robotics. Researchers explore topics such as privacy, informed consent, liability, and the ethical use of AI in healthcare [104, 105]. The Defense Advanced Research Projects Agency in the United States initiated its XAI project in 2017, aiming to develop novel techniques for explaining intelligent systems [106]. The initiative underscores the importance of explainability to enhance our understanding, trust, and control over the next generation of AI systems, in critical applications such as healthcare [70].

4.3 Future with trust-based healthcare robotics

There is a requirement to establish robust and comprehensive systematic comparisons for trust-based healthcare methodologies [107]. Moreover, developing more effective model for exploration, debugging, and validation tools, independent of shifts in data and problem at hand, will further develop trust in healthcare robotics [108]. Large sample sizes can enhance model generalization, preventing overfitting in individual cases; but, this improvement comes at the cost of increased model training expenses. Additionally, the addition of more data may require fine-tuning or training the model with the expanded dataset [109]. Explaining and fine-tuning generative models in XAI for healthcare robotics, especially for modern models like GPT-3, ChatGPT, and DALL-e 2, or stable diffusion methods, is an underexplored and interesting area [110].

Developing robust XAI tools to analyze healthcare data can reveal hidden data correlations, enhancing users' understanding of the problem under study [111]. Automated generation of decisions, arguments, explanations, and reasons is becoming more prevalent. With the rapid progress in XAI algorithmic development and robotics, healthcare professionals are also recognizing its transformative potential. All these developments are expected toward building trust in robotics and AI for healthcare in the future.

5. Conclusion

In conclusion, the possible transformation of modern healthcare through the collaborative use of robotics and AI is crucial to establishing confidence. Our study delves into important difficulties and disruptive solutions, with a particular emphasis on XAI. The initial section discusses the difficulties associated with incorporating robotics into healthcare settings, such as ethical and safety considerations, as well as the complexity of AI systems. These challenges establish the framework for understanding XAI's critical role in resolving these issues. The subsequent focus was on XAI solutions for building trust in healthcare robotics, illustrating how it facilitates transparent and reliable robot operation. Delving deeper, the chapter explored aspects such as accuracy trade-offs, explanation depth, contextual understanding, post hoc explainability techniques, evaluation metrics, and the influential power of XAI, supported by real-world examples. Subsequently, the chapter examined future developments, specifically advancements in XAI hardware for trust-based healthcare robots. This forward-looking perspective predicts a future in which healthcare robotics is based on XAI concepts. The final section looks at increasing regulatory norms and recommendations, emphasizing the safe use of AI in healthcare. This section summarizes a hopeful perspective for the future of healthcare robots, emphasizing trust as the key to their success. In conclusion, this chapter provides a practical analysis of healthcare robots, emphasizing the critical role of XAI in addressing difficulties, strengthening trust, and transforming healthcare delivery. It provides a clear and coherent framework for scholars, practitioners, and policymakers navigating the rapidly changing field of AI-integrated healthcare robotics.

References

[1] D. Kolpashchikov, O. Gerget, R. Meshcheryakov, Robotics in healthcare, in: Handbook of Artificial Intelligence in Healthcare: Vol 2: Practicalities and Prospects, Springer, 2022, pp. 281–306.

[2] F. Ju, Y. Wang, B. Xie, Y. Mi, M. Zhao, J. Cao, MDPI, The use of sports rehabilitation robotics to assist in the recovery of physical abilities in elderly patients with degenerative diseases: a literature review, Healthcare 11 (3) (2023) 326.

[3] J.-I. Yoo, M.-K. Oh, S.-U. Lee, C.H. Lee, Robot-assisted rehabilitation for total knee or hip replacement surgery patients: a systematic review and meta-analysis, Medicine 101 (40) (2022) e30852.

[4] A.J. Young, D.P. Ferris, State of the art and future directions for lower limb robotic exoskeletons, IEEE Trans. Neural Syst. Rehabil. Eng. 25 (2) (2016) 171–182.

[5] L.M. Weber, J. Stein, The use of robots in stroke rehabilitation: a narrative review, NeuroRehabilitation 43 (1) (2018) 99–110.

[6] S. Cooper, A. Di Fava, C. Vivas, L. Marchionni, F. Ferro, ARI: the social assistive robot and companion, in: 2020 29th IEEE International Conference on Robot and Human Interactive Communication (RO-MAN), IEEE, 2020, pp. 745–751.

[7] R. Yu, E. Hui, J. Lee, D. Poon, A. Ng, K. Sit, K. Ip, F. Yeung, M. Wong, T. Shibata, Use of a therapeutic, socially assistive pet robot (PARO) in improving mood and stimulating social interaction and communication for people with dementia: study protocol for a randomized controlled trial, JMIR Res. Protoc. 4 (2) (2015) e4189.

[8] S. Whelan, K. Murphy, E. Barrett, C. Krusche, A. Santorelli, D. Casey, Factors affecting the acceptability of social robots by older adults including people with dementia or cognitive impairment: a literature review, Int. J. Soc. Robot. 10 (5) (2018) 643–668.

[9] S. Reddy, J. Fox, M.P. Purohit, Artificial intelligence-enabled healthcare delivery, J. R. Soc. Med. 112 (1) (2019) 22–28.

[10] L.G. Pee, S.L. Pan, L. Cui, Artificial intelligence in healthcare robots: a social informatics study of knowledge embodiment, J. Assoc. Inf. Sci. Technol. 70 (4) (2019) 351–369.

[11] A. Langer, R. Feingold-Polak, O. Mueller, P. Kellmeyer, S. Levy-Tzedek, Trust in socially assistive robots: considerations for use in rehabilitation, Neurosci. Biobehav. Rev. 104 (2019) 231–239.

[12] Royal Society and National Academy of Sciences, The Frontiers of Machine Learning: 2017 Raymond and Beverly Sackler US-UK Scientific Forum, The National Academies Press, 2018, ISBN: 0309471958.

[13] N. Wang, D.V. Pynadath, S.G. Hill, Trust calibration within a human-robot team: comparing automatically generated explanations, in: 2016 11th ACM/IEEE International Conference on Human-Robot Interaction (HRI), IEEE, 2016, pp. 109–116.

[14] B.C. Kok, H. Soh, Trust in robots: challenges and opportunities, Curr. Robot. Rep. 1 (2020) 297–309.

[15] N. Emaminejad, R. Akhavian, Trustworthy AI and robotics: implications for the AEC industry, Autom. Constr. 139 (2022) 104298.

[16] M. Iosa, G. Morone, A. Cherubini, S. Paolucci, The three laws of neurorobotics: a review on what neurorehabilitation robots should do for patients and clinicians, J. Med. Biol. Eng. 36 (2016) 1–11.

[17] I. Asimov, Three laws of robotics, Runaround 2 (1941).

[18] F.M. Carrillo, J. Butchart, N. Kruse, A. Scheinberg, L. Wise, C. McCarthy, Physiotherapists' acceptance of a socially assistive robot in ongoing clinical deployment, in: 2018 27th IEEE International Symposium on Robot and Human Interactive Communication (RO-MAN), IEEE, 2018, pp. 850–855.

[19] A. Alaiad, L. Zhou, The determinants of home healthcare robots adoption: an empirical investigation, Int. J. Med. Inform. 83 (11) (2014) 825–840.

[20] L.M. Beuscher, J. Fan, N. Sarkar, M.S. Dietrich, P.A. Newhouse, K.F. Miller, L.C. Mion, Socially assistive robots: measuring older adults' perceptions, J. Gerontol. Nurs. 43 (12) (2017) 35–43.

[21] A. Van Maris, H. Lehmann, L. Natale, B. Grzyb, The influence of a robot's embodiment on trust: a longitudinal study, in: Proceedings of the Companion of the 2017 ACM/IEEE International Conference on Human-Robot Interaction, 2017, pp. 313–314.

[22] E. Capobianco, High-dimensional role of AI and machine learning in cancer research, Br. J. Cancer 126 (4) (2022) 523–532.

[23] R. Setchi, M.B. Dehkordi, J.S. Khan, Explainable robotics in human-robot interactions, Procedia Comput. Sci. 176 (2020) 3057–3066.

[24] M. Hossin, M.N. Sulaiman, A review on evaluation metrics for data classification evaluations, Int. J. Data Min. Knowl. Manag. Process 5 (2) (2015) 1.

[25] A. Malhi, T. Kampik, H. Pannu, M. Madhikermi, K. Främling, Explaining machine learning-based classifications of in-vivo gastral images, in: 2019 Digital Image Computing: Techniques and Applications (DICTA), IEEE, 2019, pp. 1–7.

[26] S. Nazir, D.M. Dickson, M.U. Akram, Survey of explainable artificial intelligence techniques for biomedical imaging with deep neural networks, Comput. Biol. Med. 156 (2023) 106668.

[27] A. Holzinger, G. Langs, H. Denk, K. Zatloukal, H. Müller, Causability and explainability of artificial intelligence in medicine, Wiley Interdisc. Rev. Data Min. Knowl. Discov. 9 (4) (2019) e1312.

[28] S. Ali, T. Abuhmed, S. El-Sappagh, K. Muhammad, J.M. Alonso-Moral, R. Confalonieri, R. Guidotti, J. Del Ser, N. Díaz-Rodríguez, F. Herrera, Explainable artificial intelligence (XAI): what we know and what is left to attain trustworthy artificial intelligence, Inf. Fusion 99 (2023) 101805.

[29] European Parliament, Framework of ethical aspects of artificial intelligence, robotics and related technologies, 2020. https://oeil.secure.europarl.europa.eu/oeil/popups/ficheprocedure.do?lang=en&reference=2020/2012(INL). Accessed 13.12.2023.

[30] European Parliament, Artificial Intelligence Act, 2023. https://www.europarl.europa.eu/doceo/document/TA-9-2023-0236_EN.html. Accessed 13.12.2023.

[31] A. van Wynsberghe, N. Sharkey, Special issue on responsible robotics: introduction, Ethics Inf. Technol. 22 (2020) 281–282.

[32] B.C. Stahl, M. Coeckelbergh, Ethics of healthcare robotics: towards responsible research and innovation, Robot. Auton. Syst. 86 (2016) 152–161.

[33] M. Coeckelbergh, C. Pop, R. Simut, A. Peca, S. Pintea, D. David, B. Vanderborght, A survey of expectations about the role of robots in robot-assisted therapy for children with ASD: ethical acceptability, trust, sociability, appearance, and attachment, Sci. Eng. Ethics 22 (2016) 47–65.

[34] A. Diouf, F. Sadak, I. Fassi, M. Boudaoud, G. Legnani, S. Haliyo, Automatic sorting of zebrafish embryos using deep learning, in: 2023 International Conference on Manipulation, Automation and Robotics at Small Scales (MARSS), IEEE, 2023, pp. 1–6.

[35] F. Sadak, M. Saadat, A.M. Hajiyavand, Real-time deep learning-based image recognition for applications in automated positioning and injection of biological cells, Comput. Biol. Med. 125 (2020) 103976.

[36] M. Coeckelbergh, Can we trust robots? Ethics Inf. Technol. 14 (2012) 53–60.

[37] D. Shin, The effects of explainability and causability on perception, trust, and acceptance: implications for explainable AI, Int. J. Hum. Comput. Stud. 146 (2021) 102551.

[38] C. Macrae, Governing the safety of artificial intelligence in healthcare, BMJ Qual. Saf. 28 (6) (2019) 495–498.
[39] E. Fosch-Villaronga, T. Mahler, Cybersecurity, safety and robots: strengthening the link between cybersecurity and safety in the context of care robots, Comput. Law Secur. Rev. 41 (2021) 105528.
[40] European Parliament, 2021. https://digital-strategy.ec.europa.eu/en/policies/regulatory-framework-ai. Accessed 25.12.2023.
[41] European Parliament, 2023. https://www.europarl.europa.eu/doceo/document/TA-9-2023-0236_EN.html. Accessed 25.12.2023.
[42] W. Li, C. Song, Z. Li, An accelerated recurrent neural network for visual servo control of a robotic flexible endoscope with joint limit constraint, IEEE Trans. Ind. Electron. 67 (12) (2019) 10787–10797.
[43] S. Li, S. Chen, B. Liu, Accelerating a recurrent neural network to finite-time convergence for solving time-varying Sylvester equation by using a sign-bi-power activation function, Neural Process. Lett. 37 (2013) 189–205.
[44] A.I. Chen, M.L. Balter, T.J. Maguire, M.L. Yarmush, Deep learning robotic guidance for autonomous vascular access, Nat. Mach. Intell. 2 (2) (2020) 104–115.
[45] Z. Wu, L. Li, Y. Yang, P. Hu, Y. Li, S.-Y. Yang, L.V. Wang, W. Gao, A microrobotic system guided by photoacoustic computed tomography for targeted navigation in intestines in vivo, Sci. Robot. 4 (32) (2019) eaax0613.
[46] M.R. Gubbi, M.A.L. Bell, Deep learning-based photoacoustic visual servoing: using outputs from raw sensor data as inputs to a robot controller, in: 2021 IEEE International Conference on Robotics and Automation (ICRA), IEEE, 2021, pp. 14261–14267.
[47] A.M. Antoniadi, Y. Du, Y. Guendouz, L. Wei, C. Mazo, B.A. Becker, C. Mooney, Current challenges and future opportunities for XAI in machine learning-based clinical decision support systems: a systematic review, Appl. Sci. 11 (11) (2021) 5088.
[48] G. Ras, N. Xie, M. Van Gerven, D. Doran, Explainable deep learning: a field guide for the uninitiated, J. Artif. Intell. Res. 73 (2022) 329–396.
[49] I. Choi, W. Koh, B. Koo, W.C. Kim, Network-based exploratory data analysis and explainable three-stage deep clustering for financial customer profiling, Eng. Appl. Artif. Intell. 128 (2024) 107378.
[50] S. Tilouche, V. Partovi Nia, S. Bassetto, Parallel coordinate order for high-dimensional data, Stat. Anal. Data Min. ASA Data Sci. J. 14 (5) (2021) 501–515.
[51] S. Wold, K. Esbensen, P. Geladi, Principal component analysis, Chemom. Intell. Lab. Syst. 2 (1–3) (1987) 37–52.
[52] S.G. Graabæk, E.V. Ancker, A.R. Fugl, A.L. Christensen, An experimental comparison of anomaly detection methods for collaborative robot manipulators, IEEE Access 11 (2023) 65834–65848.
[53] D.W. Hosmer Jr, S. Lemeshow, R.X. Sturdivant, Applied Logistic Regression, vol. 398, John Wiley & Sons, 2013.
[54] O.Z. Maimon, L. Rokach, Data Mining With Decision Trees: Theory and Applications, vol. 81, World Scientific, 2014.
[55] J. Jung, C. Concannon, R. Shroff, S. Goel, D.G. Goldstein, Simple rules for complex decisions, arXiv preprint arXiv:1702.04690 (2017).
[56] J. Kolodner, Case-Based Reasoning, Morgan Kaufmann, 2014.
[57] J.M. Alonso, C. Castiello, C. Mencar, Interpretability of fuzzy systems: current research trends and prospects, in: Springer Handbook of Computational Intelligence, Springer, 2015, pp. 219–237.
[58] T.J. Hastie, Generalized additive models, in: Statistical Models in S, Routledge, 2017, pp. 249–307.
[59] Z.C. Lipton, The mythos of model interpretability: in machine learning, the concept of interpretability is both important and slippery, Queue 16 (3) (2018) 31–57.
[60] T.-N. Chou, An explainable hybrid model for bankruptcy prediction based on the decision tree and deep neural network, in: 2019 IEEE 2nd International Conference on Knowledge Innovation and Invention (ICKII), IEEE, 2019, pp. 122–125.
[61] N. Papernot, P. McDaniel, Deep k-nearest neighbors: towards confident, interpretable and robust deep learning, arXiv preprint arXiv:1803.04765 (2018).

[62] D. Alvarez Melis, T. Jaakkola, Towards robust interpretability with self-explaining neural networks, Adv. Neural Inf. Process. Syst. 31 (2018) 1–10.

[63] G.E. Karniadakis, I.G. Kevrekidis, L. Lu, P. Perdikaris, S. Wang, L. Yang, Physics-informed machine learning, Nat. Rev. Phys. 3 (6) (2021) 422–440.

[64] J. Gou, B. Yu, S.J. Maybank, D. Tao, Knowledge distillation: a survey, Int. J. Comput. Vis. 129 (2021) 1789–1819.

[65] M. Sabih, F. Hannig, J. Teich, Utilizing explainable AI for quantization and pruning of deep neural networks, arXiv preprint arXiv:2008.09072 (2020).

[66] L. Weber, S. Lapuschkin, A. Binder, W. Samek, Beyond explaining: opportunities and challenges of XAI-based model improvement, Inf. Fusion 92 (2023) 154–176.

[67] X. Lu, A. Tolmachev, T. Yamamoto, K. Takeuchi, S. Okajima, T. Takebayashi, K. Maruhashi, H. Kashima, Crowdsourcing evaluation of saliency-based XAI methods, in: Machine Learning and Knowledge Discovery in Databases. Applied Data Science Track: European Conference, ECML PKDD 2021, Bilbao, Spain, September 13–17, 2021, Proceedings, Part V 21, Springer, 2021, pp. 431–446.

[68] Y.-J. Jung, S.-H. Han, H.-J. Choi, Explaining CNN and RNN using selective layer-wise relevance propagation, IEEE Access 9 (2021) 18670–18681.

[69] K. Roshan, A. Zafar, Utilizing XAI technique to improve autoencoder based model for computer network anomaly detection with SHapley Additive exPlanation (SHAP), arXiv preprint arXiv:2112.08442 (2021).

[70] D. Saraswat, P. Bhattacharya, A. Verma, V.K. Prasad, S. Tanwar, G. Sharma, P.N. Bokoro, R. Sharma, Explainable AI for Healthcare 5.0: opportunities and challenges, IEEE Access 10 (2022) 84486–84517.

[71] S. Knapič, A. Malhi, R. Saluja, K. Främling, Explainable artificial intelligence for human decision support system in the medical domain, Mach. Learn. Knowl. Extr. 3 (3) (2021) 740–770.

[72] J. Dieber, S. Kirrane, Why model why? Assessing the strengths and limitations of LIME, arXiv preprint arXiv:2012.00093 (2020).

[73] B. Crook, M. Schlüter, T. Speith, Revisiting the performance-explainability trade-off in explainable artificial intelligence (XAI), in: 2023 IEEE 31st International Requirements Engineering Conference Workshops (REW), IEEE, 2023, pp. 316–324.

[74] J. Wanner, L.-V. Herm, K. Heinrich, C. Janiesch, Stop ordering machine learning algorithms by their explainability! An empirical investigation of the tradeoff between performance and explainability, in: Conference on e-Business, e-Services and e-Society, Springer, 2021, pp. 245–258.

[75] W. Saeed, C. Omlin, Explainable AI (XAI): a systematic meta-survey of current challenges and future opportunities, Knowl. Based Syst. 263 (2023) 110273.

[76] R. Qureshi, M. Irfan, H. Ali, A. Khan, A.S. Nittala, S. Ali, A. Shah, T.M. Gondal, F. Sadak, Z. Shah, M.U. Hadi, S. Khan, Q. Al-Tashi, J. Wu, A. Bermak, T. Alam, Artificial intelligence and biosensors in healthcare and its clinical relevance: a review, IEEE Access 11 (2023) 61600–61620, https://doi.org/10.1109/ACCESS.2023.3285596.

[77] S. Buijsman, Defining explanation and explanatory depth in XAI, Minds Mach. 32 (3) (2022) 563–584.

[78] C. Bove, J. Aigrain, M.-J. Lesot, C. Tijus, M. Detynecki, Contextualization and exploration of local feature importance explanations to improve understanding and satisfaction of non-expert users, in: 27th International Conference on Intelligent User Interfaces, 2022, pp. 807–819.

[79] T. Hulsen, Explainable Artificial Intelligence (XAI) in Healthcare, CRC Press, 2024.

[80] C. Burger, L. Chen, T. Le, Are your explanations reliable?" Investigating the stability of LIME in explaining text classifiers by marrying XAI and adversarial attack, in: Proceedings of the 2023 Conference on Empirical Methods in Natural Language Processing, 2023, pp. 12831–12844.

[81] A.M. El-Gezawy, H. Abdel-Kader, A.H. Ali, A new XAI evaluation metric for classification, Int. J. Comput. Inf. 10 (3) (2023) 58–62.

[82] A.B. Arrieta, N. Díaz-Rodríguez, J. Del Ser, A. Bennetot, S. Tabik, A. Barbado, S. García, S. Gil-López, D. Molina, R. Benjamins, Explainable artificial intelligence (XAI): concepts, taxonomies, opportunities and challenges toward responsible AI, Inf. Fusion 58 (2020) 82–115.

[83] M. Langer, D. Oster, T. Speith, H. Hermanns, L. Kästner, E. Schmidt, A. Sesing, K. Baum, What do we want from explainable artificial intelligence (XAI)?—a stakeholder perspective on XAI and a conceptual model guiding interdisciplinary XAI research, Artif. Intell. 296 (2021) 103473.

[84] D. Das, S. Banerjee, S. Chernova, Explainable AI for robot failures: generating explanations that improve user assistance in fault recovery, in: Proceedings of the 2021 ACM/IEEE International Conference on Human-Robot Interaction, 2021, pp. 351–360.

[85] H. Afrisal, A.D. Setiyadi, M.A. Riyadi, R. Ismail, O. Toirov, I. Setiawan, Performance analysis of 4-DOF RPRR robot manipulator actuation strategy for pick and place application in healthcare environment, Int. J. Adv. Sci. Eng. Inf. Technol. 12 (6) (2022) 2258–2265.

[86] B. Beyret, A. Shafti, A.A. Faisal, Dot-to-Dot: explainable hierarchical reinforcement learning for robotic manipulation, in: 2019 IEEE/RSJ International Conference on Intelligent Robots and Systems (IROS), IEEE, 2019, pp. 5014–5019.

[87] S. O'Sullivan, M. Janssen, A. Holzinger, N. Nevejans, O. Eminaga, C.P. Meyer, A. Miernik, Explainable artificial intelligence (XAI): closing the gap between image analysis and navigation in complex invasive diagnostic procedures, World J. Urol. 40 (5) (2022) 1125–1134.

[88] M.L. Schrum, M. Johnson, M. Ghuy, M.C. Gombolay, Four years in review: statistical practices of Likert scales in human-robot interaction studies, in: Companion of the 2020 ACM/IEEE International Conference on Human-Robot Interaction, 2020, pp. 43–52.

[89] T.J. Wiltshire, S.F. Warta, D. Barber, S.M. Fiore, Enabling robotic social intelligence by engineering human social-cognitive mechanisms, Cogn. Syst. Res. 43 (2017) 190–207.

[90] E. Fosch-Villaronga, H. Drukarch, On healthcare robots: concepts, definitions, and considerations for healthcare robot governance, arXiv preprint arXiv:2106.03468 (2021).

[91] N. Vallès-Peris, M. Domènech, Caring in the in-between: a proposal to introduce responsible AI and robotics to healthcare, AI Soc. 38 (4) (2023) 1685–1695.

[92] C. Diaz-Arrastia, C. Jurnalov, G. Gomez, C. Townsend, Laparoscopic hysterectomy using a computer-enhanced surgical robot, Surg. Endosc. Other Interv. Tech. 16 (2002) 1271–1273.

[93] S. Panicucci, N. Nikolakis, T. Cerquitelli, F. Ventura, S. Proto, E. Macii, S. Makris, D. Bowden, P. Becker, N. O'Mahony, A cloud-to-edge approach to support predictive analytics in robotics industry, Electronics 9 (3) (2020) 492.

[94] K. Kawaharazuka, Y. Obinata, N. Kanazawa, K. Okada, M. Inaba, Robotic applications of pretrained vision-language models to various recognition behaviors, arXiv preprint arXiv:2303.05674 (2023).

[95] Z. Pan, P. Mishra, Hardware acceleration of explainable AI, in: Explainable AI for Cybersecurity, Springer, 2023, pp. 199–220.

[96] E. Tjoa, C. Guan, A survey on explainable artificial intelligence (XAI): toward medical XAI, IEEE Trans. Neural Netw. Learn. Syst. 32 (11) (2020) 4793–4813.

[97] A.B. Tosun, F. Pullara, M.J. Becich, D. Taylor, J.L. Fine, S.C. Chennubhotla, Explainable AI (XAI) for anatomic pathology, Adv. Anat. Pathol. 27 (4) (2020) 241–250.

[98] C. Rudin, J. Radin, Why are we using black box models in AI when we don't need to? A lesson from an explainable AI competition, Harv. Data Sci. Rev. 1 (2) (2019) 1–9.

[99] M. Desai, P. Kaniarasu, M. Medvedev, A. Steinfeld, H. Yanco, Impact of robot failures and feedback on real-time trust, in: 2013 8th ACM/IEEE International Conference on Human-Robot Interaction (HRI), IEEE, 2013, pp. 251–258.

[100] Q. Jenkins, X. Jiang, Measuring trust and application of eye tracking in human robotic interaction, in: IISE Annual Conference and Expo, 2010.

[101] K. Akash, W.-L. Hu, N. Jain, T. Reid, A classification model for sensing human trust in machines using EEG and GSR, ACM Trans. Interact. Intell. Syst. 8 (4) (2018) 1–20.

[102] I.B. Ajenaghughrure, S.C. Sousa, I.J. Kosunen, D. Lamas, Predictive model to assess user trust: a psycho-physiological approach, in: Proceedings of the 10th Indian Conference on Human-Computer Interaction, 2019, pp. 1–10.

[103] M.U. Hadi, R. Qureshi, A. Shah, M. Irfan, A. Zafar, M.B. Shaikh, N. Akhtar, J. Wu, S. Mirjalili, A Survey on Large Language Models: Applications, Challenges, Limitations, and Practical Usage, Authorea, 2023.

[104] M.A.R. Bak, M.C. Ploem, H.L. Tan, M.T. Blom, D.L. Willems, Towards trust-based governance of health data research, Med. Health Care Philos. 26 (2) (2023) 185–200.

[105] R. Qureshi, M. Irfan, T.M. Gondal, S. Khan, J. Wu, M.U. Hadi, J. Heymach, X. Le, H. Yan, T. Alam, AI in drug discovery and its clinical relevance, Heliyon 9 (7) (2023) e17575.

[106] D. Gunning, D. Aha, DARPA's explainable artificial intelligence (XAI) program, AI Mag. 40 (2) (2019) 44–58.

[107] B. Goodman, S. Flaxman, European Union regulations on algorithmic decision-making and a "right to explanation, AI Mag. 38 (3) (2017) 50–57.

[108] S. Amershi, M. Chickering, S.M. Drucker, B. Lee, P. Simard, J. Suh, ModelTracker: redesigning performance analysis tools for machine learning, in: Proceedings of the 33rd Annual ACM Conference on Human Factors in Computing Systems, 2015, pp. 337–346.

[109] R.M. French, Catastrophic forgetting in connectionist networks, Trends Cogn. Sci. 3 (4) (1999) 128–135.

[110] V. Nagisetty, L. Graves, J. Scott, V. Ganesh, XAI-GAN: enhancing generative adversarial networks via explainable AI systems, arXiv preprint arXiv:2002.10438 (2020).

[111] S. Krishnan, E. Wu, Palm: machine learning explanations for iterative debugging, in: Proceedings of the 2nd Workshop on Human-in-the-Loop Data Analytics, 2017, pp. 1–6.

CHAPTER 2

XAI implementation in traditional alternate medicine system

Rakesh Gnanasekaran[a], Gnanasankaran Natarajan[a], Sundaravadivazhagan Balasubramanian[b], and Shirley Chellathurai Pon Anna Bai[c]
[a]Department of Computer Science, Thiagarajar College, Madurai, Tamil Nadu, India
[b]Department of Information Technology, University of Technology and Applied Sciences, Al Mussanah, Oman
[c]Department of Computer Science and Engineering, Karunya Institute of Technology and Sciences, Coimbatore, Tamil Nadu, India

1. Traditional medicine systems

Customary medication (traditional medicine system) is depicted collectively of medical services practices and items with a long history of purpose. TM regularly alludes to clinical information created by native societies that integrate nature-based cures, otherworldly medicines and manual procedures intended to treat sickness or protect prosperity. TM is normally drilled beyond allopathic medication (regularly known as present day or Western medication), which is the prevailing arrangement of medication in the created world. A portion of the notable TM frameworks incorporate conventional Indian (Ayurveda) medication, customary Chinese medication (TCM), and customary Arabic (Unani). The Global Wellbeing Association (WHO) gives out Conventional Medication process as "The complete range of knowledge, skills, and practices influenced by different cultural backgrounds, whether scientific or applied in the context of health, prevention, diagnosis, development, or treatment of physical and mental well-being" TM rehearses, especially complete clinical frameworks, for example, customary Chinese Medication and Ayurveda, share a large number of a similar fundamental beliefs.

These practices will quite often be described by an all-encompassing and exceptionally individualized way to deal with treatment. An accentuation is generally given on augmenting the body's ability to mend. Patients are dynamic members in their own consideration, as Conventional Medication addresses physical, mental, and otherworldly traits of a sickness while underlining on counteraction and wellbeing. Rather than the all-encompassing methodology of Conventional Medications, western medication frequently centers around the unfortunate region and concealment of bothersome side effects, focusing on quick outcomes. Despite the fact that quick outcomes are understood, the impacts on long haul wellbeing, as well as consequences for different infections, essentially those of constant nature, mental, stress problems, and so on are less

Explainable AI in Healthcare Imaging for Medical Diagnoses
https://doi.org/10.1016/B978-0-443-23979-3.00002-6

29

obvious. Subsequently, by and by "elective meds," like homeopathy, naturopathy, diet or nourishment treatment, home grown, ayurvedic meds, and so forth are still popular and have all tracked down their specialties.

1.1 Informatics in health care

Medical care has turned into a data-based science. A large part of the clinical practice includes assembling, incorporating, and following up on data. This can be tended to by clinical informatics. Clinical informatics is the field that includes the mental, data handling, and correspondence assignments of clinical practice, instruction, and examination. It is the field worried about procuring, putting away, and involving data in medical services, normally including the utilization of data innovation [1].

Data in addition to correspondence innovation (ict) currently are a center component of digitized associations that object over there currently am able to work with functional viability in addition to improve upper hand. In the present fourth modern transformation (4ir) period, high level advanced innovations in addition to gadgets currently are generally applied for development in addition to worth creation across enterprises. The medical services industry currently exists no exemption. Emergency clinics in addition to care suppliers every single one over the planet, particularly in created economies, currently are forcefully conveying advanced innovations, tremendously enjoy man-made brainpower (simulated intelligence), ai, shrewd sensors in addition to robots, enormous information examination, in addition to web of things (iot), for worked on nature of care in addition to functional. It currently exists significant to examine whatever exceptional advanced gadgets currently are meaning for administration experiences among clients in addition to specialist co-ops in the medical care industry. Just as of late, there has existed far in addition to wide utilizations of simulated intelligence upheld advancements in medical services foundations for further developed care administration quality in addition to effectiveness of clinical assets. Since simulated intelligence envelops ai, normal language handling, in addition to savvy robots, computer-based intelligence-based improvement opens up multiple doors to improvement in the information of data escalated medical services industry.

1.2 Artificial intelligence

Computerized thinking (artificial intelligence) can be described as Trying again human mental procedure in the PC or ROBO—Protocol is made to arise mental powers what people use either person known for each other (like learning and critical thinking) above words and image "man-made intelligence," "AI," "Deep learning" are famous well-known words these days. Because it is AI terms which includes calculations for a wide variety of tasks, say for example matching, predictive or grouping, etc. and your calculations need to be upskilled on the information of Data. Entering more information to your calculation would make it better. Deep learning protected by fake brain networks;

it is extremely new part of Artificial intelligence. It works on same principles pattern showed by humans and also uses Information to know how to accomplish task.

Because artificial intelligence-driven improvements are being standardized into everyday life, artificial intelligence-based innovations will be essential for every association. Thus, independent of how long deep learning has progressed in handling issues in artificial intelligence subfields, enterprises must feel the computational costs of calculating the training across staggering amount of data.

Artificial intelligence-based advances are being coordinated into our regular routines. For instance, the computer-based intelligence product "Aria" delivered by SK Telecom of South Korea is a shrewd audio-initiated gadget. Aria is also able to make crisis decision when the carrier could not use other devices due to emergencies, physical incapacity, or special situations as when an elderly subject who falls and says "Aria, I need help" first calls the hospital, the predetermined relative for crisis, or ADT Covers (it's machine security in South Korea that refers to personal security guard and security services). Assuming the middle establishes that it is, as a matter of fact, a crisis circumstance it automatically direct to the number 119 (Emergency number in South Korea) [2]. Therefore it has proactively brought about saving various older individuals who live alone. Aria can likewise be utilized as a recipe help. On the off chance that the client states, "Aria, kindly assistance me with a salmon recipe," Aria AI product direct the client bit by bit through the recipe. Aria could furnish help with overseeing individual budgets. Based on client's order, Aria can propose the best Visa to utilize in light of the financing cost and yearly expenses, or help the client to remember the charge card installment due date in a given month.

As brilliant versatile applications and gadgets become generally accessible in the computerized age, purchasers are requesting separated, customized, and responsive administrations, with secure adaptability. In this unique situation, it is essential to examine certifiable instances of current simulated intelligence use by medical care associations for patient consideration and the executives of activities, it is a basic prerequisite (e.g., guideline and obligation) and backing requirements (e.g., morals, preparing programs, or counseling administrations) of cutting-edge artificial intelligence applications.

2. AI-based applications in the healthcare sectors

AI-supported intelligent robots can execute procedures and improve doctors' jobs with specialized diagnoses, treatment methods, cost and time minimization, as well as a more advanced response time to patients' needs. The total count of such rare diseases is estimated at around 6000–8000, with the care volume reaching up to 400 million people globally. The average period of diagnosing a rare disease is estimated to be five years. Hence the clinical patients with such diseases spend a huge of their time, energy, and money trying to obtain a correct diagnosis. 3Billion, a biotech start-up specializing in DNA sequencing services for complex rare oriented diseases, has reported that they have identified over 1200 patients with unique diseases using artificial intelligence. In many cases, 3Billion can run tests for around 7000 illnesses at one time. When a patient identify with certain disease get another solution from a professionalized person with the implementation of Artificial intelligence based innovation results. In the meantime, the patient can spend hours wasting time and being lost in "diagnosis wanderings" from one hospital to another. The above is a global issue for patients with some rare diseases. The count of patients that can be served by a doctor is limited. Therefore, artificial intelligence in the medical field has the potential to save millions of lives globally.

Specialists at the Moorfield's Eye Clinic in London have fostered a man-made intelligence finding framework, which can suggest medicines for in excess of approximately 50 eye illnesses with 94% precision. In China, artificial intelligence-based advances are utilized to analyze colon polyps. In a clinical review, man-made intelligence-based innovations and a gastroenterology expert cooperated to analyze a patient, while in another clinical preliminary just an expert was analyzed; when determined to have the help of man-made intelligence, the discovery pace of polyps expanded by 20% [3].

3. Explainable AI

XAI is the statutes of logic and interpretability. Both these statutes are frequently utilized reciprocally; however, they have particular contrasts. More or less, logic alludes to understanding and explaining the inner mechanics of a DL or ML model. So, they can be made sense of in human terms. Similarly, interpretability centers around foreseeing what will occur, given changes to the contributions of a model and algorithmic boundaries. In this manner, interpretability means to notice the circumstances and logical results operations of a model. It is shown in Fig. 1 [4].

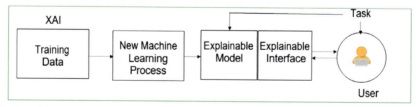

Fig. 1 Explainable AI.

On account of XAI, reasonableness depicts the dynamic properties of a learning model in light of posteriori information, which means to explain its working. Here, logic is emotional as end-clients are set in a situation to decide if the clarification gave is trustworthy. To guarantee the credibility of results produced from models, there is a need to make sense of.

- What information went into a preparation model and why? How was reasonableness evaluated? What exertion was made to lessen predisposition? (Logical information)

- What model elements were initiated and used to accomplish a given result? (Reasonable expectations)
- What individual layers make up the model, and how would they bring about a result (Reasonable calculations)

Quiet, the results of studies involving DL and ML in development are only sometimes addressed and approved by end-clients in a genuine setting, especially those connected with PC vision. Besides, the helpfulness of a DL and ML answer for an issue that tends to an industry need inside the scholarly development writing is barely at any point inspected.

3.1 Interpretability

Interpretability alludes to the uninvolved property of a learning model in light of deduced information. The objective of interpretability is to guarantee that a given model sounds good to a human onlooker utilizing language that is significant to clients. The "progress of this objective is attached to the discernment, information, and inclinations of the client." In this way, interpretability is seen as being consistent and giving a straightforward result; that is, it is promptly reasonable.

3.2 Transparency

Supporting the requirement for straightforwardness is the requirement for fair and moral direction. It is generally concurred that calculations should be effective. They additionally should be straightforward and fair and be available to ex-post and ex-bet investigation. In this manner, straightforwardness expects to beat the dark boxness related with computer-based intelligence-based models. Straightforwardness is characterized as the "capacity of a human to understand the (risk hoc) system utilized by a prescient model."

3.3 Explainability methods

AI calculations fluctuate in execution levels and are less exact than DL frameworks. In this way, the decision of calculation to apply for a given issue depends on a compromise among execution and logic. A quick survey of the writing on computer-based intelligence in development uncovers that little consideration is given to this presentation logic compromise.

1. Transparent (plan) is where a model is made sense of during its preparation. That is, the characteristic design of models fulfills one of three straightforwardness aspects.
2. Opaque, where the model is made sense of after it has been prepared. Techniques, for example, RF, Backing Vector Machine (SVM), Convolutional Brain Organizations (CNN), Multi-facet.

Strikingly, there are particular contrasts between model-freethinker and model-explicit techniques. A model-rationalist technique works for all models. Interestingly, the model-explicit techniques just work for explicit models like the DL calculations, SVM, and RF.

3.4 XAI in healthcare industries

Explainable AI is necessary because traditional machine learning models lack transparency and flexibility, making it difficult for users to understand the reasoning behind their predictions. This lack of clarity can lead to trust issues and hinder the adoption and deployment of these models. Explainable AI aims to address these challenges by providing more transparency and interpretability in machine-learning models, ultimately improving trust and understanding among users.

3.4.1 Origin of explainable AI

Explainable AI has its origins in the early 1980s, when researchers in machine learning were focused on developing algorithms and techniques for learning and making predictions. As these algorithms became more complex, there arose a need for transparency and explain ability in machine learning models. One of the pioneers in this field was Judea Pearl, who introduced the concept of cause and effect in machine learning. Pearl proposed a framework to understand and explain the key factors that influenced machine learning predictions. His groundbreaking work laid the foundation for many of the explainable AI techniques and approaches that are still in use today, enabling transparent and understandable machine learning. Another significant development in the early stages of explainable AI was the introduction of LIME (local interpretable model-agnostic explanations). This method utilizes an internal average of the model to provide insights into the factors that have the greatest similarity and influence in the model's forecasting. LIME has been widely adopted across various domains and applications.

The primary objective of explainable AI is to make machine learning models transparent and interpretable, enabling humans to comprehend and trust them. This has numerous advantages and benefits. Firstly, explainable AI enhances decision-making capabilities by providing valuable insights and information. It offers an understanding of the most relevant and influential factors in the model's predictions, enabling the identification and prioritization of actions and strategies that are likely to yield the desired outcome. Furthermore, explainable AI helps to build trust and acceptance in machine learning models. It addresses the challenges and limitations faced by traditional opaque and unpredictable models. By providing explanations and insights into the decision-making process, explainable AI can accelerate the adoption and utilization of machine learning models.

3.5 Transformation of health care industry for medical system

The healthcare sector provided within the middle of a change. The main lead of this insurgency is rising add up to health-sector fetched and a developing need of health-specialists. As an output, the healthcare sector is looking to execute modern data technology-concerned arrangements and forms that reduce amount of cost and donate arrangements to these upcoming troubles.

Healthcare frameworks around the world confront gigantic issues, counting a need of get to, tall costs, squander, and an elderly-based population. Global Pandemics just like the coronavirus (COVID-19) put a pressure on healthcare frameworks, coming about in a need of defensive gear, inadequately or wrong demonstrative tests, more vitally, a healthcare catastrophe like COVID-19.

4. Traditional applications in XAI healthcare

Here there are some applications of medical XAI application used in healthcare sectors.

4.1 XAI for drug discovery

Since the enhancement of a unused medicate could be a complex, expensive, and a very overlong handle, how to diminish costs and speed up unused sedate disclosure has turned into a compelling and quick issue within the industry. So, the significance of computer-aided medicate revelation has been relentlessly expanding in later times. Notwithstanding the expanding number of fulfilled forward-looking hones, profound learning models ought to be logical as numerical models are regularly troublesome to decipher by the human intellect. In this consider, to begin with, essential data around current logical manufactured insights (XAI) strategies such as class enactment mapping (CAM), neighborhood interpretable model-agnostic clarifications is given. A short time later, XAI approaches utilized within the setting of sedate revelation such as highlight attribution, instance-based, graph-convolution based, self-clarifying, and vulnerability forecast are said and expectations for future thinks about are made [5].

4.2 XAI for clinical outputs sample trails

A clinical sample trail may be a system in which naturally fabricated drugs are given to individualities to test working methodology. The below is an affair of a noteworthy sum of time and cash. The palm rate, in any case, is veritably moo. As an affair, clinical sample trial automation has shown to be an advantage for AI and the healthcare sector. Likewise, Fake perceptivity and healthcare help within the preemption of time-expending information data chancing strategies. Likewise, AI-supported health care trials handle extensive totalities of data and produce exceptionally perfect results. The taking later are a many of the foremost current Fake perceptivity in healthcare operations for clinical trials.

4.3 XAI clinical trials for traditional medical system

Conventional methods of clinical sampling trial methods are still the extreme best standard for guaranteeing the viability and security of unused health-based needed drugs. The long, used-and-promising strategy of unmistakable and characterized list of randomized controlled trials was created basically for assessing mass-advertise drugs and has remained for the most part unaltered in later decades. Counterfeit insights has the potential to abbreviate clinical trial cycle lengths whereas too upgrading efficiency and clinical advancement results. Conventional Sedate base have been able to get expanding volumes of logical and inquire about data from a large number of sources in later a long time used alluded to as current-world information (RWD). They have, in any case, as often as possible needed the ability and apparatuses required to effectively utilize this information. Applying prescient AI models and progressed analytics to open RWD can offer assistance analysts way better get it maladies, discover pertinent patients and imperative examiners, and empower progressive clinical think about plans. In combination with a proficient computerized framework, clinical trial information may be cleansed, amassed, coded, protected, and kept up utilizing AI calculations. Besides, progressed electronic information capture can reduce the effect of human mistake in information collection whereas too permitting for smooth framework integration.

4.4 Clinical trial and module sharing

In response to COVID-19, experimenters from some areas are rushing into an extraordinary scientific collaboration. Inflexibility of data, substantiation and law distributing styles, as well as preemption adaptation of operations to internal settings and contributing across borders, would be needed for AI bias that make a global impact. Artificial intelligence (AI) workloads bear data. A many exemplifications of COVID-19 concentrated data-participating enterprise live at the world wide, countrywide and community situations within all three operation scales. Genomic studies, inheritable sequences, protein module structure, cases' clinical instructional data, medical imaging, event data, epidemiological information, migration patterns, social media commentary, scientific literature are just a many possible exemplifications of accoutrements that was [6]. There's an implicit problem with hyperactive-fragmentation in terms of participating conditioning due to the circumstance that it can lead to changes being limited only within specific sectors or communities. The establishment of flexible means to partake information, models and canons may quicken the development and spread-eschewal novel software products. These days encyclopedically open inclusive exact transparent secure worldwide open equal data-sharing conditioning will go far in linking and enhancing cooperation among societies at this stage [7].

4.5 Patient care

Understanding results are impacted by fake insights in healthcare. Restorative Artificial intelligence organization make a framework that assist the persistent at each stage. Clinical insights too scrutinize patients' restorative data and forecast the experiences to cooperate them upgrade their best quality of life [8]. The taking after are some critical clinical insights frameworks that progress quiet care.

4.5.1 Maternal care

The taking after could be a valuable method for distinguishing huge-risk mothers and decreasing maternal mortality and issues after birth of the child.

(1) Anticipating whether hopeful moms are at critical chance of troubles amid conveyance utilizing electronic wellbeing information and counterfeit intelligence (AI).

(2) Utilizing advanced innovation to extend quiet passage to all standard and large acuity care all through their pregnancy.

The comparison to conveying in large-acuity clinics with multiple solid assets and clinical involvement, high-risk fat ladies who provide their newborn children at minimize-acuity clinics have the next hazard of creating genuine maternal horribleness [2].

4.5.2 Healthcare robotics

In addition to medical personnel, certain medical robots play a crucial role in assisting patients. For instance, exoskeleton robots are utilized to aid paralyzed patients in regaining the ability to walk and achieve independence. Another illustration of robotic technology in healthcare is the use of intelligent prostheses. These bionic limbs are equipped with sensors that enhance their responsiveness and accuracy compared to natural body parts. They can also be covered with bionic skin or connected to the user's muscles [9]. Robots are valuable in supporting rehabilitation and performing surgical procedures. An example is Cyberdyne's HAL exoskeleton, specifically designed to aid patients recovering from conditions such as spinal cord injury and stroke that result in lower limb disorders. This exoskeleton is equipped with sensors on the skin that detect electrical signals within the patient's body, enabling it to move accordingly at the joint [10].

4.5.3 Genetics XAI evidence-based medicine

Currently, healthcare consumers are getting more involved in their own medical treatment, from assaying their genome sequencing to creating a substantiated health profile using data from fitness/exertion trackers. Data-driven drug uses this massive quantum of information to gain a more precise and prophetic sapience into our health and conditions. Not only can data-driven drug enhance the delicacy and effectiveness of inheritable complaint webbing, but it can also open the door to substantiated remedial curatives [11].

5. Conclusion

Nowadays XAI is dynamically being utilized to healthcare, because it gets to be more predominant in modern undertaking and lifestyle. XAI has the content to assist healthcare suppliers in an assortment of directions, consisting of persistent treatment and authoritative assignments. The lion's share of artificial intelligence and basic healthcare advancements are valuable within the healthcare sector, but the techniques they assist can be or maybe distinctive. Whereas a few distributions on manufactured insights in wellbeing request that artificial intelligence in nowadays can able to give best as well as or more than superior to people at particular forms, Some of them are predicating ailment, These are all long time activity some time recently used AI in healthcare to alternate individuals for a wide extend of restorative employments. Hence by implementing XAI in Traditional healthcare system provide a large insight to provide access for society in a greater extent.

References

[1] N. Greenberg, M. Docherty, S. Gnanapragasam, S. Wessely, Managing mental health challenges faced by healthcare workers during COVID-19 pandemic, BMJ 368 (2020).

[2] W.T. Maphumulo, B.R. Bhengu, Challenges of quality improvement in the healthcare of South Africa post-apartheid: a critical review, Curationis 42 (1) (2019) 1–9.

[3] D.C. Angus, Randomized clinical trials of artificial intelligence, JAMA 323 (11) (2020) 1043–1045.

[4] J. Agrawal, Stethee, an AI powered electronic stethoscope, Anaesthesia Pain Intensive Care 22 (3) (2018) 412–413.

[5] O. Díaz, J.A.R. Dalton, J. Giraldo, Artificial intelligence: a novel approach for drug discovery, Trends Pharmacol. Sci. 40 (8) (2019) 550–551.

[6] J. Habermann, Psychological Impacts of COVID-19 and Preventive Strategies: A Review, 2021.

[7] M. Luengo-Oroz, K.H. Pham, J. Bullock, R. Kirkpatrick, A. Luccioni, S. Rubel, C. Wachholz, M. Chakchouk, P. Biggs, T. Nguyen, Artificial intelligence cooperation to support the global response to COVID-19, Nat. Mach. Intell. 2 (6) (2020) 295–297.

[8] S. Lip, S. Visweswaran, S. Padmanabhan, Transforming clinical trials with artificial intelligence, in: Artificial Intelligence, Productivity Press, 2020, pp. 297–306.

[9] G. Rakesh, MRI image brain tumor detection and segmentation using texture-based statistical characterization method, J. Adv. Res. Dynam. Control Syst. 12 (14) (2020).

[10] S. Harrer, P. Shah, B. Antony, J. Hu, Artificial intelligence for clinical trial design, Trends Pharmacol. Sci. 40 (8) (2019) 577–591.

[11] A. Holzinger, C. Biemann, C.S. Pattichis, D.B. Kell, What do we need to build explainable AI systems for the medical domain?, 2017. ArXiv Preprint. ArXiv:1712.09923.

CHAPTER 3

Explainable computational intelligence in bio and clinical medicine

M. Shanmuga Sundari[a] ⓘ **and Harshini Reddy Penthala**[b] ⓘ
[a]Computer Science and Engineering, BVRIT HYDERABAD College of Engineering for Women, Hyderabad, India
[b]The University of Texas at Arlington, Arlington, TX, United States

1. Introduction

1.1 Background and motivation

The field of bio and clinical medicine [1] is at the cusp of a transformative revolution, thanks to the integration of computational intelligence. Understanding the background and motivation behind this revolution is crucial to appreciating its significance and potential impact. In this section, we will delve into the historical context of medical data analysis and the driving forces propelling the adoption of computational intelligence methods in healthcare.

1.1.1 Historical context of medical data analysis

The history of medicine has been marked by continuous efforts to understand diseases, diagnose conditions, and improve patient care. Traditional medical practices relied heavily on the expertise of healthcare professionals, empirical observations, and rudimentary diagnostic tools. While these practices have undoubtedly saved countless lives, they also had limitations.

Medical data analysis [2], in its early forms, involved manual record-keeping, rudimentary statistical analyzes, and limited data-driven decision-making. Physicians and researchers faced challenges in handling the vast and complex datasets that emerged as medical science advanced. These challenges included data overload, slow processing, and a lack of tools for pattern recognition.

The turning point in medical data analysis came with the advent of computers and digital technologies. The ability to store, process, and analyze data at a scale previously unimaginable opened new avenues for understanding diseases and improving patient outcomes. However, the true revolution began with the application of computational intelligence. The Fig. 1 is showcasing the biological description with Artificial intelligence.

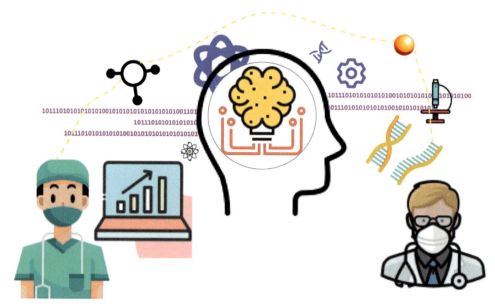

Fig. 1 Artificial intelligence in biology.

1.1.2 Driving forces behind computational intelligence in healthcare

Several factors have been instrumental in driving the adoption of computational intelligence in bio and clinical medicine:

1. **Data explosion**: The exponential growth of medical data, including electronic health records (EHRs), medical imaging, and genomic data, necessitated advanced data analysis techniques. Computational intelligence provides the means to harness the insights hidden within these vast datasets.

2. **Complexity of healthcare**: Healthcare is inherently complex, with numerous variables and factors influencing patient health. Computational models, such as artificial neural networks and machine learning algorithms, excel at handling complexity and identifying patterns in multifaceted healthcare data.

3. **Personalized medicine**: The shift toward personalized medicine, where treatment plans are tailored to individual patients, demanded sophisticated predictive models. Computational intelligence enables the development of patient-specific treatment plans, optimizing therapeutic outcomes.

4. **Early diagnosis and prevention**: Timely diagnosis and preventive measures are crucial in healthcare. Computational models can analyze patient data for early signs of diseases, facilitating early interventions and improving prognosis.

5. **Medical imaging advancements**: The field of medical imaging has witnessed remarkable advancements, and computational intelligence plays a pivotal role in image analysis, aiding in accurate diagnosis and treatment planning.

6. **Cost efficiency**: Healthcare costs have been a concern worldwide. Computational intelligence has the potential to optimize resource allocation, reduce medical errors, and enhance efficiency, leading to cost savings.

7. **Research acceleration:** Researchers benefit from computational tools that can analyze large-scale biomedical data rapidly. This accelerates the pace of medical discoveries and innovation.

The historical context of medical data analysis and the compelling factors driving the integration of computational intelligence underscore the pivotal role of this field in revolutionizing bio and clinical medicine.

1.2 Objectives of computational intelligence in bio and clinical medicine

Computational intelligence in the field of bio and clinical medicine is driven by a set of well-defined objectives aimed at enhancing healthcare outcomes, elevating diagnosis accuracy [3], and improving treatment efficacy. This section outlines the primary goals of applying computational intelligence in the realm of healthcare, underscoring its potential to revolutionize the medical landscape.

1.2.1 Advancing healthcare outcomes

One of the foremost objectives of computational intelligence [4] in bio and clinical medicine is to advance healthcare outcomes. This includes improving patient health, enhancing the quality of care, and ultimately saving lives. Computational intelligence achieves this by:

- **Early disease detection:** AI-driven algorithms can analyze vast datasets, including medical images and patient records, to identify subtle signs of diseases at their early stages. Early detection enables timely intervention and significantly improves prognosis.

- **Personalized treatment:** Computational models can generate patient-specific treatment plans based on individual health data. Tailored treatments maximize therapeutic efficacy, reduce adverse effects, and enhance patient well-being.

- **Preventive healthcare:** Predictive analytics powered by computational intelligence enable the identification of individuals at risk of certain diseases. Preventive measures can be implemented, reducing the overall disease burden.

1.2.2 Elevating diagnosis accuracy

A key objective is to elevate diagnosis accuracy, minimizing diagnostic errors and ensuring precision in healthcare practices. Computational intelligence contributes to this goal by:

- **Medical imaging enhancement**: AI-based image analysis techniques can assist radiologists and clinicians in interpreting medical images with higher accuracy. This is particularly valuable in identifying subtle abnormalities in X-rays, MRIs, and CT scans.

- **Data-driven diagnosis**: By analyzing diverse patient data, including genetic, clinical, and demographic information, computational models can provide comprehensive diagnostic insights. This holistic approach enhances diagnostic accuracy.
- **Minimizing false positives and negatives**: Machine learning algorithms can reduce false positive and false negative errors, especially in conditions where early diagnosis is critical, such as cancer screening.

1.2.3 Improving treatment efficacy

Improving treatment efficacy is central to the objectives of computational intelligence in healthcare. This objective is achieved through:

- **Treatment optimization**: Computational models can optimize treatment plans by considering various factors, including drug interactions, patient history, and genetic markers. Optimized treatments increase the likelihood of successful outcomes.
- **Monitoring and feedback**: AI-driven systems enable real-time patient monitoring. This allows healthcare providers to adjust treatment regimens promptly based on patient responses, ensuring that treatments remain effective.
- **Drug discovery**: Computational intelligence accelerates drug discovery by simulating drug interactions, identifying potential candidates, and expediting the development of new therapies.

The objectives of applying computational intelligence in bio and clinical medicine are grounded in the pursuit of better healthcare outcomes, elevated diagnosis accuracy, and improved treatment efficacy. By harnessing the power of advanced algorithms and data analysis, computational intelligence stands to revolutionize medical practices and bring about a new era of patient-centered, data-driven healthcare.

1.3 Scope and organization of the chapter

In this section, we provide readers with a clear understanding of the scope and organization of this chapter. We lay out a roadmap that highlights the various aspects of computational intelligence in healthcare [5] that will be explored, ensuring that readers are well-prepared for the journey ahead.

1.3.1 Scope of the chapter

The scope of this chapter is comprehensive, as it aims to provide a holistic view of computational intelligence's role in bio and clinical medicine. It encompasses a wide range of topics, including:

- **Fundamentals of computational intelligence**: We will start by delving into the fundamental concepts of computational intelligence, including machine learning algorithms, artificial neural networks, genetic algorithms, and data mining techniques. This foundation will prepare readers for the subsequent discussions.

- **Data acquisition and preprocessing**: Understanding the sources of biological and clinical data, data collection methods, and data preprocessing techniques is vital. This chapter will provide insights into how data is obtained, managed, and prepared for analysis.
- **Disease diagnosis and classification**: We will explore how computational intelligence is applied in disease diagnosis, both through computational models and image-based diagnosis. Additionally, we will delve into biomarker discovery and validation, a crucial aspect of modern medicine.
- **Treatment planning and personalized medicine**: This section will highlight the potential of computational intelligence in predicting treatment responses, drug discovery and optimization, and the development of patient-specific treatment plans.
- **Healthcare management and decision support**: Readers will gain an understanding of clinical decision support systems (CDSS), predictive analytics for hospital resource allocation, and the role of telemedicine and remote patient monitoring in modern healthcare.
- **Ethical and regulatory considerations**: We will explore the ethical implications of computational intelligence in medicine, including patient privacy and data security, regulatory compliance, and fairness in AI.
- **Challenges and future directions**: The chapter will discuss the challenges faced in the field, such as model interpretability and integration with traditional medical practices, and explore future directions, including the incorporation of real-time data streams.
- **Case studies and applications**: Practical case studies will be presented to illustrate how computational intelligence is making a difference in healthcare, covering areas such as cardiovascular disease, cancer diagnosis, and neurological disorders.
- **Conclusion**: The chapter will conclude by summarizing key takeaways, discussing the impact and future prospects of computational intelligence in bio and clinical medicine, and emphasizing the need for collaboration between computational experts and medical professionals.

1.3.2 Organization of the chapter

The chapter is organized in a logical sequence, moving from foundational concepts to practical applications and ethical considerations. This structured approach ensures that readers can follow the progression of ideas and insights. Each section is designed to build upon the knowledge presented in the preceding sections, offering a coherent and comprehensive understanding of computational intelligence in healthcare.

By providing a clear scope and organization, this chapter aims to guide readers through the multifaceted world of computational intelligence in bio and clinical medicine, empowering them to explore each facet and appreciate the transformative potential of this field.

2. Fundamentals of computational intelligence

2.1 Machine learning algorithms

Machine learning, a cornerstone of computational intelligence, plays a pivotal role in healthcare. This section immerses us in the core concepts of machine learning and underscores its relevance in the context of healthcare. It is essential to comprehend the diverse types of machine learning algorithms and their applications in the analysis of medical data.

2.1.1 Understanding machine learning in healthcare

Machine learning is a branch of artificial intelligence that enables computer systems to learn and make predictions or decisions based on data. In the realm of healthcare, machine learning algorithms [6] are employed to extract meaningful insights from vast datasets, such as electronic health records, medical images, and patient histories. The major classification of machine learning algorithms is depicted in Fig. 2.

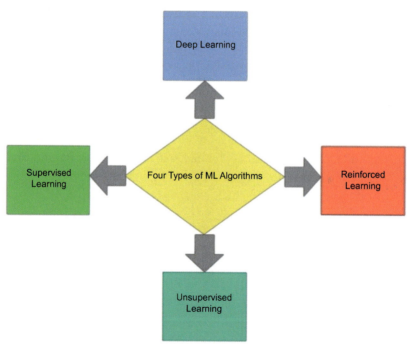

Fig. 2 Machine learning and its types.

2.1.2 Types of machine learning algorithms

This section provides an overview of various types of machine learning algorithms, including:

- **Supervised learning**: In healthcare, supervised learning algorithms are widely used for tasks like disease classification, where the algorithm learns from labeled examples (e.g., patient records with known diagnoses) to make predictions.
- **Unsupervised learning**: Unsupervised learning methods are applied to find patterns or groupings within data, aiding in tasks like patient clustering based on similar characteristics.
- **Deep learning**: Deep learning, a subset of machine learning, focuses on artificial neural networks with multiple layers. It excels in tasks like image analysis, where complex patterns need to be recognized.
- **Reinforcement learning**: Although less common in healthcare, reinforcement learning can optimize treatment plans by learning the best actions to take in response to changing patient conditions.

2.1.3 Applications in medical data analysis

Machine learning algorithms find diverse applications in medical data analysis:

- **Disease prediction**: Algorithms can predict the risk of diseases based on patient data, assisting in preventive healthcare.
- **Image analysis**: Medical image analysis, including the detection of abnormalities in X-rays or the segmentation of tumors in MRI scans [7], is significantly enhanced by machine learning.
- **Drug discovery**: Machine learning expedites drug discovery by identifying potential compounds and predicting their efficacy.
- **Patient monitoring**: Real-time patient monitoring, such as predicting deteriorations in a patient's condition, is made possible through machine learning.

2.2 Artificial neural networks

Artificial neural networks (ANNs) [8] are at the forefront of computational intelligence. This section delves into the intricacies of ANNs, exploring their architecture, training methods, and applications in the medical domain.

2.2.1 Architecture of artificial neural networks

ANNs are inspired by the human brain's structure. They consist of interconnected nodes organized into layers: an input layer, one or more hidden layers, and an output layer. ANNs excel in tasks that involve complex pattern recognition and nonlinear relationships within data. The neural network is displayed in Fig. 3.

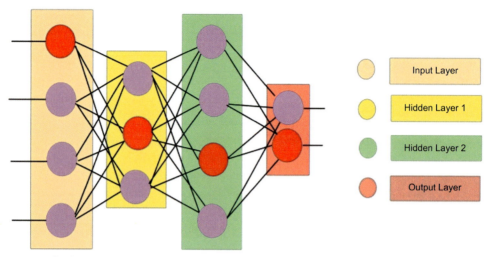

Fig. 3 Artificial neural network.

2.2.2 Training methods

Training ANNs involves adjusting the weights and biases of connections between nodes to minimize the difference between predicted and actual outcomes. Techniques like backpropagation are used for supervised learning.

2.2.3 Medical applications of ANNs

In healthcare, ANNs are harnessed for:

- **Medical imaging**: ANNs analyze medical images, detecting anomalies, segmenting structures, and aiding in disease diagnosis.
- **Disease risk assessment**: ANNs predict disease risk by considering genetic, lifestyle, and clinical data, facilitating preventive measures.
- **Drug discovery**: ANNs accelerate drug discovery by predicting the pharmacological properties of compounds.

2.3 Genetic algorithms

Genetic algorithms are optimization techniques inspired by the process of natural selection. This section introduces genetic algorithms and their applications in healthcare.

2.3.1 Optimization in healthcare

In healthcare, optimization is critical for tasks like:

- **Parameter tuning**: Genetic algorithms can optimize the parameters [9] of machine learning models to enhance their performance in tasks such as disease prediction.
- **Feature selection**: Genetic algorithms [10] aid in selecting relevant features from large datasets, reducing dimensionality while retaining essential information.

2.3.2 Genetic algorithm process

Genetic algorithms involve a population of potential solutions (chromosomes) that evolve over generations. They undergo selection, crossover (recombination), and mutation to arrive at optimal solutions. The process of genetic algorithm process is shown in Fig. 4.

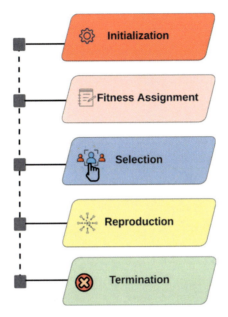

Fig. 4 Genetic algorithms.

2.3.3 Applications in healthcare

Genetic algorithms find application in:

- **Treatment planning**: Optimizing treatment plans by considering various factors, such as drug dosages and scheduling, to maximize therapeutic efficacy.
- **Disease biomarker discovery**: Identifying biomarkers associated with diseases through feature selection.

2.4 Data mining techniques

Data mining techniques are instrumental in uncovering patterns [11], associations, and insights from medical data. This section focuses on their role in bio and clinical medicine.

2.4.1 Exploring data mining

Data mining encompasses various methods which is shown in Fig. 5, including:

- **Association rule mining**: Discovering relationships between variables, such as the co-occurrence of symptoms and diseases.
- **Clustering**: Grouping similar patients based on characteristics, aiding in patient stratification.

- **Sequential pattern mining**: Identifying sequences of events, valuable for understanding disease progression.

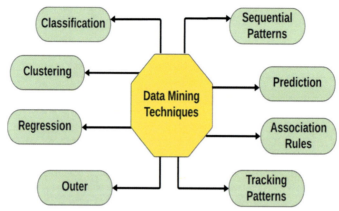

Fig. 5 Data mining techniques.

2.4.2 Applications in healthcare

Data mining techniques contribute to:

- **Clinical decision support**: Providing insights to clinicians for making informed decisions based on patient data.
- **Epidemiological studies**: Analyzing population health data to identify trends and risk factors.
- **Drug adverse event detection**: Detecting adverse drug reactions by mining large pharmacovigilance databases.

3. Data acquisition and preprocessing in bio and clinical medicine

3.1 Biological data sources

The foundation of computational intelligence in bio and clinical medicine lies in the availability of high-quality biological data. This section delves into the diverse sources of biological data, encompassing genomics, proteomics, and metabolomics. It emphasizes the critical role of collecting pristine biological data to underpin accurate analysis and medical insights.

3.1.1 Genomics data

Genomics, the study of an organism's complete set of DNA, provides a wealth of data for researchers and clinicians. Genomic data sources include:

- **DNA sequencing**: Techniques like next-generation sequencing (NGS) generate vast amounts of DNA sequence data, enabling the identification of genetic variations associated with diseases.

- **Genome-wide association studies (GWAS)**: These studies investigate genetic variants across the entire genome to identify links to specific diseases or traits.
- **Functional genomics**: Transcriptomics and epigenomics data shed light on gene expression and regulation, providing insights into disease mechanisms.

3.1.2 Proteomics data

Proteomics involves the study of proteins, their structures, functions, and interactions. The process of precision medicine is shown in Fig. 6. Sources of proteomics [12] data include:

- **Mass spectrometry**: Mass spectrometry techniques analyze proteins and peptides, offering insights into biomarker discovery and protein-protein interactions.
- **Protein expression profiling:** Techniques like 2D gel electrophoresis and protein microarrays assess protein expression levels, aiding in disease characterization.

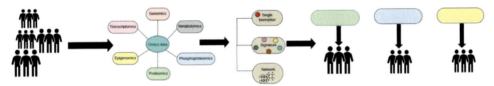

Fig. 6 An illustrative process for implementing precision medicine.

3.1.3 Metabolomics data

Metabolomics examines small molecules, metabolites, in biological systems [13]. It provides data on metabolic pathways and disease markers. Metabolomics data sources include:

- **Nuclear magnetic resonance (NMR)**: NMR spectroscopy generates data on metabolite concentrations and chemical structures.
- **Liquid chromatography-mass spectrometry (LC–MS)**: LC-MS techniques are used to profile metabolites, facilitating disease diagnosis and understanding metabolic pathways.

Collecting high-quality biological data is paramount as it forms the basis for subsequent computational analyzes. Researchers must ensure data accuracy, consistency, and integrity to derive meaningful insights and drive medical advancements.

3.2 Clinical data collection and management

Clinical data, derived from patient interactions and healthcare processes, is a cornerstone of healthcare analytics. This section explores the methods employed for clinical data collection and management, shedding light on the challenges and opportunities in handling these critical datasets.

3.2.1 Electronic health records (EHRs)

EHRs have revolutionized healthcare data collection by digitizing patient records. They contain a wealth of information, including medical history, diagnoses, medications, and treatment plans. EHRs offer real-time access to patient data, enabling data–driven decision-making and personalized medicine.

3.2.2 Wearable devices and remote monitoring

Advancements in wearable technology have introduced new opportunities for data collection. Wearable devices, such as fitness trackers and smartwatches, collect data on vital signs, physical activity, and sleep patterns. Remote patient monitoring allows continuous data collection, especially useful for chronic disease management.

3.2.3 Challenges in clinical data management

While clinical data holds immense potential, it presents challenges:

- **Data interoperability**: Integrating data from various healthcare systems and devices can be complex due to differences in data formats and standards.
- **Data privacy**: Protecting patient privacy and complying with regulations like HIPAA (Health Insurance Portability and Accountability Act) is critical.
- **Data quality**: Ensuring the accuracy and completeness of clinical data is essential for reliable analysis.
- **Data volume**: The sheer volume of clinical data generated daily requires robust storage and processing solutions.

3.3 Data preprocessing techniques

Data preprocessing [14] is a vital step in preparing healthcare data for analysis. This section covers specific data preprocessing techniques tailored to healthcare datasets, including:

- **Noise reduction**: Filtering out irrelevant or erroneous data points to enhance data quality.
- **Missing data handling**: Strategies for dealing with missing values to prevent bias in analysis results.
- **Data normalization**: Scaling data to a standard range to ensure that features with different units contribute equally to analysis.

Data preprocessing ensures that the data used for computational analysis is accurate, consistent, and devoid of artifacts, laying a solid foundation for meaningful insights and discoveries in bio and clinical medicine.

4. Disease diagnosis and classification

4.1 Computational models for disease diagnosis

Disease diagnosis is a critical aspect of healthcare, and computational intelligence offers powerful tools to automate and enhance this process. This section introduces various computational models used for disease diagnosis, underscoring their significance in improving diagnostic accuracy and efficiency.

4.1.1 Decision trees

Decision trees are a popular choice in disease diagnosis due to their simplicity and interpretability. They work by recursively splitting data into subsets based on the most informative features. Decision trees are particularly useful for tasks like identifying diseases with clear decision paths based on patient attributes.

4.1.2 Support vector machines (SVM)

SVM is a robust classification algorithm that finds optimal hyperplanes to separate data points belonging to different classes. In disease diagnosis, SVMs excel in scenarios where data is not linearly separable. They are valuable for tasks like cancer classification based on gene expression profiles.

4.1.3 Deep learning models

Deep learning, a subset of machine learning, has revolutionized disease diagnosis, especially in image-based applications. Convolutional neural networks (CNNs), a type of deep learning model, are adept at recognizing complex patterns in medical images. CNNs have become indispensable in tasks like identifying tumors in X-rays, detecting diabetic retinopathy in eye images, and segmenting brain tumors in MRI scans.

The adoption of computational models in disease diagnosis enhances the speed and accuracy of diagnoses. Moreover, these models can handle large volumes of medical data, allowing for early disease detection and proactive healthcare.

4.2 Image-based diagnosis

Medical imaging is a cornerstone of modern healthcare, and computational intelligence plays a pivotal role in analyzing and interpreting medical images. This section delves into the use of computational intelligence in medical imaging, encompassing X-rays, MRIs, CT scans, and more. The image analysis techniques are displayed in Fig. 7.

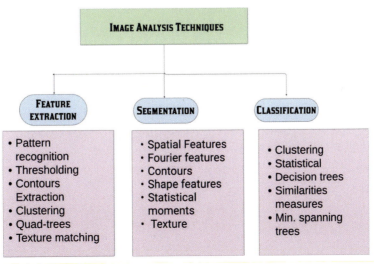

Fig. 7 Image analysis techniques.

4.2.1 Image analysis techniques

Medical image analysis involves a multitude of techniques, such as:

- **Feature extraction**: Identifying relevant features in images, such as the shape and texture of tumors or the density of lung tissue.
- **Segmentation**: Delineating structures of interest within images, crucial for tasks like tumor localization.
- **Object detection**: Detecting abnormalities or specific structures within images, aiding in early disease diagnosis.
- **Classification**: Assigning labels to images based on their content, such as classifying X-rays as normal or indicative of pathology.

Image-based diagnosis leverages these techniques to automate the identification and characterization of diseases and abnormalities. For instance, chest X-rays can be analyzed to detect pneumonia, while MRI scans of the brain can assist in identifying neurological disorders.

4.3 Biomarker discovery and validation

Biomarkers are biological indicators of disease or physiological processes. Computational approaches have revolutionized biomarker discovery and validation [14]. This section discusses the importance of biomarkers in disease diagnosis and the role of computational intelligence in this context.

4.3.1 Biomarker discovery

Computational methods, including statistical analysis and machine learning algorithms, enable the identification of potential biomarkers by analyzing large-scale biological data. These biomarkers may include genetic markers, protein levels, or metabolite concentrations. Early detection biomarkers, for example, can enable timely intervention in diseases like cancer. The drug discovery process in medical field is shown in Fig. 8.

4.3.2 Biomarker validation

Once potential biomarkers are discovered, rigorous validation is essential to ensure their reliability in clinical settings. Computational tools aid in validating biomarkers by analyzing their performance across diverse patient populations and datasets. Validated biomarkers become invaluable tools for diagnosing diseases, monitoring treatment responses, and predicting disease progression.

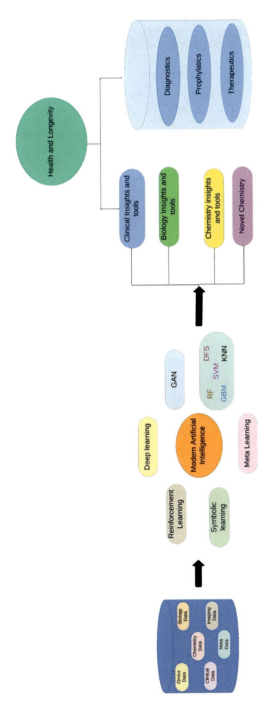

Fig. 8 The capacity of deep learning in the fields of drug discovery and the identification of biomarkers.

5. Treatment planning and personalized medicine

5.1 Treatment response prediction

Predicting how individual patients will respond to treatments is a central challenge in modern medicine. This section explores the application of computational intelligence to this critical task, highlighting how predictive models are employed to tailor treatment plans to individual patients, ultimately leading to improved therapeutic outcomes.

5.1.1 Personalized treatment plans

One-size-fits-all treatment strategies have limitations, as individuals may react differently to the same intervention. Computational intelligence, through the utilization of machine learning algorithms, helps in constructing patient-specific models. These models consider a patient's unique characteristics, including genetics, medical history, and lifestyle, to predict their response to specific treatments which is shown in Fig. 9 [15].

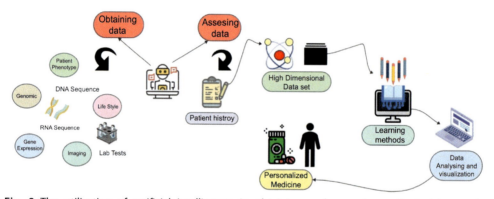

Fig. 9 The utilization of artificial intelligence in obtaining and assessing patient data to give personalized medicine.

5.1.2 Predictive models

Predictive models, often based on historical patient data, clinical trials, and genetic profiles, are pivotal in treatment response prediction. They encompass various machine learning and statistical techniques, including:

- **Regression models**: Regression analysis helps estimate how patient characteristics correlate with treatment outcomes, allowing for response prediction.
- **Random forests and ensemble methods**: These methods aggregate predictions from multiple models to enhance accuracy.
- **Deep learning**: Deep neural networks can capture complex relationships between patient attributes and treatment responses, particularly in image-based diagnoses and genomics.

The result is a treatment plan tailored to the individual, optimizing therapeutic efficacy while minimizing potential adverse effects. This personalized approach represents a significant leap forward in patient care.

5.2 Drug discovery and optimization

The process of drug discovery [15] is notoriously time-consuming and expensive. Computational intelligence offers a solution by expediting drug discovery and optimization. This section delves into the application of AI-driven methods in virtual screening, compound design, and drug optimization.

5.2.1 Virtual screening

Virtual screening employs computational models to sift through vast chemical libraries, identifying potential drug candidates. Machine learning algorithms analyze molecular structures to predict a compound's affinity for a target, such as a protein associated with a disease. This approach significantly accelerates the identification of lead compounds.

5.2.2 Compound design

AI-driven compound design tools facilitate the creation of novel drug candidates. Generative models, reinforced learning, and molecular dynamics simulations enable the generation of molecules with desired properties, enhancing the diversity and efficiency of drug development.

5.2.3 Drug optimization

Once a candidate is identified, computational methods optimize its properties. This includes fine-tuning molecular structures to improve efficacy, safety, and pharmacokinetics. In silico simulations predict how drugs interact with biological systems, guiding refinements.

These computational techniques streamline the drug development pipeline, reducing costs and time-to-market. Furthermore, they enable the exploration of novel therapeutic avenues that were previously inaccessible.

5.3 Patient-specific treatment plans

Personalized medicine is a transformative approach that tailors medical care to individual patients. This section delves into the integration of patient data, genomics, and AI to create patient-specific treatment plans that maximize therapeutic efficacy while minimizing adverse effects.

5.3.1 Genomic medicine

Understanding an individual's genetic makeup is central to personalized medicine. Genomic data, obtained through techniques like DNA sequencing, provide insights into a patient's genetic predispositions, allowing for treatments precisely aligned with their genetic profile.

5.3.2 AI-driven treatment plans

Artificial intelligence plays a pivotal role in analyzing vast genomic datasets. Machine learning algorithms identify genetic markers associated with diseases, predict disease risk, and recommend treatments tailored to an individual's genetic profile. These models consider genetic, environmental, and lifestyle factors.

5.3.3 Enhancing treatment efficacy

Personalized treatment plans enhance therapeutic outcomes by:

- **Targeted therapies**: Identifying the most effective drugs based on genetic markers.
- **Preventive measures**: Predicting disease risk and recommending preventive interventions.
- **Optimizing drug dosing**: Adjusting drug dosages to maximize efficacy while minimizing side effects.

By embracing personalized medicine, healthcare providers can significantly improve patient care, optimize resource allocation, and reduce healthcare costs.

6. Healthcare management and decision support

6.1 Clinical decision support systems (CDSS)

Clinical decision support systems (CDSS) are invaluable tools in modern healthcare and displayed in Fig. 10. This section explores their pivotal role in aiding healthcare providers in making informed decisions, optimizing patient care, and reducing medical errors through the application of computational intelligence.

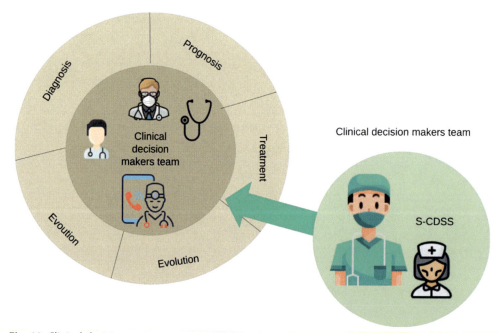

Fig. 10 Clinical decision support systems (CDSS).

6.1.1 Enhancing decision-making

CDSS leverages computational intelligence to analyze patient data, medical literature, and clinical guidelines. Machine learning algorithms process this vast information to generate recommendations for healthcare providers. These recommendations encompass:

- **Diagnosis support**: Assisting in the accurate diagnosis of diseases by analyzing patient symptoms and test results.
- **Treatment recommendations**: Suggesting evidence-based treatment options, including drug choices and dosages.
- **Alerts and warnings**: Notifying healthcare providers of potential drug interactions, allergies, or abnormal test results.

6.1.2 Reducing medical errors

One of the most significant benefits of CDSS is its ability to reduce medical errors. By providing healthcare providers with real-time information and decision support, CDSS minimizes the risk of misdiagnosis, incorrect treatments, and adverse events. This ultimately leads to improved patient safety and outcomes.

6.2 Predictive analytics for hospital resource allocation

Efficient resource allocation is critical for hospital operations and patient care. This chapter delves into the application of predictive analytics in optimizing hospital resource allocation. It highlights how AI models can forecast patient admissions, enabling hospitals to allocate resources like beds, staff, and equipment more effectively.

6.2.1 Predictive modeling

Predictive analytics models, often powered by machine learning algorithms, analyze historical patient data, seasonal trends, and other factors to forecast patient admissions. These models can predict the expected patient load with high accuracy, helping hospitals prepare for fluctuations in demand.

6.2.2 Resource optimization

Armed with accurate forecasts, hospital administrators can optimize resource allocation. For example:

- **Bed allocation**: Hospitals can proactively adjust bed availability based on expected admissions, reducing overcrowding and wait times.
- **Staff scheduling**: Predictive analytics aids in scheduling healthcare professionals according to expected patient loads, ensuring adequate staffing levels.
- **Equipment procurement**: Hospitals can plan equipment purchases and maintenance schedules based on predicted demand.

Optimized resource allocation not only enhances patient care but also contributes to cost savings and operational efficiency.

6.3 Telemedicine and remote patient monitoring

The healthcare landscape has witnessed a significant shift towards telemedicine and remote patient monitoring, especially in the context of computational intelligence. This section examines the growing importance of these technologies and how AI-driven solutions enable remote healthcare consultations and continuous patient monitoring. The telemedicine diagram is depicted in Fig. 11.

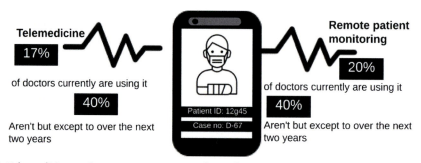

Fig. 11 Telemedicine and remote monitoring to boost home-based care.

6.3.1 Telemedicine

Telemedicine leverages video conferencing, AI-driven chatbots, and remote diagnostic tools to facilitate healthcare consultations at a distance. AI algorithms can assist in symptom assessment, offer treatment recommendations, and even support mental health services through virtual counseling.

6.3.2 Remote patient monitoring

Continuous monitoring of patients with chronic conditions or those in post-operative recovery is made possible through wearable devices and AI. These devices collect and transmit patient data, which AI algorithms analyze for abnormalities or trends. Healthcare providers receive real-time alerts and insights, allowing for timely interventions.

Computational intelligence adds value to telemedicine and remote monitoring by:

- **Enhancing diagnosis**: AI can assist in diagnosing conditions remotely, reducing the need for in-person visits.
- **Personalizing care**: AI analyzes patient data to tailor treatment plans and interventions.
- **Improving outcomes**: Remote monitoring ensures early detection of issues, leading to better patient outcomes and reduced hospital readmissions.

The integration of telemedicine and remote patient monitoring into healthcare systems has been accelerated by computational intelligence, offering more accessible and patient-centric care models.

7. Ethical and regulatory considerations

7.1 Patient privacy and data security

Patient data privacy and security are paramount in healthcare. This section addresses critical ethical considerations related to safeguarding patient information and complying with regulations like the Health Insurance Portability and Accountability Act (HIPAA).

7.1.1 Data privacy

Maintaining patient privacy is a fundamental ethical principle. Healthcare organizations and computational intelligence practitioners must implement robust data privacy measures. This includes:

- **Data encryption**: Ensuring that patient data, especially when transmitted or stored electronically, is encrypted to prevent unauthorized access.
- **Access controls**: Implementing strict access controls to limit who can view or modify patient records, ensuring that only authorized personnel have access.
- **Consent management**: Obtaining informed patient consent for data use, especially in research or when sharing data with third parties.

7.1.2 HIPAA compliance

In the United States, HIPAA sets stringent standards for protecting patient data. Ethical healthcare providers and AI developers must adhere to HIPAA regulations and compliance is shown in Fig. 12, which cover:

- **Privacy rule**: Regulating the use and disclosure of patient health information.
- **Security rule**: Mandating safeguards to protect electronic health information.
- **Breach notification rule**: Requiring notification of data breaches.

Fig. 12 HIPAA compliance.

By upholding these ethical standards and regulations, healthcare organizations ensure that patient data remains confidential and secure.

7.2 Regulatory compliance in healthcare AI

The use of computational intelligence in healthcare is subject to regulatory frameworks designed to safeguard patient safety and efficacy. This chapter outlines these regulatory frameworks and discusses the challenges and responsibilities associated with ensuring compliance.

7.2.1 FDA approvals

The U.S. Food and Drug Administration (FDA) regulates medical devices and software used in healthcare. Many AI-driven tools, especially those involved in diagnosis and treatment, require FDA approval. Ensuring compliance with FDA standards is essential for both AI developers and healthcare institutions.

7.2.2 Challenges and responsibilities

Compliance with healthcare regulations presents challenges, such as navigating complex regulatory processes and ensuring that AI algorithms meet safety and efficacy standards. The responsibility for compliance falls on both AI developers, who must rigorously test and validate their algorithms, and healthcare providers, who must implement and use AI tools in accordance with regulations.

7.3 Ethical implications of computational intelligence in medicine

The adoption of computational intelligence in medicine raises broader ethical questions beyond privacy and regulation. This section delves into the ethical implications of using AI in healthcare, including issues of bias, fairness, transparency, and accountability.

7.3.1 Algorithmic bias

AI algorithms can inadvertently perpetuate biases present in training data, leading to disparities in healthcare. Ethical considerations include identifying and mitigating bias to ensure equitable care for all patient populations.

7.3.2 Transparency and explainability

Patients and healthcare providers must understand AI-driven recommendations and decisions. Ensuring transparency and explainability in AI models is an ethical imperative. It involves making algorithms understandable and providing insights into how they arrive at specific recommendations.

7.3.3 Accountability and oversight

AI systems should have clear lines of accountability. Ethical guidelines necessitate identifying responsible parties for AI decisions and actions, ensuring that errors or adverse outcomes are addressed promptly.

Addressing these ethical considerations is crucial to building trust in AI-driven healthcare and ensuring that computational intelligence benefits all patients while minimizing harm.

8. Challenges and future directions

8.1 Interpretability and explainability of AI models

In the realm of healthcare, the interpretability and explainability of AI models are of paramount importance which is shown in Fig. 13. This section delves into the intricate challenge of making AI models transparent and interpretable in medical contexts, addressing the complexities that arise from the deployment of complex AI models and exploring techniques to enhance interpretability.

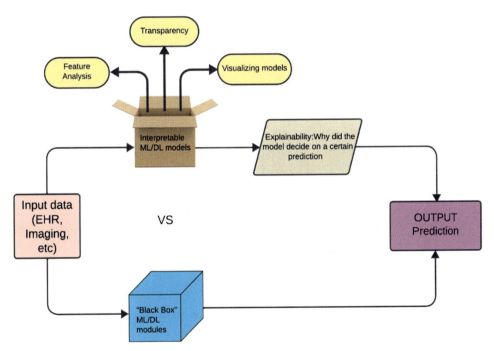

Fig. 13 Black box AI models versus interpretable and explainable AI models.

8.1.1 Complex AI models: The "black box" challenge

Modern AI models, especially deep learning models, have demonstrated remarkable predictive capabilities. However, they often operate as "black boxes," meaning that they make highly accurate predictions without providing clear explanations for their decisions. This inherent opacity presents a significant challenge in healthcare, where medical professionals and patients alike require insights into the reasoning behind AI-driven recommendations.

The Transparency Gap: The lack of transparency in complex AI models raises concerns about their trustworthiness in clinical settings. When healthcare professionals cannot understand why a particular diagnosis or treatment recommendation was made, they may hesitate to rely on AI-generated insights.

8.1.2 Interpretability techniques: Bridging the gap

Addressing the challenge of AI model interpretability involves the development and application of techniques designed to make AI decisions more transparent and interpretable:

1. **Feature importance**: One essential technique involves identifying which features or variables have the most significant impact on a model's predictions. This method allows healthcare professionals to understand which patient characteristics or clinical parameters are driving the AI's recommendations.
2. **Visualization**: Visual representations of AI model outputs and decision boundaries can simplify complex information. For instance, graphical displays can show how the model arrived at a particular conclusion, making it easier for medical professionals to grasp the reasoning.
3. **Local explanations**: Providing explanations for individual predictions is another valuable approach. By offering insights into why a specific recommendation was made for a particular patient, AI models become more transparent to healthcare practitioners. This can help build trust in AI-generated recommendations.

Improving the interpretability and explainability of AI models is essential not only for gaining the trust of medical professionals but also for ensuring that AI-driven decisions align with clinical expertise. By bridging the transparency gap and making AI more interpretable, computational intelligence can be harnessed more effectively to enhance medical diagnosis, treatment, and patient care while fostering collaboration between AI systems and healthcare professionals.

8.2 Integration with traditional medical practices

Traditional medicine encompasses a wide range of healthcare practices, including herbal medicine, acupuncture, Ayurveda, traditional Chinese medicine, and indigenous healing methods. These practices are rooted in cultural beliefs and have been handed down through generations.

8.2.1 Practices in traditional medicine

Traditional medicine often takes a holistic approach to health, considering the interconnectedness of the body, mind, and spirit. It relies on empirical observations, herbal remedies, and concepts like energy flow within the body.

8.2.2 Integration with computational intelligence

The integration of computational intelligence with traditional medicine involves merging traditional healing practices with modern technology and data-driven approaches:

1. **Data-driven insights**: Computational intelligence can analyze large datasets of traditional healthcare practices, outcomes, and patient experiences. This analysis can uncover patterns, trends, and potential areas for improvement.

2. **Personalized treatment**: AI algorithms can assist traditional healers in tailoring treatments to individual patients. By analyzing a patient's medical history, genetics, and lifestyle, AI can suggest personalized remedies and approaches.

3. **Diagnostic support**: Computational intelligence can aid in the diagnostic process by offering additional insights based on historical data and medical literature. This can complement the diagnostic expertise of traditional practitioners.

4. **Telemedicine**: Telemedicine platforms powered by AI can connect traditional healers with patients remotely, overcoming geographical barriers. Patients can receive advice and treatment recommendations from traditional healers using video consultations.

5. **Herbal medicine optimization**: AI-driven research can help identify the most effective herbal remedies for specific conditions. This can lead to the development of evidence-based herbal treatments.

6. **Public health**: Computational intelligence can contribute to public health efforts by analyzing data related to traditional medicine practices. This can inform public health policies and strategies.

7. **Ethical considerations**: Ethical frameworks should guide the integration process, ensuring that cultural sensitivities, privacy, and informed consent are respected.

8.2.3 Collaboration: A key enabler

Successful integration demands close collaboration between AI researchers, data scientists, and healthcare practitioners. AI experts must immerse themselves in the intricacies of clinical workflows, while healthcare professionals need to become familiar with AI capabilities and limitations.

8.2.4 Clinical validation: Ensuring efficacy and safety

Integrating AI into clinical practice necessitates rigorous validation to ensure alignment with established medical standards. This includes testing AI algorithms on diverse patient populations and benchmarking AI solutions against established clinical practices.

8.2.5 Workflow integration: Seamlessness is key

For AI tools to become integral to traditional medical practices, they must seamlessly integrate into existing healthcare workflows. User-friendly interfaces, integration with electronic health records (EHRs), and effective communication channels are crucial for successful integration.

The integration of computational intelligence into traditional medicine seeks to enhance the effectiveness and accessibility of traditional healing practices while respecting cultural traditions and beliefs. This fusion of traditional wisdom with modern technology holds promise for improving healthcare outcomes and expanding access to holistic healthcare solutions.

8.3 Incorporating real-time data streams

In today's healthcare landscape, the integration of real-time data streams from sources like wearable devices and the Internet of Things (IoT) is paramount. This section explores the profound impact of these streams on healthcare AI systems, discussing both their benefits and the challenges that come with utilizing real-time data for medical decision-making.

8.3.1 Benefits of real-time data

Real-time data streams offer numerous advantages that are transforming healthcare delivery and patient outcomes:

1. **Timely interventions**: Continuous monitoring allows for the early detection of health anomalies. For instance, wearable devices can instantly alert healthcare providers or patients themselves to irregular heart rhythms, indicating potential cardiac issues. This timeliness enables swift interventions, preventing adverse events and reducing the severity of conditions.

2. **Personalized care**: The dynamic nature of real-time data allows for the tailoring of treatment plans to individual patients. For example, diabetes management can be personalized based on real-time glucose levels, physical activity, and dietary habits. AI-driven systems can adapt treatment recommendations in response to changing health statuses, optimizing care plans for each patient.

3. **Remote monitoring**: Real-time data streams enable remote monitoring, reducing the need for frequent in-person visits, especially for patients with chronic conditions. This not only enhances patient comfort but also reduces healthcare costs and the burden on healthcare facilities. The real time monitoring flow is shown in Fig. 14.

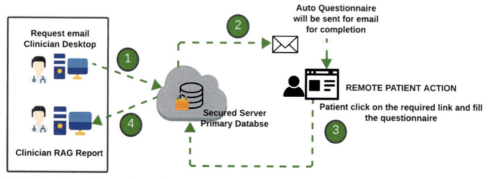

Fig. 14 Real time information exchange.

8.3.2 Challenges and data quality

While real-time data streams offer immense potential, their integration into healthcare AI systems poses several challenges:

1. **Data quality**: Real-time data must be of high quality to ensure the accuracy and reliability of AI-driven insights. Sensor accuracy, data transmission integrity, and signal noise can all impact data quality. Data validation and cleaning processes are essential to maintain data integrity.

2. **Privacy and security**: Real-time healthcare data often includes sensitive patient information. Ensuring patient privacy and data security is a paramount concern. Healthcare organizations must implement robust encryption, access controls, and data anonymization techniques to protect patient confidentiality.

3. **Scalability**: The continuous influx of real-time data can strain healthcare IT infrastructures. Scalable solutions are needed to handle the growing volume of data generated by wearables, IoT devices, and remote monitoring systems.

4. **Interoperability**: Real-time data often comes from diverse sources and devices. Ensuring interoperability between different data streams and AI systems is crucial for comprehensive patient monitoring and decision support.

Incorporating real-time data streams into healthcare AI systems requires a thoughtful approach that addresses these challenges. When successfully implemented, these systems can revolutionize patient care by providing timely, personalized, and remote monitoring solutions that enhance both patient and provider experiences while improving health outcomes.

9. Case studies and applications

This chapter delves into the practical application of computational intelligence in bio and clinical medicine through real-world case studies. Each case study highlights the transformative impact of AI on healthcare, focusing on specific medical domains and their respective challenges and outcomes.

9.1 Case study 1: Early detection of cardiovascular diseases

Introduction: Cardiovascular diseases (CVDs) are a leading cause of mortality worldwide. Early detection and intervention are crucial for improving patient outcomes. This case study illustrates how computational intelligence plays a pivotal role in the early detection and prevention of CVDs.

Methods: The case study details the methods employed, which may include machine learning algorithms, analysis of medical imaging data, and integration with patient health records. It discusses the selection of relevant features, data sources, and the development of predictive models.

Data sources: The case study outlines the sources of data used in the study. This may encompass patient medical records, cardiac imaging data, genetic information, and lifestyle factors. It emphasizes the importance of high-quality data for accurate predictions.

Outcomes: The outcomes of the case study are discussed, including the accuracy of CVD risk prediction, the reduction in false positives, and the impact on patient care. It may also touch upon the potential for early intervention to prevent cardiac events.

9.2 Case study 2: Cancer diagnosis and treatment planning

Introduction: Cancer is a complex and multifaceted disease, requiring precise diagnosis and tailored treatment plans. This case study sheds light on how computational intelligence is revolutionizing cancer diagnosis and treatment planning.

Methods: The case study elaborates on the methods employed, which may involve image analysis, genomics, and machine learning algorithms. It discusses how AI models are trained on vast datasets to recognize cancerous patterns and predict treatment responses.

Data sources: An overview of data sources is provided, which may include medical imaging data (e.g., mammograms, MRIs), genetic data, and patient clinical records. The integration of diverse data types for holistic cancer diagnosis is emphasized.

Outcomes: The case study showcases the impact of AI on cancer diagnosis accuracy, the ability to identify early-stage cancers, and the optimization of treatment plans. It may also highlight how AI contributes to personalized treatment strategies that consider individual patient characteristics.

9.3 Case study 3: Neurological disorders and computational neuromedicine

Introduction: Neurological disorders pose significant challenges in diagnosis and treatment. This case study explores the application of computational intelligence in understanding and managing neurological disorders, such as Alzheimer's disease or Parkinson's disease.

Methods: The case study delves into the methods used, which may include neuro-imaging analysis, predictive modeling, and data-driven insights. It discusses how AI aids in early disease prediction and tracking disease progression.

Data sources: An overview of data sources is provided, encompassing neuroimaging data (e.g., MRI, PET scans), genetic data, and patient clinical histories. It emphasizes the importance of longitudinal data for studying neurological disorders.

Outcomes: The case study outlines the outcomes, including advancements in early diagnosis, disease prediction, and personalized treatment plans for neurological disorders. It underscores the potential for AI to improve the quality of life for patients with such conditions.

These case studies exemplify the practical impact of computational intelligence in bio and clinical medicine, demonstrating how AI-driven approaches are revolutionizing disease detection, diagnosis, and treatment across diverse medical domains. They serve as powerful examples of the transformative potential of computational intelligence in improving patient care and outcomes.

10. Conclusion

The concluding chapter serves as the culmination of the entire book, offering a comprehensive synthesis of the key insights and implications surrounding computational intelligence in the fields of bio and clinical medicine.

10.1 Key takeaways

In this section, we revisit and distill the critical takeaways from the entire chapter. These takeaways encapsulate the overarching themes and contributions of computational intelligence to bio and clinical medicine, ensuring that readers leave with a clear understanding of the book's core messages.

Advancing healthcare: The conclusion underscores the transformative power of computational intelligence in advancing healthcare. It reiterates how AI and related technologies have the potential to reshape medical practices, from early disease detection to treatment optimization.

Data-driven medicine: A key takeaway highlights the shift towards data-driven medicine, where large datasets, AI algorithms, and predictive analytics play pivotal roles in diagnosis, treatment planning, and patient care.

Interdisciplinary collaboration: The importance of interdisciplinary collaboration is emphasized, as the book recognizes that computational experts and medical professionals must work together to harness the full potential of AI in healthcare.

10.2 Impact and future prospects

This section delves into the broader impact of computational intelligence on the healthcare landscape and its future prospects. It elaborates on the transformative effects that computational intelligence has already had and is likely to continue having on medical practices.

Improved patient outcomes: The impact of computational intelligence in improving patient outcomes is discussed in detail. The chapter highlights how AI-driven early detection, accurate diagnosis, and personalized treatment plans lead to better patient care.

Cost reduction: The potential for computational intelligence to reduce healthcare costs is explored. By streamlining processes, minimizing errors, and optimizing resource allocation, AI can contribute to cost-effective healthcare delivery.

Continued innovation: The section touches upon the ever-evolving nature of computational intelligence. It emphasizes that as technology advances, healthcare practices will continue to evolve, offering even more innovative and effective solutions.

10.3 Call for collaboration between computational experts and medical professionals

The concluding chapter concludes with a compelling call to action. It emphasizes the need for collaboration between computational experts and medical professionals to drive further advancements in healthcare through computational intelligence.

Teamwork: The chapter stresses that successful integration of computational intelligence into healthcare requires teamwork. Computational experts and medical professionals must collaborate closely to develop and implement AI-driven solutions that are both effective and ethically sound.

Ethical considerations: The call for collaboration extends to ethical considerations. The importance of addressing ethical and regulatory concerns in a concerted effort between both groups is highlighted.

Patient-centric approach: The book's conclusion reinforces the commitment to a patient-centric approach. It underscores that the ultimate goal of computational intelligence in bio and clinical medicine is to enhance patient well-being and healthcare quality.

This conclusion serves as a fitting endnote to your book, reinforcing the critical role of computational intelligence in shaping the future of bio and clinical medicine. It invites readers and professionals from various fields to join in the collective effort to harness the potential of AI and related technologies to improve healthcare on a global scale.

References

[1] M. Cui, D.Y. Zhang, Artificial intelligence and computational pathology, Lab. Investig. 101 (4) (2021) 412–422.

[2] H. Shimizu, K.I. Nakayama, Artificial intelligence in oncology, Cancer Sci. 111 (5) (2020) 1452–1460.

[3] A.C. Chang, Intelligence-Based Medicine: Artificial Intelligence and Human Cognition in Clinical Medicine and Healthcare, Academic Press, 2020.

[4] A. Bhardwaj, S. Kishore, D.K. Pandey, Artificial intelligence in biological sciences, Life 12 (9) (2022) 1430.

[5] L. Xing, M.L. Giger, J.K. Min (Eds.), Artificial Intelligence in Medicine: Technical Basis and Clinical Applications, Academic Press, 2020.

[6] M. Shanmuga Sundari, V.C. Jadala, Neurological disease prediction using impaired gait analysis for foot position in cerebellar ataxia by ensemble approach, Automatika 64 (3) (2023) 541–550.

[7] M. Shanmuga Sundari, V.C. Jadala, S.K. Pasupuleti, P. Yellamma, Deep learning analysis using ResNet for early detection of cerebellar Ataxia disease, in: 2022 International Conference on Advancements in Smart, Secure and Intelligent Computing (ASSIC), Bhubaneswar, India, 2022, pp. 1–6, https://doi.org/10.1109/ASSIC55218.2022.10088379.

[8] J.V. Chandra, S.K. Pasupuleti, Machine learning methodologies for predicting neurological disease using behavioral activity mining in health care, in: 2022 8th International Conference on Advanced Computing and Communication Systems (ICACCS), vol. 1, IEEE, 2022, March, pp. 1035–1039.

[9] L. Rundo, A. Tangherloni, C. Militello, Artificial intelligence applied to medical imaging and computational biology, Appl. Sci. 12 (18) (2022) 9052.

[10] M. Subramanian, A. Wojtusciszyn, L. Favre, S. Boughorbel, J. Shan, K.B. Letaief, et al., Precision medicine in the era of artificial intelligence: implications in chronic disease management, J. Transl. Med. 18 (1) (2020) 1–12.

[11] R. Hamamoto, Application of artificial intelligence for medical research, Biomol. Ther. 11 (1) (2021) 90.

[12] P. Mathur, S. Srivastava, X. Xu, J.L. Mehta, Artificial intelligence, machine learning, and cardiovascular disease, Clin. Med. Insights: Cardiol. 14 (2020) 1179546820927404.

[13] M.J. Iqbal, Z. Javed, H. Sadia, I.A. Qureshi, A. Irshad, R. Ahmed, et al., Clinical applications of artificial intelligence and machine learning in cancer diagnosis: looking into the future, Cancer Cell Int. 21 (1) (2021) 1–11.

[14] K. Athanasopoulou, G.N. Daneva, P.G. Adamopoulos, A. Scorilas, Artificial intelligence: the milestone in modern biomedical research, BioMedInformatics 2 (4) (2022) 727–744.

[15] Z. Ahmed, K. Mohamed, S. Zeeshan, X. Dong, Artificial intelligence with multi-functional machine learning platform development for better healthcare and precision medicine, Database 2020 (2020) baaa010.

CHAPTER 4

Enhancing medical AI interpretability using heatmap visualization techniques

Elham Moeini[a], Duygu Cakir[b], and Nafiz Arica[c]
[a]Computer Engineering, Graduate School of Engineering, Bahcesehir University, Istanbul, Turkey
[b]Software Engineering, Faculty of Engineering and Natural Sciences, Bahcesehir University, Istanbul, Turkey
[c]Information Systems Engineering, Faculty of Engineering, Piri Reis University, Istanbul, Turkey

1. Introduction

Artificial intelligence (AI) methods used to be easily interpretable until more complex deep neural networks (DNNs) came along. A DNN contains an input layer, an output layer, and hidden layers that have hundreds of layers and millions of parameters and make the model a complex black box. Black-box models are increasingly being employed to make predictions in critical contexts that have very good performance. However, as these systems get more complicated, they increasingly become more opaque and less interpretable and trustworthy. There is a need for methods that can help us peer into the mysterious black box. Interpretable deep learning (DL) or explainable artificial intelligence (XAI) are two common terms for these methods. Since these two concepts are often used interchangeably, we shall refer to them as XAI. Numerous notable XAI initiatives have received funding from the US Defense Advanced Research Projects Agency (DARPA).

XAI is a rapidly developing field of research focused on creating machine learning (ML) algorithms and models that can be easily understood and explained by humans. With the rise of ML and AI, there has been a growing concern about the *black-box* nature of these systems, meaning that the internal workings of the algorithms are often opaque and difficult for humans to interpret. This lack of transparency can lead to mistrust, as well as concerns about privacy, bias, and other ethical considerations. XAI seeks to address these concerns by developing methods that can provide precise and straightforward explanations of their decision-making processes, making them more convenient and reliable for users. XAI has the potential to remake the way we employ AI techniques, with considerable implications for domains such as healthcare, finance, and national safety.

Investigators in XAI are studying manifold methods to address this problem, including rule-based techniques, symbolic reasoning, and machine-learning approaches. While these methods have shown promise in developing explanations for simple decision-making scenarios, they face considerable challenges when dealing with complex,

Explainable AI in Healthcare Imaging for Medical Diagnoses
https://doi.org/10.1016/B978-0-443-23979-3.00004-X

DL-based methods. Similarly, there is still a lack of agreement on what comprises a good explanation, and how to evaluate the quality of explanations supplied by an AI strategy.

In computer vision tasks, especially in medical images, experts in the medical field are naturally suspicious of the opaque nature of DL, the current gold standard in medical image processing. Patients also have a right to be informed of the rationale behind a choice, as is stated in regulations such as the EU's General Data Protection Regulation (Article 15). Researchers in the field of medical imaging are increasingly turning to XAI to justify the outcomes of their algorithms. A satisfactory explanation may give insight into the neural network's reasoning and make its conclusion comprehensible. In the field of medicine, computer-aided diagnosis (CAD) is rapidly becoming a cutting-edge resource. Due to the potential risks involved with endoscopic resections, it is recommended that CAD for colon polyps be a very beneficial tool for trainee colonoscopies. Traditional CAD has been joined by a convolutional neural network (CNN) system powered by AI that has experienced remarkable growth over the last 5 years.

As the third major cause of cancer-related death in India and the fourth most common reason for cancer-related death worldwide, colorectal cancer is a serious health concern worldwide. So, to save a patient's life, a quick and correct diagnosis is needed. Polyps, which may or may not contain cancer cells, are the primary cause of cancer growth. So, the dangerous effects of cancer can be significantly decreased if cancerous polyps are found and removed early. Colorectal polyps can be found using a colonoscopy. Manual examinations by experts, on the other hand, can have many mistakes. So, some researchers have used models based on ML and DL in order to automate the process of diagnosis. However, existing models have problems with overfitting and gradients that go away. To solve these problems, a DL model that is based on CNNs is proposed.

The purpose of this research is to use a post hoc visualization method for explainability and find the best one based by calculating the Intersection over Union (IOU) between the heatmap and ground truth image and by combining some of the best methods of class activation map (CAM) and layer-wise relevance propagation (LRP), and then compare them to evaluate the heatmap and its effect on the score of IOU.

The current work provides a significant addition to the field by suggesting a method that applies CAM and LRP models after classifying images and then computing IOU between the generated heatmaps and ground truth images. This innovative strategy aims to use the IOU as an evaluation metric to show how much the heatmaps model covered the interested area. The research also suggests combining the techniques with the highest IOU values to boost the effectiveness of the heatmaps which could cover most parts of the interested area.

The remainder of this chapter is organized as follows: Section 2 contains recent advancements in the field and briefly explains ML and the history of AI. Section 3 details experiments, the work and challenges faced in the process of training, and the result achieved, highlighting the main findings and visual results. Section 4 discusses the findings and concludes the research.

2. Related work

Explanation and understanding of the decisions made by ML models are important concerns for both developers and users. The ability to comprehend how a model makes predictions is crucial to establishing trust and improving performance. XAI has been a topic of interest in recent years, addressing the issue of model interpretability.

The field of ML has recently made significant progress, not only opening a wide variety of potential applications for AI but also highlighting some of the limitations of these newer systems. Modern developments in medicine have highlighted the essential trust issues in these professions. Due of their inherent inability to provide justifications for their actions, AI systems have been widely criticized as one of the most significant barriers to realizing their full potential [1]. It is not easy to come up with convincing answers, despite the fact that there is a lot of background and effort to work with. The actions taken by AI systems increasingly become less comprehensible to human users as their power grows with millions of parameters and they get more sophisticated, and the models respond and decide more independently [2, 3].

AI systems are often built to use training data in order to discover a model that most accurately achieves a specified goal. The goal might be to determine, with the highest possible accuracy, whether a tumor is benign or malignant, given a set of diagnostic criteria. While the AI could come up with some helpful, generalizable principles like "benign tumors have a slower growth rate," it might also pick up some erroneous rules. Inappropriate connections in the training data, such as connecting a tumor's diagnosis and ID number, might increase the probability of this occurrence. If these wrong rules are discovered and subsequently used on data from the actual world, there can be severe consequences especially in the medical field because it can put human life at risk. But if AI systems could explain their results, these types of faulty rules could be simpler to identify and the model itself might be more trustworthy. Fortunately, human cognitive abilities promote the interpretation of visual input, making the route to explainability for a CNN simpler than for other kinds of models [4].

DNN models need a huge number of images and many GPUs or TPUs to train. When training on massive datasets, DNNs may take a considerable amount of time. It is possible to decrease this time by half by using model weights from models that have already been trained on common computer vision benchmark datasets like the ImageNet image recognition tasks. The model has been trained and reused for new problems known as transfer learning. Transfer learning refers to the process of applying what has been gained to a different scenario. Several state-of-the-art achievements in image classification really rely on transfer learning approaches [5–7]. Transfer learning in computer vision is often implemented by deploying previously learned models. Accordingly, due to the computational cost of training such models, it is common practice to import and use models from published literature (e.g., VGG, Inception, MobileNet). There are lots of pretrained models for image classification, but here in this research, the VGG16 model is used.

Visual Geometry Group (VGG) is a CNN architecture that was proposed in 2014 by Simonyan and Zisserman [7]. It is used for image classification tasks, which are very powerful and influential. The VGG network consists of a series of convolutional layers, followed by max-pooling layers, and at the end, includes a few fully connected layers. To detect and extract high-level features, the convolutional layers in CNNs employ small filters (3×3) with a stride of one and zero padding, while the max-pooling layers use a filter size of 2×2 and a stride of two. The network also uses rectified linear unit (ReLU) activations, which help it converge faster.

One of the main contributions of the VGG architecture was the use of deeper networks. Simonyan and Zisserman showed that increasing the depth of the network improves its performance on image classification tasks. They also found that using smaller filters with a smaller stride leads to better performance. The VGG architecture has been widely used and adapted for various computer vision tasks, such as object detection, semantic segmentation, and image captioning.

2.1 Explainable artificial intelligence

The field of XAI is rapidly expanding, with a focus on creating ML algorithms and AI systems that can offer transparent and comprehensible explanations for their decisions. As AI technologies become more sophisticated and pervasive, it is becoming increasingly important to ensure that these systems are transparent and accountable to their users and stakeholders. XAI is an interdisciplinary field that draws on computer science, statistics, psychology, and philosophy to develop new techniques and methods for creating transparent and interpretable AI systems. It aims to allow humans to comprehend the reasoning and decision-making procedures of AI methods so that we can rely and trust on them for vital tasks such as medical diagnosis, financial forecasting, and autonomous driving. In this case, XAI can transform the way we interact with AI approaches and improve our confidence in their capabilities.

One of the considerably important areas of study in XAI is the improvement and development of explainable models. These models seek to provide human-understandable explanations for their decisions by offering explanations that are obvious, straightforward, and significant to the end user. Some of the strategies employed to perform this include rule-based methods, decision trees, and linear models. While these strategies provide explanations that are easily understood, they may not be as precise or robust as more complex models.

The vision of XAI has been around for a while. XAI research goes back a significant amount of time [8, 9]. In the last decade, ML has become quite popular. Large datasets and increased computational power have allowed ML systems to surpass human performance in several tasks (deep blue surpassing the greatest chess player, IBM Watson beating the best in Jeopardy! and AlphaGo beating the best in Go). In the past, when AI algorithms were simpler to understand, there was no need to explain the reasoning behind the systems' decisions. Even the simplest DNN uses methods that are difficult to understand in today's world. As more and more of these technologies made their way into daily use, the necessity for an XAI became increasingly evident.

The complexity of these systems, however, meant that their outputs were unreliable and incomprehensible. The demand was especially great in fields like healthcare, law, military, and finance because the results of these systems had a real effect on people's lives [4].

XAI is required for several reasons [10]. First, explanations might function as a kind of system verification. When an AI system is trained on a dataset with inherent biases, it may produce incorrect results, which might be easily detected if the reasons behind those conclusions were clearly explained [11]. Second, understanding the AI system's weaknesses is crucial for making modifications to it. Black boxes' weaknesses are difficult to detect. Third, the explanations could teach us something new. Today's AI systems are able to discover new links and provide new insights since they are trained using large amounts of data that are sometimes unavailable to humans. Indeed, justifications are now a matter of law as much as they are of explanation. Every person has the right to an explanation, as stated in the new EU rule. This requires that all judgments made by AI systems also offer an explanation [12].

Both research groups and research institutions have noticed how important it is to be able to explain and understand something. As an example, which has been written in the introduction, the XAI program was established by DARPA to develop machine-learning techniques that, on the one hand, make models that are easier to understand without losing accuracy, and on the other, let people understand, trust the model, and control these AI tools [3].

2.2 Interpretability vs. explainability

The literary phrases "interpretability" and "explainability" sometimes have the same meaning in literature, but there is a distinction between them. Doshi-Velez and Kim defined *interpretability* of ML systems as their capacity to provide explanations in words that a human can comprehend [13]. Contrarily, Montavon, Samek, and Müller defined *interpretation* which is the mapping of abstract concepts into a domain that humans can understand, while *explanation* is the collection of features from easily understandable domains that have contributed to a given example in order to produce a decision [14]. *Transparency* is often used to describe that a model's interpretability is clear and simple.

Interpretability refers to the ability of an AI system to produce outputs that can be understood and interpreted by humans. An interpretable model allows users to understand how this decision has been taken, which can help build trust and confidence in the model's reliability.

Explainability, on the other hand, goes a step further by providing a clear and understandable justification for why a model or system produces a particular output or decision. Explainable frameworks are developed to generate explanations that are efficiently comprehensible and available to users, to provide insights into the model's decision-making procedure.

Explanation techniques can be classified based on the range of models they can explain and the extent of their explanations. Depending on how they are presented, these techniques can be further subdivided. This research is solely concerned with feature relevance or attribution explanations. These types of explanations illustrate how each input feature affects the model's prediction. In the case of medical imaging, the input features correspond to individual or multiple pixels in the input image, as noted by Molnar [15]. Explanation approaches are typically divided into global and local explanations, which correspond to the extent of their explanations. Global explanation methods aim to explain an entire model at once, including its decision-making process across the input space. Although such explanations can provide valuable information in cases where ML models outperform humans, they are often challenging to obtain. As noted by Molnar [15], local explanation techniques focus on explaining how a model arrives at a specific prediction. This approach can provide explanations even for models that are too complex to explain globally. Local explanation methods aim to clarify how a model arrives at a specific prediction. The approach is based on the notion that a detailed examination of a particular model prediction can provide insights even for models that are too intricate to explain globally.

Explanation techniques in XAI can be categorized based on their level of specificity, which can be either model-agnostic or model-specific. Model-agnostic techniques are preferred when explaining an alternative model because they consider the model as a "black box" and focus only on the inputs and outputs. However, model-agnostic techniques tend to be computationally complex, which limits their flexibility. In contrast, model-specific techniques are more computationally efficient and can leverage information from a model's internals, such as backpropagation of gradients, to explain a particular type of model. For instance, in CNNs, model-specific techniques can provide insights into the learned higher-level features by accessing the network's feature maps.

In the field of vision, heatmaps are a common way of representing the importance of each pixel in an input image. Heatmaps assign relevance scores to individual pixels, which indicate their contribution to a particular decision. Post hoc explanation methods are often used in the vision domain, which involves applying the explanation technique after the neural network has been trained. There are several post hoc explanation methods that can generate heatmaps in a predetermined and unambiguous manner on a previously trained neural network. Examples of such methods include Class Saliency Map, Grad-CAM, Gradient Input, Integrated Gradients, LRP, Excitation Backprop, and Guided Backpropagation. Another category of explanation techniques involves introducing randomness into the heatmap calculation. This can be achieved by generating extra perturbed training data samples or by solving an ad hoc optimization problem to produce a single heatmap. This approach is different from post hoc methods and was proposed by Zintgraf et al. [16].

Due to the limited amount of data available during prediction, the main focus of this research is on the first set of methods discussed in the study. These methods rely on a single input data point and the trained parameters of the neural network. Usually, these explanatory techniques involve performing a backward pass over the network or are based on gradient integration. According to Samek et al. [10], sensitivity analysis (SA) and LRP are two common methods for explaining the predictions of DL models. These methods, for instance, have been used to classify images and text documents and to identify human actions. Especially in image classification, a lot of progress has been made since the explanations are easy to understand. SA is a widely used method. This method computes gradients in relation to the inputs (parameters). This theory indicates that the most important input factors are those that have the greatest influence on the outcome. SA was applied to various pixels in an image by Samek et al. [10] showing that one may learn which pixels have the most impact on the decision.

As another common strategy, LRP breaks down the prediction into relevance scores and explains the decision using those scores [10]. LRP is different from previous approaches since it does not rely on gradient evaluations. In addition, GradCAM is often used to produce saliency maps, which show which pixels in the picture are the most important [17].

Previous studies that aimed to validate visual explanations focused on applying them to real-world image classification problems and analyzing the effects of pixel perturbations [10]. While evaluations based on tracking changes in model prediction when perturbing pixels or localizing an object's bounding box may be suitable in the absence of ground truth explanations for a given task, they can create a mismatch between the primary goal of the explanation (to clarify the model's current decision), which may be influenced by factors such as the image's context or dataset biases and the evaluation criteria [18].

To ensure that the interpretability of the explanations is indeed dependent on the model's parameters and the data generation process, another approach that is based on perturbation techniques, employed which involves randomly modifying the model's weights or training data [19]. In the latter case, the model is retrained on the same images with class labels randomly permuted, and the similarity between the original and the randomized explanations is measured. While such studies can be useful for verifying the coherence of the explanations, they cannot replace a comprehensive evaluation of the explanation's quality regarding a specific prediction. Another approach proposes measuring the cosine similarity of explanations throughout the lower layers of the model when the relevance is initially randomized at a given higher layer [20], to assess whether backward propagation-based explanations converge to a single direction throughout the propagation process. The disadvantage of all perturbation-based assessments is that they require using either input data or a model that is not the target model being implemented and explained at test time, and that they can be computationally expensive.

2.3 Explaining medical images by visualization methods

It has been shown through previous research that CNNs' convolutional units in the deeper layers can learn to locate objects within images despite being trained only for the task of image classification [21]. This method was first employed in the context of XAI, where object detection models may be built without the need for extensively annotated images with bounding boxes [22, 23].

This section discusses how to utilize visualization tools to explain image data. Section 2.3.1 covers the methods used for creating heatmaps. Moving on, the approach outlined as an evaluation measure is discussed in further detail in Section 2.3.2. Section 2.3.3 emphasizes this research significance in highlighting the need for mixing visualization approaches. Finally, Section 2.3.4 describes how to apply Otsu's method to binarize the resulting heatmap, which is used to calculate the IOU between the heatmap and ground truth images.

2.3.1 Visualization methods

The importance of the CAM method lies in the activation maps of the final convolutional layer that reveal the location of a particular convolutional filter in detecting the learned abstract feature of an input image. To create a saliency map, the activation maps that are most significant in recognizing a target class are selected, and their collective is magnified.

The core of CAM is the class activation maps (CAM) in the final convolutional layer, which indicate where a certain convolutional filter has located the corresponding learned abstract feature of an input image. A saliency map is produced by choosing the activation maps that were most important for identifying a target class and then scaling up the collective of those maps.

Grad-CAM

Grad–CAM is an explanation technique developed particularly for CNNs [24], which backpropagates the gradient of the model's output to the (typically) final convolutional layer rather than all the way back to the input image. The CAM approach has been enhanced with the Grad–CAM technique [21]. By using the gradient of the class score with respect to the activation maps, the Grad–CAM technique overcomes the primary apparent constraint of CAM, which is the need to train an extra linear classifier in order to create explanations. According to Selvaraju et al. [24], CAMs may be created without training a linear classifier by using the gradients of class scores with respect to the activation maps in the final convolutional layer to calculate the weights.

Grad-CAM++

A generalization of the Grad–CAM method introduced by Chattopadhyay et al. [25], where the approach is called Grad–CAM++. Its authors observed that Grad–CAM frequently fails to adequately highlight the whole object or all the objects in an image when an object has many instances of the same class. Grad-CAM++ solves this problem by providing pixel-wise weights for the activation map, class of interest, and spatial location.

Layer-wise relevance propagation

The LRP method is also a popular technique used in DL to explain the behavior of neural networks. This method is useful in understanding the factors that contribute to a neural network's output, making it a valuable tool for debugging and interpreting neural network models. This section will explore the LRP method, its key components, and its applications in DL [10].

The LRP method assesses the significance of each input feature in a neural network's output by providing significance estimates for each feature and propagating these scores back via the network layers. It depends on "layer-wise decomposition" and "top-down propagation" concepts to show relevancy scores for each input feature. The LRP strategy has three components: a predefined propagation rule, rules for nonlinear activation functions, and rules for pooling layers. The propagation rule determines how each neuron in the network is assigned relevancy scores. The rule is intended to distribute relevance scores in such a way that the sum of the relevancy scores stays consistent across the network. This confirms that the relevancy scores are preserved as they are propagated back through the network layers.

2.3.2 Intersection over Union

In IOU, we first train a model to develop a box that precisely covers an item, making it valuable in applications involving object identification. The purpose of this model is to enhance its prediction to the point where the ground truth box and detected box entirely cover, or until the IOU between the two boxes is equivalent to one. At this point, the ground truth is the mask that detects the polyp detail, and the detected box is the scope that the heatmap indicates.

2.3.3 Max-voting method

Max-voting method is one of the most popular methods for combining two heatmaps. This method applies a straightforward and effective approach for merging the two heatmaps to create a final heatmap that highlights the most relevant areas of the input image for the assigned task. To merge these two heatmaps, the max-voting approach is employed, which concerns determining the maximum value of the two heatmaps at each pixel location. This technique ensures that the most salient regions of the input image are considered in the final merged heatmap. The procedure for merging the LRP and CAM heatmaps employing the max-voting strategy can be outlined as follows:

- Ensure that the heatmaps can be approximated and merged, and then normalize them to a similar scale, such as [0, 1].

- Utilize the final integrated heatmap to interpret the neural network's decision-making procedure and to visualize the considerably appropriate regions of the input image for the assigned task.

In general, the max-voting technique presents a simple yet effective approach to integrate a couple of heatmaps and may be utilized to heatmaps developed via considerable models or procedures.

2.3.4 Otsu's binarization

Otsu's binarization technique is an efficient methodology for automatically thresholding grayscale images to create binary images. This approach, which Nobuyuki Otsu first developed in 1979 [26], is often utilized in image processing and computer vision applications. The fundamental goal of Otsu's approach is to identify the ideal threshold value that divides the pixels of the image into the foreground and background classes. The threshold value is selected such that the variance between the two classes is maximized, ensuring the greatest possible separation between them. Here are the steps of explanation of how Otsu's method works:

- Apply the Gaussian filter to delete the noises.
- Create the grayscale image's histogram. This information reveals how often each gray level in the picture occurs.
- Calculate the histogram's cumulative distribution function.
- Calculate the image's mean gray level by adding the products of the gray levels and the frequencies that correspond to them, then dividing by the total number of pixels in the image.
- Calculate the between-class variance.
- Select the threshold value that increases the variation across classes.
- Apply the threshold to the image using the chosen threshold value to obtain a binary image.

Otsu's approach is straightforward, computationally effective, and often yields positive outcomes for a variety of images. However, it may not perform well for images with intricate backgrounds or foregrounds since it might be sensitive to noise and uneven illumination.

3. Experiments

This section provides a detailed description of the experimental settings, findings, and results. Section 3.1.1 delves into the two datasets used in the research, namely Kvasir and HAM10000. Section 3.1.2 discusses the preprocessing step that was required prior to model training. Section 3.1.3 covers the implementation specifics, such as the training procedure and model finetuning. Finally, Section 3.2 presents the experimental findings, which include both classification and visualization outcomes.

3.1 Experimental settings

This section is structured to provide the details about the datasets that have been experimented with, preprocessing techniques that have been employed over the datasets, and the technical implementation details.

3.1.1 Dataset setup

Skin lesion classification is crucial for the early detection and treatment of melanoma, a potentially fatal skin cancer. Skin lesion analysis can be done visually by dermatologists or with the help of CAD systems. This research involves two medical dataset images: the Endoscopy Polyp Dataset [27] and HAM10000 [28]. The Endoscopy dataset contains ground truth images that have been defined by specialists and was used in the EndoTect 2020 Challenge. The HAM10000 ("Human Against Machine with 10000 training images") dataset is a large collection of skin lesion images, labeled with seven possible diagnostic categories, which has been widely used for developing machine-learning models for skin disease classification. The dataset has strengths such as its large size, high quality, and diverse range of lesion types, but it also has limitations such as the overrepresentation of certain diagnostic categories. This section will cover the endoscopy datasets and skin cancer datasets that have been experimented with throughout this research in detail.

Mask data

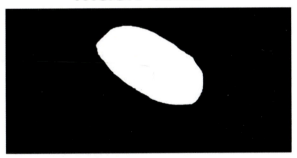

Original image

Fig. 1 Sample image of the polyp dataset and the mask image.

Endoscopy datasets

In this research, two polyp datasets have been employed: for the training part HyperK-vasir and for the test dataset Edotec 2020 datasets were used. These datasets were extracted from the video since each image has a different size, a publicly available dataset of images of gastrointestinal polyps and their corresponding segmentation masks (Fig. 1), labeled by a doctor and double-checked by a professional gastroenterologist. HyperKva-sir is an open-access dataset that can be downloaded for research purposes at: https://datasets.simula.no/hyper-kvasir/. The Edotec2020 dataset is an open-access dataset that can be downloaded at: https://endotect.com/index.html. The number of training samples for nonpolyp is 766 images and for polyp it has 1028 samples. For testing, the number of nonpolyp is 85 and the number of polyps is 200.

Skin cancer dataset

The HAM10000 dataset, published by the HARVARD Dataverse, is a large collection of skin lesion images designed to aid in the development of CAD systems for melanoma detection. The dataset consists of 10,015 images of skin lesions, which are labeled with one of seven possible diagnostic categories, and shows sample of class in Fig. 2: Actinic

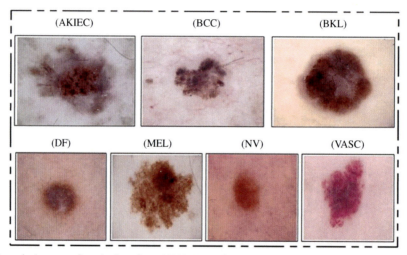

Fig. 2 Sample images of each class from HAM10000 dataset.

keratoses and intraepithelial carcinoma/Bowen's disease (akiec), basal cell carcinoma (bcc), benign keratosis-like lesions (solar lentigines/seborrheic keratoses and lichen-planus like keratoses, bkl), dermatofibroma (df), melanoma (mel), melanocytic nevi (nv), and vascular lesions (angiomas, angiokeratomas, pyogenic granulomas, hemorrhage, and vasc). The dataset is split into 0.1 for validation and 0.9 for training. As shown in Table 1, there is a severe imbalance in this dataset.

Table 1 Number of images for each class of HAM10000 dataset.

Label	Training samples	Validation samples
akiec	291	33
bcc	462	52
bkl	989	110
df	103	12
mel	1001	112
nv	6034	671
vasc	127	15

3.1.2 Preprocessing

One of the most important aspects of DL problems, especially when data imbalance is involved, is data augmentation. Its effectiveness in preventing model overfitting and improving generalization is a proven fact. In addition to the regularization function, modifications may be used to artificially increase the dataset by adding slightly changed copies of already existing images. Since the data in this research is imbalanced and is not enough to train the model appropriately, using data augmentation is essential. Here is the list of augmentation techniques used in this research:

- **Sharpness**: Change the image's sharpness in a random way with a certain possibility.
- **Invert**: Randomly inverts the image's colors with the specified probability.
- **Solarize**: Randomly solarizes the image with a specified probability by inverting all pixel values above a threshold.
- **Equalize**: Randomly equalizes the image's histogram with the specified probability.
- **Posterize**: Randomly posterizes the image with a specified probability by decreasing the number of bits in each color channel.
- **Random perspective**: Performs a random perspective change with a set probability on the input picture.
- **Vertically flip**: Vertically flips the given image randomly with a given probability.
- **Horizontal flip**: Horizontally flips the given image randomly with a given probability.
- **Colorjitter**: Randomly changes an image's brightness, contrast, saturation, and hue.
- **Compression**: Image compression is the technique of reducing a graphics file's size in bytes while maintaining an acceptable level of image quality. By decreasing the file size, more images may be stored in storage or memory.

Ideally, it is preferable for a training batch to have data for each class to have the same amount of data. For this reason, PyTorch provides a function called WeightedRandom-Sampler, which solves the problem of imbalanced data. The main idea is to draw from a multinomial distribution on the set of points. Each point (each image in this case) has a probability of being chosen as a sample. This probability is defined by its class with the given weight parameter. For each image in training, it calculates the weight. In a simple way, these weights represent the probability that an image will be selected in their class. It is necessary to be sure that the class with a high sample has less weight.

3.1.3 Implementation details

The research was implemented using the Python programming language on Google Colab and Visual Studio Code. PyTorch [29] is a popular ML framework for creating and DNNs (along with TensorFlow). It enables tensor computation with GPU acceleration and a diverse set of neural network layers, optimizers, architecture, and pretrained models. TorchCAM is a library and implementation can be obtained from GitHub. TorchCAM is used for applying the class-specific activation of a convolutional layer in PyTorch.

Since the HAM10000 dataset has the ability to leverage the LRP method, it was trained to utilize the VGG-16 architecture [7]. Subsequently, the same procedures applied to the CAM methods were implemented to generate heatmaps using the LRP method. Finally, to obtain a comprehensive understanding of the performance and coverage of the combined methods, the IOU was calculated after merging the results from both CAM and LRP methods.

The model was trained through the implementation of transfer learning, leveraging the pretrained models of EfficientNet [30] for the polyp dataset and VGG16 for the HAM10000 dataset. The experimental setup for the model involved the utilization of a learning rate of 0.00005, an image size of 256 × 256, and a batch size of 16, while Cross-EntropyLoss was employed for calculating the loss. The optimizer selected initially was Adam, which provided fast convergence. However, to identify the optimal point, it required a considerable amount of time, up to a week. As a result, the optimization algorithm was changed to Stochastic Gradient Descent after a couple of epochs, determined by a threshold epoch. The learning rate is a crucial hyperparameter in training neural networks and requires careful tuning in order to achieve optimal results. Typically, a higher learning rate is preferred during the initial training phase, with a gradual decrease to a smaller value over time. To facilitate this process, a PyTorch scheduler known as CyclicLR-triangular2 is utilized, which modulates the learning rate over the course of training in a cyclic pattern. This approach helps prevent the model from being trapped in local minima by enabling the exploration of a wider range of solutions.

3.2 Experimental results

Detailed metrics pertaining to classification performance were assessed and are presented in Tables 2 and 3, including precision, recall, and F_1 score for each respective class.

Based on the results of the classification in Table 3, it can be concluded that the model performed well in detecting polyps with high precision, recall, and F_1 score. The precision score of 1 for polyps indicates that when the model predicted the presence of polyps, it was correct every time. The recall score of 0.88 indicates that the model was able to identify a high percentage of actual polyps in the dataset. The F_1 score of 0.94 is a

Table 2 Classification results of polyp dataset.

Label	Precision	Recall	F_1 score
Nonpolyp	0.78	1.00	0.88
Polyp	1.00	0.88	0.94

proportional average of accuracy and recall, indicating the model's overall ability to iden-tify polyps. However, the accuracy score for nonpolyp was lower than that for polyps, suggesting that this group may include some misclassifications. Despite this, the findings indicate that the model has a high accuracy in recognizing polyps, which is critical in medical image analysis for early identification and treatment of cancer.

Table 3 Classification result of HAM10000 dataset.

Label	Precision	Recall	F_1 score
vasc	0.93	0.87	0.90
nv	0.94	0.80	0.86
bcc	0.66	0.94	0.78
df	0.67	0.67	0.67
akiec	0.74	0.61	0.67
bkl	0.48	0.75	0.59
mel	0.47	0.58	0.52

3.2.1 Visualization result of the polyp dataset

In the experimentation section of this part, the polyp dataset was originally trained by considering the EfficientNet network for classification objectives, and the outcomes of the findings were determined. Following that, GradCAM models were considered, and IOU values were evaluated for each model, based on the generated heatmaps and the related ground truth images.

In this stage of the study, CAM methods were employed to generate heatmaps for the images under investigation. As shown in Fig. 3A, it is apparent that the heatmap does not effectively capture all areas of interest. To address this limitation and improve the accuracy of the heatmap, additional methods were employed in combination with the initial approach.

The first step, the pixel values of the mask data were binary, with values of either 0 or 255. Similarly, the heatmap pixel values ranged from 0 to 255. To compute the IOU metric, it was necessary to binarize the pixel values by assigning them either 0 to 255, as depicted in Fig. 3B. As depicted in the figure, it is evident that the heatmap failed to cover the entire region containing polyps. Table 4 presents the mean IOU values for each model, which shows that SmoothGradCAMpp gives the best result.

The improvement of the heatmaps that represent the areas of interest, as determined by the IOU assessment metric, is the main goal of this research. The next phase involves

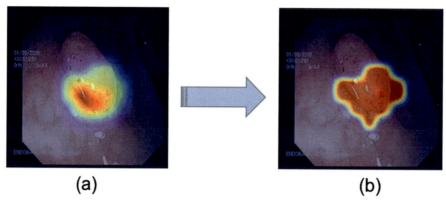

Fig. 3 Sample images show (A) heatmap and (B) binarization of heatmap.

Table 4 Mean IOU for polyp dataset.

Model	Mean IOU
GradCAM	0.37
GradCAMpp	0.35
SmoothGradCAMpp	0.38
XGradCAM	0.37
LayerCAM	0.36

merging several approaches to attain this goal, choosing and merging the strategy that produced the greatest IOU.

To further improve the IOU result and capture additional regions of interest, a max-voting technique was used to merge the outputs of two CAM models, as shown in Table 5, followed by an additional estimation of the IOU metric.

Table 5 Mean IOU of mixed models for polyp dataset.

Model	Mean IOU
SmoothGradCAMpp + GradCAM	0.40
SmoothGradCAMpp + XGradCam	0.41

Fig. 4 shows that this combination has the capacity to cover both the regions of interest and those that are not of interest; however, the areas of interest are covered more. Table 6 shows that this combination is beneficial in enhancing IOU.

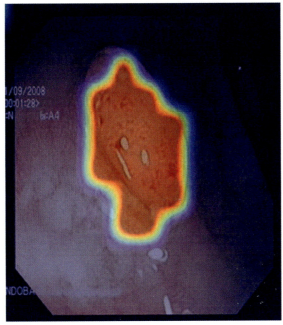

Fig. 4 Sample image of mixing methods of SmoothGradCAM and XGradCAM.

Table 6 Mean IOU for the HAM10000 dataset.

Model	vasc	nv	bcc	df	akiec	bkl	mel	Total mean
GradCAM	0.17	0.198	0.25	0.10	0.32	0.22	0.36	0.23
GradCAMpp	0.51	0.44	0.45	0.39	0.32	0.39	0.58	0.44
SmoothGradCAMpp	0.49	0.48	0.28	0.25	0.23	0.35	0.42	0.36
LayerCAM	0.56	0.46	0.45	0.49	0.30	0.40	0.55	0.46
XGradCAM	0.39	0.20	0.35	0.16	0.22	0.22	0.37	0.28
LRP-ϵ	0.22	0.25	0.26	0.31	0.37	0.30	0.31	0.29
LRP-γ	0.22	0.25	0.15	0.31	0.37	0.30	0.31	0.27
LRP$\alpha1\beta0$	0.23	0.23	0.15	0.32	0.38	0.29	0.32	0.27
LRP$\alpha2\beta1$	0.21	0.22	0.27	0.29	0.46	0.29	0.32	0.30

3.2.2 Visualization result of the HAM10000 dataset

A similar process was used in this section; however, in addition to the GradCAM methods, the LRP method was also used. Table 6 displays the calculated IOU scores for each class as well as the overall IOU average. GradCAMpp and LayerCAM have greater IOU ratings even though the table shows the IOU scores for each approach. Also for the LRP method, a relevance map is created by applying a binary threshold to the output relevance score of the input image, which is then segmented. This implies that pixels with relevance scores higher than a particular threshold are included in the segmented zone and pixels with relevance scores lower than the particular threshold excluded from the segmented zone. Using Otsu's technique, the ideal threshold value for segmentation can be determined automatically.

The heatmaps created by the GradCAMpp and LayerCAM techniques are shown in Figs. 5 and 6, respectively, and both approaches have the highest IOU ratings. These heatmaps make it clear that the red patches, which represent the most important regions using both techniques, occupy almost the same areas. This shows that the model is paying close attention to that area.

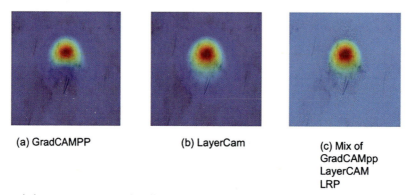

(a) GradCAMPP   (b) LayerCam   (c) Mix of GradCAMpp LayerCAM LRP

Fig. 5 Sample heatmap images with different methods. (A) GradCAMpp, (B) LayerCAM, and (C) mix of GradCAMpp, LayerCAM, LRP.

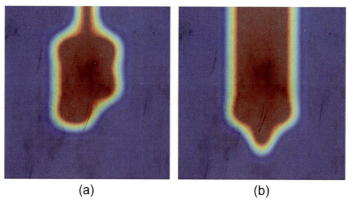

(a)   (b)

Fig. 6 Heatmap images after binarization methods. (A) Mixing the methods LayerCAM and GradCAMpp. (B) Mixing the methods LayerCAM, GradCAMpp, and LRP.

One could infer from Table 6 that the mean IOU scores for each class are almost identical across the different LRP models. Therefore, it can be concluded that there is not much difference between the LRP models in terms of their performance in this dataset. Also, as the IOU scores for different LRP models were found to be almost identical according to Table 6, it was concluded that combining one LRP method with the best-performing CAM model would be enough to mix, the results are shown in Table 7. Based on Fig. 6, the comparison of the two parts (A) and (B) indicates that the integration of LRP and CAM models is not an effective approach, as evidenced by the decreased IOU results presented in Table 7.

Table 7 Mean IOU after mixing models.

Model	vasc	nv	bcc	df	akiec	bkl	mel	Mean
GradCAMpp and LayerCAM	0.36	0.31	0.40	0.24	0.34	0.34	0.49	0.36
GradCAMpp and LayerCAM and LRP-ϵ	0.18	0.25	0.31	0.31	0.39	0.30	0.36	0.30

4. Conclusion

The use of AI in the medical field has become increasingly important in recent years. However, as AI systems get more complicated, there is rising worry regarding the interpretability and trustworthiness of their outcomes, especially in medical applications. In the field of medical images, every step would be very sensitive and needs to be clear and accurate. Each of the models needs explanations to ensure that how much the result is trustable. The XAI method that is used in this research can be helpful for explaining why the model makes the specific decision. Recent developments in the domain of XAI are shown to be positive. New explanation techniques should not be judged simply by their appearance; instead, greater effort should be put into developing more accurate quantitative ways of evaluating faithfulness. Finally, if the explanation approach is not well tested, the model might flip and provide the user with a distorted view of the model, which could have serious effects in fields such as medical images. This research shows that the evaluation of such a model is not easy and depends on lots of parameters and also can mislead the users.

The effectiveness of several XAI visualization techniques for medical pictures, notably endoscopic and skin cancer images, is investigated in this research. The Kvasir endoscopy dataset and the HAM10000 skin cancer dataset were both utilized in this investigation. Preprocessing was done to get the data ready for training, and several XAI visualization approaches were used to different models.

According to the study's findings, combining GradCAM approaches yields heatmaps that are very successful at capturing the most crucial elements of medical imaging. These conclusions were supported by the IOU assessment measure, with the best outcomes being obtained by combining the GradCAM++ and LayerCAM approaches. In addition, the research discovered that GradCAM methods beat LRP in terms of creating precise and reliable heatmaps.

The findings of this research have demonstrated the potential of attention maps in further enhancing the explainability of DL models. Attention maps enable the examination of the areas in an image that a model focuses on, thereby providing valuable insights into its decision-making process. However, despite their promise, attention maps are not without limitations, as they may fail to cover the intended regions of interest comprehensively and may occasionally cover more areas than necessary. Additionally, the results of this study have highlighted the crucial role of dataset size and visualization techniques in achieving optimal model performance. Therefore, further research is warranted to investigate and address the limitations of attention maps and to determine the optimal dataset size and visualization techniques for enhancing the interpretability of DL models.

References

[1] F.K. Došilović, M. Brčić, N. Hlupić, Explainable artificial intelligence: a survey, in: 2018 41st International Convention on Information and Communication Technology, Electronics and Microelectronics (MIPRO), IEEE, 2018, pp. 210–215.

[2] O. Biran, C.V. Cotton, Explanation and justification in machine learning: a survey or, in: IJCAI 2017 Workshop on Explainable Artificial Intelligence (XAI), 2017.

[3] D. Gunning, Explainable Artificial Intelligence (XAI), Defense Advanced Research Projects Agency (DARPA), 2017.

[4] A.B. Arrietaa, Explainable artificial intelligence (XAI): concepts, taxonomies, opportunities, and challenges toward responsible AI, Inf. Fusion 58 (2020) 82–115.

[5] K. He, X. Zhang, S. Ren, J. Sun, Deep residual learning for image recognition, in: Proceedings of the IEEE Conference on Computer Vision and Pattern Recognition, 2016, pp. 770–778.

[6] A. Krizhevsky, I. Sutskever, G.E. Hinton, ImageNet classification with deep convolutional neural networks, in: Advances in Neural Information Processing Systems, 2012, pp. 1097–1105.

[7] K. Simonyan, A. Zisserman, Very deep convolutional networks for large-scale image recognition, arXiv preprint arXiv:1409.1556 (2014).

[8] A.C. Scott, W.J. Clancey, R. Davis, E.H. Shortliffe, Explanation capabilities of production-based consultation systems, Am. J. Comput. Linguist. (1977) 1–50.

[9] W.R. Swartout, Explaining and justifying expert consulting programs, in: Computer-Assisted Medical Decision Making, Springer, 1985, pp. 254–271.

[10] W. Samek, T. Wiegand, K.R. Müller, Explainable artificial intelligence: understanding, visualizing and interpreting deep learning models, arXiv preprint arXiv:1708.08296 (2017).

[11] R. Caruana, Y. Lou, J. Gehrke, P. Koch, M. Sturm, N. Elhadad, Intelligible models for healthcare: predicting pneumonia risk and hospital 30-day readmission, in: Proceedings of the 21st ACM SIGKDD International Conference on Knowledge Discovery and Data Mining, 2015, pp. 1721–1730.

[12] B. Goodman, S. Flaxman, European Union regulations on algorithmic decision-making and a "right to explanation", AI Mag. 38 (3) (2017) 50–57.

[13] F. Doshi-Velez, B. Kim, Towards a rigorous science of interpretable machine learning, arXiv preprint arXiv:1702.08608 (2017).

[14] G. Montavon, W. Samek, K.R. Müller, Methods for interpreting and understanding deep neural networks, Digit. Signal Process. 73 (2018) 1–15.

[15] C. Molnar, Interpretable Machine Learning, 2022. https://christophm.github.io/interpretable-ml-book.

[16] L.M. Zintgraf, T.S. Cohen, T. Adel, M. Welling, Visualizing deep neural network decisions: prediction difference analysis, arXiv preprint arXiv:1702.04595 (2017).

[17] M. Harradon, J. Druce, B. Ruttenberg, Causal learning and explanation of deep neural networks via autoencoded activations, arXiv preprint arXiv:1802.00541 (2018).

[18] R. Geirhos, J.H. Jacobsen, C. Michaelis, R. Zemel, W. Brendel, M. Bethge, F.A. Wichmann, Shortcut learning in deep neural networks, Nat. Mach. Intell. 2 (11) (2020) 665–673.

[19] J. Adebayo, J. Gilmer, M. Muelly, I. Goodfellow, M. Hardt, B. Kim, Sanity checks for saliency maps, in: Advances in Neural Information Processing Systems, 2018.

[20] L. Sixt, M. Granz, T. Landgraf, When explanations lie: why many modified BP attributions fail, in: Proceedings of the International Conference on Machine Learning (ICML), 2020, pp. 9046–9057.

[21] B. Zhou, A. Khosla, A. Lapedriza, A. Oliva, A. Torralba, Learning deep features for discriminative localization, arXiv preprint arXiv:1512.04150 (2016).

[22] L. Bazzani, A. Bergamo, D. Anguelov, L. Torresani, Self-taught object localization with deep networks, in: 2016 IEEE Winter Conference on Applications of Computer Vision (WACV), IEEE, 2016, pp. 1–9.

[23] B. Zhou, A. Khosla, A. Lapedriza, A. Oliva, A. Torralba, Learning deep features for discriminative localization, in: Proceedings of the IEEE Conference on Computer Vision and Pattern Recognition, 2016, pp. 2921–2929.

[24] R.R. Selvaraju, M. Cogswell, A. Das, R. Vedantam, D. Parikh, D. Batra, Grad-CAM: visual explanations from deep networks via gradient-based localization, in: Proceedings of the IEEE International Conference on Computer Vision, 2017, pp. 618–626.

[25] A. Chattopadhyay, A. Sarkar, Grad-CAM++: generalized gradient-based visual explanations for deep convolutional networks, CoRR abs/1710.11063 (2017). http://arxiv.org/abs/1710.11063.

[26] N. Otsu, A threshold selection method from gray-level histograms, IEEE Trans. Syst. Man Cybern. 6 (3) (1979) 62–66.

[27] S.A. Hicks, D. Jha, V. Thambawita, P. Halvorsen, H.L. Hammer, M.A. Riegler, The EndoTect 2020 challenge: evaluation and comparison of classification, segmentation and inference time for endoscopy, in: Pattern Recognition. ICPR International Workshops and Challenges: Virtual Event, January 10–15, 2021, Proceedings, Part VIII, Springer, 2021, pp. 263–274.

[28] H. Borgli, V. Thambawita, P.H. Smedsrud, S. Hicks, D. Jha, S.L. Eskeland, K.R. Randel, K. Pogorelov, M. Lux, D.T.D. Nguyen, D. Johansen, C. Griwodz, H.K. Stensland, E. Garcia-Ceja, P.T. Schmidt, H.L. Hammer, M.A. Riegler, P. Halvorsen, T de Lange, HyperKvasir, a comprehensive multi-class image and video dataset for gastrointestinal endoscopy, Sci. Data 7 (1) (2020) 283, https://doi.org/10.1038/s41597-020-00622-y.

[29] A. Paszke, S. Gross, F. Massa, A. Lerer, J. Bradbury, G. Chanan, T. Killeen, Z. Lin, N. Gimelshein, L. Antiga, PyTorch: an imperative style, high-performance deep learning library, in: Advances in Neural Information Processing Systems, vol. 32, 2019.

[30] M. Tan, Q. Le, EfficientNet: rethinking model scaling for convolutional neural networks, in: Proceedings of the International Conference on Machine Learning, PMLR, 2019, pp. 6105–6114.

CHAPTER 5

An interpretation-model-guided classification method for malignant pulmonary nodule

Bin Li and Weilin He
School of Automation Science and Engineering, South China University of Technology, Guangzhou, China

1. Introduction

Lung cancer is one of the most common cancer types worldwide. Generally, patients diagnosed with lung cancer are already at late stages as lung cancer is unlikely to present any apparent symptoms until the disease is advanced, resulting in a 5-year survival rate of only 10–20% [1]. Low-dose computed tomography (LDCT) is one of the most effective methods for early lung cancer diagnoses [2]. Pulmonary nodules are small, round or oval lesions in the lung that range in size from 3 to 30 mm. Pulmonary nodules are broadly classified as malignant/benign and considered potential candidates for cancer. It is more likely to be effectively treated if lung cancer is discovered at an early stage, when it is small and has not spread. Therefore, lung cancer screening using LDCT has become standard practice to determine the presence and malignancy of a pulmonary nodule. However, according to the National Lung Screening Trial, only one in every 20 nodules detected with LDCT screening is pulmonary cancer, creating a difficult environment for accurate diagnosis [3].

In clinical practice, malignancy of pulmonary nodules is still mainly judged by radiologists based on their expertise and experience. However, considering the difficulty of classifying pulmonary nodules and the large number of patients encountered by radiologists, there is constant pressure on them to analyze a massive amount of data and decide as quickly as possible based on the analysis, leading to mistakes due to the overwhelming number of cases. So it is an emerging interest in computer–aided diagnosis (CAD) systems to improve malignant pulmonary nodules detection [4], especially classification of early stage lung cancer with less obvious symptoms [5]. Radiomics refers to the high-throughput extraction and analysis of large amounts of quantitative imaging features from regions of interest in radiographic images [6]. Radiomics could correlate quantitative imaging features with patient data and clinical results, which not only helps make lung

Explainable AI in Healthcare Imaging for Medical Diagnoses
https://doi.org/10.1016/B978-0-443-23979-3.00005-1

cancer diagnosis but provides information about prognosis and response to treatment [7]. Furthermore, radiomics features can capture associations between image voxels that skilled radiologists may miss, allowing for a comprehensive quantitation of nodules [6]. Besides radiomics features, some radiologist-assessed semantic features had shown improved performance for diagnosing pulmonary nodules malignancy, such as spiculation, lobulation, etc. [8].

Although radiomics has shown significant advantages in the fields of diagnostics and treatment of lung cancer [9], radiomics and its learning models are mostly considered to be a black box that provides predictive results for specific clinical outcomes, making them a less accepted approach [10–12]. Though some models are considered interpretable due to their simple structure, such as sparse linear models and short decision trees, it should be noted that the quantity and complexity of model features have a direct effect on the model's interpretability. While high-performance learning classifiers can be obtained by training hundreds of parameters, they can also become hard to understand [13]. Therefore, a particular method must be devised and applied to the learning model to explain and guide its decisions. Despite many efforts have been made to enhance the interpretability of traditional complex black-box machine learning models [14], there is still a shortage of appropriate interpretable models that have both good interpretability and high performance (including good guiding function).

In this chapter, an interpretation-model-guided classification method for malignant pulmonary nodule in LDCT is proposed for classification of malignant pulmonary nodules, whose main contributions are summarized as follows.

(1) Optimal features, classifiers and their parameters are adaptively selected and set based on the proposed interpretation-model guiding and explanation algorithm. The proposed Improved Shapley Additive exPlanations (ISHAP) method automatically selects best subset of features based on ISHAP explanation and recursive feature elimination algorithm. Moreover, each iterative process incorporates the successive halving random hyperparameter optimization to obtain the optimal classification performance. The importance and preference of features for classifying the malignancy of pulmonary nodules are explained in our method.

(2) Semantic and radiomics features are extracted by using medical priori knowledge and image understanding in the lung images dataset.

2. Related works

2.1 Machine-learning-based methods for classification of malignant and benign pulmonary nodules

Machine learning is used in computer-aided diagnosis of medical imaging, and today's radiomics systems are proving to be reliable in real-world situations [15]. As a result, researchers are increasingly interested in machine-learning-based radiomics models for classification of malignant pulmonary nodules [16]. To categorize malignant and benign nodules, Orozco et al. [17] used 11 wavelet transform-based features and Support Vector Machine (SVM) classifier, getting a sensitivity of 90.90%, a specificity of 73.91%, and a total accuracy of 82.2%. Taşcı et al. [18] suggested a CAD system that provided automatic detection of juxta-pleural nodules utilizing a generalized linear model regression classifiers, with an accuracy of 92.91% using a testing dataset of 124 juxta-pleural nodules. They offered seven new nodule candidate attributes that improved detection performance. To identify malignant nodules, we computed shape, intensity, and texture characteristics to define pulmonary nodules and applied them to an enhanced random forest (RF) algorithm [19]. Gong et al. [20] and Wu et al. [21] discovered that a combination of shape, intensity, and texture characteristics performed exceptionally well in identifying nodules.

Deep learning techniques, specifically convolutional neural networks (CNNs), have recently been utilized to classify pulmonary nodules with promising results. These deep learning models learn optimal features from images in a purely data-driven manner. Kumar et al. [22], for example, used an unsupervised deep autoencoder to extract possible features from 2D CT patches. These retrieved deep features were then put into lung cancer prediction decision trees. Tajbakhsh and Suzuki [23] examined the ability of massive-training ANNs (MTANNs) and CNNs to identify malignant from benign nodules. The AUC of MTANNs was 88.06%, which was higher than the AUC of CNNs, which was 77.55%. More and more network models are optimized based on CNNs as deep learning techniques advance, such as 3D multi-view convolutional neural networks (MV-CNN) [24] and 3D multi-crop convolutional neural networks (MC-CNN) [25]. 3D CNNs can encode higher spatial information from CT images than 2D networks, allowing them to learn more distinct characteristics. In addition, Afshar et al. [26] suggested a 3D Multi-scale Capsule Network (3D-MCN) framework for predicting lung tumor aggressiveness. When tested on the Lung Image Database Consortium (LIDC) dataset, the suggested 3D-MCN architecture predicted pulmonary nodule malignancy with an accuracy of 93.12%, a sensitivity of 94.94%, an AUC of 96.41%, and a specificity of 90%. It is worth noting that these cited works treat deep learning as a "black box" from which only classification results are expected, with no explanation of how the model uses the features to produce a given prediction. This low level of interpretability arguably prevents radiologists and clinicians from understanding how the model works and ultimately hinders its clinical application.

2.2 Classification of pulmonary nodules based on interpretable methods

As discussed above, interpretability is essential to facilitate the interaction between radiologists and radiomics models by providing transparent and trustworthy predictions. Although machine learning/deep learning based methods for classifying benign and malignant pulmonary nodules can achieve high accuracy, their clinical application in radiology is still not satisfactory. The reason behind this is the "black box" of deep learning models, which means that the model cannot provide reasons for making such judgments and lacks interpretability. This is unacceptable for fields with high decision risks, especially clinical application. Therefore, with the rapid development of machine learning, interpretability methods have also emerged. From a model perspective, interpretable methods developed for medical image analysis can be divided into model-based methods and post hoc explanation methods [27]. Among them, model-based interpretable methods are usually relatively simple traditional machine learning models, such as support vector machines, linear regression, and other machine learning models, which themselves have sparsity, that is, they only use a limited number of features, so that the users of the model can understand their decision-making process. Relatively speaking, the post explanation method focuses on analyzing pre-trained models to understand their decision-making process. This type of method mainly includes representing the learned features, their importance, and visual interpretation of saliency maps. There have been considerable interpretable works conducted by previous researchers in the analysis of pulmonary nodule images. For example, we proposed a lung nodule benign and malignant classification method based on the Improved Shapley Additive exPlanations (ISHAP) model and ensemble learning classifiers. This chapter is an extension of the article [28]. More researchers are adopting methods based on convolutional neural networks for interpretable classification research. Zhou et al. [29] proposed a localization method called CAM (Class Activation Map) in 2016, which aims to use the feature map of the last layer of the neural network to output the region of interest of the neural network. Jiang et al. [30] proposed a classification method for benign and malignant pulmonary nodules using neural architecture search and convolutional block attention module, and used attention visualization to understand the inference process of the model. The attention maps of the proposed method focus on the boundaries and interior of lung nodules, which is consistent with the clinical diagnostic process of physicians, therefore the model has a certain interpretability. Shen et al. [31] proposed an interpretable method based on a hierarchical semantic model, which provides outputs at two levels: low-level semantic features and high-level malignant features of nodules. The low-level semantic feature output reflects the diagnostic evaluation indicators of radiologists and can interpret images in an expert knowledge manner.

2.3 Shapley Additive exPlanations (SHAP) for machine learning interpretation

Recently, SHAP (Shapley Additive exPlanations) has gained increasing attention in machine learning interpretation domains due to its effectiveness and strong mathematical foundation. Lundberg et al. [32] first proposed a new framework called SHAP

[33] to interpret machine learning models, providing help to improve the interpretability and trustworthiness of learning models in clinical applications. SHAP is based on the Shapley value [34] from game theory and its related extensions to link optimal credit allocation to local explanations. In the SHAP framework, each feature is treated as a player and by calculating the player's Shapley value, which is also the importance of a feature in the model prediction. Since SHAP has a solid theoretical foundation in game theory, boosting confidence in the interpretation results, recently several studies have successfully used SHAP in the field of machine learning radiomics models for model interpretation [35–37]. However, much of the research to date has been limited to treating the SHAP method as a separate interpretive module of the machine learning process, ignoring the feedback role of SHAP for classifier training.

3. The proposed interpretation-model-guided classification method for malignant lung nodule

An interpretation-model-guided classification method shown in Fig. 1 is proposed for the classification of malignant pulmonary nodules. In the proposed classification method, semantic features assessed by radiologists and multiple types of radiomics features are extracted from the lung images dataset, and furthermore, optimal features, classifiers and their parameters are guided to select and set adaptively based on the proposed ISHAP explanation and recursive feature elimination algorithm.

3.1 Semantic and radiomics features extraction using medical priori knowledge and image understanding

In this chapter, both semantic features from radiologist annotations and radiomics features from lung CT images are extracted and combined to improve diagnostic power.

3.1.1 Image preprocessing

Laplacian of Gaussian (LoG) filter is used to find areas of gray level change in CT images [38]. Besides that, 3D wavelet transform [17,39] is used to extract image radiomics features in this work. The 3D wavelet transform decomposes the 3D volume (CT image sequences) into eight sub-volumes using a high-pass filter and a low-pass filter.

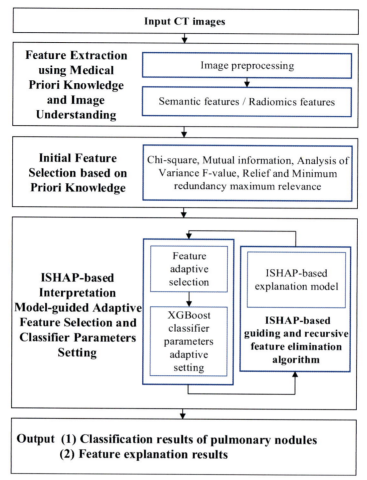

Fig. 1 Flowchart of the proposed interpretation-model-guided classification method for malignant pulmonary nodule.

3.1.2 Feature extraction based on medical priori knowledge and image understanding

In this chapter, both semantic features and radiomics features [6] are extracted from lung CT images to characterize volumes of interest qualitatively and quantitatively, which are shown in Tables 1 and 2. The following subsections describe these features in more detail.

(1) Semantic features

Semantic features are medical empirical features proposed by radiologists to describe volumes of interest qualitatively. These characteristics are difficult to describe mathematically, but they are instrumental in clinical settings. In this chapter, we use semantic features obtained in two different ways. First, for each nodule

≥ 3 mm in the LIDC dataset [40], we can obtain a set of semantic descriptors assessed by four radiologists. Secondly, shape-based features are extracted from lung CT images, which are often used to describe geometric features of volumes of interest, such as perimeter, elongation, and surface area.

(2) Radiomics features

Radiomics features could be defined as mathematically extracted quantitative descriptors. These can be divided into first-order statistical (histogram) features and higher-order statistical (texture) features. The distribution of voxel values within the volumes of interest is described using first-order features. The kurtosis, entropy, energy, range, mean and intensity maximum are all examples of first-order features. The objective assessment of the entire volume of interest is greatly aided by histogram analysis. Second-order features describe the statistical and spatial relationship between voxels with similar/dissimilar contrast values, which numerically characterize the image's texture. Statistical features based on Gray Level Co-occurrence Matrix (GLCM), such as contrast, correlation and variance, can characterize the texture of images. In addition, Gray Level Dependence Matrix (GLDM), Gray Level Run Length Matrix (GLRLM), Gray Level Size Zone Matrix (GLSZM) [41] and Neighbouring Gray Tone Difference Matrix (NGTDM) are also typical methods of texture analysis.

3.2 The proposed ISHAP-based model-guided algorithm for adaptive feature selection and classifier parameters setting

3.2.1 Initial feature and classifier selection
Initial feature selection based on priori knowledge

By extracting a large number of quantitative features, radiomics analysis underlies the curse of dimensionality because most of the features are highly relevant. There is a danger of overfitting analyses with such considerable complexity, and hence, dimensionality must be reduced by feature selection [42]. In addition, the use of feature selection methods can also bring other advantages, such as increasing the time efficiency of model training, reducing measurement and storage requirements, reducing the complexity of the model and improving interpretability [43]. Filter methods, wrapper methods, and embedded methods are the three types of feature selection methods used in radiomics studies [44–46].

The search for feature subsets is integrated into the classification model construction when using embedded methods. It's a search for feature subsets and hypotheses in a combined space. Embedded methods are also specific to classification models, just like wrapper methods are. Embedded methods, on the other hand, reduce the time required to reclassify different feature subsets. As a result, embedded methods are computationally faster and less prone to overfitting than wrapper methods [47]. In this chapter, five feature selection methods based on filter approaches are used in the analysis, including

Chi-square (CHSQ), Mutual information (MUIF), Analysis of Variance (ANOVA) F-value, Relief (RELF) [48] and Minimum redundancy maximum relevance (MRMR) [49]. We choose these methods primarily because they are well-established in the literature, computationally efficient and reproducible [50]. The following subsections provide additional information about the feature selection methods used in this chapter.

Table 1 Semantic features.

Radiologist-assesed features			Radiologist-assesed features				
Spiculation	Texture	Malignancy	Elongation	Flatness	Mesh Volume	SurfaceArea	SurfaceVolumeRatio
Calcification	Margin	Sphericity	LeastAxisLength	VoxelVolume	Maximum2DDiameterSlice		MinorAxisLength
Internal Structure	Subtley	Lobulation	Maximum2DDiameterColumn		Maximum2DDiameterRow		Maximum3Ddiameter

Table 2 Radiomics features.

First order-based features							
10th Percentile	Skewness		Total Energy	Minimum		Kurtosis	Maximum
90th Percentile	Interquartile Range		Uniformity	Median	Energy	Variance	Range
Mean Absolute Deviation	Entropy		Root Mean Squared Error	Mean		Robust Mean Absolute Deviation	

GLCM features				
Sum Average	Autocorrelation	Joint Average	Correlation	Inverse Difference Moment Normalized
Difference Variance	Cluster Prominence	Joint Energy	Contrast	Informational Measure of Correlation 1
Sum Squares	Cluster Shade	Maximum Probability	Sum Entropy	Informational Measure of Correlation 2
Difference Average	Cluster Tendency	Inverse Difference	Joint Entropy	Inverse Difference Normalized
Difference Entropy	Inverse Variance	Inverse Difference Moment		Maximal Correlation Coefficient

NGTDM features				
Busyness	Coarseness	Complexity	Contrast	Strength

GLDM features			
Dependence Entropy	Dependence Variance	GrayLevel NonUniformity	Large Dependence Low Gray Level Emphasis
Dependence NonUniformity	Small Dependence High Gray Level Emphasis		Small Dependence Low Gray Level Emphasis
Dependence NonUniformity Normalized	Small Dependence Emphasis	Large Dependence Emphasis	
Large Dependence High Gray level Emphasis	Low Gray Level Emphasis	High Gray Level Variance	High Gray Level Emphasis

GLRLM features				
Long Run High Gray Level Emphasis	Gray Level NonUniformity	Run Variance	Run Entropy	Run Percentage
Long Run Low Gray Level Emphasis	Gray Level Variance	High Gray Level Run Emphasis		Run Length NonUniformity
Low Gray Level Run Emphasis	Long Run Emphasis	Short Run Emphasis		Short Run High Gray Level Emphasis
Giray Level NonUniformity Normalized	Run Length NonUniformity Normalized			Short Run Low Gray Level Emphasis

GLSZM features				
Gray Level NonUniformity	Small Area Emphasis	Size Zone NonUniformity	Zone Entropy	Large Area Emphasis
Gray Level NonUniformity Normalized	Size Zone NonUniformity Normalized		Large Area High Gray Level Emphasis	
Gray Level Variance	Zone Percentage	Small Area Low Gray Level Emphasis	Large Area Low Gray Level Emphasis	
High Gray Level Zone Emphasis	Small Area High Gray Level Emphasis		Zone Variance	Low Gray Level Zone Emphasis

(1) Chi-square (CHSQ)

CHSQ is a statistical technique used to measure the independence between features and classes. Usually, the Chi-Square test starts with the assumption that the features and classes are independent and then observes the degree of deviation between the actual value and theoretical value. If the deviation is sufficiently small, the error is considered to be a natural sample error caused by a lack of precision in measurement or by chance. Therefore, the feature and the class are considered independent and the original hypothesis is accepted. If the deviation is large enough to make it unlikely that such an error is due to chance or inaccurate measurement, we consider that the two are actually correlated, i.e. the original hypothesis is rejected. The lower the Chi-Square value, the greater the independence between a feature and a class, which means we could remove this feature. Chi-square score χ^2 is defined as Eq. (1).

$$\chi^2 = \sum \frac{(O_i - E_i)^2}{E_i} \qquad (1)$$

where O_i is number of actual observation value in class i, E_i is number of theoretical observation value in class i if there is no relationship between the feature and the class.

(2) Mutual information

When used as a scoring criterion for feature selection, it is equal to zero if the feature and class are independent, and a higher value means a higher dependency. Mutual information (MUIF) $MI(F; Y)$ with features F and classes Y is defined as Eq. (2).

$$MI(F; Y) = \sum_{y \in Y} \sum_{f \in F} p(f, y) \log\left(\frac{p(f, y)}{p(f)p(y)}\right) \qquad (2)$$

where $p(f, y)$ is the joint probability mass function of feature f and class y. $p(f)$ and $p(y)$ are the marginal probability mass functions of feature f and class y.

(3) Analysis of variance (ANOVA) F-value

The ANOVA F-value method applies an ANOVA test to each feature with respect to the target labels in order to study the difference between the means of various data groups. A higher F-value of a feature means a higher importance in prediction.

(4) Relief method

Relief (Relevant Features) [48] is a well-known filtering feature selection method designed initially to apply binary classification problems with discrete or numerical features. Relief is designed to measure the importance of features by using a correlation statistic, which is a vector and each component of this vector corresponds to a feature. The importance of a feature subset is determined by the sum of correlation statistics for each feature in the subset. We could select the features with the largest correlation statistic components by specifying the number of features to be selected. Given a training set D including (x_1, y_1), (x_2, y_2), ..., (x_m, y_m), for each sample x_i, relief first finds its nearest neighbors $x_{i,nh}$ among samples of same class of x_i, called near-hit, and then finds its nearest neighbors $x_{i,nm}$ among different class samples of x_i, called near-miss. Correlation statistic component δ^f of feature f is defined as Eq. (3).

$$\delta^f = \sum_i - \text{diff}\left(x_i^f, x_{i,nh}^f\right)^2 + \text{diff}\left(x_i^f, x_{i,nm}^f\right)^2 \tag{3}$$

where x_i^f represents the value of the sample x_i on the feature f. For discrete (e.g. categorical or nominal) features, $\text{diff}(x_a^f, x_b^f)$ is defined as Eq. (4).

$$\text{diff}\left(x_a^f, x_b^f\right) = \begin{cases} 0, & x_a^f = x_b^f \\ 1, & \text{otherwise} \end{cases} \tag{4}$$

and for continuous features, $\text{diff}(x_a^f, x_b^f)$ is defined as Eq. (5).

$$\text{diff}\left(x_a^f, x_b^f\right) = \left| x_a^f - x_b^f \right| \tag{5}$$

If the distance between x_i^f and its near-hit $x_{i,nh}^f$ is less than the distance between x_i^f and its near-miss $x_{i,nm}^f$, it means that f is suitable for distinguishing similar and heterogeneous samples, and vice versa. The larger the value of δ^f, the stronger the classification ability of f.

(5) Minimum redundancy maximum relevance

The MRMR (max-relevancy, min-redundancy) [49] is another filter method based on mutual information. The core idea of MRMR is to maximize the relevance between features and classes, while minimizing the relevance between features and features, which can reduce the redundancy of features. Let S represents the subset of features and $|S|$ is the number of features in S. The maximal relevance is as Eq. (6).

$$\max D(S, c), \quad D = \frac{1}{|S|} \Sigma_{f_i \in S} MI(f_i; c) \tag{6}$$

where f_i is the ith feature, c is the class. The minimal redundancy is as Eq. (7).

$$\min R(S, c), \quad R = \frac{1}{|S|^2} \Sigma_{f_i, f_j \in S} MI\left(f_i; f_j\right) \tag{7}$$

Initial classifier selection and their parameters setting

In order to select the best machine learning method for differentiating benign and malignancy of pulmonary nodules, we test eight commonly used classifiers. They involve Logistic Regression (LG) [51], Support Vector Machine (SVM) [52], K-nearest Neighbours (KNN) [53], Naïve Bayes (NB) [54], Decision Tree (DT), Random Forest (RF)

[55], Multi-layer Perceptron (MLP) [56], Extreme Gradient Boosting (XGBoost) [57]. LG is a simple but effective method for supervised classification. However, when the feature space is large, LG does not perform very well and can easily be underfitting. SVM algorithm is an advanced supervised algorithm that can deal with regression and classification tasks. Yet there is no universal solution to non-linear problems, and sometimes it is difficult to find a suitable kernel function. KNN is one of the simplest and earliest classification algorithms. One big drawback of KNN is for each new sample, all the distances from K neighbors need to be calculated again and again, resulting in a significant consumption of computational time. NB requires a small amount of training data to estimate the necessary parameters and it does not work well if the sample attributes are correlated. DT is a non-parametric supervised learning method used for classification and regression. However, an overly complex DT also means a greater risk of overfitting. RF is an ensemble learning algorithm, which consists of multiple decision trees. While an RF often achieves higher accuracy than a single decision tree, it sacrifices the intrinsic interpretability in decision trees, which is vital in clinical applications. An MLP is a class of feedforward artificial neural networks (ANN). One flaw is that MLP has many hyperparameters to tune. Moreover, MLP has multiple non-convex functions, which means that there is a risk of falling into the local minimum during training, affecting computational efficiency and classification accuracy.

Extreme Gradient Boosting (XGBoost) [57] is a machine learning technique for regression and classification problems (GBDT). Multiple Classification And Regression Trees (CART) are built using feature splitting nodes. The objective function is reduced by fitting the last model's residuals to the CART tree. Finally, a strong classifier is built from many CART weak classifiers, with each leaf node representing a score. After predicting the sample, the model finds the corresponding leaf node in each tree. The sample's predicted value is the sum of all leaf node scores. The XGBoost model's optimal parameters are obtained through training samples using the principle of minimization of the objective function. So new samples are predicted using optimal parameters. To avoid overfitting, XGBoost reduces the model's variance by adding a regularization term to the objective function. XGBoost also considers disk efficiency, especially when data is large and memory is limited. A multi-threaded approach greatly improves efficiency by combining data compression and fragmentation. Like other tree-based algorithms, XGBoost struggles with extrapolation tasks.

In this chapter, by 10-fold cross-validation in terms of predictive performance and stability, we compare the classification performance of different feature selection methods and classifiers combinations to demonstrate the best performance of XGBoost and select it as the classifier for implementing the proposed improved SHAP algorithm. A well-tuned XGBoost can achieve state of art prediction accuracy compared to a not properly configured XGBoost. We optimize some of the most commonly used hyperparameters of XGBoost [58], which are listed in Table 3.

Table 3 The hyperparameters of XGBoost used in this chapter.

Hyperparameter type	Hyperparameter name	Description
Booster parameters	Learning_rate	Step size shrinkage used in update
	Max depth	Maximum depth of a tree
	Min_child_weight	Minimum sum of weights in a child
	Subsample	Subsample ratio of the training instances
	Col sample by a tree	Subsample ratio of columns for each tree
	Gamma	Minimum loss reduction
	Alpha	L1 regularization term
	Lambda	L2 regularization term
Command line parameters	Number of estimators	Number of estimators

3.2.2 The proposed ISHAP guiding classification and explanation algorithm SHAP-based explanation model

In the task of classifying benign and malignant pulmonary nodules, machine learning models have achieved good results. However, most machine learning models are lacking in explanatory aspects due to the complexity of their principles and parameters, making them more of a black box. In medical diagnostic systems, patients and doctors are not only interested in the results of the diagnosis and classification, but also in why the models make the predictions and whether the models are trustworthy.

In response to this, we use SHAP a game-theoretic approach to explain the output of any machine learning model. Let $\{f_1, f_2, \ldots, f_p\}$ represents the features used in the model, p is the number of features and S represents a subset of the features. The Shapley value $\phi_j(val)$ of feature f_j can be calculated as Eq. (8).

$$\phi_j(val) = \sum_{S \subseteq \{f_1, \ldots, f_p\} \{f_j\}} \frac{|S|!(p-|S|-1)!}{p!} \left(val\left(S \cup \{f_j\} \right) - val(S) \right) \quad (8)$$

where $val(S)$ is the prediction for feature values in set S that are performed multiple integrations for each feature that is not included in set S. For a given prediction, the further the value of $\phi_j(val)$ is from 0, the more important the feature x_j is, and the sign of the value indicates the class toward which the prediction is pushed by the feature.

The SHAP importance of a given feature is calculated by averaging the absolute Shapley values calculated on a given dataset. It represents the average strength of the feature's impact on the predictions on a given dataset.

The proposed ISHAP-based model-guided algorithm for adaptive feature selection and classifier parameters setting

(1) The proposed Improved–SHAP method

Although extensive research has been carried out on machine learning model interpretation using SHAP, most studies just remove the least important features based on SHAP importance with a given threshold. The top-ranked features, on the other hand, are not always the most important for each individual. In some ways, only the features of a subset when taken together are optimal [59]. To address this drawback, we propose an improved SHAP method for the model explanation and model-guided algorithm. An iterative method called recursive feature elimination is introduced for feature subset ranking. At each iteration, the classifier is first trained on a set of features and mean SHAP importance of each feature is calculated. Then, the least important features are pruned from this set of features. However, the number of features to be pruned at each iteration is a parameter that needs to be given. To enhance the computational efficiency, we use an adaptive coefficient α to control how many features are removed at each iteration.

(2) The proposed ISHAP-based guiding algorithm for adaptive feature selection and classifier parameters setting

In practice, researchers usually consider the machine learning model hyperparameters optimization and model interpretation as two separate steps. Therefore, the choice of features may have an impact on which hyperparameters for the model are optimal, and which hyperparameters we select for the model may have an impact on which choice of features would be optimal. In this study, we optimize the hyperparameters of machine learning model at each iteration before calculating feature importance to ensure that the model fits the data well throughout this process.

Grid search and manual search have been frequently used in the optimization of hyperparameters. Grid search, on the other hand, is not feasible when the number of hyperparameters is increased, as with XGBoost. Random search has been shown to be more efficient at optimizing hyperparameters than grid search [60]. In order to further improve the performance of hyperparameter optimization, we conduct the random search with successive halving method [61]. Successive halving is an iterative selection process where all candidates (the hyperparameter combinations) are evaluated with a small amount of data at the first iteration. Only some of these candidates are selected for the next iteration, which will be allocated more data. The detailed methods are presented in Algorithm 1.

Optimal features, classifiers and their parameters are adaptively selected and set based on the proposed ISHAP-based interpretation-model guiding and explanation algorithm. The proposed ISHAP method automatically selects the best subset of features through an iterative approach. Moreover, each iterative process incorporates the successive halving random hyperparameter optimization to obtain the optimal classification performance. The importance and preference of features for the classification of malignancy of pulmonary nodules are explained in our method.

The benefits of the proposed method are as follows. (1) The introduction of recursive feature elimination (RFE): could automatically select the best feature subset with best CV performance. (2) The introduction of adaptive coefficient: greatly reduce computation time without any loss in performance. (3) The introduction of hyperparameters optimization: ensure that the model fits the data well throughout the whole process (every iteration), avoid both overfitting and underfitting.

The proposed ISHAP-based model-guided algorithm implementation

The detailed implementation for the proposed ISHAP-based guiding algorithm for adaptive feature selection and classifier parameters setting is shown in Algorithm 1.

Algorithm 1. ISHAP-based guiding algorithm for adaptive feature selection and classifier parameters setting

Input:

Training dataset D_{train}; Testing dataset D_{test}; Set of p features $\{f_1, f_2, ..., f_p\}$;
Classifier A; Number of iterations R; Number of sampling settings of hyperparameter N;
Adaptive coefficient α; Cross-validation fold k

Output:

Testing pulmonary nodule classification results; Testing dataset feature explanation results

partition D_{train} into $D_1, D_2, ..., D_k$
initialize $i \leftarrow p$
initialize subset of features $S_i \leftarrow \{f_1, f_2, ..., f_i\}$
for $r = 1$ to R **do**
  for $n = 1$ to N **do** (hyperparameter optimization)
    successive halving random search the setting of hyperparameter H_n
    for $j = 1$ to k **do** (cross-validation for hyperparameter optimization)
      set D_j as the validation set; train classifier A with H_n on remaining $k-1$ folds;
      evaluate classification performance on D_j
    end
    calculate the mean performance of H_n
  end
  select the hyperparameter with the best performance H_r
  for $c = 1$ to k **do** (cross-validation for SHAP calculation)
    set D_c as the validation set; train classifier A with H_r on remaining $k-1$ folds;
    calculate SHAP value of each feature in S_i on D_c
  end
  calculate the mean SHAP value of each feature in S_i
  rank S_i using the mean SHAP value
  $m \leftarrow \alpha \times i$
  remove the last m features from S_i
  $i \leftarrow i - m$
end

select the subset of features with best validation performance from $\{S_1, S_2, ..., S_p\}$ as S_{best} and the corresponding optimal hyperparameters H_{best}

train classifier A with H_{best} on D_{train}

predict D_{test} with trained classifier **(classification results output)**

calculate the mean SHAP value of each feature in S_{best} on D_{test} **(feature explanation results output)**

4. Experiments

4.1 Lung nodule dataset

In this chapter, we evaluate the proposed method for pulmonary nodule classification using the LIDC dataset of thoracic CT scans and provide a quantitative explanation for how the method makes a particular prediction. LIDC [40] is the world's largest freely accessible collection of 1018 thoracic CT scans gathered from various universities and organizations. The database contains 2669 lesions that have been classified as nodules ≥3 mm in diameter by at least one of four radiologists. Each of these lesions is annotated with nodule contours, quantified values for eight nodule characteristics, and a quantified value for the nodule's malignancy estimation. The resulting collection contains CT images in Digital Imaging and Communications in Medicine (DICOM) format, as well as annotations in the Extensible Markup Language (XML) format. Without publicly available tools to accompany the dataset, converting XML contours to alternative representations is error-prone. Nonetheless, until recently, the only representation accessible to users of the LIDC collection and widely used by previous works was the XML annotations. Fedorov et al. [62] present a standardized DICOM representation of the annotations associated with the LIDC-generated pulmonary nodules ≥3 mm. This standardized DICOM representation benefits the research community by making it more FAIR (Findable, Accessible, Interoperable, and Reusable) and allowing for integration with other standardized data collections. As a result, this chapter uses the standardized DICOM representation of annotations rather than XML annotations. Naidich et al. [63] and Manos et al. [64] recommend that thin-slice CT scans be used to manage pulmonary nodules. As a result, scans with a slice thickness of ≥3 mm are discarded, as are scans with inconsistent slice spacing or missing slices, resulting in the final list of 2255 scans.

In the LIDC dataset, malignancy refers to the possibility of a pulmonary nodule developing into a malignant nodule. In this chapter, all samples are classified as malignant or benign nodules based on radiologist–assessed malignancy values. Values less than 3 (i.e., 1 and 2) are classified as benign, while values greater than 3 (i.e., 4 and 5) are classified as malignant. To avoid ambiguity in nodule samples, we remove pulmonary nodules with a score of 3 in malignancy. This results in a collection of 1829 nodule scans with 3595 annotations. Additionally, we consider nodules from the same patient to be completely distinct between the training and testing datasets. To ensure that the training and testing datasets are balanced, the same number of nodules are randomly chosen for each class, resulting in a training dataset of 1500 nodules with 2356 annotations and a testing dataset of 300 nodules with 434 annotations. Fig. 2 shows examples of benign and malignant pulmonary nodules with different ranks of malignancy from the LIDC dataset.

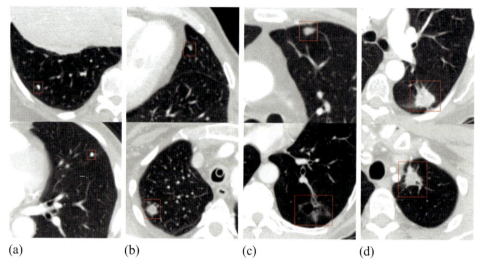

(a)   (b)   (c)   (d)

Fig. 2 Examples of rated benign and malignant pulmonary nodules from the LIDC radiologist's marks. (A) The benign pulmonary nodules with the rank of malignancy "1." (B) The benign pulmonary nodules with the rank of malignancy "2." (C) The malignant pulmonary nodules with the rank of malignancy "4." (D) The malignant pulmonary nodules with the rank of malignancy "5."

4.2 Evaluation metrics

There are four metrics to evaluate the performance of pulmonary nodules classification model, including sensitivity (SEN), specificity (SPE), accuracy (ACC) and Area Under the Receiver Operating Characteristic (ROC) Curve (AUC). Sensitivity and specificity are calculated as Eqs. (10), (11), respectively.

$$SEN = \frac{TP}{TP + FN} \tag{9}$$

$$SPE = \frac{TN}{TN + FP} \tag{10}$$

where TP is true positive; TN is true negative; FP is False positive; FN is False negatives. ACC is the proportion of correct predictions (both True positives and True negatives) among the total number of nodules, which is formulated as Eq. (11).

$$ACC = \frac{TP + TN}{TP + TN + FP + FN} \tag{11}$$

The predictive performance of the classification models and their stabilities are evaluated by the mean value and relative standard deviation (RSD) of the four metrics used in this chapter, respectively. Here, the RSD is defined as Eq. (12).

$$\text{RSD} = \frac{\text{std}_{\text{metric}}}{\text{mean}_{\text{metric}}} \times 100 \tag{12}$$

where $\text{std}_{\text{metric}}$ and $\text{mean}_{\text{metric}}$ are the standard deviation and mean of the evaluation metric. A lower RSD value corresponds to the higher stability of the predicting model.

4.3 Experimental configuration

The experimental environment is as follows: windows 10, Central Processing Unit (CPU) Intel Core i7-10875 H, 2.30 GHz, and 16 GB of Random-Access Memory (RAM). All image preprocessing and radiomics feature extraction are conducted in Python by the "PyRadiomics" package [65]. The radiomics features calculated by this package comply with the feature definitions which are described by the Image Biomarker Standardization Initiative [66]. Therefore, the reproducibility and repeatability of the result in this study are effectively guaranteed. Standardization, discretization and feature selection methods including CHSQ, MUIF and ANOVA F-value are performed by a Python machine learning library named "scikit-learn" [67]. RELF is conducted using a Python implementation of Relief-based methods named "scikit-rebate" [68]. And MRMR algorithm is implemented by importing the "pymrmr" package [49] in Python. Classifiers including LG, SVM, KNN, NB, DT, RF and MLP are imported from "scikit-learn." XGB is performed using an open-source software library named "XGBoost" [57]. Evaluation and visualization of the model performance are conducted by Python packages "scikit-learn" and "Seaborn" [69]. A Python package named "SHAP" [33,70] is used to calculate Shapley value and interpret the radiomics machine learning model. Our ISHAP algorithm is implemented by "scikit-learn," "XGBoost" and "SHAP."

4.4 Implementation details and experimental results

4.4.1 Performance validation of initial feature and classifier selection

In order to validate the performance of different initial feature selection methods and classifiers, we conduct feature extraction of CT images and evaluate the model performance using 10-fold cross-validation with multiple evaluation metrics.

(1) Implementation details

The CT images of pulmonary nodules are first filtered by a 3D wavelet transform and multiple 3D LoG filters (σ values range from 0.5 to 3 with 0.5 steps), respectively. Radiomics features are extracted from original CT images, wavelet images and LoG images, including shape-based, first-order, GLCM, GLDM, GLRLM, GLSZM and NGTDM features. Each feature is named by underlining image type from which it was extracted, feature type and feature name. For instance, original_shape_Elongation is a shape-based feature extracted from the original image with the name Elongation. After extracting eight radiologist-assessed nodule semantic features from the standardized DICOM representation of the annotations and combining them with imaging radiomics features, we get a total of 1417 features for each annotation.

Standardization (or Z-score normalization) are applied on all radiomics features so that we are able to compare features that have different units. In order to avoid information leakage from the testing dataset, standardization of the training dataset and testing dataset are both performed with the mean and standard deviation of the training dataset [71]. In addition, radiomics features in continuous values are partitioned into discrete values by discretization. In this chapter, the quantile strategy is used for discretization and each of the five bins had the same number of values split based on percentiles. Therefore, this is equivalent to evaluating radiomics features using a rating system similar to radiologist-assessed nodule semantic features, ranging from one to five. The features under the same scoring system allow us to compare each feature fairly, which is also conducive to the interpretation of the model.

Five feature selection methods based on filter approaches are used in this study, including Chi-square (CHSQ), Mutual information (MUIF), Analysis of Variance (ANOVA) F-value, ReliefF (RELF), and Minimum redundancy maximum relevance (MRMR). After selecting the top 500 ranked features for each feature selection method, we apply eight supervised machine learning algorithms, which are Logistic Regression (LG), Support Vector Machine (SVM), K-nearest Neighbours (KNN), Naïve Bayes (NB), Decision Tree (DT), Random Forest (RF), Multi-layer Perceptron (MLP) and XGBoost (XGB). In combination with five feature selection methods and eight classifiers, we build 40 models and compare their performance using the 10-fold cross-validation.

(2) Validation results

In Fig. 3, the mean values and RSD values of sensitivity, specificity, ACC and AUC of the 40 models using 10-fold cross-validation on the training dataset are presented respectively in the heatmap. As shown in Fig. 3, the MRMR_XGB model obtains the highest mean values of specificity (0.872), ACC (0.882), AUC (0.946) and a second highest value of sensitivity (0.892). In addition, the MRMR_XGB model obtains low RSD values of sensitivity (2.779), specificity (2.145), ACC (1.479) and AUC (1.174), which demonstrates the high stability of the model. Therefore, in terms of predictive performance and stability, we select the MRMR and XGBoost as our initial feature selection method and classifier.

4.4.2 Performance validation of the proposed ISHAP-based interpretation-model-guided classification method for malignant pulmonary nodule

In order to demonstrate the effectiveness of the proposed ISHAP-based interpretation-model-guided classification method in terms of pulmonary nodule classification performance and model interpretation, we conduct a comparative experiment with respect to different feature selection methods, as well as interpretation and analysis of the most important features obtained based on the ISHAP method.

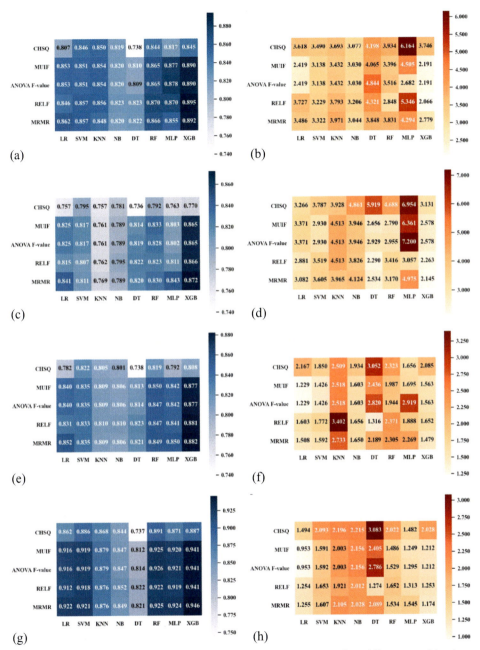

Fig. 3 Heatmaps illustrating the sensitivity, specificity, ACC and AUC of 40 different combinations of feature selection methods (in rows) and classification algorithms (in columns) using 10-fold cross-validation on the training dataset. (A) Mean values of sensitivity. (B) RSD values of sensitivity. (C) Mean values of specificity. (D) RSD values of specificity. (E) Mean values of ACC. (F) RSD values of ACC. (G) Mean values of AUC. (H) RSD values of AUC.

Classification results using the proposed ISHAP-based model-guided algorithm for adaptive feature selection and classifier parameters setting

After comparing the performance of different combination models and selecting the MRMR_XGB as the initial model, 500 features are selected from all the 1417 features and the proposed ISHAP-based method is conducted to get the best feature subset and interpret the model. At each iteration, we randomly search the hyperparameters of XGBoost with successive halving method and evaluate the AUC performance using 10-fold cross-validation. Then, we select the hyperparameter with the best performance and calculate the mean SHAP values of each feature using 10-fold cross-validation. The adaptive coefficient is set to be 0.1 and about 10% of the features would be eliminated at each iteration. In addition, by eliminating a portion of the features at each iteration, we change the search space of the hyperparameters in each iteration. As a result, the ISHAP method uses the SHAP values calculated in the previous iteration to guide the hyperparameter optimization for this round.

The AUC score for train and validation splits of our model for different numbers of features is shown in Fig. 4. We conduct a comparative experiment that only removes fixed 5 features at each iteration and Fig. 5 shows the performance of the comparative experiment. To better compare the two methods, the two validation performance curves are placed together which is shown in Fig. 6. The line represents the CV-averaged AUC score, while the shaded region behind shows the standard deviation of the CV scores. As shown in Figs. 4–6, we see that the model AUC performance on validation dataset starts dropping around 20 features with both two methods. However, the model with an adaptive coefficient is more computationally efficient as it greatly reduces the number of iterations while maintaining the performance of the model. By using the ISHAP method, we obtain both the best subset of features with 54 features and the best hyperparameters, which has the highest mean value (0.954) and lowest standard deviation value (0.008) of model AUC performance on validation dataset.

To validate the effect of our improved SHAP, a comparative experiment is conducted on different subsets of features. The 500 features subset selected by MRMR is considered as the baseline. In addition, we carry out the CHSQ, MUIF, ANOVA F-value, MRMR, RELF and the original SHAP methods to separately select 54 most important features using the same 10-fold cross-validation. To better compare the best performance of different feature subsets and hyperparameters, the same hyperparameter optimization is performed for all feature subsets except those 54 features and hyperparameters obtained by the ISHAP method. Fig. 7 shows the model sensitivity, specificity, ACC and AUC performance comparison of eight different feature subsets on the testing dataset.

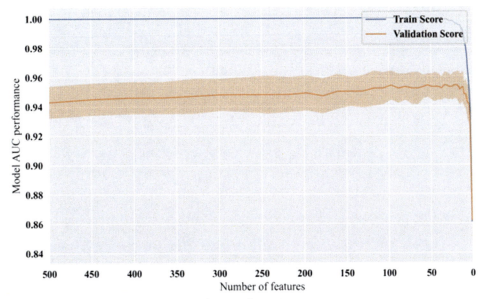

Fig. 4 Model performance with adaptive feature elimination.

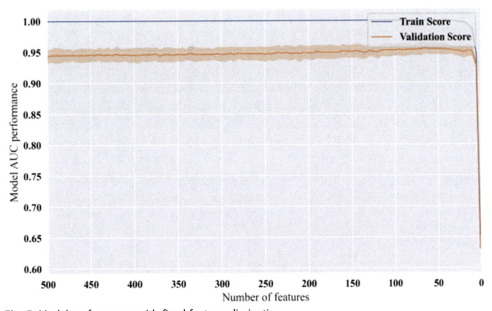

Fig. 5 Model performance with fixed feature elimination.

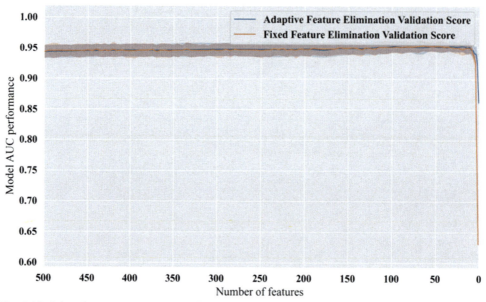

Fig. 6 Model performance comparison of adaptive feature elimination and fixed feature elimination.

As shown in Fig. 7, the improved SHAP obtains the highest sensitivity (0.862), ACC (0.873) and AUC (0.941) on the testing dataset. We can see that the baseline performance is much higher than most other methods, except for the original SHAP method, indicating that excellent classification performance can be obtained using SHAP for feature selection. Compared to the baseline, we have reduced the number of features from 500 to 54 without any loss of model predictive performance, which significantly reduces the difficulty of model interpretation. In addition, compared to conducting the feature selection and hyperparameter optimization respectively, the improved SHAP method guides the hyperparameter optimization through the SHAP values in each iteration, which results in faster computational efficiency and better classification performance. Moreover, the introduction of the adaptive recursive approach allows us to automatically select the optimal subset of features by comparing the performance of each feature subset on the validation dataset, rather than manually selecting features. Therefore, we consider the 54 features obtained from the improved SHAP as the best subset of features.

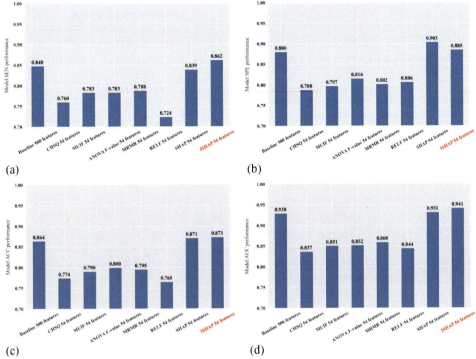

Fig. 7 Model performance comparison of eight different feature subsets. (A) Model sensitivity performance. (B) Model specificity performance. (C) Model ACC performance. (D) Model AUC performance.

Feature and model explanation results using the proposed ISHAP-based model-guided algorithm

The SHAP values of each feature in the best feature subset are calculated on testing dataset for model interpretation. A positive SHAP value for each prediction suggests an increased risk of the malignant nodule and vice versa. The SHAP summary graphic of the 20 most relevant features is shown in Fig. 8. The SHAP values are represented on the X-axis, while the features are represented on the Y-axis, which are ranked in descending order by feature relevance. A SHAP value for a prediction and feature is assigned to each point. High SHAP values are represented by red spots, whereas low SHAP values are represented by blue points. As shown in Fig. 8, the lobulation and spiculation medical semantic features are the most critical factors for classifying malignant and benign pulmonary nodules. Other radiologist-assessed semantic features like margin, sphericity and texture also demonstrate the significant impact on model prediction. Moreover, Fig. 8 demonstrates that some radiomics features extracted from CT images are more important than the radiologist-assessed features, such as the Surface Volume Ratio extracted from original images and the Large Area Low Gray Level Emphasis of GLSZM features extracted from the images after the 3D LoG filter with $\sigma=1.0$.

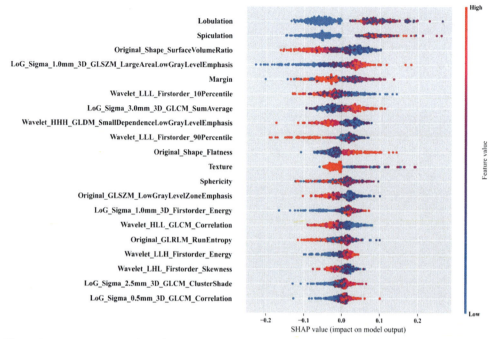

Fig. 8 SHAP summary graphic of top 20 features.

Figs. 9 and 10 show the feature attributions for two individual predictions, one of which is predicted to be a benign nodule and the other is a malignant nodule. In Figs. 9 and 10, f(x) represents the prediction of the classification model, which takes on values ranging from 0 to 1. For a given input, the closer the value of f(x) is to 1, the more likely the nodule is to be malignant. The baseline of SHAP values E[f(X)] is on the bottom of the plots, which is the average of all predictions f(X). As shown in Figs. 9 and 10, the baseline of SHAP values is 0.538, a value approximately equal to 0.5, which is because the training dataset is balanced. Fig. 9 illustrates a prediction for a benign nodule in the testing dataset. It starts with the base value of 0.538 and receives the final prediction of 0.13 by adding (red) and subtracting (blue) values. Positive and negative feature contributions are represented by the red and blue arrows. Each arrow's length is proportional to the feature's SHAP value. We can find that the texture, Surface Volume Ratio extracted from original images, lobulation, spiculation and 90Percentile of first-order features extracted from the images after wavelet filter have a considerable contribution to identifying benign nodules. Fig. 10 indicates that the nodule is identified as malignant (final prediction of 0.996), mainly due to the lobulation, spiculation, texture and Small Dependence Low Gray Level Emphasis of GLDM features extracted from the images after wavelet filter.

The relationship between the SHAP values of the features and the values of the features is depicted in Fig. 11. In this study, the two most essential features lobulation and spiculation are used to reveal this relationship. As shown in Fig. 11, the SHAP values for points with a lobulation and spiculation of 1 are both negative, while the SHAP values for points with a lobulation and spiculation of greater than 1 are both positive. Therefore, the model is more likely to predict the nodules with no lobulation and spiculation as benign nodules. As the score of lobulation and spiculation increases, the likelihood of the nodule being predicted as malignant also increases.

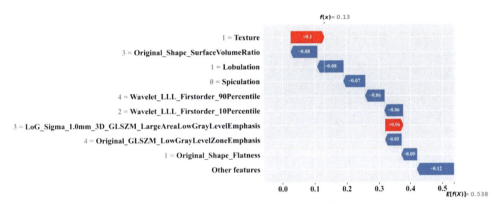

Fig. 9 SHAP explains how the model discriminates a benign nodule.

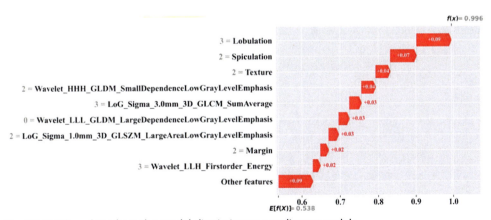

Fig. 10 SHAP explains how the model discriminates a malignant nodule.

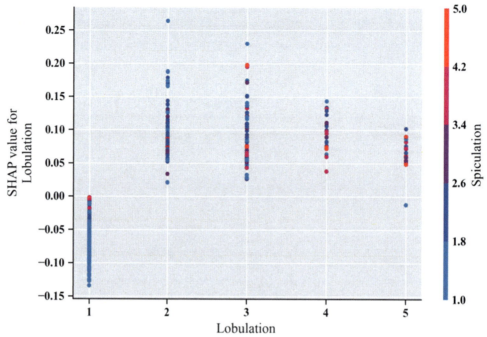

Fig. 11 SHAP dependence graphic of lobulation and spiculation.

5. Discussion

In this chapter, we first evaluate the predictive performance and stability of 40 combined models of feature selection and classification methods. Filter-based feature selection methods are preferred over wrapper and embedded methods because they are more computationally efficient and less prone to overfitting [43]. Additionally, filter methods are classifier-independent and thus enable the predictive analysis's modeling and feature selection components to be separated, increasing the generality of each component and thus the analysis as a whole. Notably, we use classifiers from different classifier families, and the large differences in their parameter configurations and tuning pose a challenge for unbiased parameter optimization of the classifiers.

Comparing the classifiers belonging to different families by using a limited number of parameter configurations and optimization methods may bias the results in favor of a particular classifier [72]. For all of the classifiers in our study, we use the same implementation tools ("scikit-learn" and "XGBoost"). Furthermore, to ensure unbiased classifier usage, we use default parameter configurations previously defined by open-source software libraries. Furthermore, the models in our study are only evaluated using 10-fold cross-validation of the training dataset.

As a result, our experimental design enables us to conduct unbiased evaluations of various combination models of feature selection methods and classifiers. The MRMR_XGB model has the best predictive performance on the training dataset, according to our findings. It's worth noting that among the eight different classifier combinations, MRMR has the highest mean AUC (0.898). Most feature selection methods do not take into account the redundancy of selected features, which could explain this result. As a result, MRMR outperforms other methods because it reduces feature redundancy. Furthermore, when combined with the XGBoost classifier, most feature selection methods provide the best predictive performance. This finding is consistent with previous research, which found that XGBoost has a high predictive performance in a variety of biomedical and other domain applications [73,74]. These findings suggest that using the MRMR feature selection method and the XGBoost classification method in radiomics improves predictive performance.

A major problem with radiomics models based on machine learning algorithms is the interpretability of the models. While the use of complex machine learning algorithms allows models to have good classification performance, clinicians are unable to understand or interpret some of the model's predictions. As a result, for the sake of interpretability, clinical models are frequently limited to simple linear modeling methods such as logistic regression, which can result in lower classification accuracy. We propose an improved SHAP method for interpreting the model to address this issue. SHAP has the advantage of being able to uncover patterns discovered by complex prediction models without being limited to simple modeling methods. What is more, we combine radiologist-assessed semantic features and radiomics features coming from lung CT images together. By discretizing the radiomics features, we can think of them as having a rating system similar to the semantic features, ranging from one to five, which allows us to compare the semantic features with the radiomics features fairly.

Instead of manually selecting a specific number of features based on the SHAP values, we introduce an adaptive recursive feature elimination method to automatically select the subset of features with the best performance. In addition, hyperparameter optimization is incorporated into the recursive process and guided by SHAP values, so that the model could fit the data better while calculating SHAP values. The comparative experiment among the ISHAP method (54 features), the baseline (500 features) and other feature selection methods (54 features) demonstrates the effectiveness of our proposed method for pulmonary nodule classification.

To further demonstrate the effectiveness of our ISHAP method, we compare the ISHAP method with other State-of-the-Art (SOTA) classification methods. However, a fair comparison of the different methods is very difficult, especially when specific codes for these SOTA methods are not available. Further, even if the publicly available LIDC dataset is used in those studies, differences in the way the dataset are divided into training and test sets can result in different results. In Table 4, we summarize several existing methods which trained and validated on the LIDC dataset in order to provide a rough estimate of the ISHAP method's performance.

Table 4 Comparisons of the classification performance between the proposed ISHAP method and other State-of-the-Art methods.

Method	Sensitivity	Specificity	Accuracy	AUC
Orozco et al. [17]	0.909	0.739	0.82	–
Li et al. [19]	0.92	0.83	0.90	0.95
Shen et al. [25]	0.77	0.93	0.87	0.93
Afshar et al. [26]	0.949	0.90	0.931	0.964
The proposed method	**0.862**	**0.885**	**0.873**	**0.941**

Orozco's method [17] employed the Daubechies db1, db2, and db4 wavelet transforms and 19 features were computed from each wavelet sub-band. Then, 11 features were selected by using the sub-band and attribute selection. The SVM classifier was employed to classify malignant and benign lung nodules. As shown in Table 4, Orozco's method [17] obtained a sensitivity of 0.909, a specificity of 0.7391 and a total accuracy of 0.82. Li's method [19] represented the lung nodules in terms of shape, intensity, and texture features. An improved random forest (RF) algorithm was proposed to differentiate malignant lung nodules from benign lung nodules. Li's method [19] obtained a sensitivity of 0.92, a specificity of 0.83, an accuracy of 0.82 and an AUC of 0.95, which yielded a significant performance improvement compared to Orozco's method [17]. Shen et al. proposed a multi-crop CNN (MC-CNN) to extract multi-scale nodule salient features for malignant and benign nodules classification [25]. Shen's method [25] obtained a sensitivity of 0.77, a specificity of 0.93, an accuracy of 0.87 and an AUC of 0.93. The specificity of Shen's method [25] is significantly higher than the other methods listed, but the sensitivity, accuracy and AUC of Shen's method [25] are relatively lower than those of Li's method [19]. Afshar et al. [26] proposed a 3D Multi-scale Capsule Network framework for lung tumor malignancy prediction. Afshar's method [26] predicted lung nodule malignancy with a sensitivity of 0.949, a specificity of 0.90, an accuracy of 0.931 and an AUC of 0.964. Although Afshar's method [26] obtained a slightly lower specificity than Shen's method [25], Afshar's method obtained the highest scores in all three remaining evaluation metrics. The proposed ISHAP method achieved an improvement over Shen's method [25] in terms of sensitivity, accuracy and AUC. The AUC reflects the classifier's trade-off between hit rate and false alarm rate. Our method yielded a slightly lower AUC than that of Li's method [19] and Afshar's method [26]. As mentioned above, differences in the class distribution of the data, the number of samples and the way in which the training and test datasets are divided can result in different scores for the evaluation metrics. There may be some gaps in classification performance between our method and the latest methods, however, our method can provide not only classification results, but also results that provide explanation of features, which are even more critical for doctors in clinical applications.

With the proposed ISHAP method, we have obtained a lot of information that helps to interpret the features of lung nodules. Some semantic features, such as spiculation and lobulation, have been used by radiologists to distinguish benign from malignant nodules with favorable clinical results [75]. Spiculation is caused by septal thickening, fibrosis, or tumor-filled lymphatic channels [76]. It has a high predictive value for malignancy, up to 90% [77]. Additionally, lobulation within a nodule is a result of variable or uneven growth rates, which is strongly associated with malignancy [78]. Our analysis indicates that the most critical factors in pulmonary nodule classification are lobulation and spiculation. This finding is consistent with that of Hancock et al. [79], who determined that spiculation and lobulation are the two most important features. According to data from the Dutch-Belgian randomized lung cancer screening trial (NELSON), spiculation (as well as lobulation) is associated with a higher risk of lung cancer than smooth, round, or polygonal shapes [80]. This is supported by our result that nodules without spiculation and lobulation are likely to be predicted as benign nodules. Another exciting finding is that two radiomics features, the Surface Volume Ratio (SVR) extracted from original images and the Large Area Low Gray Level Emphasis (LALGLE) of GLSZM features extracted from the images after 3D LoG filter with $\sigma=1.0$, have a more significant impact on model prediction than many semantic features such as margin, texture and sphericity. The Surface Volume Ratio is defined as Eq. (13).

$$SVR = \frac{A}{V} \tag{13}$$

Here A is the surface area and V denotes the volume of the pulmonary nodule. Our results show that nodules with smaller Surface Volume Ratio are more likely to be predicted as malignant by the model, while nodules with small LALGLE are more likely to be predicted as benign. The pulmonary nodule with the smallest surface area (and therefore with the smallest SVR) is a sphere for a given volume. In contrast, a pulmonary nodule with sharp corners will have an extensive surface area for a given volume. That is, SVR, like the semantic feature sphericity, characterizes the three-dimensional shape of the nodule in terms of its roundness. Interestingly, SVR shows more contribution to the model and thus has higher predictive power than sphericity. A possible explanation for this might be that the SVR quantifies the roundness of the nodes based on information from the CT images, while the radiologist judges the roundness of the nodes by experience. The LALGLE measures the proportion in the image of the joint distribution of larger size zones with lower gray-level values. However, LALGLE is not as intuitive as SVR, which measures the properties of the image itself. In general, the imaging radiomics features extracted in this study can indeed be used well to differentiate between benign and malignant pulmonary nodules. Further work is required to help explore the relationship between radiomics features, radiologist-assessed semantic features and the malignancy of pulmonary nodules.

The most important limitation lies in the fact that the SHAP value is calculated in relation to the machine learning algorithm used by the model. If there is a pattern in the data that cannot be learned by the machine learning algorithm, then this pattern will not be reflected in the SHAP value of the feature. In addition, outliers in the data cannot be interpreted using SHAP. Despite its limitations, the findings of this work imply that explanatory machine learning techniques like SHAP may be useful in comprehending predictions generated by machine learning-based radiological or clinical pulmonary nodule classification models.

6. Conclusion

In this chapter, we proposed an interpretation-model-guided classification method for the classification of malignant pulmonary nodules. We have developed an improved SHAP method for the radiomics machine learning model, aiming at malignant pulmonary nodules classification and model interpretation. By evaluating the different combination models of five feature selection methods and eight classifiers in terms of their predictive performance and stability, the MRMR method and XGBoost classifier are selected for their excellent performance on 10-fold cross-validation. In addition, a comparative experiment is conducted among the ISHAP method and other feature selection methods, which demonstrates the effectiveness of our proposed method for pulmonary nodule classification. Furthermore, we explore the most critical features of the model classification process using the improved SHAP method, which allows us to identify the factors that make a nodule malignant. Therefore, our model is no longer a black box, but an interpretable model that can provide doctors with trustworthy results in clinical applications.

Further research should be carried out to improve the malignant pulmonary nodule classification model's performance and optimize our improved SHAP method. Moreover, our further work might incorporate transfer learning methods, with the purpose of learning transferable information from radiologist-assessed semantic features and radiomics features to predict the insufficient imaging samples with truly pathological-proven labels, which could be valuable for the diagnosis and treatment of lung cancer in clinical applications.

References

[1] C. Wild, E. Weiderpass, B. Stewart, World Cancer Report: Cancer Research for Cancer Prevention, International Agency for Research on Cancer, Lyon, 2020, pp. 23–33.
[2] P.B. Bach, J.N. Mirkin, T.K. Oliver, C.G. Azzoli, D.A. Berry, O.W. Brawley, T. Byers, G.A. Colditz, M.K. Gould, J.R. Jett, Benefits and harms of CT screening for lung cancer: a systematic review, JAMA 307 (2012) 2418–2429.
[3] N.L.S.T.R. Team, Reduced lung-cancer mortality with low-dose computed tomographic screening, N. Engl. J. Med. 365 (2011) 395–409.

[4] P. Lambin, E. Rios-Velazquez, R. Leijenaar, S. Carvalho, R.G. Van Stiphout, P. Granton, C.M. Zegers, R. Gillies, R. Boellard, A. Dekker, Radiomics: extracting more information from medical images using advanced feature analysis, Eur. J. Cancer 48 (2012) 441–446.

[5] A. El-Baz, G.M. Beache, G. Gimel'farb, K. Suzuki, K. Okada, A. Elnakib, A. Soliman, B. Abdollahi, Computer-aided diagnosis systems for lung cancer: challenges and methodologies, Int. J. Biomed. Imaging 2013 (2013) 1–46. 942353.

[6] R.J. Gillies, P.E. Kinahan, H. Hricak, Radiomics: images are more than pictures, they are data, Radiology 278 (2016) 563–577.

[7] P. Lambin, R.T. Leijenaar, T.M. Deist, J. Peerlings, E.E. De Jong, J. Van Timmeren, S. Sanduleanu, R.T. Larue, A.J. Even, A. Jochems, Radiomics: the bridge between medical imaging and personalized medicine, Nat. Rev. Clin. Oncol. 14 (2017) 749–762.

[8] Q. Li, Y. Balagurunathan, Y. Liu, J. Qi, M.B. Schabath, Z. Ye, R.J. Gillies, Comparison between radiological semantic features and lung-RADS in predicting malignancy of screen-detected lung nodules in the National Lung Screening Trial, Clin. Lung Cancer 19 (2018) 148–156.

[9] B. Chen, R. Zhang, Y. Gan, L. Yang, W. Li, Development and clinical application of radiomics in lung cancer, Radiat. Oncol. 12 (2017) 1–8.

[10] V. Kumar, Y. Gu, S. Basu, A. Berglund, S.A. Eschrich, M.B. Schabath, K. Forster, H.J. Aerts, A. Dekker, D. Fenstermacher, Radiomics: the process and the challenges, Magn. Reson. Imaging 30 (2012) 1234–1248.

[11] V. Verma, C.B. Simone, S. Krishnan, S.H. Lin, J. Yang, S.M. Hahn, The rise of radiomics and implications for oncologic management, JNCI J. Natl. Cancer Inst. 109 (7) (2017) 1–3, https://doi.org/10.1093/jnci/djx055.

[12] E. Limkin, R. Sun, L. Dercle, E. Zacharaki, C. Robert, S. Reuzé, A. Schernberg, N. Paragios, E. Deutsch, C. Ferté, Promises and challenges for the implementation of computational medical imaging (radiomics) in oncology, Ann. Oncol. 28 (2017) 1191–1206.

[13] M. Reyes, R. Meier, S. Pereira, C.A. Silva, F.-M. Dahlweid, H.V. Tengg-Kobligk, R.M. Summers, R. Wiest, On the interpretability of artificial intelligence in radiology: challenges and opportunities, Radiol. Artif. Intell. 2 (2020) e190043.

[14] A. Singh, S. Sengupta, V. Lakshminarayanan, Explainable deep learning models in medical image analysis, J. Imaging 6 (2020) 52.

[15] B.J. Erickson, P. Korfiatis, Z. Akkus, T.L. Kline, Machine learning for medical imaging, Radiographics 37 (2017) 505–515.

[16] R. Thawani, M. McLane, N. Beig, S. Ghose, P. Prasanna, V. Velcheti, A. Madabhushi, Radiomics and radiogenomics in lung cancer: a review for the clinician, Lung Cancer 115 (2018) 34–41.

[17] H.M. Orozco, O.O.V. Villegas, V.G.C. Sánchez, H. de Jesús Ochoa Domínguez, M. de Jesús Nandayapa Alfaro, Automated system for lung nodules classification based on wavelet feature descriptor and support vector machine, Biomed. Eng. Online 14 (2015) 1–20.

[18] E. Taşcı, A. Uğur, Shape and texture based novel features for automated juxtapleural nodule detection in lung CTs, J. Med. Syst. 39 (2015) 1–13.

[19] X.-X. Li, B. Li, L.-F. Tian, L. Zhang, Automatic benign and malignant classification of pulmonary nodules in thoracic computed tomography based on RF algorithm, IET Image Process. 12 (2018) 1253–1264.

[20] J. Gong, J.-Y. Liu, X.-W. Sun, B. Zheng, S.-D. Nie, Computer-aided diagnosis of lung cancer: the effect of training data sets on classification accuracy of lung nodules, Phys. Med. Biol. 63 (2018) 035036.

[21] W. Wu, H. Hu, J. Gong, X. Li, G. Huang, S. Nie, Malignant-benign classification of pulmonary nodules based on random forest aided by clustering analysis, Phys. Med. Biol. 64 (2019) 035017.

[22] D. Kumar, A. Wong, D.A. Clausi, Lung nodule classification using deep features in CT images, in: 2015 12th Conference on Computer and Robot Vision, IEEE, 2015, pp. 133–138.

[23] N. Tajbakhsh, K. Suzuki, Comparing two classes of end-to-end machine-learning models in lung nodule detection and classification: MTANNs vs CNNs, Pattern Recogn. 63 (2017) 476–486.

[24] G. Kang, K. Liu, B. Hou, N. Zhang, 3D multi-view convolutional neural networks for lung nodule classification, PLoS One 12 (2017) e0188290.

[25] W. Shen, M. Zhou, F. Yang, D. Yu, D. Dong, C. Yang, Y. Zang, J. Tian, Multi-crop convolutional neural networks for lung nodule malignancy suspiciousness classification, Pattern Recogn. 61 (2017) 663–673.

[26] P. Afshar, A. Oikonomou, F. Naderkhani, P.N. Tyrrell, K.N. Plataniotis, K. Farahani, A. Mohammadi, 3D-MCN: a 3D multi-scale capsule network for lung nodule malignancy prediction, Sci. Rep. 10 (2020) 1–11.

[27] S. Ali, T. Abuhmed, S. El-Sappagh, et al., Explainable Artificial Intelligence (XAI): what we know and what is left to attain Trustworthy Artificial Intelligence, Inf. Fusion 99 (2023) 101805.

[28] W. He, B. Li, R. Liao, et al., An ISHAP-based interpretation-model-guided classification method for malignant pulmonary nodule, Knowl.-Based Syst. 237 (2022) 107778.

[29] B. Zhou, A. Khosla, A. Lapedriza, et al., Learning deep features for discriminative localization, in: Proceedings of the IEEE Conference on Computer Vision and Pattern Recognition, 2016, pp. 2921–2929.

[30] H. Jiang, F. Shen, F. Gao, et al., Learning efficient, explainable and discriminative representations for pulmonary nodules classification, Pattern Recogn. 113 (2021) 107825.

[31] S. Shen, S.X. Han, D.R. Aberle, et al., Explainable hierarchical semantic convolutional neural network for lung cancer diagnosis, in: CVPR Workshops, 2019, pp. 63–66.

[32] S.M. Lundberg, B. Nair, M.S. Vavilala, M. Horibe, M.J. Eisses, T. Adams, D.E. Liston, D.K.-W. Low, S.-F. Newman, J. Kim, Explainable machine-learning predictions for the prevention of hypoxaemia during surgery, Nat. Biomed. Eng. 2 (2018) 749–760.

[33] S. Lundberg, S.-I. Lee, A unified approach to interpreting model predictions, arXiv preprint arXiv:1705.07874, 2017.

[34] L.S. Shapley, A Value for N-Person Games, Princeton University Press, 2016.

[35] R.P. Shah, H.M. Selby, P. Mukherjee, S. Verma, P. Xie, Q. Xu, M. Das, S. Malik, O. Gevaert, S. Napel, Machine learning radiomics model for early identification of small-cell lung cancer on computed tomography scans, JCO Clin. Cancer Inf. 5 (2021) 746–757.

[36] R. Du, V.H. Lee, H. Yuan, K.-O. Lam, H.H. Pang, Y. Chen, E.Y. Lam, P.-L. Khong, A.W. Lee, D.L. Kwong, Radiomics model to predict early progression of nonmetastatic nasopharyngeal carcinoma after intensity modulation radiation therapy: a multicenter study, Radiol. Artif. Intell. 1 (2019) e180075.

[37] L. Ma, Z. Xiao, K. Li, S. Li, J. Li, X. Yi, Game theoretic interpretability for learning based preoperative gliomas grading, Futur. Gener. Comput. Syst. 112 (2020) 1–10.

[38] M. Diwakar, M. Kumar, A review on CT image noise and its denoising, Biomed. Signal Process. Control 42 (2018) 73–88.

[39] H. Mahersia, H. Boulehmi, K. Hamrouni, CAD system for lung nodules detection using wavelet-based approach and intelligent classifiers, in: 2020 17th International Multi-Conference on Systems Signals & Devices (SSD), IEEE, 2020, pp. 173–178.

[40] M.F. McNitt-Gray, S.G. Armato III, C.R. Meyer, A.P. Reeves, G. McLennan, R.C. Pais, J. Freymann, M.S. Brown, R.M. Engelmann, P.H. Bland, The Lung Image Database Consortium (LIDC) data collection process for nodule detection and annotation, Acad. Radiol. 14 (2007) 1464–1474.

[41] G. Thibault, B. Fertil, C. Navarro, S. Pereira, P. Cau, N. Levy, J. Sequeira, J.-L. Mari, Shape and texture indexes application to cell nuclei classification, Int. J. Pattern Recognit. Artif. Intell. 27 (2013) 1357002.

[42] Y. Balagurunathan, V. Kumar, Y. Gu, J. Kim, H. Wang, Y. Liu, D.B. Goldgof, L.O. Hall, R. Korn, B. Zhao, Test–retest reproducibility analysis of lung CT image features, J. Digit. Imaging 27 (2014) 805–823.

[43] I. Guyon, A. Elisseeff, An introduction to variable and feature selection, J. Mach. Learn. Res. 3 (2003) 1157–1182.

[44] G. Chandrashekar, F. Sahin, A survey on feature selection methods, Comput. Electr. Eng. 40 (2014) 16–28.

[45] Z. Liu, S. Wang, J.W. Di Dong, C. Fang, X. Zhou, K. Sun, L. Li, B. Li, M. Wang, J. Tian, The applications of radiomics in precision diagnosis and treatment of oncology: opportunities and challenges, Theranostics 9 (2019) 1303.

[46] J. Li, K. Cheng, S. Wang, F. Morstatter, R.P. Trevino, J. Tang, H. Liu, Feature selection: a data perspective, ACM Comput. Surv. 50 (2017) 1–45.

[47] Y. Saeys, I. Inza, P. Larranaga, A review of feature selection techniques in bioinformatics, Bioinformatics 23 (2007) 2507–2517.

[48] R.J. Urbanowicz, M. Meeker, W. La Cava, R.S. Olson, J.H. Moore, Relief-based feature selection: introduction and review, J. Biomed. Inform. 85 (2018) 189–203.

[49] H. Peng, F. Long, C. Ding, Feature selection based on mutual information criteria of max-dependency, max-relevance, and min-redundancy, IEEE Trans. Pattern Anal. Mach. Intell. 27 (2005) 1226–1238.

[50] C. Parmar, P. Grossmann, J. Bussink, P. Lambin, H.J. Aerts, Machine learning methods for quantitative radiomic biomarkers, Sci. Rep. 5 (2015) 1–11.

[51] R.-E. Fan, K.-W. Chang, C.-J. Hsieh, X.-R. Wang, C.-J. Lin, LIBLINEAR: a library for large linear classification, J. Mach. Learn. Res. 9 (2008) 1871–1874.

[52] J.A. Suykens, J. Vandewalle, Least squares support vector machine classifiers, Neural. Process. Lett. 9 (1999) 293–300.

[53] K. Murphy, B. van Ginneken, A.M. Schilham, B. De Hoop, H.A. Gietema, M. Prokop, A large-scale evaluation of automatic pulmonary nodule detection in chest CT using local image features and k-nearest-neighbour classification, Med. Image Anal. 13 (2009) 757–770.

[54] I. Rish, An empirical study of the naive Bayes classifier, in: IJCAI 2001 Workshop on Empirical Methods in Artificial Intelligence, 2001, pp. 41–46.

[55] L. Breiman, Random forests, Mach. Learn. 45 (2001) 5–32.

[56] G.E. Hinton, Connectionist learning procedures, Mach. Learn. (1990) 555–610. Elsevier.

[57] T. Chen, C. Guestrin, Xgboost: a scalable tree boosting system, in: Proceedings of the 22nd ACM SIGKDD International Conference on Knowledge Discovery and Data Mining, 2016, pp. 785–794.

[58] S. Putatunda, K. Rama, A comparative analysis of hyperopt as against other approaches for hyper-parameter optimization of XGBoost, in: Proceedings of the 2018 International Conference on Signal Processing and Machine Learning, 2018, pp. 6–10.

[59] I. Guyon, J. Weston, S. Barnhill, V. Vapnik, Gene selection for cancer classification using support vector machines, Mach. Learn. 46 (2002) 389–422.

[60] J. Bergstra, Y. Bengio, Random search for hyper-parameter optimization, J. Mach. Learn. Res. 13 (2012) 281–305.

[61] L. Li, K. Jamieson, G. DeSalvo, A. Rostamizadeh, A. Talwalkar, Hyperband: a novel bandit-based approach to hyperparameter optimization, J. Mach. Learn. Res. 18 (2017) 6765–6816.

[62] A. Fedorov, M. Hancock, D. Clunie, M. Brochhausen, J. Bona, J. Kirby, J. Freymann, S. Pieper, H.-J.W.L. Aerts, R. Kikinis, DICOM re-encoding of volumetrically annotated Lung Imaging Database Consortium (LIDC) nodules, Med. Phys. 47 (2020) 5953–5965.

[63] D.P. Naidich, A.A. Bankier, H. MacMahon, C.M. Schaefer-Prokop, M. Pistolesi, J.M. Goo, P. Macchiarini, J.D. Crapo, C.J. Herold, J.H. Austin, Recommendations for the management of subsolid pulmonary nodules detected at CT: a statement from the Fleischner Society, Radiology 266 (2013) 304–317.

[64] D. Manos, J.M. Seely, J. Taylor, J. Borgaonkar, H.C. Roberts, J.R. Mayo, The Lung Reporting and Data System (LU-RADS): a proposal for computed tomography screening, Can. Assoc. Radiol. J. 65 (2014) 121–134.

[65] J.J. Van Griethuysen, A. Fedorov, C. Parmar, A. Hosny, N. Aucoin, V. Narayan, R.G. Beets-Tan, J.-C. Fillion-Robin, S. Pieper, H.J. Aerts, Computational radiomics system to decode the radiographic phenotype, Cancer Res. 77 (2017) e104–e107.

[66] A. Zwanenburg, M. Vallières, M.A. Abdalah, H.J. Aerts, V. Andrearczyk, A. Apte, S. Ashrafinia, S. Bakas, R.J. Beukinga, R. Boellaard, The image biomarker standardization initiative: standardized quantitative radiomics for high-throughput image-based phenotyping, Radiology 295 (2020) 328–338.

[67] F. Pedregosa, G. Varoquaux, A. Gramfort, V. Michel, B. Thirion, O. Grisel, M. Blondel, P. Prettenhofer, R. Weiss, V. Dubourg, Scikit-learn: machine learning in Python, J. Mach. Learn. Res. 12 (2011) 2825–2830.

[68] R.J. Urbanowicz, R.S. Olson, P. Schmitt, M. Meeker, J.H. Moore, Benchmarking relief-based feature selection methods for bioinformatics data mining, J. Biomed. Inform. 85 (2018) 168–188.

[69] M.L. Waskom, Seaborn: statistical data visualization, J. Open Source Softw. 6 (2021) 3021.

[70] S.M. Lundberg, G. Erion, H. Chen, A. DeGrave, J.M. Prutkin, B. Nair, R. Katz, J. Himmelfarb, N. Bansal, S.-I. Lee, From local explanations to global understanding with explainable AI for trees, Nat. Mach. Intell. 2 (2020) 56–67.

[71] T. Hastie, R. Tibshirani, J. Friedman, The Elements of Statistical Learning: Data Mining, Inference, and Prediction, Springer Science & Business Media, 2009.

[72] M. Fernández-Delgado, E. Cernadas, S. Barro, D. Amorim, Do we need hundreds of classifiers to solve real world classification problems? J. Mach. Learn. Res. 15 (2014) 3133–3181.

[73] R.S. Olson, W.L. Cava, Z. Mustahsan, A. Varik, J.H. Moore, Data-driven advice for applying machine learning to bioinformatics problems, in: Proceedings of the Pacific Symposium, 2018, World Scientific, 2018, pp. 192–203.

[74] S. Ramraj, N. Uzir, R. Sunil, S. Banerjee, Experimenting XGBoost algorithm for prediction and classification of different datasets, Int. J. Control Theory Appl. 9 (2016) 651–662.

[75] J.J. Erasmus, J.E. Connolly, H.P. McAdams, V.L. Roggli, Solitary pulmonary nodules: part I. Morphologic evaluation for differentiation of benign and malignant lesions, Radiographics 20 (2000) 43–58.

[76] C. Zwirewich, S. Vedal, R. Miller, N. Müller, Solitary pulmonary nodule: high-resolution CT and radiologic-pathologic correlation, Radiology 179 (1991) 469–476.

[77] J.L. Leef 3rd, J.S. Klein, The solitary pulmonary nodule, Radiol. Clin. North Am. 40 (2002) 123–143.

[78] J. Gurney, Determining the likelihood of malignancy in solitary pulmonary nodules with Bayesian analysis. Part I. Theory, Radiology 186 (1993) 405–413.

[79] M.C. Hancock, J.F. Magnan, Lung nodule malignancy classification using only radiologist-quantified image features as inputs to statistical learning algorithms: probing the Lung Image Database Consortium dataset with two statistical learning methods, J. Med. Imaging 3 (2016) 044504.

[80] A. Snoeckx, P. Reyntiens, D. Desbuquoit, M.J. Spinhoven, P.E. Van Schil, J.P. van Meerbeeck, P.M. Parizel, Evaluation of the solitary pulmonary nodule: size matters, but do not ignore the power of morphology, Insights Imaging 9 (2018) 73–86.

CHAPTER 6

Case studies: Explainable AI for Healthcare 5.0

M. Shanmuga Sundari[a] ⓘ **and Mayukha Mandya Ammangatambu**[b] ⓘ
aComputer Science and Engineering, BVRIT HYDERABAD College of Engineering for Women, Hyderabad, India
bThe University of Utah, Salt Lake City, UT, United States

1. Introduction to Explainable AI

1.1 Overview of AI in healthcare

The introduction of artificial intelligence (AI) into the healthcare industry has brought about a significant shift in recent years [1], challenging established ideas and providing creative solutions in a range of fields. This chapter aims to offer a comprehensive analysis of artificial intelligence (AI) applications in the healthcare industry, examining how these applications affect drug development, virtual health assistants, diagnostics, personalized treatment, predictive analytics, and administrative effectiveness. We can see how AI is more than just a technological advancement and represents a fundamental change in how healthcare is conceived and provided by comprehending the complexities of these applications.

The introduction of artificial intelligence (AI) into the healthcare industry [2] has brought about a significant shift in recent years, challenging established ideas and providing creative solutions in a range of fields. This chapter aims to offer a comprehensive analysis of artificial intelligence (AI) applications in the healthcare industry, examining how these applications affect drug development, virtual health assistants, diagnostics, personalized treatment, predictive analytics, and administrative effectiveness. We can see how AI is more than just a technological advancement and represents a fundamental change in how healthcare is conceived and provided by comprehending the complexities of these applications (Fig. 1).

Explainable AI in Healthcare Imaging for Medical Diagnoses
https://doi.org/10.1016/B978-0-443-23979-3.00006-3

Fig. 1 Explainable AI in medicine.

The era of personalized medicine [3] has been ushered in by AI, revolutionizing the way healthcare is delivered on an individual level. By meticulously analyzing patient data, including genetic markers, lifestyle choices, and medical histories, AI facilitates the development of treatment plans tailored to the unique characteristics of each patient. This departure from a standardized treatment model allows for a more targeted and effective approach, optimizing the chances of treatment success. As AI continues to evolve, the landscape of healthcare is shifting toward a more patient-centric model, where interventions are not only evidence-based but also finely tuned to the specific needs of each individual.

The integration of AI into healthcare is prominently manifested in the domain of predictive analytics. By harnessing machine learning algorithms, healthcare providers can analyze vast datasets to identify potential risks and predict adverse events. This proactive approach to patient care represents a paradigm shift from reactive to preventive medicine. The ability to foresee health complications before they escalate allows for timely interventions, reducing the burden on healthcare systems and significantly improving patient outcomes. As AI continues to refine predictive analytics, the potential for identifying subtle patterns and associations in patient data becomes increasingly promising.

AI-powered virtual health assistants offer a substantial leap in patient involvement and healthcare communication. These digital entities, which are frequently in the form of chatbots, offer patients immediate assistance by answering questions, booking appointments, and providing important health information. The use of virtual health assistants simplifies communication routes between healthcare providers and patients, resulting in a scalable and efficient method of giving information and support. This improves not only the patient experience but also the overall efficiency of healthcare delivery systems.

AI is speeding up the traditionally difficult drug discovery and development process in the pharmaceutical industry. Machine learning algorithms process the large amount of biological and chemical data required to find possible drug candidates efficiently. This improves the accuracy of medication development while also hastening the identification of interesting molecules. There is great potential for meeting unmet medical needs and enhancing therapeutic results with the development of more effective and tailored pharmaceuticals. The way AI and drug development are working together more and more is an indication of a paradigm shift that will make the pharmaceutical industry more effective and significant.

Beyond its impact on clinical aspects, AI is enhancing administrative efficiency within healthcare systems. Automation of tasks such as billing, appointment scheduling, and record-keeping streamlines the operational aspects of healthcare delivery. By reducing the administrative burden on healthcare professionals, AI ensures that valuable time and resources are redirected toward direct patient care. The increased efficiency in administrative processes not only enhances the overall quality of healthcare delivery but also contributes to cost savings and resource optimization.

The integration of AI into healthcare is not a mere technological augmentation but a fundamental reimagining of the entire healthcare ecosystem. From redefining diagnostic precision to enabling personalized medicine, predicting health risks [4], engaging patients through virtual assistants, expediting drug discovery, and optimizing administrative processes, AI is at the core of a healthcare revolution. As these applications continue to mature and evolve, the transformative potential of AI in healthcare [5] becomes increasingly evident, paving the way for a future where technology and human expertise converge to deliver more effective, personalized, and efficient healthcare solutions.

1.2 Importance of explainability

The emergence of recent advances like artificial intelligence (AI) and machine learning have resulted in an innovative period in the field of healthcare today. Although these innovations have great potential to improve patient care, questions are raised regarding the interpretability and transparency of the underlying algorithms. It is impossible to overestimate the importance of explainability in healthcare since it has a significant impact on people's acceptance, trust, and ethical considerations surrounding these technologies. Establishing a strong foundation on which the healthcare community can confidently embrace these advancements and ensure a harmonious integration of technology and human expertise requires a nuanced understanding of how AI systems work.

Explainability is foundational in fostering trust among healthcare professionals, patients, and other stakeholders. A clear understanding of how AI systems arrive at decisions enhances confidence in these technologies [6]. Trust is pivotal in the intricate relationship between physicians and patients, and transparent AI algorithms contribute to a collaborative healthcare environment where stakeholders feel supported and informed. The ability to decipher and trust the decision-making process of AI systems creates a sense of partnership, where healthcare professionals view these technologies as valuable aids rather than opaque entities.

The high-stakes nature of healthcare decisions necessitates a commitment to ethical standards [7]. Explainability plays a crucial role in ensuring that AI-driven decisions align with established guidelines. Transparent AI models empower healthcare providers to scrutinize the reasoning behind specific recommendations, allowing for informed decision-making that prioritizes the ethical dimensions of patient care. This transparency not only aligns with the moral imperatives of the healthcare profession but also establishes a framework where ethical considerations are woven into the fabric of technological advancements.

The healthcare industry operates within stringent regulatory frameworks designed to safeguard patient rights and uphold care quality. Explainability is a linchpin in meeting these regulatory requirements. Transparent AI systems enable healthcare organizations to demonstrate accountability, and adherence to standards, and thereby steer clear of legal and ethical complications. By providing a clear trail of how decisions are reached, these systems contribute to a regulatory landscape that can adapt to the evolving nature of healthcare technologies, ensuring that patient welfare remains at the forefront.

Explainable AI contributes to more informed clinical decision-making by providing healthcare professionals with insights into algorithmic recommendations. The ability to comprehend the rationale behind AI-generated diagnoses and treatment plans allows clinicians to validate or challenge these suggestions, ultimately elevating the overall quality of patient care. With this depth of understanding, healthcare professionals are better equipped to leverage AI as a collaborative tool, incorporating its insights seamlessly into their decision-making processes.

A significant challenge in AI applications within healthcare is the potential for bias and discrimination. Explainability emerges as a crucial tool for identifying and rectifying biases within algorithms. Healthcare professionals, armed with an understanding of how AI systems reach conclusions, can actively intervene to mitigate biases, ensuring fair and equitable treatment across diverse patient populations [8]. This proactive approach not only aligns with principles of social justice but also contributes to a healthcare ecosystem where technological advancements actively work toward eliminating disparities in patient care.

Explainability not only enhances trust but also serves as an educational tool, empowering healthcare professionals with a deeper understanding of AI technologies. This knowledge fosters continuous learning and skill development, enabling clinicians to effectively collaborate with AI systems seamlessly integrated into healthcare workflows. The educational aspect of explainability extends beyond mere comprehension; it cultivates a mindset of adaptability and ongoing learning, ensuring that healthcare professionals remain at the forefront of technological advancements (Fig. 2).

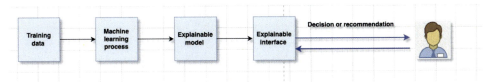

Fig. 2 White box AI.

In navigating the dynamic terrain of healthcare technology, the paramount importance of explainability becomes evident. Transparent AI systems not only establish trust but also contribute to ethical decision-making, regulatory compliance, improved clinical outcomes, bias mitigation, and ongoing education for healthcare professionals. Prioritizing explainability ensures the healthcare industry can harness the full potential of AI, fostering a balance between technological innovation and the well-being and trust of patients and practitioners alike. The journey toward a seamlessly integrated future of healthcare and AI relies on the foundation of transparency, ensuring that every stakeholder is not only a beneficiary but an informed collaborator in this transformative journey [9].

2. Case study 1: Interpretable models for diagnosing X

2.1 Problem statement and data

In delving into the intricate landscape of diagnosing Type 2 Diabetes, the problem statement serves as the compass guiding this case study. The challenge is articulated not merely as a quest for predictive accuracy but as a nuanced exploration into interpretability. Recognizing the complexity of Type 2 Diabetes diagnosis, the problem statement underscores the imperative of unveiling the decision-making process [10]. This transparency is viewed not just as a desirable feature but as an essential component in fostering a more collaborative and informed relationship between technology and healthcare practitioners.

The dataset, meticulously curated to encapsulate the breadth of factors influencing Type 2 Diabetes [11], extends beyond a compilation of clinical data. It's a dynamic compilation that weaves together comprehensive clinical histories, diverse glucose levels, cholesterol readings, and a myriad of lifestyle factors [12]. The preprocessing endeavor is not a routine data-cleaning task but an intricate choreography, addressing missing values, standardizing formats, and ensuring data quality. This meticulous curation is akin to preparing fertile soil, creating a robust foundation for subsequent model training. The richness and diversity of this dataset are intentional, aiming to capture the multifaceted nature of Type 2 Diabetes [13] and ensuring the model is exposed to a representative spectrum of patient scenarios (Fig. 3).

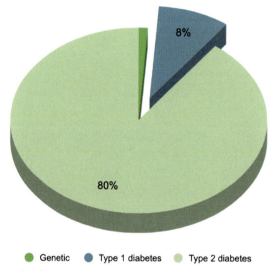

Fig. 3 Types of diabetes distribution.

2.2 Model selection and training

The journey into model selection and training is not merely a technical expedition but a strategic voyage where precision meets purpose. The chosen model, a carefully selected companion in this diagnostic quest, undergoes an extensive training regimen. The intricacies of this phase involve not only technical considerations like hyperparameter optimization and cross-validation but also a deliberate commitment to setting a benchmark that exceeds conventional clinical expectations. The emphasis on interpretability is not an afterthought; it is an intentional design choice. It is a conscious decision to steer away from the allure of complex black-box models toward a model that not only predicts Type 2 Diabetes accurately but does so in a way that aligns seamlessly with the day-to-day realities of healthcare practices.

The model training phase becomes a symphony where each hyperparameter adjustment and each validation iteration harmonize to create a model that not only meets the clinical gold standard but surpasses it. The goal is a model that doesn't merely offer predictions but does so in a manner that healthcare professionals can understand and trust. This emphasis on interpretability becomes a bridge, connecting the technical sophistication of the model with the pragmatic needs of clinical workflows. It is not just about achieving accuracy; it is about achieving a balance where accuracy and interpretability coalesce, ensuring the model becomes a valuable asset in the hands of healthcare practitioners.

2.3 Interpretability techniques applied

The section dedicated to interpretability techniques is akin to unlocking the secrets of the diagnostic model. Feature importance analysis emerges as the first act, where the spotlight shines on critical variables such as glucose levels, BMI, and physical activity. This analysis goes beyond a mere list of significant features; it delves into their contextual relevance, unraveling the intricate web of factors that influence the model's predictions. Decision boundary visualization techniques emerge as the visual narrative, offering clinicians more than just an abstract understanding. These techniques paint a vivid picture of how the model navigates through the diagnostic landscape, providing valuable insights into its decision-making process.

The storyline leaps advanced interpretability techniques, with SHAP becoming a protagonist. SHAP doesn't merely provide explanations; it crafts a personalized narrative for each patient. It's an intricate dance of Shapley values, assigning contributions to each feature, and creating a mosaic of insights into the specific factors influencing each patient's likelihood of having Type 2 Diabetes. These techniques, far from being mere add-ons, become integral chapters in the interpretability narrative. They empower healthcare professionals with not just an understanding of the model's decisions but a profound comprehension of the nuanced factors shaping those decisions. This interpretability is not just a feature; it's the key to unlocking a collaborative dialog between the model and the healthcare practitioner.

2.4 Results and clinical insights

The section dedicated to results and clinical insights is the outcome of this case study, where the impact of interpretable models for Type 2 Diabetes unfolds. The results extend beyond the numerical metrics, painting a comprehensive picture of the model's efficacy. It's not just about accuracy; it's about the model's capacity to offer actionable clinical insights. The model's commendable sensitivity and specificity aren't just figures on a chart; they set a new standard for diagnostic excellence, challenging and reshaping the existing benchmarks. Interpretability becomes the transformative force, elevating the diagnostic process from a mere prediction to a comprehensible and transparent decision-making tool [14].

Clinicians, in this narrative, aren't just passive recipients of predictions; they become active participants in understanding the underlying rationale for each diagnosis. The transparency facilitated by interpretability fosters trust and confidence among healthcare professionals. The model is not positioned as a replacement for clinical expertise; it becomes an invaluable ally that enriches the decision-making process. The success of this case study centered on Type 2 Diabetes, is not just a triumph of numerical achievements; it resonates with the profound impact that interpretability can wield in medical AI applications [15]. It's a stride toward a future where technological advancements seamlessly integrate with human expertise, promising not just accurate predictions but a holistic enhancement of patient outcomes. The denouement, therefore, isn't just a conclusion; it's an invitation to envision a healthcare landscape where interpretability becomes the linchpin in the collaboration between technology and human care.

3. Case study 2: Explainable AI in drug discovery

3.1 Drug discovery process overview

The drug discovery process is an intricate journey characterized by multiple interlinked phases, each demanding precision and innovation. Initiated with target identification, researchers delve into the intricate world of molecular structures and proteins associated with a specific disease [16]. This phase involves leveraging vast datasets and biological knowledge to pinpoint potential targets for therapeutic intervention. Following target identification, the process advances to lead compound identification, where researchers sift through compound libraries, employing advanced screening techniques to identify promising drug candidates. The subsequent hit-to-lead optimization phase involves refining and enhancing the selected candidates, focusing on improving their efficacy, safety profiles, and overall drug-like properties. Moving into preclinical testing, the selected compounds undergo rigorous evaluation for toxicity and pharmacokinetics, providing critical insights before advancing to the complex and resource-intensive clinical trial phases. The drug discovery process, therefore, entails navigating a labyrinth of biological data, chemical structures, and experimental validations to unveil potential therapeutic breakthroughs.

3.2 Application of Explainable AI

Explainable AI (XAI) emerges as a revolutionary tool within the drug discovery landscape, introducing transparency into the decision-making processes of intricate machine learning models. In the realm of target identification, XAI techniques offer a comprehensive understanding of the molecular features and interactions influencing a model's predictions, thereby providing researchers with valuable insights into the biological rationale underlying potential drug targets [17]. During lead compound identification and optimization, XAI elucidates the critical molecular descriptors and structural motifs driving a compound's predicted activity and safety profile. This transparency not only expedites decision-making but also empowers researchers to make informed choices in prioritizing compounds that align closely with therapeutic goals. In the preclinical and clinical phases, XAI facilitates the interpretation of model-driven predictions, offering insights into factors influencing toxicity, efficacy, and pharmacokinetics. By demystifying the black box of machine learning models, XAI enhances researchers' ability to make informed decisions about advancing promising compounds through the drug development pipeline, thereby optimizing resources and expediting the path to therapeutic breakthroughs.

3.3 Case study results and insights

Within the domain of drug discovery, a compelling case study underscores the tangible impact of integrating Explainable AI (XAI) into the research pipeline. This case study, focused on a specific therapeutic area, showcased the transformative potential of XAI in elucidating complex biological interactions. By applying XAI techniques during target identification, researchers gained granular insights into the model's decision-making process, unveiling the specific molecular features influencing predictions [18]. This newfound transparency allowed for a more focused and efficient target selection process, streamlining the identification of potential therapeutic candidates. Moreover, during lead compound identification and optimization, XAI played a pivotal role in unraveling the intricacies of molecular interactions, providing researchers with a roadmap for refining compounds based on specific structural motifs and functional groups. This transparency not only expedited the optimization phase but also contributed to a more rational and informed selection of lead compounds for further development.

As the case study progressed through preclinical testing and clinical phases, XAI continued to be a guiding force. It facilitated a deeper understanding of model predictions related to compound toxicity, efficacy, and pharmacokinetics. Researchers were equipped with nuanced insights, enabling them to make informed decisions on whether to advance a compound to the next stage of development. The interpretability provided by XAI proved instrumental in steering the drug discovery process, fostering a more dynamic and collaborative relationship between computational models and experimental validation. Ultimately, the case study showcased that the integration of XAI not only expedites decision-making but also elevates the quality and efficiency of the drug discovery process, marking a significant stride toward more effective and transparent therapeutic development.

4. Ethical considerations in Explainable AI

4.1 Bias and fairness

In the ethical landscape of Explainable AI (XAI), the intricate challenge of bias and fairness demands meticulous consideration. Bias in AI models, often rooted in the data used for training, can perpetuate and exacerbate existing societal inequities. Particularly concerning is the potential for disparate impacts on various demographic groups, amplifying historical biases. In XAI, recognizing and mitigating bias goes beyond technical intricacies; it necessitates a comprehensive approach that includes transparent communication about how decisions are made. Fairness-aware machine learning techniques and regular audits of model predictions are essential tools in the developer's arsenal to identify and rectify biases. Striving for fairness in XAI systems is not just a technical endeavor but a fundamental ethical imperative, requiring ongoing vigilance, commitment, and a relentless pursuit of creating AI systems that serve diverse populations equitably [19]. As the field evolves, a continuous dialog about the ethical dimensions of bias and fairness is crucial, ensuring that XAI contributes to a more just and equitable future.

4.2 Privacy concerns

Privacy stands as a cornerstone of ethical considerations in the deployment of Explainable AI, especially in contexts where these systems operate on vast datasets. The potential revelation of sensitive information poses significant challenges, particularly in healthcare, where XAI is increasingly employed for diagnostics and treatment decisions. Patient privacy becomes paramount, and the transparency provided by XAI must be delicately balanced with the imperative to safeguard personal information. As models unravel intricate details of individuals' health conditions, the need for robust data security measures becomes pronounced. Privacy-preserving techniques, such as federated learning, offer a pathway by allowing models to be trained across decentralized data sources without directly sharing raw data. Ethical frameworks and legal standards must evolve hand in hand with technological advancements to address the dynamic landscape of AI and privacy. Individuals' control over their personal information should remain sacrosanct, necessitating ongoing ethical reflections and proactive measures to align XAI practices with evolving standards, fostering public trust and ethical use of these transformative technologies.

4.3 Regulatory compliance

In the ethical dimension of Explainable AI, regulatory compliance serves as a linchpin, shaping responsible and lawful deployment. The varied standards and regulations across industries and sectors necessitate a nuanced understanding and meticulous adherence to established frameworks. In healthcare, for instance, rigorous regulations like the Health Insurance Portability and Accountability Act (HIPAA) set stringent standards for protecting patient data. Ethical considerations in XAI extend beyond technical facets to encompass the alignment of AI practices with existing legal frameworks. Staying informed about evolving regulations, transparently communicating the capabilities and limitations of AI systems, and proactively working toward compliance are essential steps. This not only mitigates legal risks but also contributes to building public trust. As the regulatory landscape evolves, organizations must engage in continuous ethical reflection, ensuring that the deployment of XAI is not just technically sound but ethically robust, accountable, and aligned with societal expectations [20].

5. Case study 3: Predicting patient outcomes with interpretable models

5.1 Data collection and preprocessing

Imagine a scenario where a healthcare system is seeking to predict patient outcomes, such as the likelihood of readmission following a hospital stay. In this case, the first crucial step is data collection. Diverse datasets are accumulated, spanning a multitude of patient-related variables, including demographics, medical histories, laboratory results, and details of previous treatments. For instance, patient age, gender, chronic conditions, and medication history are among the myriad factors considered. This wealth of data is akin to gathering puzzle pieces, each carrying a unique aspect of the patient's health journey.

Once collected, the data undergoes meticulous preprocessing. Missing values are addressed, ensuring that the information used to train the model is complete. Formats are standardized to create a cohesive dataset, and data quality is ensured by identifying and correcting anomalies. Imagine this phase as assembling the puzzle pieces—organizing and refining the data to construct a comprehensive picture of each patient's health profile. Patient privacy and confidentiality are paramount during this process, aligning with ethical standards like HIPAA to safeguard sensitive health information. The resulting dataset becomes the foundation for the subsequent steps in predicting patient outcomes.

5.2 Model development and explainability

With the curated dataset in hand, the next phase involves model development and explainability. Picture selecting a powerful yet interpretable model, such as a decision tree. This model is trained using the assembled dataset, with its parameters fine-tuned through iterative processes to enhance predictive accuracy. Unlike complex black-box models, the chosen model is designed to be interpretable from the outset. Now, envision the model as a medical detective, scrutinizing the patient data to uncover patterns and relationships.

Explainability techniques come into play, unveiling the model's decision-making process. Feature importance analysis identifies which puzzle pieces (variables) significantly influence predictions. For instance, the model might recognize that a patient's history of previous hospitalizations and specific chronic conditions strongly correlates with the likelihood of readmission. Visualization tools, like decision boundary maps, provide clinicians with a visual representation of how the model distinguishes between different outcomes. This transparency is akin to the detective sharing the clues and logic behind their conclusions. The goal is not just accurate predictions but insights that clinicians can trust and comprehend.

5.3 Clinical impact and feedback

Now, imagine integrating the interpretable model into real-world clinical settings. Clinicians use the model's predictions as an additional tool in their decision-making toolkit. Let's say a physician receives a prediction of a high likelihood of patient readmission. The interpretability of the model allows the clinician to understand the specific factors contributing to this prediction, such as the patient's medical history and recent treatment.

This transparency fosters trust, transforming the model from a mysterious algorithm to a collaborator. Continuous feedback loops are established, resembling regular check-ins between the detective and the healthcare team. Clinicians provide insights into the model's performance based on their real-world experiences, refining the model's parameters accordingly. The iterative nature of this process ensures the model remains adaptive to the evolving dynamics of patient populations and healthcare practices.

Ultimately, the clinical impact is profound. The model, with its transparency and continual refinement, becomes an integral part of the healthcare team, contributing valuable insights and augmenting the decision-making process. This symbiotic relationship between technology and human expertise reshapes how patient outcomes are predicted and managed, ensuring a balance of accuracy, interpretability, and clinical relevance.

6. Integration of Explainable AI into clinical practice

6.1 Challenges and solutions

The integration of Explainable AI (XAI) into clinical practice is not without its intricacies, presenting challenges that necessitate innovative solutions. One primary challenge lies in the inherent complexity of healthcare data, where multifaceted variables and intricate relationships often defy straightforward interpretation. Addressing this challenge involves the development of advanced XAI techniques specifically tailored to the nuances of healthcare data. Enhanced visualization tools, such as interactive dashboards and dynamic graphs, can offer a more intuitive understanding of complex data structures. Feature importance techniques can be refined to highlight the variables most influential in predictive outcomes, aiding clinicians in grasping the essential factors guiding AI predictions [21]. Collaborative model development, involving close collaboration between data scientists and healthcare professionals, becomes pivotal to ensure that AI models align with the intricacies of real-world clinical scenarios.

The interpretability-accuracy trade-off introduces another layer of complexity. As models become more interpretable, there may be concerns about potential compromises in predictive performance. Innovative solutions involve the exploration of hybrid models that strike a delicate balance between interpretability and accuracy. By leveraging ensemble methods or incorporating model-agnostic techniques, it becomes possible to harness the advantages of interpretability without sacrificing predictive power. Interdisciplinary collaborations that bring together data scientists, clinicians, and ethicists play a crucial role in navigating the intricate landscape of healthcare decision-making [22], ensuring that XAI is not just a technical tool but a clinically relevant and ethically sound asset.

6.2 Training and adoption by healthcare professionals

The successful integration of XAI into clinical practice hinges on robust training programs and widespread adoption by healthcare professionals. Training initiatives need to transcend the mere imparting of technical knowledge; they must familiarize clinicians with the interpretability of AI models, their implications for patient care, and the ethical considerations surrounding their use. Bridging the gap between technical expertise and clinical understanding is paramount, fostering trust and acceptance among healthcare professionals. Intuitive visualization tools, resembling user-friendly interfaces and interactive displays, can significantly enhance the adoption of XAI by making the technology accessible to healthcare professionals without extensive technical backgrounds. Continuous education programs, including workshops, webinars, and real-world case studies, contribute to building a strong foundation for understanding and leveraging XAI in the daily workflow of healthcare providers.

The feedback loop between clinicians and AI models is a critical component of successful adoption. Clinicians must not only interpret AI-driven predictions but actively engage in a collaborative dialog with the models. This involves seeking insights into the decision-making process, questioning assumptions, and providing context-specific feedback. Establishing a culture of continuous learning and adaptation ensures that AI becomes an integral part of the decision-making toolkit for healthcare professionals.

7. Future directions and advancements in Explainable AI for healthcare

7.1 Emerging technologies and trends

The future of XAI in healthcare holds the promise of numerous emerging technologies and trends, each poised to enhance interpretability and impact. Natural Language Processing (NLP) stands out as a transformative force, enabling AI models to communicate their findings in a more human-readable format. Imagine AI systems generating explanations in plain language, elucidating complex medical predictions for clinicians and patients alike. The integration of NLP not only enhances interpretability but also contributes to building trust and transparency in the AI decision-making process.

Advancements in model-agnostic techniques and ensemble methods offer versatility and adaptability across various types of AI models. Picture a healthcare system leveraging a combination of decision trees, neural networks, and ensemble methods, providing a comprehensive and nuanced understanding of patient outcomes. This diversification of interpretability techniques ensures that healthcare professionals have a suite of tools at their disposal, each tailored to specific clinical scenarios.

Additionally, the incorporation of uncertainty quantification methods emerges as a pivotal advancement. This involves acknowledging and quantifying the inherent uncertainties in medical predictions. Healthcare is inherently uncertain, and AI models that can provide probabilistic estimates rather than deterministic predictions offer a more realistic and actionable understanding of patient outcomes. Imagine a predictive model indicating not only the likelihood of a specific outcome but also expressing the degree of uncertainty associated with that prediction. This nuanced approach allows clinicians to make more informed decisions, especially in scenarios where the clinical landscape is complex and dynamic.

7.2 Potential impact on healthcare

The potential impact of XAI on healthcare extends far beyond accurate predictions; it encompasses a transformative redefinition of how patient care is delivered. The transparent and interpretable nature of these models empowers clinicians to make more informed decisions, leading to optimized treatment [23] plans, personalized interventions, and ultimately, improved patient outcomes.

Consider the potential revolution in diagnostic processes. Imagine a clinician leveraging an XAI model to guide them through intricate decision-making pathways. The model not only predicts outcomes but provides transparent insights into the factors driving those predictions. Visualize a collaborative environment where clinicians and AI systems work synergistically, each contributing unique insights to the diagnostic process [24]. This collaborative approach doesn't replace human expertise; rather, it amplifies and refines it, leading to enhanced diagnostic accuracy and more tailored patient care.

Furthermore, the continuous refinement of models based on real-world feedback ensures that XAI becomes a dynamic ally in navigating the complexities of healthcare. The iterative nature of this process envisions a future where AI models evolve with the changing dynamics of patient populations, emerging medical research, and evolving healthcare practices.

8. Conclusion

In conclusion, the integration of Explainable AI (XAI) into clinical practice represents a pivotal advancement in healthcare, promising to redefine how decisions are made and patient outcomes are managed. As we navigate this transformative landscape, it is evident that challenges exist, ranging from the inherent complexity of healthcare data to the delicate balance between interpretability and accuracy. However, innovative solutions, such as advanced XAI techniques tailored to healthcare nuances and interdisciplinary collaborations, illuminate the path forward. These solutions ensure that XAI not only meets the technical demands of predictive modeling but also aligns seamlessly with the intricate realities of clinical decision-making.

The training and adoption of XAI by healthcare professionals emerge as linchpins in this transformative journey. Robust training programs, transcending technical knowledge, foster a mutual understanding between data scientists and clinicians. The adoption of XAI relies on intuitive visualization tools, user-friendly interfaces, and a continuous feedback loop that encourages clinicians to actively engage with AI models. This collaborative approach ensures that XAI becomes a trusted ally, augmenting clinical expertise rather than supplanting it. The successful integration of XAI into clinical practice hinges on nurturing a culture of continuous learning, where the dynamic relationship between technology and healthcare expertise thrives.

Looking to the future, emerging technologies and trends in XAI promise even greater interpretability and impact on healthcare decision-making. The integration of Natural Language Processing (NLP), advancements in model-agnostic techniques, and the incorporation of uncertainty quantification methods offer a glimpse into a future where AI communicates seamlessly with healthcare professionals and provides nuanced insights into patient outcomes. This evolution not only enhances the interpretability of AI models but also positions XAI as a dynamic force capable of adapting to the complexities of evolving healthcare landscapes. As XAI becomes a cornerstone in healthcare decision-making, it holds the potential to revolutionize diagnostics, foster a culture of data-driven decision-making, and ultimately elevate the standard of patient care.

References

[1] R.P. Ekins, A.J. Williams, How will artificial intelligence change the drug discovery process? J. Med. Chem. 62 (11) (2019) 5836–5850.
[2] S. Mullard, Artificial intelligence in drug discovery, Clin. Pharmacol. Ther. 101 (3) (2019) 320–324.
[3] A.K. Jha, A. Kumar, Y.K. Dwivedi, Acceptance of artificial intelligence (AI) powered healthcare chatbots by patients and healthcare professionals: an empirical study, J. Healthcare Eng. 2020 (2020) 1–10.
[4] B.G. Tait, E. Holz, S. Emery, Virtual assistants: opportunities and challenges in healthcare, J. Am. Med. Inform. Assoc. 26 (6) (2019) 981–985.
[5] V. Gulshan, M. Wulkenberg, C. Funk, Machine learning in medical imaging: a review, Annu. Rev. Biomed. Eng. 21 (1) (2019) 747–774.
[6] D.S. Char, N.H. Shah, D. Magnus, M.A. Hlatky, J. Tsai, Implementing machine learning in health care—addressing ethical challenges, N. Engl. J. Med. 378 (11) (2018) 981–983.
[7] European Commission, White Paper on Artificial Intelligence—A European Approach to Excellence and Trust, European Commission, 2020.
[8] A. Rajkomar, E. Oren, K. Chen, A.M. Dai, N. Hajaj, M. Hardt, et al., Scalable and accurate deep learning for electronic health records, npj Digital Med. 2 (1) (2019) 1–10.
[9] Z. Obermeyer, B. Powers, C. Vogeli, S. Mullainathan, Dissecting racial bias in an algorithm used to manage the health of populations, Science 366 (6464) (2019) 447–453.
[10] M.T. Ribeiro, S. Singh, C. Guestrin, "Why should I trust you?" Explaining the predictions of any classifier, in: Proceedings of the 22nd ACM SIGKDD International Conference on Knowledge Discovery and Data Mining, 2016, pp. 1135–1144.
[11] American Diabetes Association (ADA), Standards of medical care in diabetes, Diabetes Care 42 (Supplement 1) (2019) S29–S33.
[12] R. Bellazzi, B. Zupan, Predictive data mining in clinical medicine: current issues and guidelines, Int. J. Med. Inform. 77 (2) (2008) 81–97.
[13] P. Srinivasan, A. Karthikeyan, K.K. Singh, A review on detection of diabetes using machine learning techniques, Mater. Today Proc. 18 (2019) 2363–2368.
[14] T. Hastie, R. Tibshirani, J. Friedman, The Elements of Statistical Learning, Springer, 2009.
[15] J. Bergstra, Y. Bengio, Random search for hyper-parameter optimization, J. Mach. Learn. Res. 13 (February) (2012) 281–305.

[16] D.C. Swinney, J. Anthony, How were new medicines discovered? Nat. Rev. Drug Discov. 10 (7) (2011) 507–519.

[17] T. Ching, D.S. Himmelstein, B.K. Beaulieu-Jones, A.A. Kalinin, B.T. Do, G.P. Way, et al., Opportunities and obstacles for deep learning in biology and medicine, J. R. Soc. Interface 15 (141) (2018) 20170387.

[18] D. Duvenaud, D. Maclaurin, J. Iparraguirre, R. Bombarell, T. Hirzel, A. Aspuru-Guzik, R.P. Adams, Convolutional networks on graphs for learning molecular fingerprints, in: Advances in Neural Information Processing Systems, NeurIPS Proceedings, 2015, pp. 2224–2232.

[19] M. Hardt, E. Price, N. Srebro, Equality of opportunity in supervised learning, in: Advances in Neural Information Processing Systems, NeurIPS Proceedings, 2016, pp. 3315–3323.

[20] R. Caruana, Y. Lou, J. Gehrke, P. Koch, Intelligible models for healthcare: predicting pneumonia risk and hospital 30-day readmission, in: Proceedings of the 21st ACM SIGKDD International Conference on Knowledge Discovery and Data Mining, 2015, pp. 1721–1730.

[21] Z.C. Lipton, The mythos of model interpretability, in: Proceedings of the 2016 ICML Workshop on Human Interpretability in Machine Learning, 2016, pp. 1–6.

[22] K. Padmanandam, M.V. Rajesh, A.N. Upadhyaya, B. Chandrashekar, S. Sah, Artificial intelligence biosensing system on hand gesture recognition for the hearing impaired, Int. J. Oper. Res. Inf. Syst. 13 (2) (2022) 1–13.

[23] S. Sundari, Y. Divya, K.B.K.S. Durga, V. Sukhavasi, M.D. Sugnana Rao, M.S. Rani, A stable method for brain tumor prediction in magnetic resonance images using Finetuned XceptionNet, Int. J. Comput. Digit. Syst. 14 (1) (2023) 1–14.

[24] M.K. Doma, K. Padmanandam, S. Tambvekar, K. Kumar, B. Abdualgalil, R.N. Thakur, Artificial intelligence-based breast cancer detection using WPSO, Int. J. Oper. Res. Inf. Syst. 13 (2) (2022) 1–16.

CHAPTER 7

OML-GANs: An optimized multi-level generative adversarial networks model for multi-omics cancer subtype classification

Nabaa Abd Mohammed[a] and Mazin Abed Mohammed[b]
aDepartment of Computer Science, College of Computer Science and Information Technology, University of Anbar, Anbar, Iraq
bDepartment of Artificial Intelligence, College of Computer Science and Information Technology, University of Anbar, Anbar, Iraq

1. Introduction

Omics comprises all studies in biological sciences that finish with "omics" such as transcriptomics, genomics, metabolomics, or proteomics. The suffix "ome" is associated with the research subjects in fields like the metabolome, transcriptome, proteome, or genome [1]. Omics studies are principally designed to search, to describe, and to quantify all these varieties of biomolecules which are the underlying components of the cell, tissue, or organism. Multi-omics research is thriving more in the areas of microbiome, radiomics, transcriptomics, genomes, metabolomics, pathomics, and proteomics [2], while the advent of the most recent high-through techniques have made it possible to assemble omics data from numerous sources like epigenome, genome, proteome, metabolome, transcriptome, etc. [3]. Omics data represent the biomolecules from various levels of cells that contain many details that complete and respond to each other [3]. The big joint scientific projects like the cancer genome atlas (TCGA) [4], the International Cancer Genome Consortium (ICGC) [5], and Cancer Analysis of Whole Genomes (PCAWG) [6] have collected a large number of biological specimens, which include multi-level molecular profiles as shown by various measuring technologies. Since the molecular diversities of cancer etiology (the origin of cancer) are at multiple levels, the multi-omics data analysis gives a deep view of the basic processes of cancers. Consequently, the compilation of omics data from which the relationships between the molecules can be visualized has turned out to be essential and, in fact, the analysis of these data has been possible [7]. This is beneficial in many ways, such as understanding complex systems of biology in a holistic manner [8]. Multidimensional scaling is a very popular method in statistics and some other areas of mathematics [9] which is used in

Explainable AI in Healthcare Imaging for Medical Diagnoses
https://doi.org/10.1016/B978-0-443-23979-3.00007-5

143

the evaluation and creation of replicable data. And it is more and more used in multivariate analysis of variance and scale-invariant approaches and also in certain dimensional data and regression. Analyses in psychoacoustics, a types of multisensory tools, are usually performed using this method. Each method has its advantages and uses which can be either qualitative or quantitative. Besides these, the identical tools such as AND/OR systems and High-Level Petri Nets can be extensively employed to enumerate the perceptual system gauging the visualization algorithms. Such visual conditions are generally created in two basic forms distinct or integral. Scientific characteristics are important conditions that have been foreseen by this authentication process. Inseparable representation mechanisms have been employed to establish query designs and exploit data correlations. A successful clustering analysis (MDS process), that is the outcome of developing a dichotomous model, can be grounded in the adjustment of the eases in such a way that the total power is maximized [10].

Recent machine learning achievements in the scientific community have delivered a vivid representation of where deep learning could be applied to cancer identification. For illustration, deep learning models have been proved to bring about the quality and efficiency of tumor identification according to the radiographic images or histopathological of tumors [11,12]. Well, items such as deep learning systems including [13–15], have been built to identify gene expression data or the effects of noncoding gene mutations in the genome based on high-dimensional genomic or epigenomic profiles. In addition to this, the possibility of cancer detection and diagnosis by supervised and unsupervised deep learning has been highlighted and it is predicted that these methods will be more effective in comparison to other traditional ones, for example, some which are already developed like support vector machine (SVM) and random forests (RF) [16,17]. Gene expression profiles reengineering of the pathways have been proven to provide less informative and less stable results for disease classification than actuals [18]. However, the majority of the current publications have addressed the investigation of molecular subgroups or assisting prognosis investigation using single-omics data [15]. The other factor that has been described as a challenge for analyzing multi-omics data is the availability of efficient and suitable methodological approaches for analyzing such complex data. It has been shown that integrating multi-omics datasets is imperative especially with the increasing volume of omics data. However, high computational costs are apparently a problem that has not yet been solved [16]. There is a high need to adopt new approaches that capitalize on the available computing tools and capabilities fully to analyze multi-omics data integration.

The purpose of this study is to propose an efficient multi-omics integration model that is capable of learning the features of cancer subtypes and classify them in a precise, robust and efficient manner within a reduced timeframe. This has to be a developed method intelligent and robust with the capacity to handle different data sources and yield coherent and credible results. The objective of this study is to provide further insights of biological phenomena that contribute to cancer and find better solutions on diagnostic, prognostic, and therapeutic approaches. The aim of this study is to develop a deep learning model that accurately classify cancer subtypes based on multi-omics data, improving diagnosis and prognostic for cancer patients. The proposed method improves the accuracy of cancer subtype diagnosis based on multi-genome data. The main contributions can be summarized as:

- To propose and develop a optmized multi-level generative adversarial networks model (ML-GANs) for multi-omics cancer subtype classification.
- To develop and prepare multi-omics cancer data a number of preprocessing techniques are used including remove outliers, handling missing values, normalization, and standardized features.
- To develop dimensionality reduction method of high dimensional and heterogeneity integrated multi-omics data while considering preserving biological meanings for enhance the cancer classification models using K means clustering.
- This ML-GANs model is created by combining an integrated GANs-based detector and a fine-tuned balanced classifier for Multi-Omics Cancer Subtype Classification.

The rest organization of this study are: Section 2 presented the related works on multi-omics cancer classification with AI methods. The proposed methodology for optimized multi-level GAN in Section 3. Section 4 presented the results and discussion with features selected of multi omics data. Finally, the conclusion and future works presented in Section 5.

2. Related works

Studies in the literature have illustrated the usability of methods in the analysis of multi-omics data. A cascade deep forest model to classify breast cancer subtypes (IntClust and Pam50) with the use of multi-omics datasets was proposed in the work by **Ala'a El-Nabawy et al. (2021)**. The preprocessing stage involved cleaning data and applying statistical analysis feature engineering to create integrated data profiles from clinical and gene expression features. The cascade deep forest model was then used to classify the cancer subtypes, achieving accuracies of 83.45% for five subtypes and 77.55% for ten subtypes. A challenge in the study was the high dimensionality of multi-genome data, which can lead to overfitting, especially with limited training data [19]. Also, **Reinel Tabares-Soto et al. (2020)** applied machine learning and deep learning methods to classify 11 cancer types using a microarray dataset through two unsupervised methods: K–means clustering and hierarchical clustering. Various classification algorithms, including logistic regression and convolutional neural networks (CNN), were tested, achieving tumor identification accuracies between 90.6% and 94.43%. The study highlighted that tuning the models had limited impact on accuracy. A key limitation was the imbalance between cancer classes, which could lead to biased predictions favoring majority classes [20]. According to **Bingjun Li et al. (2021)** proposed a multifaceted model based on deep learning that integrates both previous biological knowledge and assembled data from different genomes to classify molecular cancer subtypes. This architecture leverages a graph convolution network (GCN) to model the relationships between the genes, and a fully connected neural network for global feature extraction. The study investigated different input graphs and evaluated model performance against baseline methods (CNN, FC-NN, RF, SVM). The proposed GCN model achieved an accuracy of 81.4%, outperforming most baseline models. Limitations include the exclusion of non-coding region CNV data and handling of imbalanced datasets [21].

Also, **Yuqi Lin et al. (2020)** proposed DeepMO, a deep neural network model for classifying breast cancer subtypes using multi-genome data, including mRNA, DNA methylation, and CNV data from TCGA. After preprocessing and feature selection, genomic data were input into a deep neural network with coding and classification subnetworks. DeepMO achieved high accuracy in binary classification (90.8%) and multi-genome classification (96.2%), outperforming other methods. However, data imbalance and unoptimized hyperparameters (e.g., hidden layers, learning rates) were limitations that could impact the model's performance. Optimizing these parameters may further enhance its accuracy. Baoshan Ma et al. (2020) proposed an XGBoost-based classification model to distinguish early and late-stage cancers using multi-genome data from TCGA. The model achieved high accuracy rates across four cancer types (KIRC, KIRP, HNSC, LUSC), outperforming or matching other machine learning methods. The integration of multi-genome data via autoencoder improved classification accuracy, and bioinformatics analysis identified key genes linked to cancer stages. However, the small sample sizes from TCGA may limit result generalizability [22]. **Wei Dai et al. (2021)** introduced a cancer subclassification method using a residual graph convolutional network (GCN) and a sample similarity network. The network was built based on gene co-expression patterns, with gene expression profiles as initial features to prevent over-smoothing in the GCN model. The model, applied to BRCA, GBM, and LUNG datasets, achieved accuracy rates of 82.58%, 85.13%, and 79.18%, respectively, outperforming existing methods. However, setting the appropriate threshold for the similarity network remains challenging, and only gene expression data was used, limiting the integration of other genomic data [23]. **Chaoyi Yin et al. (2022)** proposed the multi-omics graph convolutional network (M-GCN), integrating multi-omics data for molecular classification. The framework uses HSIC Lasso to select molecular transcriptional features and builds a sample-sample similarity graph. It processes gene expression, SNVs, and CNVs to learn multi-view representations, which are further refined by a robust GCN model. Experimental results on breast and gastric cancer show M-GCN outperforms existing methods, achieving 94% accuracy. However, the imbalance in dataset sizes between BRCA and STAD may limit the model's generalizability across different cancers [24].

The recent studies have observed that the DL approaches can help in modifying the multi-omics data into more generalized models and with the help of non-linear patterns, it is possible to get better prognosis models. The major advantages of the DL models over the standard ML models are the following; automatic creation of high quality features; capability to handle large data sets. However, the most of current DL-based solutions use only one model, and it is generally not rational and efficient in dealing with big data in complex environment. Moreover, the training of multiple learning many DL models is mainly relies on the backpropagation method that is operated with a gradient descent optimization. It is also well understood that the algorithm of BP has some drawbacks, for example, converging to the local suboptimal solutions and ignoring the trial-and-error process in the parameter tuning. Therefore, the motivation for the research that is the subject of the present paper is the prospect of development and testing of ensemble methods for multi-omics data analysis and cancer subtype recognition and classification, as indeed it can cover features of several types of data for effective integrating of multi-omics data.

3. The proposed methodology

In this study, a multi-omics cancer subtype classification system is proposed using a CNN model. The proposed CNN has several advantages when dealing with datasets: the CNN algorithm can perform some tasks faster, which can lead to faster processing times, and the CNN can provide a more compact feature representation, which reduces the complexity of the model and improves its performance. Once the multi-omics cancer subtypes dataset is obtained, the phases of the proposed system are implemented. Before entering the classification step, the multi-omics cancer subtypes dataset undergoes pre-processing, followed by feature extraction from the multi-omics cancer subtypes dataset. The next stage is feature selection, where optimal features are selected from the list of features extracted from the feature extraction process. All the steps and procedures of the proposed multi-omic cancer subtype classification system are explained in detail in the next subsection. Fig. 1 also provides a brief overview of the steps of the proposed system.

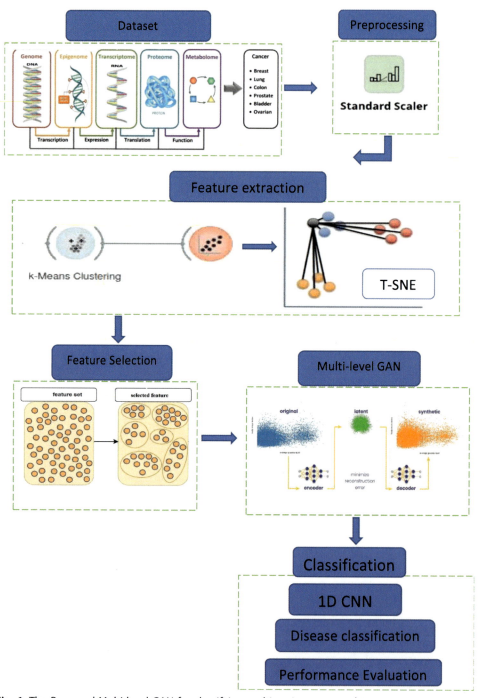

Fig. 1 The Proposed Multi level GAN for classifying multi-omics cancer subtypes.

3.1 OMICS dataset

The dataset is accessible for scholarly purposes and may be obtained from the Internet. The combined dataset we have contains a vast amount of information on various types of cancer. Each row in the dataset represents a unique patient sample, while each column provides valuable information on the molecular factors contributing to cancer [25]. A comprehensive overview of many types of cancers has been presented, including hepatocellular carcinoma (HCC), brain cancer (GBM), pancreatic cancer (PaCa), breast cancer (BrCa), colon cancer (CoCa), gastric cancer (GaCa), and lung cancer. A detailed molecular blueprint of these disorders is now being constructed, with a particular focus on the cancer cells themselves. The dataset consists of 20,244 samples and 69 features. The features in the dataset for multi-omics cancer subtypes are described and explained in Table 1.

Table 1 The details of OMICS dataset used.

NO.	Cancer type/healthy	Number of features
1	Healthy	8
2	Lung cancer	15
3	Hepatocellular carcinoma	10
4	Pancreatic cancer	7
5	Breast cancer	4
6	Colon cancer	4
7	Gastric cancer	5
8	Brain cancer	4
9	Hepatitis B, (HBV)	7
10	Blood	4
	Total	68

3.2 The preprocessing using Standard Scaler

Converting numerical input from their original ranges to a single common range is a technique that greatly improves the performance of various deep learning procedures. As for the standardization, it is in essence the procedure of adjusting every single variable at a time. This is done by taking away the mean (alternatively known as centering) and then dividing by the standard deviation. The objective of this treatment is to reposition the distribution so that it has a mean of zero and a standard deviation of one. The multi-omics cancer dataset is characterized by the fact that it contains 20,244 rows (samples) and 69 columns. The first column ("Gene Name") contains the names of the genes, while the remaining 68 columns are numerical and contain expressive data for different samples, representing different health conditions and experimental conditions. When applying Stenderscaler to a multi-omics cancer dataset, the "genename" column will first be dropped from the dataset to focus on the numeric data for measurement. The target features and columns representing gene expression levels for each sample are then defined, and the rows represent observations (individual genes). Then the average is calculated for each feature in the data set, after which the standard deviation for each feature is calculated. Finally, normalization is calculated by subtracting the arithmetic mean from each feature and then dividing it by the standard deviation. This involves scaling features (columns) to a standard range, between 0 and 1, or using z-score normalization (mean = 0, standard deviation = 1). This Stenderscaler application will therefore scale each feature individually so that it is in a certain range, which is between 0 and 1. Although the initial scan did not show missing values, it is necessary to ensure that there are no NaNs in the dataset before applying the scaling. The utilization of the Standard Scaler method to our dataset through it may fit perfectly to a zero mean and a standard deviation of one and, consequently, achieving a distribution that is uniformly spread out. By the normalization of the features and subtraction of the respective mean value of the features, the obtained value is divided by the standard deviation of the features. The standardization formula, represented by Algorithm 1, is

$$Z = \frac{(X - \mu)}{\sigma} \tag{1}$$

where Z indicates the scaled data, x is the data to scale (van Diest) and σ represents the standard deviation, μ is still the mean of the original training data set, which is a set of samples of the training features. Standardization preserves the presence of outliers, enhancing the resistance of the final algorithm to external influences compared to an algorithm that has not undergone standardization.

Algorithm 1: Standardization

Input: Dataset features

Output: Scaled data features

Start

Step 1: Define matrix X as the dataset, where each row represents a sample and there are N columns representing the features.

Step 2: For a feature: Confirm that the mean (μ) is equal to 0 and the standard deviation (σ) is equal to 1.

Step 3: For $i = 1$ to N in X matrix do:

1. Calculate mean (μ) by $\mu = \frac{1}{N}\sum_{i=1}^{N} X_i$

2. Calculate standard deviation (σ) by $\sigma = \sqrt{\frac{1}{N}\sum_{i=1}^{N}(X_i - \mu)^2}$

3. Calculate StandardScaler by $Z = \frac{(X - \mu)}{\sigma}$

   End for

Step 4: Retrieve the updated value Z

End

3.3 Feature extraction using K-means

The feature extraction process is an important stage in classifying cancer types, because it helps identify the most important and useful features for training deep learning models. K-means clustering is an important step because it is a basic introduction to feature extraction [26]. Source data is the input for feature extraction. With the help of clustering, it is achievable to find the points that are alike even without knowing the labels. Consequently, the features responsible for the similarity of the points to the others as well as the points which make them different from others can be discovered as well. Since the multi-omics cancer dataset is unlabeled, it will be labeled using K-means. We can use the group labels assigned to data points as new features. These labels represent the group to which each data point belongs, adding categorical information to the data set.

In the first step, known as initialization, (k) the initial cluster centroids are randomly determined from the multi-omics cancer data set. Here $k = 10$ is set, so the clustering process will start with 10 randomly chosen central cores. In the second step, each data point is assigned to the closest centroid based on the Euclidean distance. Each data point is then assigned to one of the ten groups whose centroid is the closest. After all data points are assigned to the groups, the centroids for each group are updated, where the centroid is calculated New to a group is defined as the average of all data points assigned to that group as shown in Fig. 2. This process is repeated until the maximum amount of data is reached, at which point we group similar data points into 10 groups. Eventually all samples in the multi-omics cancer dataset will be labeled. Algorithm 2 provides a brief description of the K-means algorithm.

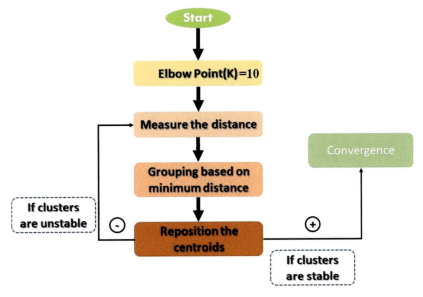

Fig. 2 Steps of the K-mean algorithm.

Algorithm 2: K-means algorithm
Input: D = preprocessing data, K = number of clusters
Output = Clustering data
Start
Step 1: Randomchoose K data item from D as initial centroids.
Step2: Repeat
Assign each item d1 to the clusters which has centroids.
Calculate new mean for each cluster.
Step3: Until convergence criteria is met.

3.3.1 T-SNE algorithm

T-SNE enables us to get deeper insights into the inherent patterns and interconnections within the data. The t-SNE technique is used to compute the similarity measure between pairs of cases in both the highest and lowest dimensions space of the multi-omics cancer dataset. Furthermore, it aims to improve the two similarity measurements [27]. A Gaussian kernel is used to compute the pairwise similarity of data points in a space with a large number of dimensions. It calculates the arrangement of points in both upper and lower dimensions, assigning a greater likelihood to points that are closer together and a lower likelihood to ones that are further away. Subsequently, we use the technique to convert the data points from a higher-dimensional space to a lower-dimensional one, while maintaining the shared characteristics between pairs of points. This is accomplished by reducing the difference between the original probability distribution in a space with many dimensions and the probability distribution in a space with fewer dimensions, as seen in Flowchart 1. The method use gradient descent to minimize variance. Algorithm 3 provides a detailed explanation of how the t-SNE algorithm works.

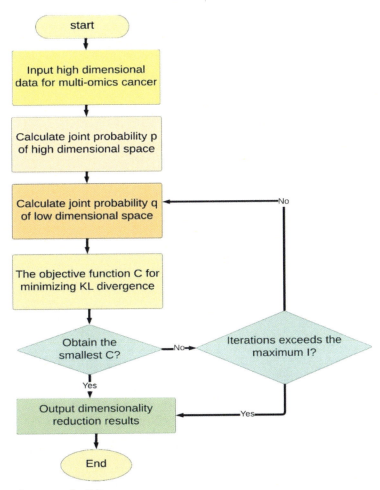

Flowchart 1 The steps of t-SNE.

Algorithm 3: T-SNE algorithm

Input: High-dimensional data X, number of dimensions for output d = 2, perplexity P, number of iterations N

Output: low-dimensional Y

Start

Step 1. Normalize the data X

Step 2. Compute pairwise affinities in high-dimensional space:

For each data point x_i in X:

- Compute the conditional probability $p\_j|i$ that point x_i would pick x_j as its neighbor using a Gaussian distribution with mean x_i and variance σ_i^2
- Adjust σ_i to match the desired perplexity P using a binary search
- Symmetrize the affinities:

$$p_{ij} = \frac{p_{i|j} + p_{i|j}}{2n} \text{ where } n \text{ is the number of data points}$$

Step 3. Initialize low-dimensional map Y randomly.

Step 4. Compute pairwise affinities in low-dimensional space:

- For each pair of points y_i and y_j in Y:
- Compute the similarity q_{ij} using a Student's t-distribution with one degree of freedom:

$$q_{ij} = (1 + ||y_i - y_j||^2)^{-1}$$

- Normalize qij to ensure that sum(q_{ij}) = 1

Step 5. Minimize the Kullback-Leibler (KL) divergence between the high-dimensional and low-dimensional distributions:

- For iteration t in range (1, N + 1):
- Compute the gradient of the KL divergence with respect to y_i:

$$\frac{dKL}{dy_i} = 4 * \text{sum}((p_{ij} - q_{ij}) * (y_i - y_j) * (1 + ||y_i - y_j||^2)^{-1})$$

- Update the positions of the points in the low-dimensional map:

$$y_i = y_i - \eta * \frac{dKL}{dy_i}$$

Step 6. Return the low-dimensional representation Y

3.4 Feature selection using SelectKBest

The multi-omics cancer dataset is large, and classifying it using DL algorithms requires a significant amount of time and computing resources. Therefore, it is necessary to carefully select essential features and exclude irrelevant features. Feature selection in Python involves calling the "SelectKBest" function with the "k" argument and a chi-square. This method detects whether a feature has an impact on the class by first calculating the chi-square. The chi-square test measures the dependence between two variables of a multi-omics cancer dataset as it measures the dependency between each feature and the target variable. For each feature, a chi-square statistic is calculated based on the frequency of occurrence of each feature value with respect to each classification of the multi-omics cancer data class. The observed frequency of each feature value is compared with the expected frequency assuming independence between the feature and the target variable, as shown in Fig. 3.

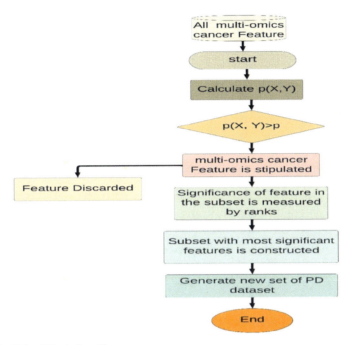

Fig. 3 Steps for SelectKBest algorithm.

Later, when each features was processed through the chi-square stat, the "SelectKBest" classified the feature list based on their chi-square scores. The features that scored higher chi-square scores are recognized as they are the more important for the prediction of the target variable or useful for it. Lastly, "SelectKBest" selects the k-best features with the highest chi-square scores. It is the way the value of k is specified, that these are the specific features used to train the deep learning model. The value of "k" is the number of features that the chi-square selects in order to rank and classify them based on their connection to the output variable. After this, the K features with the best scores are the only ones that hold the final feature subset. Where $k=60$ is set, which indicates the number of features to be selected that are related to the target value. In this case, after specifying a value for K, we can calculate the cost function for each value of K, and then choose a value corresponding to the smallest value of the cost function. SelectKBest is described in Algorithm 4.

Algorithm 4: SelectKBest algorithm steps
Input:
Feature (X): where each column represents a feature, Target (Y): The target variable used for feature selection, k: The number of features to select.
Output: Selected feature (X_selected)
Start
1. Compute a univariate statistical score for each feature using a chi-squared test for categorical data for numerical data.
2. Rank the features based on their statistical scores.
3. Select the top k features with the highest scores.
4. Return the selected feature matrix.

3.5 Data splitting

The process of data preparation and feature extraction has a very vital and important significance in the completion of the model. It requires the data sets to be split into train and test sets by the hold-out method which is evident from the classification. The multi-omics cancer datasets are divided as follows: the training set contains 80% of the original dataset while the test set comprises 20% of it, illustrated in Fig. 4.

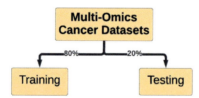

Fig. 4 Dataset splitting.

3.6 Classification based on multi-level GAN

Generative adversarial networks (GANs) are a type of deep learning models that have been specifically designed to produce data which is almost identical to a given dataset [28]. GANs are brought into the world through two interconnected neural networks: a generator and a discriminator [29,30]. The generator creates artificial data samples. Moreover, the discriminator confirms that the sample is authentic multi-omics cancer dataset else generated they by the generator. The operation of the networks is initiated through the adversarial training, i.e., the generator's aim is to replicate original data, and the discriminator's mission is to improve its power of discernment. This adversarial process repeats until the generator produces data that is almost indistinguishable from real data. In this study, developed and trained a multi-level GAN system utilizing the TensorFlow and Keras frameworks. A multi-level GAN system was constructed, comprising of three tiers, each comprising of a generator and a discriminator. The generators generate artificial data by utilizing random noise from pre-processed multi-omics cancer data. The generator includes a dense layer with 128 units and ReLU activation that is next to a dense output layer with a specified number of units and sigmoid activation. The first generative model is identified. When the generator model produces a sample in the domain from a random vector with a specified length is as input, the generator model is thus mentioned. In Dual GAN, the generator is responsible for generating sentences from a given specific latent space, therefore, the topics be assigned to the space and based on these facts the generator can develop new and diverse samples through the execution of a network. It is trying to obtain a real, valid output by using accurate propagation data that will confuse the discriminator, which is the one used as an example to distinguish the actual data produced by the generator from it. The generator employs the input as a random sample and then leads to the production of many new samples. This process is done by a three-layer generator model. The initial layer is the input and the size of the sample is obvious. Using a convolution layer, the kernel size, network learning rate, and stride are represented. On the contrast, rectified linear unit (ReLU) is added as an activation function. Softmax is an activation function for the normalization purpose. Moreover, the classification layer classifies them into specific categories by applying probabilities (Fig. 5).

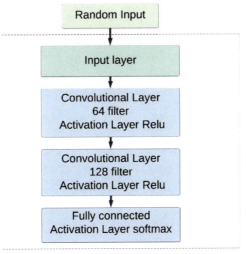

Fig. 5 Generator model description.

There are two phases for the multi-level GAN model in the second part, namely the discriminator model. The true and fabricated samples are passed to the discriminator in order to make a correct judgement of the data. The model inputs the domain sample and predicts whether the category label is true or false. In the construction of the model, three layers are used. The first layer is the input layer, which has the addition of the restraint such as weight, height and the channel's size. Whereas for the convolutional layer, the number of filters, size, step, and kernel value hosts the input of the layer. Moreover, in the form of the rectified linear unit function (ReLU) an activation function is added and the max pooling layer calculates the peak for each region. The general alignment of the layers is demonstrated in Fig. 6, where the fully connected layer is the end condenser for all the information the previous layers create. It would not be wrong saying that it communicates every feature that was taught by the classes. Another special role is also played by the sigmoid– it also acts as an activation function.

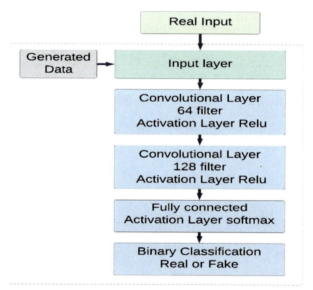

Fig. 6 Discriminator model description.

Multi-omics data is processed via a dense layer containing 128 modules and a ReLU activation function. The output layer is also dense, containing a single unit and a sigmoid activation function, which produces a probability indicating whether the data is valid or fraudulent. Multi GAN models undergo an iterative training process where the discriminator is first taught on both true and false data, and after that, the generator is trained to generate improved false data that may fool the discriminator. The procedure is repeated for 100 periods, thus enhancing the generator's ability to generate original data. Data produced using multi GAN is used to train convolutional neural networks, enabling us to independently detect patterns in the input data. Algorithm 5 explains the details of how the multi-level GAN algorithm works.

Algorithm 5: The multi-level GAN model

Input: Best feature

Output: new feature

Start
Initialize generator network G and discriminator network D
Initialize learning rate α and number of Level N
for iteration in range(N):
Step 1: Train the Discriminator
  for k steps:
    # Sample minibatch of m real examples from real data distribution
    real_data_batch = sample_real_data(m)
    # Generate m fake examples using the generator
    noise = sample_noise(m)
    fake_data_batch = G(noise)

    # Compute the discriminator loss on real and fake data

$$real_{loss} = -log\left(D(real_{data_{batch}})\right)$$
$$fake_{loss} = -log\left(1 - D(fake_{data_{batch}})\right)$$

$$D_{loss} = real_{loss} + fake_{loss}$$

    # Update the discriminator parameters
    D_optimizer
    D_loss
Step 2: Train the Generator
# Sample a new batch of noise
noise = sample_noise(m)
  # Generate fake examples using the generator
fake_data_batch = G(noise)
  # Compute the generator loss
G_loss = -log(D(fake_data_batch))
  # Update the generator parameters
G_optimizer
G_loss
  # Optionally: Print/log losses and other metrics for monitoring
if iteration % log_interval == 0:
  return D_loss, G_loss
Save the final trained generator model
End

GAN models are trained in a loop three times. Each time, a new GAN level is generated and generates new data different from the previous level, as shown in Fig. 7, where the discriminator is first trained on real and fake data, followed by training the generator to produce fake data each time of the Multi-omics dataset so that it is better can fool the discriminator. The process is repeated for each level for 100 periods, gradually improving the generator's ability to produce realistic data.

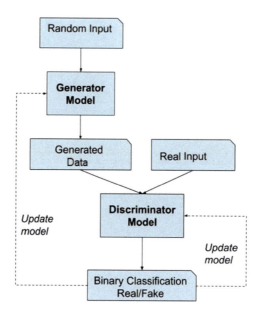

Fig. 7 Multi-level GAN.

3.7 Evaluation stage

The metrics used to assess the performance of the deep learning algorithms are included in this section. The metrics accuracy, recall, precision, and f1_score are used to assess classification methods. As seen in Table 2, the confusion matrix, which consists of four predictions, is the basis for calculating these measures.

Table 2 The confusion matrix.

Confusion matrix	Positive	Negative
Positive	True positive (TP)	False negative (FN)
Negative	False positive (FP)	True negative (TN)

Each component within the matrix denotes the following [2]:
True positives (TP): Cases in which observed cases were positive and expected cases were also positive.

True negatives (TN): Cases in which observed cases were negative and expected cases were also negative.

False positives (FP): Cases in which observed cases were actually negative but expected cases were positive.

False negatives (FN): Cases in which observed cases were positive but expected cases were negative.

Based on the value of the confusion matrix predictions, the evaluation metrics are computed [3,4]. The way the measurements are described is as.

1. **Accuracy**: The percentage of samples accurately categorized from the tested samples is known as the accuracy rate. In other words, accurate categorizations of negative and positive specimens are identified. Eq. (2) provides a formal definition of it

$$Accuracy = \frac{TN + TP}{TN + TP + FN + FP} \tag{2}$$

2. **Recall**: A measure of sensitivity that counts the number of accurately categorized positive samples; a high recall rate indicates a low number of positively classified as negative samples. Eq. (3) provides a formal definition of it

$$Recall = \frac{TP}{FN + TP} \tag{3}$$

3. **Precision**: To calculate the quantity of undesired positives, divide the true positives by the total of the true positives and false positives. Eq. (4) has a formal definition of it

$$Precision = \frac{TP}{FP + TP} \tag{4}$$

4. **f1_score:** officially stated in Eq. (5), it is the rate at which accuracy and recall are combined and scaled to yield accurate prediction accuracy

$$f1\_scoreRate = \frac{2*(Precision)*(Recall)}{(Recall + Precision)} \tag{5}$$

4. Results and discussions

This section presents the results of the implementation stages of the proposed approach for an improved multi-level GAN model for multi-level cancer subtype classification. Furthermore, this chapter presents the results of dataset preprocessing, feature extraction, feature selection, and multi-level cancer subtype classification using the CNN model. The performance metrics for multi-level cancer subtype classification using the CNN model are calculated. The confusion matrix, accuracy, precision rate, recall rate, and F1 score of the CNN used in this study are presented. This chapter also presents a comparative analysis of the results generated by the used algorithms. Furthermore, the results achieved by the proposed approach when implementing an improved multi-level GAN model for multi-level cancer subtype classification with the CNN algorithm and the results achieved without using multi-level GAN are presented. In addition, a comparison is made between the proposed approach and other classification methods in other researchers' studies.

4.1 The results of preprocessing stage

The first stage of the proposed approach involves preprocessing, which is applied to a multi-level cancer subtype dataset. The importance of the preprocessing step and its impact on the accuracy and efficiency of the system are discussed in Section 3. Preprocessing plays a crucial role in multi-level cancer subtype classification. Preprocessing involves a series of stages that provide accurate data to handle throughout the classification stage.

4.1.1 StandardScaler

In the dataset, certain characteristics have large numerical magnitudes, whereas other features have lesser values. Scaling the dataset and ensuring size consistency are essential to reduce the influence on deep learning performance. The Python "scikit-learn" package's "StandardScaler" function is used to standardize the dataset's features. This is important for the clustering process that will be explained in the next step because it ensures that each feature contributes equally to the distance calculations. All feature sizes will align after the features in the dataset are standardized, as shown in Fig. 8.

althy_10	0.319371	0.056045	-0.770931	-0.312356	0.411140	0.265339	-0.866614	-0.122457	0.676843	-0.169825	...	-0.645279	0.241984	0.546736	0.043907
althy_11	-0.771447	-0.428438	-0.429695	0.303430	0.029697	0.881671	-0.923687	0.278453	-0.392353	-0.832780	...	0.257404	-0.768499	0.542601	-0.568909
althy_27	-0.139653	0.084283	-0.192835	0.232896	-0.138500	0.154733	0.418261	-0.066226	-0.102361	-0.295328	...	0.017914	0.656527	-0.612958	-0.527963
thy_35-1	-0.019463	-0.827820	-0.248695	0.097097	0.430397	0.211874	0.154564	1.000532	-0.692231	-0.684346	...	0.077333	1.448417	0.703076	0.122331
thy_36-1	-0.376713	-0.432248	-0.414847	-0.160162	-0.082100	-0.308705	0.809914	0.000596	-0.086795	-1.086819	...	0.371855	1.613471	0.646298	0.209978
...	...	...	...	...	...	...	...	...	...	...	...	...	...	...	...
_HBV646	0.301886	1.216917	-0.148765	1.861554	3.195776	-0.895567	-1.349643	0.451172	-0.364756	0.395591	...	0.365619	-0.564556	0.572438	0.215474
lood_35-blood	-0.646803	0.261048	-0.856859	-1.106893	-0.814245	1.410606	-1.858800	-1.896090	3.055235	-0.994415	...	2.550495	0.251957	-0.981273	1.541941
lood_36-blood	-0.548598	-0.098000	-0.659194	-0.925545	-0.701871	1.032562	-1.693901	-1.740837	2.245136	0.014322	...	2.063888	0.156592	-0.727644	1.543271
ut_input	-2.131707	-2.214888	1.387722	-0.079344	-0.959800	-1.231178	1.549898	-2.101280	-0.792680	1.246837	...	1.210747	-0.029181	-0.812702	-2.580126
Blood-t_blood-input	-1.998366	-1.816805	1.516625	-0.180719	-1.838610	-1.393978	0.939322	-2.138789	-0.499462	1.942035	...	1.199856	0.470295	-0.091346	-2.566446

Fig. 8 Dataset after applying StandardScaler.

4.2 The results of feature extraction stage

The second stage of the proposed system involves feature extraction, which is applied to a multi-level cancer subset after pre-processing. The feature extraction stage involves a series of steps that provide accurate data to handle throughout the classification stage.

4.2.1 K-means clustering

Feature extraction is the second stage in the proposed multi-level cancer subtype classification system, and it was applied to the multi-level cancer subtype dataset resulting from preprocessing to extract features using feature extraction algorithm. The cluster K-means algorithm was used to extract features from the multi-level cancer subtype dataset. It is known that the multi-level cancer subtype dataset is unlabeled, so it is necessary to label this dataset to be classified. This is done by dividing the data into clusters based on their features. The K-means clustering process involves initializing the weight points, each data point is assigned to the nearest center of gravity based on the Euclidean distance measure, where the data is divided into ten random clusters where a cluster is named for each feature such that the data points within the same cluster are more similar to each other than those in other clusters. These labels indicate the cluster to which each feature is assigned based on the K-means model. Then, a dictionary is created that maps each feature (column name) to its assigned cluster. Each feature is then paired with its corresponding cluster label, and the dictionary comprehension generates the final mapping. Thus, the data set is divided into 10 labeled groups as shown in Table 3.

Table 3 Dataset labeling.

Features	Label
Healthy_10 , Healthy_11, Healthy_35-1, Healthy_36-1, Healthy_38, Healthy_39o, BrCa_BR13, BrCa_BR14, BrCa_BR5-1, CoCa_colon16, GaCa_stomach2, GaCa_stomach3, GaCa_stomach8, GBM_GBM57, GBM_GBM58, GBM_GBM66, GBM_GBM76.	0
LuCa-M0_lung395, LuCa-M0_lung492, LuCa-M0_lung493, LuCa-M0_lung512, LuCa-M0_lung513, LuCa-M0_lung514, LuCa-M0_lung515, LuCa-M1_lung517, HCC_HCC320, PaCa_pancreatic9, GaCa_stomach4, HBV_HBV374.	1
HCC_HCC150, HCC_HCC195, HCC_HCC237, HCC_HCC256, HCC_HCC260,HCC_HCC290, HCC_HCC46, HBV_HBV455.	2
HCC_HCC628, PaCa_pancreatic15, PaCa_pancreatic27, PaCa_pancreatic68.	3
Input_input, Blood-input_blood-input, LuCa-M1_lung323.	4
Healthy_27, Healthy_38o, PaCa_pancreatic22, BrCa_BR7-1, CoCa_colon13,CoCa_colon17, CoCa_colon19, GaCa_stomach1, HBV_HBV334, HBV_HBV640, HBV_HBV646.	5
HCC_HCC285, HBV_HBV268	6
LuCa-M0_lung419, LuCa-M0_lung496, LuCa-M1_lung293, LuCa-M1_lung324, LuCa-M1_lung417, LuCa-M1_lung418.	7
PaCa_pancreatic69, PaCa_pancreatic75.	8
HBV_HBV397, Blood_35-blood, Blood_36-blood.	9

To increase the system's capacity to differentiate between the various cancer states and their normal equivalents, features that enhance this capability are determined throughout the feature extraction process. There are a total of 68 features that were retrieved from both the healthy samples and the cancer group. Clustering and classification of the multi-omics data becomes effective after this complete feature extraction procedure, which helps identify the multiple biological patterns and assists in the interpretation of cancer subtypes. The number of features retrieved for each form of cancer is summarized in the Table 4.

Table 4 Extracted features of cancer type.

Types of cancer	Extracted features
Healthy	8
Lung cancer	15
Hepatocellular carcinoma	10
Pancreatic cancer	7
Breast cancer	4
Colon cancer	4
Gastric cancer	5
Brain cancer	4
Hepatitis B, (HBV)	7
Blood	4

4.2.2 T-SNE

T-SNE, a technique for reducing dimensionality, is best used to reduce high-dimensional data to lower-dimensional space while preserving the local structure of the data. After performing a K-means clustering step t-SNE is applied to the dataset. In the high-dimensional space of a multi-omics cancer dataset, t-SNE determines pairwise similarities between the data points. It then arrives at a low-dimensional representation that most accurately preserves the similarities. The algorithm converts the precise Euclidean distances of the ten data points from the clustering step into joint probabilities of pairwise data points that are more representational of the similarities in the data. Next, the algorithm makes adjustments in the low-dimensional space in order to reduce divergence between the joint probabilities of the high-dimensional space and the low-dimensional embedding. This loop is repeated for each cluster (from 0 to 9) where cluster data is extracted for each cluster based on the cluster labels (row clusters). This extraction involves selecting rows from the reduced dataset where the cluster label matches the current cluster index. Next, a 2D representation of the multi-omics cancer dataset is obtained that preserves the local structure and similarities of the original high-dimensional dataset. Each cluster is then plotted in a 2D scatter plot that makes it easier to observe the clustering and separation of data points as determined by K–means clustering as shown in Fig. 9.

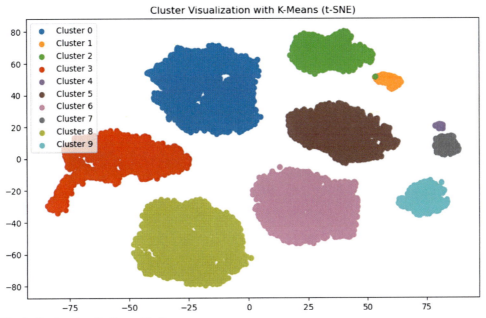

Fig. 9 Cluster visualization with K-means.

4.3 The results of feature selection stage

The multi–omics cancer dataset can be classified using CNN when combined with the multilevel GAN method. Feature selection describes the essential features and their correlations, allowing for a more thorough knowledge of the data in order to construct a robust deep learning model. The ability to calculate the ideal feature count depending on K value is the foundation of Kselectbest. A random number for K is chosen each time since there is no statistical approach to identify the ideal K for the feature selection process. Several trials are then carried out to choose the best K. The trial and error method is used while using the dataset during the training phase. From a total of 68 features acquired from the feature extraction process, as shown in Table 5, the top 60 features were chosen for this thesis.

Table 5 Names of selected features.

No.	Feature selected	No.	Feature selected
1.	Healthy_10	2.	PaCa_pancreatic15
3.	Healthy_11	4.	PaCa_pancreatic22
5.	Healthy_27	6.	PaCa_pancreatic27
7.	Healthy_35-1	8.	PaCa_pancreatic68
9.	Healthy_36-1	10.	PaCa_pancreatic9
11.	Healthy_38	12.	BrCa_BR13
13.	Healthy_38o	14.	BrCa_BR14
15.	Healthy_39o	16.	BrCa_BR5-1
17.	LuCa-M0_lung395	18.	BrCa_BR7-1
19.	LuCa-M0_lung419	20.	CoCa_colon13
21.	LuCa-M0_lung492	22.	CoCa_colon16
23.	LuCa-M0_lung493	24.	CoCa_colon17
25.	LuCa-M0_lung496	26.	CoCa_colon19
27.	LuCa-M0_lung512	28.	GaCa_stomach1
29.	LuCa-M0_lung513	30.	GaCa_stomach2
31.	LuCa-M0_lung514	32.	GaCa_stomach3
33.	LuCa-M0_lung515	34.	GaCa_stomach4
35.	LuCa-M1_lung293	36.	GaCa_stomach8
37.	LuCa-M1_lung418	38.	GBM_GBM57
39.	LuCa-M1_lung517	40.	GBM_GBM58
41.	HCC_HCC150	42.	GBM_GBM6
43.	HCC_HCC195	44.	GBM_GBM76
45.	HCC_HCC237	46.	HBV_HBV268
47.	HCC_HCC256	48.	HBV_HBV334
49.	HCC_HCC260	50.	HBV_HBV374
51.	HCC_HCC285	52.	HBV_HBV397
53.	HCC_HCC290	54.	HBV_HBV455
55.	HCC_HCC320	56.	HBV_HBV640
57.	HCC_HCC46	58.	HBV_HBV646
59.	HCC_HCC628	60.	Blood_35-blood

4.4 The results of classification stage based CNN with multi-level GAN

The multilevel GAN architecture is achieved by training three levels of GANs sequentially. The idea behind this setup is to gradually improve the quality of the generated data by feeding the output of one GAN to another GAN at the next level. In a level-1 multilevel GAN, generator1 learns how to generate data similar to the multi-omics cancer dataset. Discriminator1 tries to distinguish between the real and generated data. After training, it produces a new dataset named generated_data1. In a level-2 multilevel GAN, generator2 uses level-1's generated_data1 as the real data to train. Discriminator2 distinguishes between generated_data1 and the new data generated by generator2. After training, generator2 produces a new dataset named generated_data2. Then comes the level 3 of the multi-level GAN where the third generator (generator3) uses generated_data2 from level 2 as the real data to train it. The discriminator (discriminator3) distinguishes between generated_data2 and the new data generated by (generator3). After training, (generator3) produces a new dataset named generated_data3. At the end of the process, the new dataset generated_data3 is produced consisting of 16,195 samples, whose shape matches the shape of the original multi-omics cancer dataset that will be the input to the CNN algorithm.

The proposed CNN model in this thesis consists of three convolutional layers followed by pooling layers, flattening layers, and dense layers. The first convolutional layer applies 32 filters to the input data, each of which performs a convolutional operation that scans the input with a kernel size of 3. This operation helps in extracting low-level features. After the first convolution, max pooling is applied to reduce the spatial dimensionality and computational load while retaining the most important features. The second convolutional layer processes the pooled feature maps using 64 filters, capturing more complex patterns as the subsequent max pooling layer again reduces the dimensionality, ensuring that the most important features are retained. The third convolutional layer applies 128 filters to capture more complex and abstract features. The flattening layer then converts the 2D feature maps into a 1D vector, making it suitable for dense layers performing the classification task. The first dense layer, which contains 128 ReLU activation units, is applied to introduce nonlinearity. Then, the output layer consisting of two softmax activation units is applied to produce the probability distribution of ten probabilities for multi-omics cancer. Fig. 10 shows detailed information about each layer in the proposed CNN algorithm.

Layer (type)	Output Shape	Param #
conv1d_18 (Conv1D)	(None, 58, 32)	128
max_pooling1d_18 (MaxPooling1D)	(None, 29, 32)	0
conv1d_19 (Conv1D)	(None, 27, 64)	6,208
max_pooling1d_19 (MaxPooling1D)	(None, 13, 64)	0
conv1d_20 (Conv1D)	(None, 11, 128)	24,704
max_pooling1d_20 (MaxPooling1D)	(None, 5, 128)	0
flatten_2 (Flatten)	(None, 640)	0
dense_50 (Dense)	(None, 128)	82,048
dense_51 (Dense)	(None, 10)	1,290

Total params: 114,378 (446.79 KB)

Trainable params: 114,378 (446.79 KB)

Non-trainable params: 0 (0.00 B)

Fig. 10 Details of the proposed CNN algorithm.

sparse_categorical_crossentropy is used as the loss function and the Adam optimizer is used to train the model across 100 epochs. Testing 16, 64, and 128 led to the determination of this value via experimental epochs with a batch size of 32. The number of iterations per epoch for bigger batches of 64 and 128 increases the number of iterations per epoch based on experimental findings which indicate that 32 is the best batch size. Validation is a step in the training process that tracks how well the model performs with unknown data. After several experiences over various epochs. To reduce overfitting and enhance experimental outcomes, it is essential to choose the ideal number of epochs. These settings are established empirically to avoid overfitting. Table 6 shows the classifier's performance using the multi-omics cancer data.

Table 6 The evaluation metric on multi-omics cancer dataset.

Class	Precision	Recall	F1-measure	Accuracy
Cluster 0	99%	99%	99%	
Cluster 1	93%	89%	91%	
Cluster 2	98%	98%	98%	
Cluster 3	100%	98%	99%	
Cluster 4	100%	87%	93%	98.4
Cluster 5	99%	98%	98%	
Cluster 6	98%	100%	99%	
Cluster 7	93%	99%	96%	
Cluster 8	98%	99%	99%	
Cluster 9	98%	96%	97%	

The accuracy, precision, recall, and F1 score values for each class are shown in Table 6 and the outcomes are as follows: accuracy 98.4%, precision 98.5%, recall 98.4%, and F1 score 98.4% on the multi-omics cancer dataset, as shown in Fig. 11.

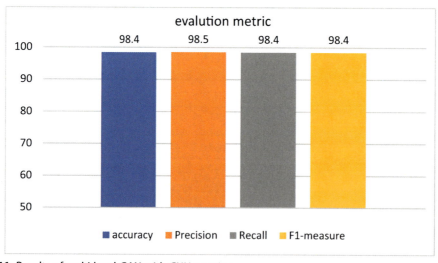

Fig. 11 Results of multi-level GAN with CNN metrics.

A confusion matrix is a graph that evaluates the performance of a classification method. A confusion matrix is created to thoroughly evaluate the model's performance. The efficacy of the multi-level GAN with CNN approach is assessed by computing its confusion matrix, seen in Fig. 12, after the implementation of the multi-level GAN with CNN using multi-omics cancer dataset and attaining notable accuracy.

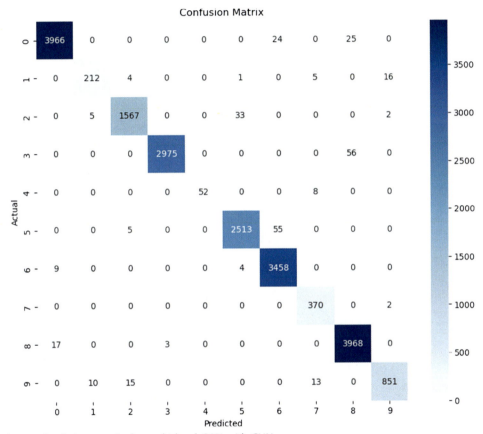

Fig. 12 Confusion matrix the multi-level GAN with CNN.

Following the completion of training, the model undergoes evaluation on the test set in order to quantify its level of accuracy. The multi-level GAN using CNN algorithm, used to deep learning of multi-omics cancer data, achieves an accuracy of up to 98.4%. The training and assessment accuracy are predicted using an equal number of epochs. The loss function is optimized via the "Adam" optimizer. The training and validation accuracy of the suggested model. The architecture of the multi-level GAN using the CNN method is shown in Fig. 13, whilst the training and validation loss are represented in Fig. 14.

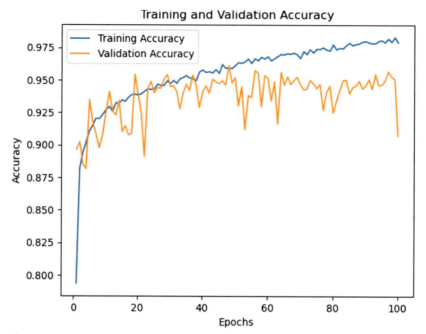

Fig. 13 The training and validation accuracy with multi-level GAN.

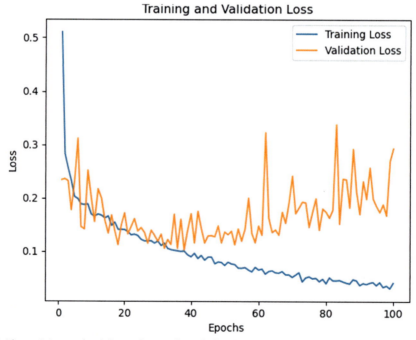

Fig. 14 The training and validation loss with multi-level GAN.

These results represent performance metrics for the multi-level GAN with CNN classification model across ten Cluster or classes. In Cluster 0, the model performs exceptionally well at classifying cases in Cluster 0, with high accuracy, recall, and F1 score, indicating balanced performance in both detecting true positives and avoiding false positives. The slight drop in accuracy indicates a very small number of misclassifications (Accuracy: 99%, Recall: 99%, F1 score: 99%). For Cluster 1, accuracy is relatively high at 93%, but recall is slightly lower at 89%, resulting in a lower F1 score of 91%. This indicates that while the model is good at correctly identifying Cluster 1 cases, it misses a few of them, resulting in low recall. The low recall indicates that some true positives are missed. The model performs equally well at classifying Cluster 2 cases, with high accuracy, recall, and F1 score of 98%. This reflects a well-balanced model for this set, effectively identifying true positives while avoiding false positives. The precision is perfect at 100%, meaning that all predicted instances for Cluster 3 are correct. However, the recall is slightly lower at 98%, indicating that a small number of true positives are missed. The F1 score is still very high at 99%, reflecting strong performance overall. While precision is perfect at 100% for Cluster 4, indicating that there are no false positives, recall is relatively low at 87%. This suggests that the model is missing many true positives, impacting the F1 score of 93%. The gap between precision and recall indicates an imbalance, perhaps due to the model being more conservative in its predictions for this set. The model performs well with a high precision of 99% and recall of 98%, resulting in a strong F1 score of 98%. This indicates that the model is accurate in its predictions and comprehensive in identifying actual cases for Cluster 5. The high recall of 100% also indicates that all actual positives for Cluster 6 were correctly identified. The precision is slightly lower at 98%, meaning that there are few false positives, but the overall performance is strong, as reflected in the high F1 score of 99%. The model has a high recall of 99% for Cluster 7, correctly identifying most of the Cluster 7 cases, but the precision is slightly lower at 93%, leading to some false positives. The F1 score of 96% shows that the model maintains a good balance between precision and recall. The model performs very well on Cluster 8 classification, where both precision and recall are high (precision 98%, recall 99%). This results in a strong F1 score of 99%, indicating a balanced and effective model for this cluster. On Cluster 9, the precision is high at 98%, indicating that most of the predicted positives are correct, but the recall is slightly lower at 96%, meaning that some cases are missed. The F1 score of 97% still indicates strong overall performance for this ensemble. The high scores on most metrics across most ensembles indicate a well-rounded model that is capable of handling the classification task effectively.

The findings demonstrated the superior performance of the suggested multi-level GAN technique across all measures, surpassing the performance of all other researchers that used the same metric to assess their models. The acquired findings demonstrate the efficacy of the suggested methodology in discovering fundamental data that can be used to categories multi-omics cancer. The issue of lacking dataset labelling for multi-omics cancer, a common challenge faced by researchers in this field, is resolved by the use of clustering algorithms. These approaches provide labels to the dataset by dividing it into 10 distinct groups. The comparison with state of arts provided in Table 7.

Table 7 Comparison with previous studies.

Author	Year	Method	Accuracy	Precision	Recall	F1-score
Reinel Tabares–Soto et al. [1]	2020	KNN	88.57%	–	–	–
		SVC	94.29%			
		RF	85.71%			
		LR	90.60%			
		DT	51.14%			
		CNN	94.43%			
Bingjun Li et al. [2]	2024	graph convolutional network (GCN), graph attention network (GAT)	77.9% 82.5%	–	–	76% 82%
Bingjun Li et al. [3]	2021	CNN	81.4%	80%	79%	79%
		FC–NN	82.3%	81%	80%	80%
		RF	78.7%	73%	77%	73%
		SVM	81.5%	74%	78%	75%
		GCN	81.4%	81%	81%	81%
Baoshan Ma et al. [4]	2020	XGBoost	85.9%	–	–	–
Hao Li et al. [5]	2022	GNN	87%	–	–	–
Our work	2024	Multi-level GAN with CNN	98.4%	98.5%	98.4%	98.4%

5. Conclusion

To improve cancer subtype classification with multi–omics data a novel optimized multi-level generative adversarial networks (OML-GANs) model is proposed and developed. This model utilizes a multi–continuous dataset, that is GANs, to create artificial data. This method was employed to overcome problems of balanced and sparse data. The experiment results confirm the thesis that the OML-GANs technique outperforms the other methods like machine learning and deep learning in terms of accuracy, precision, recall, and F1–score across many subtypes. The OML-GANs structure takes the tedious task of addressing the interconnected pathways between various omics data points into a simple level where a more comprehensive view of the cancer biology is developed. Moreover, the optimization strategies built into the GAN model decreases the chance of particular states of the model of becoming dominant over other states and thus enables the model to be applied in real-challenges situations, e.g., clinical diagnosis. This study is another contribution tinging the solid collection of research concerning the virtues of multi-omics in identifying different cancer subtypes. It is a scalable tool which can be used for clinical diagnosis and personalized treatment. Results demonstrated a significant improvement in detection accuracy compared to traditional methods, with an accuracy of 98.5%, precision of 98.5%, recall 98.4%, and F1 score 98.4%. Moreover, for the future studies, possible avenues may embrace the By adding new omics data units such as epigenomics and proteomics to represent the entire range of cancer subtypes and thus to improve the classification accuracy. It will be critical to data pre-processing in the form of feature extraction and dimensionality reduction, which needs to be optimized. Which were in place efficient feature extraction and dimensionality reduction will be key to handling the high dimensionality of these datasets. Besides, the approach also involves the usage of pretrained GAN models from larger multi-omics datasets through transfer learning, which may significantly decrease training time and also bring about better performance, especially for hard to detect cancer subtypes. On the other hand, this practice can be employed to tailor the model to different cancer types and platforms.

References

[1] A.M. Ali, M.A. Mohammed, A comprehensive review of artificial intelligence approaches in omics data processing: evaluating progress and challenges, Int. J. Math. Stat. Comput. Sci. 2 (2024) 114–167.

[2] S.B. Zaghlool, O. Attallah, A review of deep learning methods for multi-omics integration in precision medicine, in: 2022 IEEE International Conference on Bioinformatics and Biomedicine (BIBM), IEEE, 2022, December, pp. 2208–2215.

[3] M.A. Mohammed, K.H. Abdulkareem, A.M. Dinar, B.G. Zapirain, Rise of deep learning clinical applications and challenges in omics data: a systematic review, Diagnostics 13 (4) (2023) 664.

[4] M.A. Mohammed, A. Lakhan, K.H. Abdulkareem, B. Garcia-Zapirain, Federated auto-encoder and XGBoost schemes for multi-omics cancer detection in distributed fog computing paradigm, Chemom. Intell. Lab. Syst. 241 (2023) 104932.

[5] M. Mohammed, Enhanced cancer subclassification using multi-omics clustering and quantum cat swarm optimization, Iraqi J. Comput. Sci. Math. 5 (3) (2024) 552–582.

[6] M.A. Mohammed, A. Lakhan, K.H. Abdulkareem, B. Garcia-Zapirain, A hybrid cancer prediction based on multi-omics data and reinforcement learning state action reward state action (SARSA), Comput. Biol. Med. 154 (2023) 106617.

[7] M.S. Karthika, H. Rajaguru, A.R. Nair, Performance enhancement of classifiers through Bio inspired feature selection methods for early detection of lung cancer from microarray genes, Heliyon 10 (16) (2024).

[8] Z. Yang, G. Michailidis, A non-negative matrix factorization method for detecting modules in heterogeneous omics multi-modal data, Bioinformatics 32 (1) (2016) 1–8.

[9] C. Chauvel, A. Novoloaca, P. Veyre, F. Reynier, J. Becker, Evaluation of integrative clustering methods for the analysis of multi-omics data, Brief. Bioinform. 21 (2) (2020) 541–552.

[10] S. Canzler, J. Schor, W. Busch, et al., Prospects and challenges of multi-omics data integration in toxicology, Arch. Toxicol. 94 (2) (2020) 371–388.

[11] K. Yan, C. Li, X. Wang, et al., Comprehensive Autoencoder for Prostate Recognition on Mr Images, School of Information Technologies, University of Sydney Department of Nuclear Medicine, Ruijin Hospital, Shanghai Jiaotong University School of Medicine, 2016, pp. 1190–1194.

[12] F.G. Ebg, Breast Cancer Histopathological Image Classification Using Convolutional Neural Networks, IEEE, 2016, pp. 2560–2567.

[13] Y. Chen, Y. Li, R. Narayan, A. Subramanian, X. Xie, Gene expression inference with deep learning, Bioinformatics 32 (February) (2016) 1832–1839.

[14] R. Singh, J. Lanchantin, G. Robins, Y. Qi, DeepChrome: deep-learning for predicting gene expression from histone modifications, Bioinformatics 32 (17) (2016) i639–i648.

[15] J. Zhou, O.G. Troyanskaya, Predicting effects of noncoding variants with deep learning-based sequence model, Nat. Methods 12 (10) (2015).

[16] E.M. Karabulut, T. Ibrikci, Discriminative deep belief networks for microarray based cancer classification, Biomed. Res. 28 (3) (2017) 1016–1024.

[17] R. Ibrahim, N.A. Yousri, M.A. Ismail, N.M. El-makky, Multi-level gene/MiRNA feature selection using deep belief nets and active learning, Annu. Int. Conf. IEEE Eng. Med. Biol. Soc. (2014) 3957–3960.

[18] M.R. Young, D.L. Craft, Pathway-informed classification system (PICS) for cancer analysis using gene expression data, Cancer Inform. (2016) 151–161.

[19] A. El-Nabawy, N.A. Belal, N. El-Bendary, A cascade deep forest model for breast cancer subtype classification using multi-omics data, Mathematics 9 (13) (2021) 1–14.

[20] R. Tabares-Soto, S. Orozco-Arias, V. Romero-Cano, V.S. Bucheli, J.L. Rodríguez-Sotelo, C.F. Jiménez-Varón, A comparative study of machine learning and deep learning algorithms to classify cancer types based on microarray gene expression data, PeerJ Comput. Sci. 2020 (4) (2020) 1–22.

[21] B. Li, T. Wang, S. Nabavi, Cancer molecular subtype classification by graph convolutional networks on multi-omics data, in: Proc. 12th ACM Conf. Bioinformatics, Comput. Biol. Heal. Informatics, BCB 2021, 2021.

[22] Y. Lin, W. Zhang, H. Cao, G. Li, W. Du, Classifying breast cancer subtypes using deep neural networks based on multi-omics data, Genes (Basel) 11 (8) (2020) 1–18.

[23] W. Dai, W. Yue, W. Peng, X. Fu, L. Liu, L. Liu, Identifying cancer subtypes using a residual graph convolution model on a sample similarity network, Genes (Basel) 13 (1) (2022).

[24] C. Yin, Y. Cao, P. Sun, et al., Molecular subtyping of cancer based on robust graph neural network and multi-omics data integration, Front. Genet. 13 (May) (2022) 1–14.

[25] CfOmics, A Cell-Free MultiOmics Database for Diseases. https://cfomics.ncrnalab.org.

[26] S. Khadirnaikar, S. Shukla, S.R.M. Prasanna, Machine learning based combination of multi-omics data for subgroup identification in non-small cell lung cancer, Sci. Rep. 13 (1) (2023) 4636.

[27] X. Zhang, J. Zhang, K. Sun, X. Yang, C. Dai, Y. Guo, Integrated multi-omics analysis using variational autoencoders: application to pan-cancer classification, in: 2019 IEEE International Conference on Bioinformatics and Biomedicine (BIBM), IEEE, 2019, November, pp. 765–769.

[28] K.T. Ahmed, J. Sun, S. Cheng, J. Yong, W. Zhang, Multi-omics data integration by generative adversarial network, Bioinformatics 38 (1) (2022) 179–186.

[29] S. Afroz, N. Islam, M.A. Habib, M.S. Reza, M.A. Alam, Multi-omics data integration and drug screening of AML cancer using Generative Adversarial Network, Methods 226 (2024) 138–150.

[30] H. Yang, R. Chen, D. Li, Z. Wang, Subtype-GAN: a deep learning approach for integrative cancer subtyping of multi-omics data, Bioinformatics 37 (16) (2021) 2231–2237.

CHAPTER 8

Explainable artificial intelligence in epilepsy management: Unveiling the model interpretability

Najmusseher[a], P.K. Nizar Banu[a], Ahmad Taher Azar[b,c,d], and Nashwa Ahmad Kamal[e]

[a]Department of Computer Science, CHRIST (Deemed to be University), Central Campus, Bangalore, Karnataka, India
[b]College of Computer and Information Sciences, Prince Sultan University, Riyadh, Saudi Arabia
[c]Automated Systems and Soft Computing Lab (ASSCL), Prince Sultan University, Riyadh, Saudi Arabia
[d]Faculty of Computers and Artificial Intelligence, Benha University, Benha, Egypt
[e]Faculty of Engineering, Cairo University, Giza, Egypt

1. Introduction

Epilepsy, a neurological condition impacting approximately 1% individuals worldwide as reported by the World Health Organization, is recognized as a hypersensitive disease by clinical associations [1]. Electroencephalography, commonly abbreviated as EEG, is a fundamental tool in the field of neuroscience research. It is highly esteemed for its exceptional capability to map out the complex electrical activities within the brain. Utilizing a non-invasive approach, EEG captures the synchronized activity of millions of neurons, offering crucial insights into cognitive functions, sleep cycles, and various neurological conditions. Its extensive application in both clinical and experimental environments highlight its indispensable contribution to understanding the intricate workings of the human brain. EEG, with its unique and universally stable characteristics, holds significant credibility for researchers. It is a technique used to capture the electrical activity inside the brain [2]. Consequently, it has become the most widely utilized test for the detection and prediction of epileptic seizures. In recent years, the field of epileptic seizure classification has undergone a remarkable evolution, driven by advancements in leveraging EEG data for precise and timely diagnoses. This chapter seeks to elucidate the advancements made in harnessing EEG technology for detecting, predicting, and monitoring epileptic seizures. Specifically, it delves into the utilization of the Xgboost Machine learning (ML) model. Emphasizing the significance of employing interpretable models like XAI, it aims to offer transparent explanations for the predictions generated within this framework. Our specific focus lies in the utilization of the XGBoost ML model, a choice made after careful consideration of its numerous merits. XGBoost stands out for its exceptional ability to manage intricate datasets with efficiency, its robustness in guarding against overfitting, and its unique capacity to provide interpretable results. This emphasis

Explainable AI in Healthcare Imaging for Medical Diagnoses
https://doi.org/10.1016/B978-0-443-23979-3.00008-7

on interpretability aligns seamlessly with our overarching aim of offering transparent and comprehensible explanations for the predictions generated within the framework of seizure detection and prediction.

Machine learning is crucial across various fields due to its ability to enhance decision-making, efficiency, and innovation. In classification tasks, it enables accurate sorting and categorization of data, which is vital for applications like spam detection, medical diagnosis, and image recognition [3–7]. For prediction, machine learning models analyze historical data to forecast future events, aiding in financial market analysis, weather forecasting, and demand planning [8–12]. In robotics, machine learning empowers robots with advanced perception, decision-making, and adaptation capabilities, facilitating their use in manufacturing, healthcare, and autonomous vehicles, thereby revolutionizing industries with improved precision, safety, and productivity [13–16].

ML methods have demonstrated considerable promise in binary classification tasks, particularly in the challenging endeavor of distinguishing between seizure and healthy EEG signals. Despite achieving notable accuracy, a critical necessity has emerged for a robust explanation of results and decision-making processes. This requirement is particularly crucial for technical verification and clinical validation, highlighting the significance of XAI in the domain. Fig. 1 illustrates the overview of XAI, which focuses on creating models and algorithms that can provide transparent and understandable explanations for their decisions, while ML primarily focuses on developing systems that can learn from data and make predictions or decisions. XAI enables users to understand the inner workings of ML models, facilitating trust, interpretability, and accountability in decision-making processes. In EEG analysis, XAI can provide crucial insights by explaining how the ML algorithms interpret brain activity patterns, helping researchers and clinicians understand the reasoning behind the results, identify potential biases, and improve the accuracy and reliability of diagnoses or predictions.

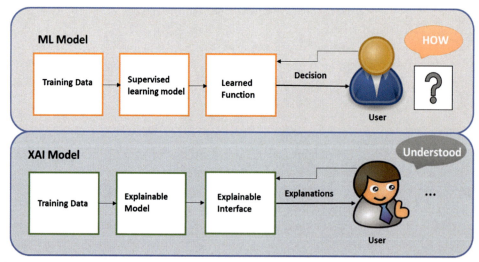

Fig. 1 XAI overview.

This chapter introduces a pivotal component in epilepsy diagnosis: a binary classification model leveraging XGBoost to accurately categorize EEG signals as either seizure or normal. As EEG data plays a crucial role in identifying epileptic episodes, the development of a reliable classification model holds significant promise for enhancing diagnostic accuracy and timely intervention. To ensure robust understanding and trustworthiness of the model's predictions, advanced techniques such as LIME and SHAP are incorporated. LIME facilitates local interpretability by elucidating individual predictions, while SHAP provides a comprehensive understanding of feature importance, aiding in the identification of EEG patterns indicative of seizures. Through the integration of these cutting-edge XAI techniques, this study endeavors to enhance the transparency and reliability of epilepsy diagnosis using EEG data.

This research highlights the transformative impact of LIME and SHAP in EEG-based seizure classification. Beyond enhancing the transparency and interpretability of the model, the integration of XAI techniques equips clinicians and researchers with the tools to make more informed decisions. Ultimately, this contributes to the improvement of patient care and outcomes in epilepsy management. By bridging the gap between complex EEG data and actionable insights, this study signifies a significant paradigm shift in the application of XAI techniques in medical diagnostics. It foresees a new era in epilepsy diagnosis and management, where advanced ML models guided by LIME and SHAP play a pivotal role in revolutionizing healthcare practices [17]. In the field of healthcare, there is a growing trend toward the incorporation of Artificial Intelligence (AI) systems, encompassing applications from surgical robots to automated medical diagnostics. While a ML engineer may concentrate on the parameters determining the performance and accuracy of these AI systems, it is hypothesized that medical practitioners prioritize considerations related to the practicality and usefulness of these systems within a medical context. Nevertheless, medical professionals often lack the necessary skills for a meaningful interpretation of AI-based systems. This poses a dual concern. Firstly, it hinders the widespread adoption of systems capable of automating routine analytical tasks, thereby impeding potential productivity gains. Secondly, and of potentially greater significance, it limits the pool of expertise available to contribute to the validation, refinement, and enhancement of AI-based systems designed to address healthcare challenges.

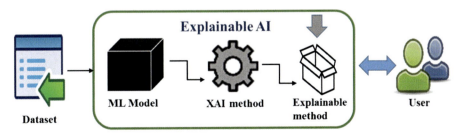

Fig. 2 XAI process.

Fig. 2 describes the general process of XAI, where feature importance analysis identifies influential factors in model predictions. Explanation generation techniques elucidate model decisions through understandable formats. Human–AI interaction facilitates user understanding and trust in model outputs through interactive visualization and dialog. XAI stands at the forefront of innovative developments in the field of artificial intelligence. As the complexity of AI systems continues to grow, the ability to comprehend and interpret their decision-making processes becomes increasingly crucial [18]. XAI aims to bridge the gap between the inherent opacity of advanced AI models and the need for transparency in understanding their outputs. XAI encompasses a collection of approaches and techniques designed to offer transparent and easily understandable explanations for the decisions produced by AI and ML models.

By incorporating an explainability layer into these models, Data Scientists and ML practitioners have the capability to develop systems that are not only more reliable but also transparent. This transparency benefits various stakeholders, including developers, regulators, and end-users, fostering trust and understanding in the decision-making processes of AI systems [19]. Fig. 3 illustrates the spectrum of XAI highlighting the key principles. Here are key principles of XAI:

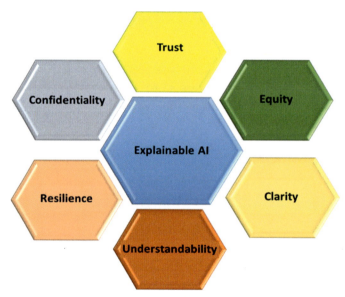

Fig. 3 XAI spectrum.

1. **Trust:** This involves assessing the level of confidence human users have in their interactions with the AI system, promoting reliability and user trust.
2. **Equity:** This principle emphasizes the importance of ensuring that the decisions made by the models are fair and just for all individuals, regardless of race, religion, gender, disability, or ethnicity.
3. **Clarity:** It ensures that stakeholders have a clear understanding of the decision-making process utilized by the models, promoting transparency and comprehension.

4. Understandability: This principle stresses the significance of providing human-friendly explanations for the predictions and outcomes generated by the AI models, enhancing user comprehension and enabling informed decision-making.

5. Resilience: This principle highlights the importance of developing models that can withstand changes in input data or modifications in model parameters, ensuring consistent and reliable performance in uncertain or unexpected circumstances.

6. Confidentiality: It focuses on guaranteeing the protection of sensitive user information, demonstrating a commitment to ethical and secure AI practices.

In this chapter, we delve into the critical methodologies within the realm of XAI that significantly contribute to enhancing our understanding of complex ML models. Two key players in this domain are LIME and SHAP. These methodologies serve as essential tools for interpreting the decisions made by advanced AI systems. LIME is a noteworthy methodology designed to address the challenges associated with comprehending the decisions of complex ML models. Recognizing the need for interpretability, LIME offers a versatile and model-agnostic approach, providing explanations that are both understandable and insightful. By acting as a bridge between intricate algorithms and human understanding, LIME significantly enhances the accessibility and trustworthiness of AI systems. SHAP stands as a pivotal advancement in the field of XAI. Developed on the basis of cooperative game theory, SHAP provides a comprehensive framework for understanding and attributing the impact of each feature on the predictions made by a model. Through the assignment of unique Shapley values to individual features, SHAP enables a sophisticated investigation into the impact of each variable on shaping the model's outcomes. This approach not only improves interpretability but also fosters a deeper and refined understanding of complex AI models, promoting trust and accountability in their decision-making processes.

In the subsequent discussion of this chapter, we will delve into an in-depth exploration of LIME and SHAP specifically in the context of seizure classification, unraveling their roles in enhancing the interpretability and reliability of ML models in the domain of epilepsy diagnosis.

2. Exploring XAI in healthcare

Artificial intelligence (AI) has found diverse applications in healthcare, including health services management, predictive medicine, clinical decision-making, and patient data diagnostics [20–26]. Despite achieving impressive performance levels, AI models are often underutilized due to their perceived black-box nature, leading to a lack of trust. Introducing XAI addresses this issue by offering transparency in model predictions, explaining the rationale behind predictions [27]. This enhances confidence in AI systems, particularly in healthcare, where trust is paramount. This review aims to identify healthcare areas that necessitate increased focus from the XAI research community to foster greater transparency and trust in AI applications. The integration of ML techniques in healthcare, particularly in the context of predicting epileptic seizures presents a unique challenge due to the inherent black-box nature of these models. The opacity of traditional ML and deep learning (DL) approaches raises concerns among both medical professionals and patients, leading to a diminished level of trust in the predictions generated. In the critical domain of healthcare, where the consequences of decision-making are profound, the need for transparent and comprehensible predictions becomes paramount. The reliability of epileptic seizure prediction using ML and DL methods is often a subject of skepticism among general users, as the intricacies of these models are not readily interpretable. The desire for a predictive technique that not only delivers accurate results but also provides clear and understandable reasons for its predictions has led to the emergence of XAI in the healthcare landscape.

2.1 Ethics of AI in healthcare

The ethical use of AI in healthcare, particularly in EEG analysis, is paramount to ensure patient well-being, privacy, and trust. For instance, AI-powered EEG systems should adhere to strict privacy protocols to safeguard patient data. Additionally, transparency in AI algorithm's decision-making processes is crucial, ensuring that healthcare professionals can validate and understand the recommendations provided. Ultimately, ethical considerations should guide the development and deployment of AI technologies in healthcare, fostering patient-centric care and maintaining the integrity of medical practice. For example, in EEG analysis, an ethical use of AI would involve ensuring that patient consent is obtained for data collection and analysis. The AI system should be transparent about how it processes EEG data and provide clear explanations for its findings. Moreover, the AI model should be trained on diverse datasets to avoid biases and ensure equitable treatment for all patients, regardless of demographic factors. Additionally, patient confidentiality must be maintained by securely storing and handling EEG data to prevent unauthorized access or breaches of privacy. By adhering to these ethical principles, AI-driven EEG analysis can contribute to improved diagnosis and treatment while upholding patient rights and well-being.

2.2 XAI diversity

XAI plays a transformative role by meeting the increasing demand for transparency in the decision-making processes of ML models. In contrast to conventional black-box models, XAI strives to demystify the intricate inner workings of these algorithms, making their predictions accessible and understandable for both medical professionals and patients. Through the application of XAI techniques, users gain the ability to explore the nuanced details and rationales behind specific predictions, fostering a deeper understanding of the decision-making process. XAI represents a dedicated field focused on researching methods and strategies that facilitate the comprehension and interpretation of ML models

[28]. The surge in the adoption of AI-based solutions, coupled with evolving regulatory requirements, has generated widespread interest in XAI [29]. Traditionally, the understanding of ML models has been predominantly approached from the perspective of computer scientists, with limited emphasis on supporting non-computer science domains [30]. Fig. 4 illustrates the diversity of XAI in various fields. XAI exhibits versatility across domains, adapting its transparent decision-making to diverse applications from healthcare to finance. Its flexibility empowers stakeholders to comprehend and trust AI systems' outputs, fostering inclusivity and innovation across industries.

Fig. 4 XAI diversity.

These domains may include healthcare, finance, biology, psychology, sociology, and many others. For example, in healthcare, ML models are used to analyze medical images, signals etc. for diagnosis or predict patient outcomes based on clinical data. In finance, these models can be employed for risk assessment, fraud detection, or algorithmic trading. In biology, ML techniques are applied to analyze genomic data or model protein structures. In psychology, they might be utilized to predict behavior patterns or analyze sentiment from text data. In each of these examples, experts from fields like medicine, finance, biology, psychology, or sociology may not have a background in computer science but still utilize ML models to enhance their research, analysis, or decision-making processes. Therefore, integrating XAI with ML models offers valuable insights into understanding and explaining ML models and their workflows, seamlessly blending into typical ML algorithms. This helps bridge the divide between technical complexities and the comprehension requirements of various user groups, promoting a wider and more inclusive grasp of AI-driven decision-making processes.

2.3 XAI in EEG

As EEG data is multi-dimensional and often high-dimensional, analyzing it requires sophisticated computational techniques, including machine learning and artificial intelligence. However, traditional AI models such as deep neural networks are often regarded as "black boxes," meaning they provide predictions without transparently revealing the reasoning behind those predictions. In medical contexts like EEG analysis, transparency and interpretability are paramount for clinical decision-making and understanding brain function.

XAI techniques aim to bridge the gap between complex AI models and human understanding by providing explanations for their predictions. In EEG analysis, XAI methods can elucidate the features, patterns, and relationships in the data that contribute to specific outcomes, such as diagnosing neurological disorders or decoding cognitive states. Here are some ways XAI is applied in EEG.

XAI in EEG involves delving into the complexities of EEG data, the challenges in interpreting it, and how XAI methods address these challenges. EEG, a non-invasive technique for recording brain activity, captures the intricate interplay of neural processes through voltage fluctuations in the brain's neurons. However, the multi-dimensional and high-dimensional nature of EEG signals presents significant challenges in analysis, particularly in understanding the underlying neural mechanisms and translating findings into actionable insights for clinical practice.

Traditionally, EEG analysis relied on manual interpretation by experts, which is time-consuming, subjective, and prone to errors. With the advent of machine learning and artificial intelligence, automated methods have been developed to analyze EEG data, offering the promise of faster and more objective assessments. However, many AI models, particularly deep neural networks, are often regarded as "black boxes" due to their opacity, providing predictions without transparently revealing the reasoning behind them. In medical contexts like EEG analysis, where transparency and interpretability are crucial for clinical decision-making, this lack of explainability poses significant challenges.

This is where XAI techniques come into play. XAI methods aim to bridge the gap between complex AI models and human understanding by providing explanations for their predictions. In the context of EEG analysis, XAI techniques offer several avenues for enhancing interpretability. For instance, they can identify which features or electrodes in the EEG signal are most influential in making a prediction, highlighting specific brain regions or frequency bands associated with particular cognitive tasks or pathological conditions. Additionally, XAI methods visualize how the AI model processes EEG data over time or across different brain regions, providing insights into the internal workings of the model and the underlying brain processes it captures.

Moreover, XAI enables the extraction of interpretable rules or decision trees from complex EEG data, offering actionable insights into the relationships between different EEG features and clinical outcomes. By generating counterfactual explanations, XAI methods show how changes in EEG features would alter the model's predictions, helping clinicians assess the robustness of the model and explore hypothetical scenarios for personalized treatment strategies. Integrating XAI techniques into EEG analysis not only enhances clinical interpretability but also improves quality control by identifying and mitigating biases, errors, or artifacts in EEG recordings.

Overall, XAI plays a vital role in advancing EEG analysis by providing transparent and interpretable insights into complex brain activity patterns. By empowering clinicians with actionable insights and explanations, XAI enables more informed clinical decisions, fosters a deeper understanding of brain function, and ultimately improves patient outcomes in neurological and psychiatric disorders. As research in XAI continues to evolve, its integration into EEG analysis holds great promise for revolutionizing our understanding and treatment of brain-related conditions.

XAI is a game-changer in healthcare, revolutionizing decision-making processes with transparency and interpretability for advanced AI models. As the healthcare industry embraces AI, the demand for trustworthy and understandable systems becomes paramount. XAI, focusing on enhancing the interpretability of ML models, emerges as a crucial solution for addressing challenges in healthcare. The rising adoption of AI-driven tools in healthcare aims at better diagnostics, personalized treatment, and enhanced patient care. However, the complexity of advanced ML models presents a hurdle in understanding predictions, especially for non-technical stakeholders like healthcare practitioners. In healthcare, XAI plays a pivotal role, ensuring clinicians can trust and comprehend the AI systems integrated into their workflows. By providing interpretable insights into AI models' decision-making, XAI bridges the gap between sophisticated algorithms and real-world healthcare applications.

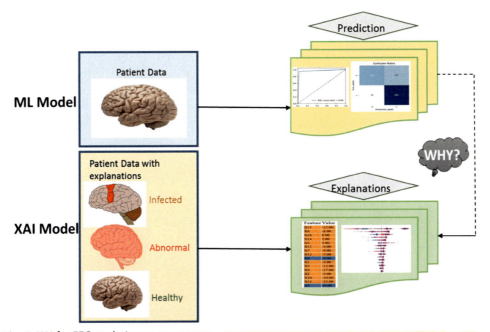

Fig. 5 XAI for EEG analysis.

Fig. 5 illustrates XAI for EEG data analysis, In EEG data brain classification, ML utilizes algorithms to analyze patterns in brainwave signals, identifying features that distinguish between different states or conditions. ML models, such as neural networks or support vector machines, learn from labeled data to make predictions about new EEG recordings. XAI enhances this process by providing insights into how the ML model reaches its decisions, highlighting the specific features or signals that contribute most to classifications. This transparency aids researchers and clinicians in understanding the underlying mechanisms driving EEG classifications, ensuring more accurate and interpretable results for brain analysis tasks. In the field of epileptic seizure classification, XAI significantly impacts by providing transparent insights into ML model's decision-making processes. Techniques like LIME and SHAP enable a clear understanding of features influencing seizure predictions, empowering clinicians with actionable insights for reliable diagnoses. This transformative role bridges the gap between complex ML algorithms and critical medical applications, enhancing patient care in epilepsy management. Exploring XAI in healthcare equips medical practitioners with tools for understanding factors influencing AI-based diagnoses and treatment recommendations, fostering trust and collaborative decision-making. XAI's contribution to meeting regulatory and ethical standards through transparency and accountability underscores its importance in successfully integrating AI technologies into medical practices. It addresses interpretability challenges, promoting an ethical approach to leveraging AI for improved patient outcomes and healthcare delivery.

2.4 The importance of XAI in healthcare

Interpretability, defined as the degree to which the rationale behind a decision is observable within a system, plays a crucial role in the acceptance and trustworthiness of ML models [31]. When the operation of an ML model is easily comprehensible, it is deemed interpretable. Explainability, on the other hand, goes a step further by addressing the extent to which the internal workings of a system can be articulated in human terms. XAI serves as a comprehensive methodology designed to render AI systems both interpretable and explainable [29]. XAI holds profound significance in healthcare by addressing the inherent challenge of interpretability in advanced ML models. In the context of healthcare applications, understanding and trusting the decisions made by AI systems are paramount for successful integration into clinical practice. One crucial area where XAI plays a transformative role is in the classification of EEG based epileptic seizures.

In healthcare, the context of interpretability and explainability takes on a domain-specific nature, varying between different stakeholders such as ML engineers and medical practitioners. From the perspective of an ML engineer, interpretability involves understanding the internal workings of a system to fine-tune technical parameters for enhanced overall performance. In contrast, a medical practitioner seeks a higher-level understanding of a system's internal operation as it pertains to the medical functions it serves.

Furthermore, explainability for an ML engineer revolves around presenting technical information in a format that enables effective system evaluation, while for medical practitioners, it relates to understanding the rationale behind prescribed courses of action for patients.

In the healthcare setting, it is postulated that AI-based systems must accommodate the perspectives of medical practitioners to be considered truly explainable and interpretable. This assertion introduces several challenges that are addressed in the course of this work. Designing XAI systems that are domain-agnostic and simultaneously accommodate multiple perspectives proves to be complex, as explanations need contextual relevance within specific domains such as engineering, medicine, or healthcare. For example, interactive visualizations explaining neural network layers may be beneficial for ML engineers but less crucial for radiologists using the network for analyzing MRI scans.

The scope of interpretability and explainability extends beyond the operation of an ML model to encompass the entire workflow employed in training these models. This workflow encompasses critical technical knowledge, from pre-processing steps and ML model details for ML engineers to an overview of underlying data, the model's interpretation of the data, and performance metrics for medical practitioners involved in diagnostics. Grasping the complexities of XAI in healthcare presents several challenges. These include the necessity for interpretability metrics tailored to medical practitioners, integrating clinical expertise into feature engineering, ensuring clear explanations of both data and ML models, and adapting models to operate effectively across diverse healthcare settings.

XAI holds immense significance in healthcare by unraveling the complexities of AI models, offering transparent insights into decision-making processes. In epileptic seizure classification, traditional ML models often operate as opaque black boxes, hindering medical practitioner's understanding of predictive factors. Through methods like LIME and SHAP, XAI clarifies these complexities by furnishing clear explanations for predictions. For instance, LIME unveils essential EEG features driving individual classification decisions, while SHAP provides a broader view of feature importance, illuminating the collective impact on classification outcomes. This empowers clinicians with a deeper comprehension of seizure prediction, enhancing diagnostic accuracy and patient care.

Through the utilization of XAI in epileptic seizure classification, healthcare practitioners acquire a deeper understanding of the intricacies behind the AI model's decision-making process. This enhances their trust in the system's accuracy and dependability. This transparency is crucial in clinical settings, where decisions based on AI predictions have direct implications for patient care. Understanding why a particular prediction is made allows clinicians to validate and refine the model's recommendations, fostering a collaborative approach between human expertise and machine intelligence. However, the importance of XAI in healthcare, especially in epileptic seizure classification, cannot be overstated. XAI brings transparency to complex AI models, enabling healthcare practitioners to comprehend and trust the decisions made by these systems. This not only enhances the acceptance and adoption of AI technologies in healthcare but also ensures that the integration of these systems results in improved patient outcomes and more informed clinical decision-making [32]. While this chapter does not comprehensively address all challenges, it takes initial steps toward a solution. Acknowledging the complexities, it emphasizes the necessity for XAI approaches to seamlessly integrate interpretability and explainability into the standard ML workflow, encompassing stages from data collection and preparation to modeling, training, evaluation, parameter tuning, and deployment in healthcare applications.

3. Architectural Framework for XAI

The Architectural Framework for XAI pertains to the structured design and organization of components and processes that facilitate the development and implementation of explainable AI systems. This framework is crucial for delivering transparency, interpretability, and accountability in the decision-making processes of intricate ML models [33]. The workflow for XAI involves a systematic and comprehensive approach to ensure transparency and interpretability in the decision-making process of ML models as illustrated in Fig. 6. The framework consists of the following key steps:

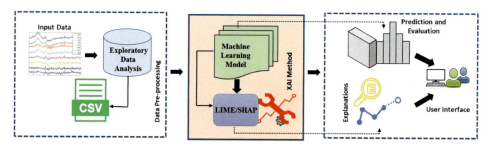

Fig. 6 XAI workflow.

1. **Input data & preprocessing**: Foundational to any ML model, The framework initiates with the collection and pre-processing of data. This stage involves ensuring data quality, handling missing values, and preparing the dataset for model training.

2. **Model building**: During this phase, ML models are constructed. The framework incorporates considerations for selecting models that inherently support interpretability and explainability, along with the choice of model architecture, algorithms, and parameters. XGBoost is a powerful machine learning algorithm widely employed for seizure classification tasks. Its ability to handle complex datasets, optimize feature selection, and provide robust predictive modeling makes it particularly suitable for detecting patterns in EEG data associated with seizures [34]. Boost's ensemble learning approach enhances the accuracy of seizure classification, contributing to improved diagnostic capabilities in the field of epilepsy research and healthcare.

3. **Prediction and evaluation**: The trained model predicts outcomes for new data, with accuracy quantifying overall correctness. A Confusion Matrix provides a detailed breakdown, distinguishing between true positives, true negatives, false positives, and false negatives.

4. **XAI methods (LIME or SHAP)**: Incorporates Explainable AI methods like LIME and SHAP into the framework. LIME's surrogate models offer localized interpretability, revealing features influencing individual predictions. SHAP provides a global perspective on feature importance, elucidating collective impacts on classification outcomes.

5. **User interface**: This final step involves a purposeful design of a user interface, for presenting explanations to end-users. This interface is designed to be user-friendly and accessible to individuals who may not have a technical background, such as healthcare professionals or business stakeholders.

By integrating these steps, XAI establishes a systematic and transparent approach. This enables stakeholders not only to trust the model's predictions but also to comprehend the influential factors behind them. Such transparency is pivotal for bolstering the acceptance and usability of machine learning models, especially in critical domains like healthcare [35].

3.1 Varieties of XAI

XAI encompasses a range of approaches designed to augment the interpretability and transparency of machine learning models. Gaining insight into the subtleties of XAI is essential for constructing reliable and interpretable AI systems. The primary types such as model specific, model agnostic, post-hoc and surrogate models, along with their respective subtypes [36], are illustrated in Fig. 7 respectively.

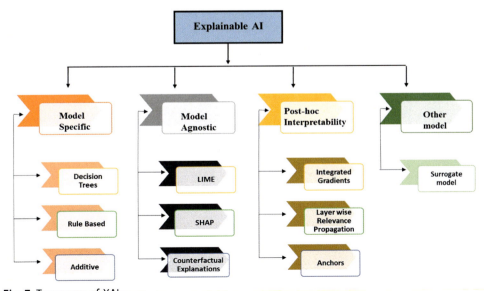

Fig. 7 Taxonomy of XAI.

1. **Model-specific approaches**: Tailored methods that focus on explicit rule formulation within models like decision trees and additive models for enhanced interpretability, which includes the following.
   a. Decision trees: Recognized for their inherent interpretability, decision trees formulate explicit rules within their branches, offering a clear understanding of decision-making processes.
   b. Rule-based models: Represented explicitly as sets of rules, these models provide transparency by outlining the decision logic in a comprehensible manner.
   c. Additive models: Represent predictions as a sum of contributions from individual features, facilitating a straightforward interpretation of feature impacts.
2. **Model-agnostic approaches**: Techniques, such as LIME and SHAP, provide interpretability irrespective of the underlying model architecture, ensuring broad applicability, which includes the following.
   a. Local Interpretable Model-agnostic Explanations (LIME): This method generates interpretable models tailored for specific instances, providing localized insights into model behavior.
   b. Shapley Additive explanation (SHAP): Leveraging Shapley values, SHAP offers a global perspective on feature importance, aiding in understanding the overall impact of features on predictions.
   c. Counterfactual Explanations: By presenting alternative scenarios, counterfactual explanations offer insights into how changes in input variables influence model predictions.
3. **Post hoc interpretability techniques**: Strategies like Integrated Gradients, LRP, and Anchors that assess models after training, offering insights into feature importance and decision rationale.
   a. Integrated gradients: This technique measures the impact of each feature by integrating gradients along the path from a baseline to the input, offering a comprehensive view of feature importance.
   b. Layer-wise relevance propagation (LRP): Assigning relevance scores to each input feature, LRP provides a layer-wise understanding of feature contributions in complex neural network models.
   c. Anchors: Anchors identify minimal input features necessary for a consistent model prediction, offering clear and concise rules for specific decisions.
4. **Other model**: Miscellaneous approaches that may not fit neatly into the previous categories, encompassing innovative methods for interpretability enhancement.
   a. Surrogate models: These are simplified versions of complex models that maintain similar decision boundaries, facilitating a more accessible understanding of the overall model behavior.

Table 1 Examples for varieties of XAI.

SL. No.	Approach	Technique	Scenario	Example
1	Model specific	Decision trees	Loan approval system	A bank uses a decision tree model to automate loan approvals. If a loan applicant has a credit score above 700 and an income below $50,000, the model might automatically deny the loan
		Rule based	Loan approval system	A hospital employs a rule-based model to prioritize emergency room patients. If a patient presents with symptoms like chest pain and shortness of breath, the model may prioritize them for immediate attention
		Additive	Loan approval system	A retail store uses an additive model to predict daily sales. The model calculates the expected sales by summing up contributions from factors like promotions, weather conditions, and historical sales data
2	Model agnostic	LIME	Image classification	A photo-sharing app uses LIME to explain why a particular image was classified as "cat." LIME highlights the pixels in the image that contributed most to the model's classification decision
		SHAP	Image classification	An insurance company uses SHAP values to understand why certain customers were classified as high-risk. SHAP reveals that factors like age and medical history had the greatest impact on the model's risk assessment
		Counterfactual	Image classification	An insurance company uses SHAP values to understand why certain customers were classified as high-risk. SHAP reveals that factors like age and medical history had the greatest impact on the model's risk assessment

Continued

Table 1 Examples for varieties of XAI.—cont'd

SL. No.	Approach	Technique	Scenario	Example
3	Post hoc	Integrated gradients	Sentiment analysis system	A movie review website uses integrated gradients to explain why a review was classified as "positive." The technique highlights the words in the review that had the greatest influence on the sentiment classification
		Layer-wise relevance propagation (LRP)	Sentiment analysis system	A medical imaging system uses LRP to interpret why a certain region in an MRI scan was flagged as abnormal. LRP identifies which pixels in the scan contributed most to the abnormality detection
		Anchors	Sentiment analysis system	A credit card fraud detection system employs anchors to explain why a transaction was flagged as suspicious. Anchors identify specific features of the transaction, such as location and transaction amount, that led to the fraud alert
4	Other model	Surrogate models	Stock price prediction system	A financial institution uses a simpler linear regression model as a surrogate for a complex deep learning model predicting stock prices. The surrogate model provides insights into which economic indicators, such as interest rates and GDP growth, are most influential in predicting stock movements

Understanding the different varieties of XAI is essential for choosing the most suitable approach according to the specific needs of a particular application. Whether focusing on local explanations, feature importance, or rule-based models, the goal remains consistent. Table 1 provides suitable examples for different varieties of XAI. This comprehensive array of types and subtypes forms a diverse toolkit for XAI techniques [37]. Practitioners have the flexibility to select or combine methods based on the unique characteristics of their model and the interpretability requirements inherent to a specific task. This versatility ensures an adaptable and effective approach to enhancing interpretability in machine learning models.

3.2 XAI applications in healthcare

XAI has emerged as a transformative force in healthcare, offering transparency and interpretability in complex decision-making processes. By providing insights into the inner workings of machine learning models, XAI plays a crucial role in enhancing trust and facilitating the integration of AI technologies in healthcare settings. Below, we explore the diverse applications of XAI in healthcare:

(i) XAI plays a vital role in enhancing diagnostic accuracy through transparent explanations for model predictions. In medical imaging, XAI techniques reveal the features influencing a model's decision, empowering clinicians to validate and trust AI-driven diagnoses [38]. For example, in a mammogram analysis, XAI can highlight areas of concern, helping radiologists interpret the image and make informed decisions about patient care. This transparency aids in improving diagnostic accuracy and patient outcomes.

(ii) AI models frequently assist in recommending personalized treatment plans for patients. With XAI, these recommendations become not only accurate but also explainable. For instance, in cancer treatment, an AI model might suggest a specific chemotherapy regimen based on a patient's genetic profile. XAI techniques can provide insights into why this particular treatment is recommended, such as highlighting genetic mutations or biomarkers influencing the decision. This transparency enables clinicians to understand the rationale behind the treatment suggestion, fostering collaboration between AI systems and healthcare professionals to optimize patient care.

(iii) XAI helps doctors understand why a patient might need frequent check-ups. For instance, in diabetes management, it can reveal that high blood sugar levels and missed insulin doses increase the risk of complications, prompting doctors to adjust treatment plans, accordingly, ensuring better patient outcomes.

(iv) In pharmaceutical research, XAI clarifies why certain compounds are promising for drug development. For instance, it might reveal that compounds with specific chemical structures are more effective in targeting disease proteins, guiding researchers to focus on those compounds for further testing, ultimately leading to the development of safer and more effective drugs.

(v) In Clinical Decision Support Systems (CDSS), XAI helps doctors understand why a certain treatment is recommended for a patient. For example, if a CDSS suggests a specific medication for a patient with diabetes, XAI might reveal that the recommendation is based on factors like the patient's blood sugar levels and medical history, enabling doctors to make informed treatment decisions with confidence.

(vi) In Electronic Health Records (EHR), XAI helps doctors understand why certain patients are at risk for certain conditions. For instance, if an AI model predicts a high risk of heart disease for a patient, XAI might reveal that the prediction is based on factors like cholesterol levels, blood pressure, and family history of heart disease, empowering doctors to intervene early and provide targeted care.

(vii) In healthcare, XAI ensures regulatory compliance and ethical transparency by explaining AI model decisions. For example, in a diagnostic AI system for breast cancer detection, XAI might reveal which features of a mammogram image led to the model's classification, helping regulatory authorities verify the system's accuracy and ensuring patients understand the basis of their diagnoses, thereby addressing ethical concerns and fostering trust in AI applications.

(viii) Transparent AI models empower patients by explaining medical decisions in understandable terms. For instance, in a diabetes management app, an AI-driven recommendation to adjust insulin dosage might be accompanied by an explanation that high blood sugar levels indicate the need for increased insulin to maintain stable glucose levels. This transparency helps patients understand the reasoning behind treatment recommendations, empowering them to actively participate in their healthcare decisions. This fosters trust between patients and healthcare providers, enabling more informed discussions about treatment options and fostering a collaborative approach to healthcare decision-making [39].

(ix) In healthcare, XAI contributes to the creation of explainable Chabot's that interact with patients. For instance, a Chabot providing mental health support might explain relaxation techniques by detailing how deep breathing can reduce stress and anxiety. This clear explanation enhances patient engagement and understanding, fostering effective communication and promoting overall well-being.

(x) In healthcare, XAI helps identify and address biases in AI models to ensure fairness. For example, in a skin cancer detection app, XAI might reveal that the model disproportionately misclassifies darker skin tones due to insufficient representation in the training data. By recognizing and explaining these biases, XAI enables developers to improve model accuracy and equity, ensuring that healthcare decisions are fair and unbiased for all patients.

Fig. 8 illustrates the various applications of XAI, beyond healthcare, XAI finds diverse applications in finance, aiding in risk assessment and fraud detection, while also revolutionizing customer service and personalization in industries like retail and e-commerce. Its transparent decision-making processes foster trust and efficiency across various sectors, driving innovation and informed decision-making. Hence, the applications of XAI in healthcare are diverse and impactful. From improving diagnostics to fostering patient trust and facilitating regulatory compliance, XAI is a key enabler in the evolution of healthcare toward more transparent, accountable, and patient-centric practices. As the healthcare industry continues to embrace AI technologies, the integration of XAI will be pivotal in realizing the full potential of these advancements while ensuring ethical and patient-focused healthcare delivery.

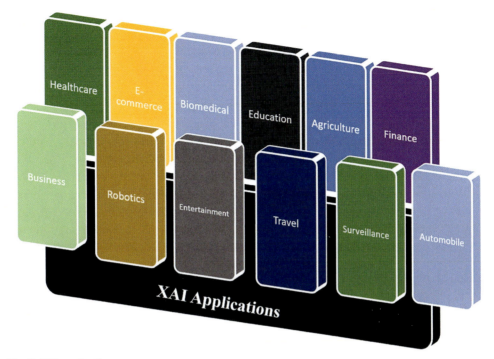

Fig. 8 XAI applications.

4. Understanding LIME

LIME, stands for Local Interpretable Model-agnostic Explanations, is a highly effective and commonly used method in the field of XAI. Originating from the work of Marco Tulio Ribeiro and his research team [40], LIME offers a technique for making black-box machine learning models interpretable on a local scale [41]. The primary goal of LIME is to offer human-understandable insights into the decision-making processes of complex models, particularly in scenarios where the models themselves lack inherent interpretability. Instead of offering a comprehensive overview of the model across the entire dataset, LIME concentrates on elucidating the model's predictions for individual instances [42].

4.1 Key principles of LIME

1. **Local approximation**: LIME operates on the principle of creating local approximations of a black-box model's behavior. Instead of attempting to interpret the entire model globally, LIME focuses on explaining individual predictions by generating a simplified, interpretable model in the vicinity of the specific instance under consideration.

2. **Instance perturbation**: To create these local approximations, LIME introduces the concept of instance perturbation. It involves perturbing the input features of a particular instance and observing the resulting changes in the model's predictions. By generating a dataset of perturbed instances and their corresponding predictions, LIME constructs a local neighborhood for the interpretability task.

3. **Interpretable model construction**: Within the defined local neighborhood, LIME employs an interpretable model, such as a linear or decision tree model, to approximate the behavior of the black-box model. The interpretable model is trained to mimic the predictions of the complex model within the local context, providing a simplified and understandable representation.

4. **Weights and importance**: LIME assigns weights to each perturbed instance based on their proximity to the original instance. Instances that are closer in feature space receive higher weights, emphasizing their importance in constructing the local approximation. This ensures that the interpretable model focuses on capturing the relevant features for the specific prediction under investigation.

In the context of ML and data analysis, perturbation involves modifying input data or features in order to observe how these changes affect the output or predictions of a model. In the given context of LIME, perturbation is used to create local approximations of a black-box model's behavior. This involves intentionally altering the input features of a particular instance in the dataset and observing the resulting changes in the model's predictions. By systematically perturbing the input features and recording the corresponding changes in predictions, LIME constructs a dataset of perturbed instances. These perturbed instances, along with their predictions, form a local neighborhood around the specific instance under consideration. Hence, perturbation in of LIME is a methodical approach to exploring how variations in input features impact the predictions of a machine learning model, ultimately facilitating the generation of interpretable explanations for individual predictions.

4.2 Applications of LIME

(i) In the domain of computer vision, LIME is extensively applied to interpret image, signal and text classification models. For an image analysis, by perturbing pixels in an image and observing the changes in predictions, LIME can highlight the regions and features crucial for the model's decision, offering insights into why a particular classification was made.

(ii) LIME finds application in explaining predictions made by natural language processing models. By perturbing words or phrases in a text, LIME can reveal the key linguistic elements influencing the model's output, aiding in the interpretation of sentiment analysis or text classification models.

(iii) LIME is valuable in healthcare applications where complex predictive models are employed. By explaining the predictions of models that assess patient risk or diagnose diseases, LIME provides clinicians with insights into the features contributing to individual patient assessments, fostering trust and aiding in decision-making.

(iv) In fraud detection models within the finance sector, LIME helps explain why certain transactions are flagged as fraudulent. By perturbing transaction features, LIME can elucidate the factors driving the model's decisions, offering transparency and supporting financial analysts in understanding and validating model outputs.

(v) While LIME is a potent tool for local interpretability, it is essential to note that it provides simplified approximations and might not capture the entire complexity of the global model. Additionally, the choice of the interpretable model and the perturbation strategy can impact the quality of explanations. Researchers and practitioners using LIME must carefully consider these factors to ensure meaningful and accurate interpretations.

Therefore, LIME stands as a pivotal technique in the XAI landscape, offering a pragmatic approach to understanding the decisions of black-box models. Its versatility and applicability across various domains make it a valuable tool for practitioners seeking transparent and interpretable AI systems in specific, localized contexts. Fig. 9 provides he workflow of LIME.

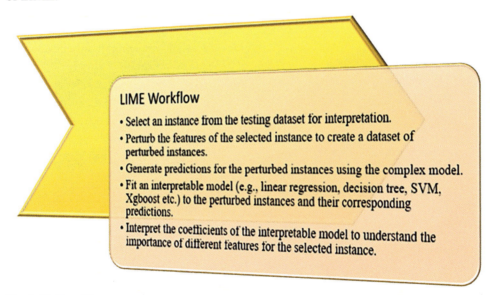

LIME Workflow
- Select an instance from the testing dataset for interpretation.
- Perturb the features of the selected instance to create a dataset of perturbed instances.
- Generate predictions for the perturbed instances using the complex model.
- Fit an interpretable model (e.g., linear regression, decision tree, SVM, Xgboost etc.) to the perturbed instances and their corresponding predictions.
- Interpret the coefficients of the interpretable model to understand the importance of different features for the selected instance.

Fig. 9 LIMEworkflow.

5. Understanding SHAP

SHAP, which stands for Shapely Additive Explanations, is a groundbreaking method in the field of XAI [43]. Developed based on cooperative game theory and named after Shapley values, SHAP offers a robust framework for interpreting complex machine learning models. The central objective of SHAP is to provide a fair and consistent way of attributing contributions to each feature in a model's prediction, thereby enhancing transparency and interpretability [44].

5.1 Key principles of SHAP

1. **Shapley values**: At the core of SHAP lies the concept of Shapley values, borrowed from cooperative game theory. Shapley values represent a unique way of fairly distributing the contribution of each player in a cooperative game. In the context of machine learning, features are treated as players, and Shapley values allocate the contribution of each feature to the final prediction in a way that is both mathematically rigorous and intuitively appealing.
2. **Coalitional game framework**: SHAP extends the concept of cooperative games to the realm of machine learning by considering all possible combinations (coalitions) of features as potential contributors to the prediction. This approach addresses the challenge of feature interactions and ensures a comprehensive evaluation of each feature's impact within the context of different feature subsets.
3. **Shapley additive explanations**: SHAP values provide additive explanations for individual predictions. This means that the sum of Shapley values assigned to each feature equals the difference between the model's prediction for a specific instance and the average prediction across all instances. This additive property ensures consistency and reliability in attributing contributions to features.
4. **Kernel SHAP**: To compute Shapley values efficiently, SHAP introduces the concept of Kernel SHAP. Kernel SHAP uses a weighted linear regression approach, leveraging a weighted combination of model predictions on different subsets of features. This method ensures a scalable and computationally feasible way of calculating Shapley values, even for complex models.

SHAP, draws upon the concept of Shapley values from cooperative game theory to fairly attribute the contribution of each feature to a machine learning model's prediction. For example, in a credit scoring model, SHAP assigns Shapley values to features like income and credit history, indicating their impact on predicting creditworthiness. By considering all possible feature combinations, SHAP's coalitional game framework comprehensively evaluates each feature's influence, addressing interactions between features. The additive property of SHAP values ensures consistency, with the sum of Shapley values equating to the difference between a specific prediction and the model's average prediction. Furthermore, SHAP introduces Kernel SHAP, a method that efficiently computes Shapley values by using a weighted linear regression approach. This enables scalable and computationally feasible interpretation of feature contributions, even in complex models.

5.2 Applications of SHAP

(i) SHAP is particularly powerful in explaining predictions made by black-box models, such as deep neural networks or ensemble methods. By assigning Shapley values to each feature, SHAP unveils the contribution of individual features to the model's output, enabling users to understand the decision-making process.

(ii) SHAP offers a refined approach to feature importance analysis. Rather than providing a static ranking of features, SHAP values capture the dynamic influence of each feature based on the context of other features. This results in a more accurate and context-aware assessment of feature importance.

(iii) SHAP has been applied to assess and mitigate biases in machine learning models. By providing insights into the contribution of each feature, SHAP facilitates the identification of discriminatory factors, promoting fairness and equity in model predictions.

(iv) SHAP aids in model debugging by revealing unexpected or counterintuitive feature contributions. It serves as a valuable tool for validating models, ensuring that predictions align with domain knowledge and expectations.

(v) While SHAP is a powerful tool for interpretability, users must be mindful of certain challenges. In particular, the interpretation of high-dimensional data and the potential for complex interactions among features may pose challenges in applying SHAP effectively. Additionally, users should be cautious in interpreting Shapley values in isolation, as the context of feature interactions is crucial for accurate interpretation.

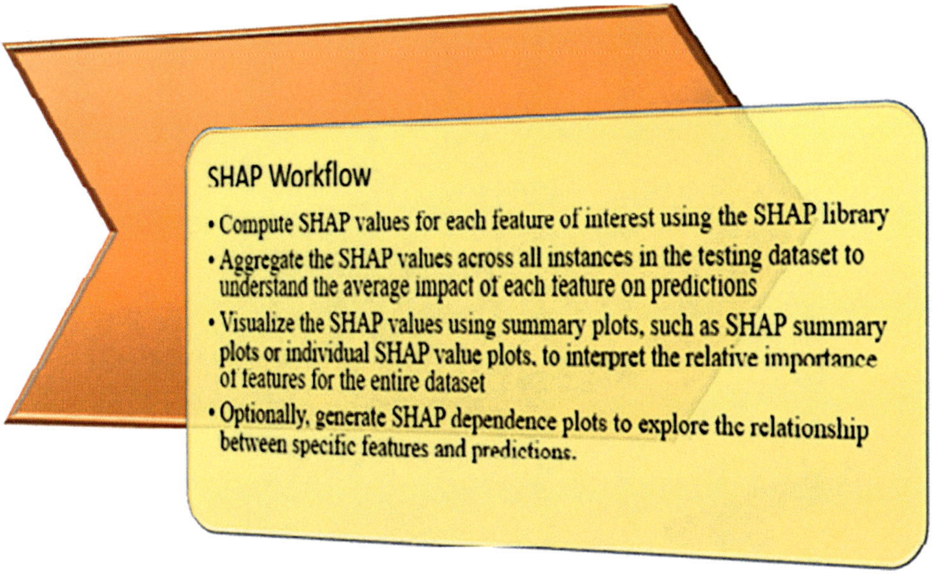

SHAP Workflow

- Compute SHAP values for each feature of interest using the SHAP library
- Aggregate the SHAP values across all instances in the testing dataset to understand the average impact of each feature on predictions
- Visualize the SHAP values using summary plots, such as SHAP summary plots or individual SHAP value plots, to interpret the relative importance of features for the entire dataset
- Optionally, generate SHAP dependence plots to explore the relationship between specific features and predictions.

Fig. 10 SHAP workflow.

Fig. 10 provides the workflow for SHAP, which represents a significant advancement in the field of XAI, providing a principled and versatile framework for understanding complex machine learning models. Its application in diverse domains, from model interpretation to fairness evaluation, underscores its importance in creating transparent and accountable AI systems [45]. As the field of XAI continues to evolve, SHAP stands out as a cornerstone in fostering trust and comprehension in the age of sophisticated machine learning models.

6. A case study on seizures using EEG data

The diagnosis and treatment of epileptic seizures present considerable challenges, and the integration of XAI has demonstrated potential in enhancing patient outcomes. However, the intricate nature of ML models utilized for seizure prediction and detection necessitates transparent and interpretable explanations. This case study delves into the application of XAI techniques, specifically LIME and SHAP, to unravel the predictions of a seizure prediction model. In this investigation, the Bangalore EEG Epilepsy Dataset (BEED) was employed, encompassing EEG recordings from both seizure patients and healthy subjects, featuring 16 EEG channels. The ML model, such as Xgboost, was chosen for classification, aiming to predict the likelihood of a seizure based on EEG signals. Despite the model's high predictive accuracy, comprehending the factors that contribute to its decisions is paramount to garnering trust from medical practitioners and ensuring the secure and effective implementation of clinical strategies.

6.1 LIME analysis

LIME is employed to provide local interpretations for individual predictions made by the seizure prediction model. Fig. 11 illustrates the LIME functionality.

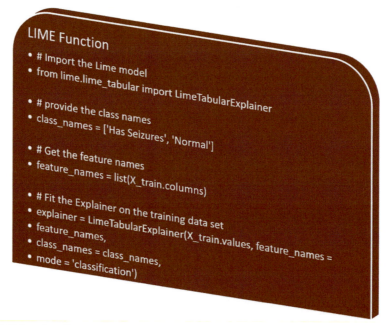

```
LIME Function
• # Import the Lime model
• from lime.lime_tabular import LimeTabularExplainer

• # provide the class names
• class_names = ['Has Seizures', 'Normal']

• # Get the feature names
• feature_names = list(X_train.columns)

• # Fit the Explainer on the training data set
• explainer = LimeTabularExplainer(X_train.values, feature_names =
• feature_names,
• class_names = class_names,
• mode = 'classification')
```

Fig. 11 LIME functionality.

A sample EEG recording that resulted in a positive seizure prediction was selected. LIME generated a surrogate interpretable model, such as a linear model, trained on perturbed instances of the original EEG data. The interpretable model aimed to mimic the behavior of the black-box model locally. The LIME interpretation revealed specific patterns and features in the EEG data that strongly influenced the model's prediction for a seizure. Clinicians could now grasp the relevance of certain signal patterns, such as high-frequency spikes or abnormal waveforms, in the context of a positive seizure prediction. This local interpretability facilitated a more transparent understanding of the model's decision for this particular instance. LIME explainer involves two primary steps, importing the LIME module, and fitting the explainer with the training data and targets. In this process, the model is configured for classification, aligning with the specific task at hand [46]. LIME results for seizure classification typically include three key components as illustrated in Fig. 12.

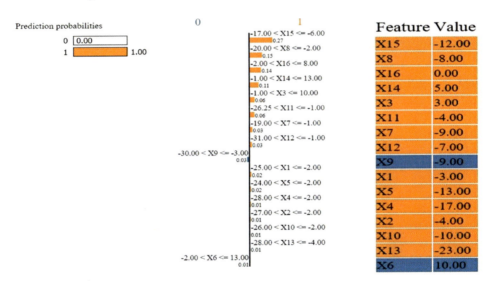

Fig. 12 LIME explanations.

In LIME (Local Interpretable Model-agnostic Explanations), feature values are calculated based on the perturbed instances generated around the instance of interest.

1. Model's predictions: This refers to the output of the ML model for a specific instance, indicating whether it predicts a seizure or not. The prediction is a binary outcome, where the model assigns a probability or a label (e.g., "Seizure" as class 1 and "non-seizure" as class 0) based on the input features.

2. Feature contributions: LIME provides an explanation for the model's prediction by highlighting the contributions of individual features. It identifies the impact of each

feature on the model's decision for the given instance. Positive and negative contributions indicate how each feature influences the prediction toward or against the seizure classification. The coefficients or feature weights of the interpretable model represent the importance of each feature in predicting the target outcome. These coefficients indicate how much each feature contributes to the model's predictions for the instance of interest.

3. Actual value for each feature: This component involves displaying the actual values of the features for the specific instance under consideration. Knowing the values of input features allows a deeper understanding of why the model made a particular prediction. For seizure classification, relevant features include EEG signals data that the model uses to make predictions. The feature values in the LIME output represent the coefficients or weights assigned to each feature by the interpretable model. Higher positive values indicate features that positively influence the prediction, while lower or negative values suggest features that have a negative impact.

Examining Fig. 12 reveals a prediction of seizures with a high confidence level of 100%. The model arrived at this decision due to specific patient feature values such as X1, X2, X3...X16, notably, X9 being less than −3, and X6 being less than 13. The interpretation of these features is substantiated by referring to the accompanying table on the right, where the actual values for each feature can be verified. This detailed analysis helps elucidate the rationale behind the model's prediction, connecting the observed feature values to the high confidence in predicting the occurrence of seizures in the given instance.

LIME calculates feature values by fitting an interpretable model to perturbed instances and their corresponding predictions, with the feature weights indicating the importance of each feature in explaining the model's behavior for the instance of interest. Hence, LIME results for seizure classification offer insights into the model's predictions, break down the contributions of individual features, and present the actual values of these features for a given instance. This interpretability is crucial in understanding the decision-making process of the model and can aid in building trust and confidence in the model's predictions, especially in critical applications like healthcare.

6.2 SHAP analysis

SHAP is then applied to provide a holistic understanding of feature importance across multiple instances. SHAP values were calculated for each feature in the EEG data, considering all possible combinations of features. Fig. 13 illustrates the SHAP functionality.

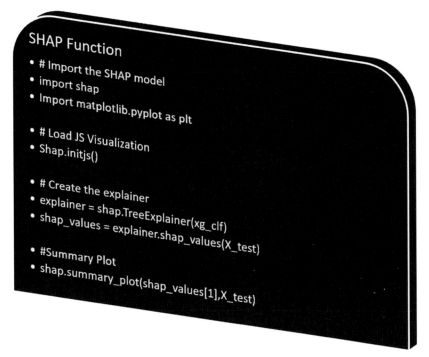

Fig. 13 SHAP functionality.

The resulting SHAP values highlighted the individual contributions of each feature to the overall seizure prediction, considering interactions and dependencies among features. The SHAP analysis demonstrated that certain frequency components and spatial patterns in the EEG data had consistently high Shapley values across multiple instances, indicating their significant influence on seizure predictions. This global interpretability offered valuable insights into the model's general decision-making process, aiding clinicians in identifying consistent biomarkers for seizure risk. SHAP analysis for seizure prediction comprises of two different plots such as Variable Importance Plot and Summary plot.

1. **Variable importance plot**: The Variable Importance Plot in SHAP aims to highlight the significance of each feature in influencing the model's predictions. Features with higher absolute SHAP values contribute more substantially to the model's output. Positive SHAP values push the prediction in favor of the target class, while negative values push it against the target class. The longer the bars in the plot, the more influential the corresponding features are in determining the model's predictions. Fig. 14 illustrates the variable importance plot.

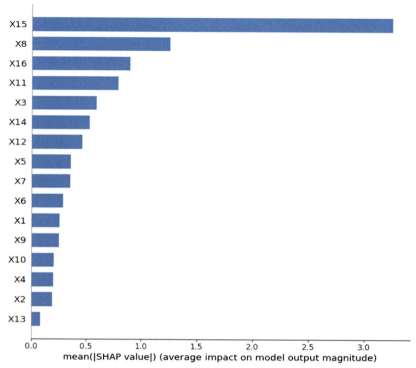

Fig. 14 Variable importance plot.

2. **Summary plot**: The Summary Plot provides an overview of the impact of each feature on a set of predictions.

For each instance in the dataset, the plot displays a dot representing the SHAP value for a specific feature. The color of the dot indicates the feature value (red for high values, blue for low values). The spread of dots along the y-axis shows the distribution of SHAP values, revealing how different feature values contribute to predictions across multiple instances [46]. Fig. 15 illustrates the Summary plot. This plot aids in understanding the consistency and directionality of feature impacts.

Hence, the Variable Importance Plot emphasizes the individual contribution of features to model predictions, while the Summary Plot provides a broader view, illustrating how the impact of features varies across different instances. Together, these visualizations offer valuable insights into the importance and role of each feature in the model's decision-making process.

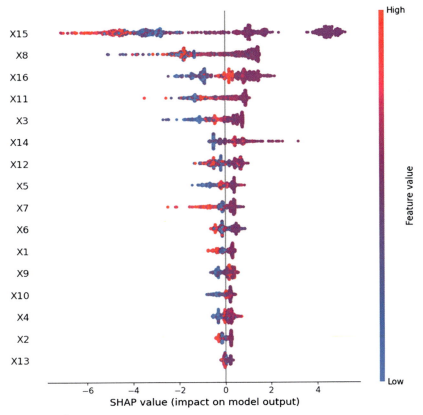

Fig. 15 Summary plot.

In Python, generating Variable Importance Plots and Summary Plots for SHAP analysis can be achieved using the SHAP library's built-in functions. Variable Importance Plots provide a visual representation of feature importance by displaying the magnitude and direction of each feature's impact on model predictions. On the other hand, Summary Plots offers a comprehensive overview of feature contributions across the dataset, showcasing both the average impact and variability of features. While Variable Importance Plots focus on individual feature importance, Summary Plots provide insights into the overall behavior of features, aiding in understanding their collective influence on model predictions. Both plots are valuable tools for interpreting SHAP values and gaining insights into the factors driving model predictions, offering complementary perspectives on feature importance.

6.3 Examples for LIME and SHAP

Suppose we have ML model that predicts house prices based on various features such as square footage, number of bedrooms, and location. We want to understand how the model makes predictions for a specific house. With LIME, we select the specific house we want to interpret and create perturbed instances by randomly perturbing its features within a certain range. We then observe how the model's predictions change for these perturbed instances. Based on this, LIME generates an interpretable model (e.g., linear regression, SVM, XGBoost, etc.) that approximates the behavior of the complex model locally around the selected house. This interpretable model helps us understand which features contribute the most to the prediction for that particular house. Using SHAP, we can calculate the SHAP values for each feature of the house. For instance, if the model predicts a higher price for a house with a larger square footage, SHAP will attribute a positive SHAP value to the square footage feature, indicating its positive impact on the prediction. Similarly, if the model predicts a lower price for a house in a particular location, SHAP will assign a negative SHAP value to the location feature, highlighting its negative influence on the prediction.

The key difference between SHAP and LIME lies in their approach to interpretability. SHAP provides both local and global interpretability by attributing feature importance to individual predictions as well as across the entire dataset. It offers a holistic view of feature contributions, aiding in understanding model behavior at both local and global levels. On the other hand, LIME focuses on creating local approximations of black-box models, providing interpretable explanations for individual predictions only. It offers insights into how features influence specific predictions within a localized context. In summary, SHAP offers a broader scope of interpretability, while LIME provides more localized insights into model predictions.

7. Key insights from the study

The integration of XAI techniques, specifically LIME and SHAP, in interpreting a seizure prediction model, offers numerous benefits. This case study underscores the importance of these techniques in understanding complex ML models for epileptic seizure prediction. By providing both local and global explanations, LIME and SHAP empower clinicians with valuable insights into the decision-making process of the model. Clinicians gain transparency into the black-box model, comprehending the critical features and patterns influencing specific predictions. This interpretability not only enhances trust in the AI system but also enables medical professionals to validate model predictions against their domain expertise. The study has revealed consistent biomarkers through LIME and SHAP analysis, potentially contributing to a deeper understanding of epileptic seizure mechanisms.

This discovery may pave the way for targeted interventions and personalized treatment plans for patients with epilepsy. As the healthcare industry increasingly embraces AI, the application of XAI in critical domains like epilepsy is poised to have a lasting impact on patient care and treatment strategies. Furthermore, the study delves into the nuances of LIME and SHAP, emphasizing their roles in providing localized expla-nations and offering a comprehensive view of feature significance. The transparent and interpretable nature of XAI is crucial for fostering trust, validating model predictions, and advancing the understanding of epileptic seizures in the medical community. The study also underscores ethical considerations in the application of XAI in healthcare, ensuring responsible and ethically sound practices. In essence, the integration of LIME and SHAP not only enhances the transparency of the seizure classification model but also provides actionable insights for future refinements, contributing to the responsible deployment of AI in clinical settings [47]. Table 2 describes the characteristics of LIME and SHAP used in this research study.

Table 2 Comparative characteristics of LIME and SHAP.

Characteristic	LIME	SHAP
Interpretability	LIME focuses on creating local approximations of black-box models, providing interpretable explanations for individual predictions. It offers insights into how features influence specific predictions, enhancing local interpretability	SHAP provides both local and global interpretability by attributing feature importance to individual predictions as well as across the entire dataset. It offers a holistic view of feature contributions, aiding in understanding model behavior at both local and global levels
Model agnosticism	LIME is model-agnostic, meaning it can be applied to any machine learning model without relying on specific assumptions about model architecture	Similarly, SHAP is also model-agnostic, making it compatible with various machine learning models, including black-box and white-box models
Computation complexity	LIME's local approximation approach involves generating perturbed instances and fitting interpretable models, which can be computationally intensive, especially for large datasets or complex models	SHAP's computation complexity depends on the method used for calculating SHAP values. Kernel SHAP, for example, offers efficient computation of SHAP values for complex models, making it suitable for large datasets and computationally demanding scenarios
Scope of interpretability	LIME provides insights into the importance of features for individual predictions, offering local interpretability within a specific instance's context	SHAP offers both local and global interpretability, allowing users to understand feature contributions across the entire dataset and their impact on individual predictions

Continued

Table 2 Comparative characteristics of LIME and SHAP.—cont'd

Characteristic	LIME	SHAP
Ease of use	LIME is relatively straightforward to implement and understand, making it accessible to users with basic machine learning knowledge	SHAP requires a deeper understanding of concepts such as Shapley values and cooperative game theory, which may pose a learning curve for some users. However, SHAP libraries provide user-friendly implementations and visualization tools to facilitate interpretation

Therefore, LIME and SHAP offer complementary strengths for seizure classification scenarios. LIME excels in providing local interpretability for individual predictions, while SHAP offers both local and global interpretability, making it suitable for understanding feature contributions across the entire dataset. The choice between LIME and SHAP depends on the specific requirements of the seizure classification task, considering factors such as interpretability needs, computational resources, and user expertise.

8. Wrapping up and looking ahead

As the healthcare industry continues to integrate AI into various facets of diagnosis, treatment, and patient care, the demand for transparency and interpretability in AI models becomes increasingly crucial. XAI has emerged as a transformative field, providing solutions to demystify complex machine learning models and foster trust among healthcare professionals, patients, and regulatory bodies. As we wrap up current achievements and challenges in XAI for healthcare, it's essential to reflect on the progress made and consider the future directions for this evolving field.

   (i) **Improved trust and adoption**: XAI has played a pivotal role in enhancing the trustworthiness of AI models in healthcare. By providing understandable explanations for model predictions, clinicians and healthcare practitioners are more likely to adopt and integrate AI technologies into their workflows.

  (ii) **Validation and compliance**: XAI has facilitated the validation of AI models in healthcare settings. Regulatory bodies often require transparency in decision-making processes, and XAI techniques such as LIME and SHAP have enabled model validation, ensuring compliance with healthcare standards and regulations.

 (iii) **Enhanced clinician understanding**: XAI has empowered clinicians to better understand the predictions made by AI models. By offering insights into the features and factors influencing a particular decision, XAI facilitates collaboration between machine intelligence and human expertise in delivering optimal patient care.

(iv) Identification of biomarkers: XAI has contributed to the identification of significant biomarkers and features within medical data. This has implications for early disease detection, personalized treatment plans, and a deeper understanding of the physiological factors contributing to specific medical conditions.

(v) Reduced bias and disparities: XAI has been instrumental in identifying and mitigating biases within healthcare AI models. By providing explanations for predictions, XAI helps ensure fairness and equity in the delivery of healthcare services, reducing the potential for biases that may disproportionately impact certain demographic groups.

9. Challenges and opportunities

(i) The increasing complexity of AI models, particularly deep learning models, poses challenges for interpretability. Future research in XAI should focus on developing techniques that effectively explain the decisions of intricate models while maintaining accuracy.

(ii) As healthcare datasets continue to grow in size and complexity, scalability becomes a critical consideration for XAI techniques. Future developments should address the scalability of interpretability methods to handle large and diverse healthcare datasets.

(iii) The success of XAI in healthcare relies on effective collaboration between computer scientists, clinicians, ethicists, and other stakeholders. Future endeavors should encourage interdisciplinary research to address the unique challenges and requirements of the healthcare domain.

(iv) Future applications of XAI in healthcare should prioritize patient-centric approaches. Providing patients with understandable explanations for AI-driven decisions promotes shared decision-making and fosters a sense of empowerment and trust in AI technologies.

In the future, researchers are expected to work on making XAI more advanced and specific to different medical fields, tailoring methods for various specialties and diagnostic tools. It's important for these techniques to seamlessly fit into existing clinical workflows, ensuring they complement current practices and provide understandable insights without disrupting healthcare efficiency. Ethical considerations, such as privacy and security, will continue to be a top priority in the development and use of XAI in healthcare. Future efforts should focus on creating ethical frameworks to guide the responsible deployment of AI technologies. Patient education about the role of AI in healthcare is crucial, and clear communication strategies should be implemented to help patients understand and accept AI-driven interventions. Additionally, as the field progresses, establishing regulatory standards for XAI in healthcare will be essential. Collaboration between regulatory bodies, researchers, and industry stakeholders is necessary to develop guidelines that ensure the responsible and ethical use of XAI in medical settings.

As AI progresses and becomes more intricate, the task of comprehending and inter-preting algorithms to discern their decision-making processes is growing more intricate. This complexity prompts researchers to explore novel approaches and refine existing ones continually. Many XAI models necessitate simplifying the underlying model, often resulting in a trade-off with predictive performance. Furthermore, current explainability methods may not encompass all facets of the decision-making process, particularly in dealing with more intricate models, thereby limiting the explanatory benefits. Emerging research methodologies are concentrating on improving explainable AI techniques by devising more efficient algorithms that not only address ethical concerns but also generate user-friendly explanations.

In conclusion, this study introduces a groundbreaking approach to EEG-based seizure classification, employing the Xgboost classifier within a comprehensive framework. The incorporation of XAI techniques, namely LIME and SHAP, addresses the crucial need for transparency in model decision-making processes. By offering both localized inter-pretability through LIME and a global perspective on feature importance through SHAP, our binary classification model not only demonstrates high accuracy but also provides invaluable insights into the intricate patterns of EEG data. This transformative integration of XAI techniques marks a paradigm shift in epilepsy diagnostics, enhancing clinician decision-making and ultimately improving patient care and outcomes. Further, ongoing research endeavors are poised to yield more sophisticated methods that enhance transpar-ency, trustworthiness, and fairness in the field of artificial intelligence.

XAI transcends the confines of the medical sphere, finding indispensable utility across a multitude of domains. In legal contexts, XAI illuminates intricate legal documents and elucidates the factors shaping judicial decisions, ensuring transparency and fairness in legal proceedings. Similarly, in finance and banking, XAI empowers risk assessment and fraud detection while fostering trust in financial decision-making by providing clear insights into the rationale behind credit scoring and investment recommendations. Moreover, XAI enhances customer service and marketing endeavors by elucidating personalized recommendations and shedding light on customer behavior patterns, thereby refining service offerings and optimizing marketing strategies. In the educational arena, XAI facil-itates personalized learning experiences, predicts student performance, and informs cur-riculum development through comprehensive explanations of learning patterns and teaching efficacy. Furthermore, XAI plays a pivotal role in manufacturing processes by optimizing operations, predicting equipment failures, and streamlining supply chain management, thus driving efficiency and cost-effectiveness. In environmental science, XAI aids in analyzing vast datasets, forecasting climate change impacts, and guiding con-servation efforts by providing interpretable insights into environmental trends and eco-system dynamics. Across these diverse sectors, XAI serves as a cornerstone for transparency, accountability, and trust in decision-making processes, fostering progress and innovation in myriad fields.

Acknowledgment

This work is derived from a research grant funded by the Research, Development, and Innovation Authority (RDIA)—Kingdom of Saudi Arabia—with grant number (13382-psu-2023-PSNU-R-3-1-EI-). The authors would like to thank Prince Sultan University, Riyadh, Saudi Arabia, for support with the article processing charges (APC) of this publication. The authors specially acknowledge the Automated Systems and Soft Computing Lab (ASSCL) at Prince Sultan University, Riyadh, Saudi Arabia.

References

[1] N. McCallan, S. Davidson, K.Y. Ng, P. Biglarbeigi, D. Finlay, B.L. Lan, J. McLaughlin, Epileptic multi-seizure type classification using electroencephalogram signals from the Temple University Hospital Seizure Corpus: a review, Expert Syst. Appl. (2023) 121040.

[2] A. Capurro, L. Diambra, D. Lorenzo, O. Macadar, M.T. Martin, C. Mostaccio, et al., Human brain dynamics: the analysis of EEG signals with Tsallis information measure, Phys. A: Stat. Mech. Appl. 265 (1–2) (1999) 235–254.

[3] A.M. Anter, A.T. Azar, N. El-Bendary, A.E. Hassanien, M.A. El Soud, Automatic computer aided segmentation for liver and hepatic lesions using hybrid segmentations techniques, in: 2013 Federated Conference on Computer Science and Information Systems (FedCSIS), Kraków, Poland, September 8-11, 2013, 2013.

[4] E. Emary, H. Zawbaa, A.E. Hassanien, G. Schaefer, A.T. Azar, Retinal blood vessel segmentation using bee colony optimisation and pattern search, in: IEEE 2014 International Joint Conference on Neural Networks (IJCNN 2014), July 6–11, Beijing International Convention Center, Beijing, China, 2014.

[5] E. Emary, H. Zawbaa, A.E. Hassanien, G. Schaefer, A.T. Azar, Retinal vessel segmentation based on possibilistic fuzzy c-means clustering optimised with cuckoo search, in: IEEE 2014 International Joint Conference on Neural Networks (IJCNN 2014), July 6–11, Beijing International Convention Center, Beijing, China, 2014.

[6] G. Jothi, H.H. Inbarani, A.T. Azar, Hybrid tolerance rough set: PSO based supervised feature selection for digital mammogram images, Int. J. Fuzzy Syst. Appl. 3 (4) (2013) 15–30.

[7] H.I. Elshazly, A.M. Elkorany, A.E. Hassanien, A.T. Azar, Ensemble classifiers for biomedical data: performance evaluation, in: IEEE 8th International Conference on Computer Engineering & Systems (ICCES), 26 Nov-28 Nov 2013, Ain Shams University, 2013, pp. 184–189, ISBN: 978-1-4799-0078-7, https://doi.org/10.1109/ICCES.2013.6707198.

[8] A. ASA, A.E. Hassanien, A.T. Azar, S.E. Hanafy, Genetic algorithm with different feature selection techniques for anomaly detectors generation, in: 2013 Federated Conference on Computer Science and Information Systems (FedCSIS), Kraków, Poland, September 8-11, 2013, 2013.

[9] H.H. Inbarani, P.K.N. Banu, A.T. Azar, Feature selection using swarm-based relative reduct technique for fetal heart rate, Neural Comput. Appl. 25 (3–4) (2014) 793–806. Springer https://doi.org/10.1007/s00521-014-1552-x.

[10] A.T. Azar, S.A. El-Said, A.E. Hassanien, Fuzzy and hard clustering analysis for thyroid disease, Comput. Methods Prog. Biomed. 111 (1) (2013) 1–16.

[11] P.K.N. Banu, A.T. Azar, H.H. Inbarani, Fuzzy firefly clustering for tumor and cancer analysis, Int. J. Model. Identif. Control 27 (2) (2017) 92–103.

[12] T. Ashfaq, R. Khalid, A.S. Yahaya, S. Aslam, A.T. Azar, S. Alsafari, I.A. Hameed, A machine learning and blockchain based efficient fraud detection mechanism, Sensors 22 (19) (2022) 7162, https://doi.org/10.3390/s22197162.

[13] F. Ajeil, I.K. Ibraheem, A.T. Azar, A.J. Humaidi, Autonomous navigation and obstacle avoidance of an omnidirectional mobile robot using swarm optimization and sensors deployment, Int. J. Adv. Robot. Syst. 17 (3) (2020) 1–15, https://doi.org/10.1177/1729881420929498.

[14] A.A. Najm, I.K. Ibraheem, A.T. Azar, A.J. Humaidi, Genetic optimization-based consensus control of multi-agent 6-DoF UAV system, Sensors 20 (12) (2020) 3576, https://doi.org/10.3390/s20123576.

[15] A.A. Al-Qassar, A.I. Abdulkareem, A.F. Hasan, A.J. Humaidi, I.K. Ibraheem, A.T. Azar, A.H. Hameed, Grey-wolf optimization better enhances the dynamic performance of roll motion for tail-sitter VTOL aircraft guided and controlled by STSMC, J. Eng. Sci. Technol. 16 (3) (2021) 1932–1950.

[16] A.J. Humaidi, H.T. Najem, A.Q. Al-Dujaili, D.A. Pereira, I.K. Ibraheem, A.T. Azar, Social spider optimization algorithm for tuning parameters in PD-like interval type-2 fuzzy logic controller applied to a parallel robot, Meas. Control 54 (3–4) (2021) 303–323, https://doi.org/10.1177/0020294021997483.

[17] B.P. Evans, B. Xue, M. Zhang, What's inside the black-box? A genetic programming method for interpreting complex machine learning models, in: Proceedings of the Genetic and Evolutionary Computation Conference, 2019, July, pp. 1012–1020.

[18] D. Raab, A. Theissler, M. Spiliopoulou, XAI4EEG: spectral and spatio-temporal explanation of deep learning-based seizure detection in EEG time series, Neural Comput. Appl. 35 (14) (2023) 10051–10068.

[19] T. Miller, Explanation in artificial intelligence: insights from the social sciences, Artif. Intell. 267 (2019) 1–38.

[20] G. Atteia, E.M. El-Kenawy, N.A. Samee, M.M. Jamjoom, A. Ibrahim, A.A. Abdelhamid, A.T. Azar, N. Khodadadi, R.A. Ghanem, M.Y. Shams, Adaptive dynamic dipper throated optimization for feature selection in medical data, Comput. Mater. Continua 75 (1) (2023) 1883–1900.

[21] H.H. Inbarani, A.T. Azar, G. Jothi, Leukemia image segmentation using a hybrid histogram-based soft covering rough K-means clustering algorithm, Electronics 9 (1) (2020) 188, https://doi.org/10.3390/electronics9010188.

[22] S.S. Kumar, H.H. Inbarani, A.T. Azar, A.E. Hassanien, Rough set-based meta-heuristic clustering approach for social E-learning systems, Int. J. Intell. Eng. Inf. 3 (1) (2015) 23–41.

[23] A. Koubaa, A. Ammar, M. Alahdab, A. Kanhouch, A.T. Azar, DeepBrain: experimental evaluation of cloud-based computation offloading and edge computing in the internet-of-drones for deep learning applications, Sensors 20 (18) (2020) 1–25, https://doi.org/10.3390/s20185240. 5240.

[24] H.H. Ammar, A.T. Azar, R. Shalaby, M.I. Mahmoud, Metaheuristic optimization of fractional order incremental conductance (FO-INC) maximum power point tracking (MPPT), Complexity 2019 (2019) 1–13, https://doi.org/10.1155/2019/7687891. Article ID: 7687891.

[25] A.T. Azar, A.E. Hassanien, T.H. Kim, Expert system based on neural-fuzzy rules for thyroid diseases diagnosis, in: International Conference on Bio-Science and Bio-Technology (BSBT 2012), December 16-19, 2012, Korea. Vol. 353 of the Communications in Computer and Information Science Series, Springer, 2012, pp. 94–105, ISBN: 978-3-642-35520-2, https://doi.org/10.1007/978-3-642-35521-9_13.

[26] T.S. Gorripotu, H. Samalla, C. Jagan Mohana Rao, A.T. Azar, D. Pelusi, TLBO algorithm optimized fractional-order PID controller for AGC of interconnected power system, in: Soft Computing in Data Analytics. Advances in Intelligent Systems and Computing, vol. 758, Springer, Singapore, 2019.

[27] J. Gerlings, M.S. Jensen, A. Shollo, Explainable AI, but explainable to whom? An exploratory case study of XAI in healthcare, in: Handbook of Artificial Intelligence in Healthcare: Vol 2: Practicalities and Prospects, Springer International Publishing, Switzerland, 2022, pp. 169–198.

[28] C. Manresa-Yee, M.F. Roig-Maimó, S. Ramis, R. Mas-Sansó, Advances in XAI: explanation interfaces in healthcare, in: Handbook of Artificial Intelligence in Healthcare: Vol 2: Practicalities and Prospects, Springer International Publishing, Cham, 2021, pp. 357–369.

[29] A. Holzinger, C. Biemann, C.S. Pattichis, D.B. Kell, What do we need to build explainable AI systems for the medical domain?, 2017. *arXiv preprint arXiv:1712.09923.*

[30] D. Wang, Q. Yang, A. Abdul, B.Y. Lim, Designing theory-driven user-centric explainable AI, in: Proceedings of the 2019 CHI Conference on Human Factors in Computing Systems, 2019, May, pp. 1–15.

[31] P. Rathod, S. Naik, Review on epilepsy detection with explainable artificial intelligence, in: 2022 10th International Conference on Emerging Trends in Engineering and Technology-Signal and Information Processing (ICETET-SIP-22), IEEE, 2022, April, pp. 1–6.

[32] E. Tjoa, C. Guan, A survey on explainable artificial intelligence (XAI): toward medical XAI, IEEE Trans. Neural Networks Learn. Syst. 32 (11) (2020) 4793–4813.

[33] R. Calegari, G. Ciatto, J. Dellaluce, A. Omicini, Interpretable narrative explanation for ML predictors with LP: a case study for XAI, in: CEUR Workshop Proceedings, vol. 2404, Sun SITE Central Europe, RWTH Aachen University, 2019, pp. 105–112.

[34] M. Rashed-Al-Mahfuz, M.A. Moni, S. Uddin, S.A. Alyami, M.A. Summers, V. Eapen, A deep convolutional neural network method to detect seizures and characteristic frequencies using epileptic electroencephalogram (EEG) data, IEEE J. Transl. Eng. Health Med. 9 (2021) 1–12.

[35] A.M. Antoniadi, Y. Du, Y. Guendouz, L. Wei, C. Mazo, B.A. Becker, C. Mooney, Current challenges and future opportunities for XAI in machine learning-based clinical decision support systems: a systematic review, Appl. Sci. 11 (11) (2021) 5088.

[36] R. Dwivedi, D. Dave, H. Naik, S. Singhal, R. Omer, P. Patel, et al., Explainable AI (XAI): core ideas, techniques, and solutions, ACM Comput. Surv. 55 (9) (2023) 1–33.

[37] A.B. Arrieta, N. Díaz-Rodríguez, J. Del Ser, A. Bennetot, S. Tabik, A. Barbado, et al., Explainable artificial intelligence (XAI): concepts, taxonomies, opportunities and challenges toward responsible AI, Inf. Fusion 58 (2020) 82–115.

[38] W. Saeed, C. Omlin, Explainable AI (XAI): a systematic meta-survey of current challenges and future opportunities, Knowl.-Based Syst. 263 (2023) 110273.

[39] U. Pawar, D. O'Shea, S. Rea, R. O'Reilly, Incorporating Explainable Artificial Intelligence (XAI) to Aid the Understanding of Machine Learning in the Healthcare Domain, AICS, 2020, December, pp. 169–180.

[40] S.E. Sánchez-Hernández, R.A. Salido-Ruiz, S. Torres-Ramos, I. Román-Godínez, Evaluation of feature selection methods for classification of epileptic seizure EEG signals, Sensors 22 (8) (2022) 3066.

[41] D.T. Barus, F. Masri, A. Rizal, NGBoost interpretation using LIME for alcoholic EEG signal based on GLDM feature extraction, in: Proceedings of the Computational Methods in Systems and Software, Springer International Publishing, Cham, 2020, pp. 894–904.

[42] E.H. Bijoy, M.H. Rahman, S. Ahmed, M.S. Laskor, An Approach to Detect Epileptic Seizure Using XAI and Machine Learning (Doctoral dissertation), Brac University, 2022.

[43] J.C. Vieira, L.A. Guedes, M.R. Santos, I. Sanchez-Gendriz, Using explainable artificial intelligence to obtain efficient seizure-detection models based on electroencephalography signals, Sensors 23 (24) (2023) 9871.

[44] V. Gabeff, T. Teijeiro, M. Zapater, L. Cammoun, S. Rheims, P. Ryvlin, D. Atienza, Interpreting deep learning models for epileptic seizure detection on EEG signals, Artif. Intell. Med. 117 (2021) 102084.

[45] A. Tahmassebi, J. Martin, A. Meyer-Baese, A.H. Gandomi, An interpretable deep learning framework for health monitoring systems: a case study of eye state detection using EEG signals, in: 2020 IEEE Symposium Series on Computational Intelligence (SSCI) (pp., IEEE, 2020, December, pp. 211–218.

[46] J. Breitenbach, D. Raab, E. Fezer, D. Sauter, H. Baumgartl, R. Buettner, Automatic diagnosis of intellectual and developmental disorder using machine learning based on resting-state EEG recordings, in: 2021 17th International Conference on Wireless and Mobile Computing, Networking and Communications (WiMob), IEEE, 2021, October, pp. 7–12.

[47] A. Pradhan, S. Prabhu, K. Chadaga, S. Sengupta, G. Nath, Supervised learning models for the preliminary detection of COVID-19 in patients using demographic and epidemiological parameters, Information 13 (7) (2022) 330.

CHAPTER 9

Revolutionizing cancer diagnosis with AI-enhanced histopathology and deep learning: A study on enhanced image analysis and model explainability

Seema Kashyap[a], Arvind Kumar Shukla[a], and Iram Naim[b]
[a]School of Computer Science and Engineering, IFTM University, Moradabad, Uttar Pradesh, India
[b]Faculty of Engineering and Technology, MJP Rohilkhand University, Bareilly, India

1. Introduction

An estimated 238,340 newly diagnosed cases and 127,070 fatalities of lung cancer are projected by the American Cancer Society for 2023, constituting a significant public health concern in the United States. It remains the leading cause of cancer-related mortality, responsible for almost three times the number of fatalities compared to colorectal cancer. The projected overall cancer mortality in the United States is anticipated to reach 609,820, with lung and bronchus cancer accounting for the highest number of deaths. Colorectal cancer is the third most common kind of cancer worldwide, accounting for more than 10% of all cancer cases. Additionally, it is the second leading factor contributing to cancer-related deaths globally.

A histopathological study not only aids in the confirmation of the cancer diagnosis but also yields details on the kind, grade, and stage of the disease. In addition, molecular testing can be conducted to discover precise genetic alterations that might inform the use of targeted medicines. The histopathological results are essential in identifying the optimal treatment regimens for individuals with lung and colon cancer. Deep learning algorithms have gained prominence in the detection of lung and colon cancer in recent years. These algorithms utilize histology slides, X-rays, MRIs, endoscopic pictures, and CT scans. CNNs revolutionized the field of medical image categorization. Traditional and separable convolutional neural networks (CNNs) have distinct operational characteristics. Convolutional neural networks (CNNs) have the capability to identify cancer in medical imagery.

In the course of our study, we have developed a deep-learning model for the early detection of lung and colon cancer by using histopathological data. To fabricate a Deep Learning model capable of precisely classifying nascent forms of colon and lung cancer, we utilize histopathological images acquired from the LC25000 dataset.

Explainable AI in Healthcare Imaging for Medical Diagnoses
https://doi.org/10.1016/B978-0-443-23979-3.00009-9

In the field of Explainable AI (XAI), LIME (Local Interpretable Model-agnostic Explanations) and SHAP (SHapley Additive exPlanations) are widely used approaches to understand complex machine learning models like convolutional neural networks (CNNs). LIME provides localized insights into individual predictions, making it highly suitable for applications such as the analysis of medical imaging for the detection of lung cancer. Conversely, SHAP provides explanations at both the local and global levels, facilitating comprehension of the significance of features over the whole dataset. By integrating LIME with SHAP, researchers may attain a holistic comprehension of a model's decision-making process, enabling model enhancement, bias identification, and overall interpretability. This methodology guarantees openness and reliability, which are essential in applications such as medical diagnostics.

1.1 Contributions

The subsequent section delineates the principal contributions of our paper:

- The suggested model excels in detecting and categorizing lung anomalies, reaching the highest level of performance. The model enhances accuracy and efficiency in comparison with present methods.
- This study creates a sophisticated deep-learning algorithm to detect early stages of lung and colon cancer by analyzing histopathology pictures. Deep learning algorithms are becoming increasingly popular for detecting lung and colon cancer using histology slides and medical photos. Convolutional neural networks (CNNs) have transformed medical picture classification, although their opaque nature hinders interpretability.
- LIME and SHAP are utilized for model interpretation to comprehend convolutional neural network predictions. LIME offers specific insights into individual forecasts that are appropriate for medical imaging. SHAP provides explanations at both the local and global levels to comprehend the overall value of features.
- Integrating LIME and SHAP allows for a thorough comprehension of model decisions. These methods guarantee transparency and dependability in medical diagnostics by ensuring model interpretability.
- Overall, the study shows that that accuracy in detecting and categorizing lung cancer is greatly improved, surpassing state-of-the-art models.

1.2 Organization

The paper's organization is now summarized as follows: Section 2 talks about past work that used deep learning, machine learning, and XAI to find and sort lung problems. Methodology, including model architecture, dataset, XAI approaches, and preprocessing, is presented in Section 3. In Section 4, the accuracy, sensitivity, and specificity of the proposed model are assessed, along with its comparison to other deep learning models that are already available. Section 5 examining the benefits and drawbacks of the suggested method. In Section 6, we have outlined the main points, consequences, and potential avenues for further study.

2. Related works

Early cancer detection is crucial, especially in low- and middle-income nations where the disease has a high death rate. Patients encounter difficulties in getting vital healthcare treatments due to problems such as insufficient financial resources and undeveloped medical infrastructure. To address these problems and improve patient survival rates, the passage suggests the implementation of early diagnosis programs. These programs are specifically developed to accurately identify and detect the early indications and manifestations of cancer. The initiative's primary objective is to improve the survival rates of cancer patients in low-resource regions by emphasizing early detection. This approach intends to address the inequalities in healthcare access and eventually better results in those with economic disadvantages.

Hadiyoso et al. [1] suggested CNN-based VGG16 architecture and CLAHE classify histopathological cancer pictures into five groups, including lung and colon cancer. The model has 98.96% accuracy, Precision, False Positive Rate, and F1 score in experiments. Early diagnostic findings were 98.96% sensitive and 99.74% specific.

Ibrahim et al. [2] proposed a method to enhance images using DWT and Double-CLAHE stages, resulting in 99.5% accuracy, 99.6% precision, 99.4% recall, and 99.6% F1-score, proving its effectiveness in colon and lung cancer classification.

Bukhari et al. [3] employed the ResNet18, ResNet30, and ResNet50 models to train and develop algorithms for the purpose of detecting colonic cancer. The LC25000 and CRAG image datasets were utilized for this task. They achieved validation accuracies of 93.91%, 93.04%, and 93.91% using three different CNN architectures.

Fatoki et al. [4] deploy a computational model for lung cancer prediction using machine learning. Various algorithms were evaluated, with support vector machine (SVM) proving to be the most effective, indicating its potential for real-world lung cancer prediction.

In their research, Nishio et al. [5] investigated the use of homology-based image processing and texture analysis. The researchers used conventional machine learning methods such as Perceptron, Logistic Regression, KNN, SVM, Decision Tree, Random Forest, and Gradient Tree Boosting. Achieved a precision rate of 78.33% and an accuracy rate of 99.43% for both private and public datasets.

The authors Sikder et al. [6] used pathological image analysis and supervised learning based on deep learning. Identified five discrete tissue categories, with three being carcinogenic and two being non-malignant, inside tumors located in the lungs and colon. In addition, I attained a remarkable F-measure score of 96.38% and a maximum classification accuracy of 96.33% for the detection of lung and colon cancer.

Abbas and Bukhari et al. [7] used many pre-trained convolutional neural networks (namely AlexNet, VGG-19, ResNet-18, ResNet-34, ResNet-50, and ResNet-101) that were trained on the ImageNet dataset. Classified histological slides into categories of benign lung tissue, squamous cell carcinoma, and adenocarcinoma. Achieved outstanding scores of 0.973, 0.997, 0.986, 0.992, 0.999, and 0.999 for the test dataset.

Hatuwal et al. [8] developed a convolutional neural network (CNN) model that achieved high levels of precision. The training accuracy was 96.11% and the validation accuracy was 97.20%.

Bukhari et al. [3] investigated the suitability of three CNN architectures (ResNet-18, ResNet-30, and ResNet-50) for the task of categorizing colonic tissue images. The sensitivity of ResNet-50 is the greatest, measuring at 96.77%. It is followed by ResNet-30 with a sensitivity of 95.74% and ResNet-18 with a sensitivity of 94.79%. More complex models, such as ResNet-50, provide superior performance in categorizing images of colonic tissue, highlighting the impact of architectural depth on the sensitivity of medical image classification.

Dyah Irawati et al. [9] suggest that FFT outperforms IFFT, DWT, and non-sparse methods in terms of precision. KNN classifies the "N" image as having uniquely extracted attributes with 100% accuracy. Adenocarcinoma (ACA) and squamous cell carcinoma (SCC) images had 70% accuracy, showing that extracted features may not be enough to distinguish these cancers. Strong image classifiers like FFT and KNN may assist in identifying cancer types.

A model that is based on DenseNet and includes pooling kernels and three-dimensional filters was proposed by Zhang et al. [10]. In terms of classification accuracy, this model achieved 92.4% on the LUNA16 dataset.

The "Adaptive Hierarchical Heuristic Mathematical Model (AHHMM)" is a method for detecting lung cancer that was proposed by Yu et al. [11]. The suggested method includes capturing images, processing them, converting them to binary, establishing a threshold, dividing them into segments, extracting characteristics, and finally, identifying them with a deep neural network (DNN). Test findings showed that the system has a 96.67% accuracy rate in predicting lung cancer.

A "vesselness filter" was applied to the MSDLF by Zhang et al. [12] to detect large lung nodes accurately and with little error. An LIDC-IDRI dataset-based four-channel CNN model is used for segmentation and classification, along with data preparation, lung contour correction, tissue segmentation, vascular removal, and dataset standardization. Messalas et al. [13] suggest enhancing ML model explanations by fusing the interpretability of surrogate models with the accuracy of Shapley values. They use "Top j Similarity" to assess how consistent various approaches are with one another. Fryer et al. [14] also criticize Explainable AI's (XAI) use of Shapley values for feature selection.

To automatically detect lung cancer in CT scans, Lin et al. [15] introduce a 2D convolutional neural network (CNN) with Taguchi parametric optimization. CNN outperforms AlexNet in terms of accuracy because of the Taguchi method.

Cai et al. [16] used volume rendering with Mask R-CNN to segment 3D lung nodules. Preprocessing, segmentation, detection, and three-dimensional reconstruction are all part of the model. The algorithm achieved an accuracy rate of 88.7% in detecting lung cancer using the LUNA-16 and Ali Tian Chi [17] datasets.

The 3D deep convolutional neural networks (3DDCNN) demonstrated by Masood et al. [18] can detect lung nodules by combining deep learning with cloud computing. The mRPN had a 98.5% success rate in detecting lung cancer. Nohara et al. [19] highlighted how important model interpretability is to machine learning, particularly when it comes to decision-making processes. Using hospital data, they developed the Shapley Additive Explanation (SHAP) approach for interpreting gradient-boosting decision tree models.

The KAMP-Net was demonstrated by Guo et al. [20]. By continuously adding new input, this network trains neural networks with multiple layers. The technique outperforms physicians by 28.1% (AUC = 0.82) when CNN and SVM are combined to assess mortality risk using the National Pulmonary Evaluation Study (NLST) dataset. The results of applying the different machine learning and deep learning models to the datasets are displayed in Table 1.

Table 1 Comparative analysis of various studies.

Author	Image type/dataset image	Number of classes	Method	ACC (%)
Hadiyoso et al. [1]	LC25000 Lung and Colon Histopathology	Five classes	CNN-CLAHE-VGG16	98.96
Ibrahim et al. [2]	LC25000 Lung and Colon Histopathology	Lung and colon classes	DWT and Double-CLAHE	99.5%
Bukhari et al. [3]	LC25000 Lung and Colon Histopathology	Colon dataset	ResNet18, ResNet30, and ResNet50	validation accuracies of 93.91%, 93.04%, and 93.91%
Fatoki et al. [4]	LC25000 Lung and Colon Histopathology	Lung dataset	CNN + SVM	
Nishio et al. 2021 [5]	LC25000 Lung and Colon Histopathology	Colon and pulmonary class	Homology-based techniques combined with machine learning	78.3/99.33
Bukhari et al. [7]	LC25000 Lung and Colon Histopathology	lung classes	CNN-ResNet	99.8
Sikder et al., 2021 [6]	LC25000 Lung and Colon Histopathology	Colon and pulmonary class	CNN	96.33
Hatuwal et al. [8]	LC25000 Lung and Colon Histopathology	Lung classes	CNN	97.2

Continued

Table 1 Comparative analysis of various studies—cont'd

Author	Image type/dataset image	Number of classes	Method	ACC (%)
Khalid Bukhari et al. [3]	LC25000 Lung and Colon Histopathology	Colon classes	CNN–ResNet	93.91
Dyah et al. [9]	LC25000 Lung and Colon Histopathology	Lung classes	Compressive sensing—KNN 88	
Zhang et al. [10]	LUNA16 dataset	Lung classes	DenseNet-based model with 3D filters and pooling kernels	92.4% accurately
Yu et al. [11]	Ct scan dataset	Lung classes	Adaptive Hierarchical Heuristic Mathematical Model (AHHMM)"	96.67% accuracy
Zhang et al. [12]	LIDC-IDRI dataset	Lung classes	MSDLF with a "vesselness filter"	High accuracy
Cai et al. [16]	LUNA-16 and Ali TianChi datasets	Lung classes	Mask R–CNN and volume rendering	88.7% accuracy
Guo et al. [20]		Lung classes	National Pulmonary Evaluation Study (NLST) dataset	AUC = 0.82

3. Proposed methodology

A systematic approach was adopted for this research, which includes the accumulation of datasets, preprocessing, model development, training, and evaluation (Fig. 1). After gathering data, Fig. 2 shows the steps that were taken in the right order, from using explainable AI methods (LIME and SHAP) for local and global reasons to showing the results. This methodology guarantees openness and repeatability in our approach, hence improving the interpretability of our model's predictions.

Label Name	image type	number of samples
Colon_aca	colon Adenocarcinoma	5000
colon_n	colon Benign tissue	5000
Lung_aca	lung Adenocarcinoma	5000
Lung_n	lung Benign tissue	5000
Lung_scc	lung Squamous-cell carcinoma	5000

Fig. 1 Number of each data sample in the LC25000 dataset.

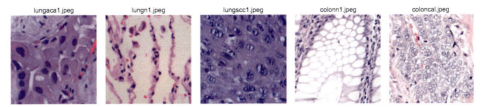

Fig. 2 Sample image from the LC25000 dataset.

3.1 Dataset

The collection consists of 25,000 histopathological pictures, with each image having dimensions of 768 by 768 pixels and being in JPEG format. The dataset was created by Brokowski et al. [21] initially comprised of 750 images depicting lung tissue (250 benign, 250 adenocarcinomas, and 250 squamous cell carcinomas) and 500 images representing colon tissue (250 benign and 250 adenocarcinomas), all of which originated from a validated and HIPAA-compliant source. By employing the Augmenter program for augmentation, the dataset was enlarged to include 25,000 photos. Within the collection, there are five distinguishable categories, each of which has 5000 images:

- Non-cancerous lung tissue
- Lung adenocarcinoma refers to a type of cancer that originates in the glandular cells of the lungs
- Squamous cell cancer of the lung
- Adenocarcinoma of the colon
- Noncancerous tissue of the colon

This dataset is ideal for histopathological image classification tasks, especially for lung and colon tissue, due to its balanced class distribution and various representations. It offers a wonderful asset for instruction and education.

3.2 Data preprocessing

Image preprocessing function:
- The scalar function is designed to preprocess an input image.
- It converts the image from RGB to LAB color space.
- It separates the LAB channels into L (luminance), A (green to magenta), and B (blue to yellow).
- Histogram equalization is applied to the L channel to enhance the contrast of the image.
- Contrast Limited Adaptive Histogram Equalization (CLAHE) is applied to the L channel for further contrast enhancement.
- The LAB channels are merged back, and the image is converted back to RGB.
- The pixel values are scaled between −1 and 1.

Data augmentation and generators:
* The ImageDataGenerator from TensorFlow is used for data augmentation.
* The preprocessing function parameter of the generator is set to the scalar function for real-time data preprocessing during training.
* Separate generators (train_gen, test_gen, and valid_gen) are created for the training, testing, and validation datasets.
* The flow_from_dataframe method is used to generate batches of augmented images from Data Frame containing file paths and labels.
* The target size, class mode, color mode, shuffle, and batch size parameters are appropriately configured for each generator.

A model applies color spatial manipulations and data augmentation techniques to the images before classification to classify them using a deep learning pipeline (Figs. 3 and 4). While the model is being trained, it is set up in such a way that the producers can

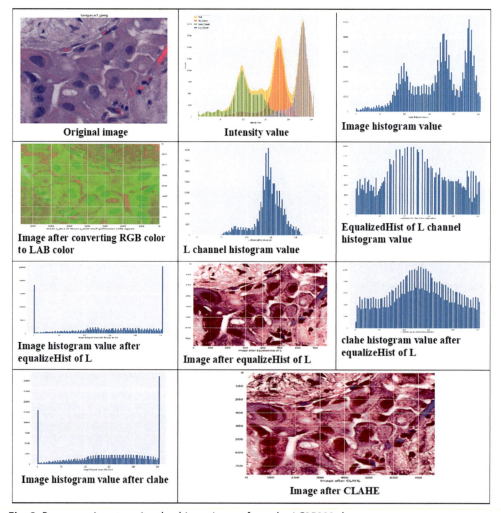

Fig. 3 Preprocessing steps involved in an image from the LC25000 dataset.

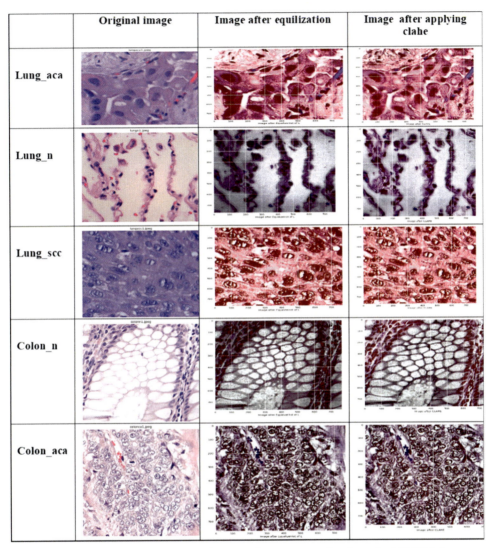

Fig. 4 Original and improved each type of image after preprocessing from the LC25000 dataset.

provide groups of enhanced images for training, testing, and validation. It executes critical image preparation operations in support of a deep learning model. Utilizing the ImageDataGenerator to supplement data, converting color spaces from RGB to LAB, and increasing contrast through histogram equalization and CLAHE are all steps involved in this process. In addition to accelerating the convergence of training, the purpose of these operations is to improve the acquisition of features, as well as to increase the model's durability. In addition, the algorithm creates data generators for training, testing, and validation, which ensures that the preprocessing is consistent and encourages the generalization of the model. Fig. 5 shows the flow diagram for proposed model. Fig. 6 shows the Technique of explainable artificial intelligence (XAI).

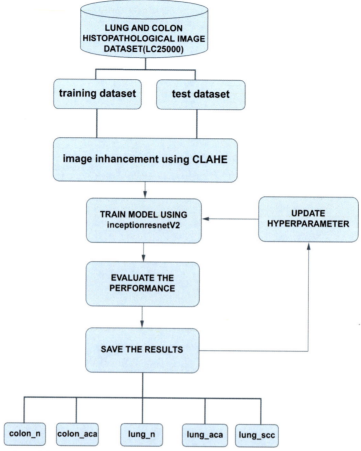

Fig. 5 Proposed model.

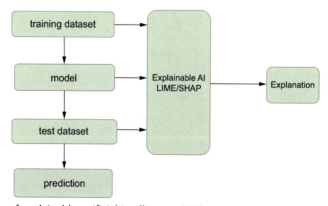

Fig. 6 Technique of explainable artificial intelligence (XAI).

3.3 Model architecture

3.3.1 Inception-ResNet-v2

Inception-ResNet-v2 is an architectural design that integrates concepts from both the Inception and ResNet systems. The purpose of Inception-ResNet-v2 is to leverage residual connections to speed up training while preserving the advantages of the inception modules for collecting a variety of characteristics. Inception-ResNet-v2 replaces the pooling process inside the main Inception modules with residual links. Also, after each Inception block comes a filter extension layer (1×1 convolution without activation) that helps make the filters bigger before they are added to the residuals. The input size must be matched, and the measurements must be compatible with the residual link. Inception-ResNet-v2's incorporation of residual connections enables the optimization-enhanced training of extremely deep networks, which is especially advantageous for difficult tasks and massive datasets. This architecture has been proven to exhibit exceptional performance on diverse computer vision tasks. It is important to consider that various designs may exhibit varying performance on distinct datasets. This network is selected for the research due to its high performance on the utilized dataset.

The system incorporates stem, reduction modules, Inception-ResNet modules, pooling layers, and linear classification layers. The stem consists of three layers with 32, 32, and 64 filters respectively, using 3×3 convolution, Batch Normalization, and Rectified Linear Unit activation function. Reduction modules decrease the resolution via strided convolution. Inception-ResNet modules consist of 1×1, 3×3, and 5×5 convolution branches that are combined by concatenating them channel-wise. A 1×1 convolution is applied before adding the residual to ensure that the dimensions match. Apply the Rectified Linear Unit (ReLU) activation function after each convolution operation and concatenation. Loading Using ImageNet weights circumvents random initialization and enables the utilization of acquired hierarchical features. Model freeze the lower levels of the original model to only train the new classifier layers. Lower learning rates are employed to refine the patterns that have already been acquired by the base model. The 256-unit Rectified Linear Unit (ReLU) dense layer acquires novel feature combinations tailored to our dataset. Batch normalization aids in training by standardizing activations. Alleviates covariate shift. Regularization techniques such as L1 and L2 on dense layer weights aid in preventing overfitting to the training data. During training, dropout randomly assigns inputs to 0 to decrease dependence on certain weights. The Adamax optimizer employs adaptive learning rates and momentum to mitigate sudden shifts. The architecture effectively utilizes transfer learning and enables dataset-specific customization through the classifier layers and training parameters.

3.4 Explainable artificial intelligence (AI) approaches

The main objective of this research is to make deep learning model predictions on lung images more understandable. To accomplish this, we employ explainable AI methods, namely LIME and SHAP.

3.4.1 LIME (Local Interpretable Model-agnostic Explanations)

LIME is a well-known method that is utilized for the purpose of producing local explanations for machine learning models. It works by changing predictions based on perturbed input cases, allowing it to infer the relevance of certain attributes. Using LIME, this study creates picture segmentations of lungs, drawing attention to the areas that are most relevant for classification. The model's decision-making process can be better understood with the help of these segmentations. Based on black-box models, LIME is a model-independent technique that produces explanations for specific predictions. It sets up a local linear model around the prediction point and gives input trait weights to figure out how important they are to the forecast process. The Python Lime module was employed to create explanations for the model's predictions on both the validation and test datasets.

3.4.2 SHAP (SHapley Additive exPlanations)

SHAP is another popular tool for describing ML model results. It uses Shapley values from cooperative game theory to figure out how much each trait contributed to the model's prediction. By computing SHAP values for a group of photos, it is possible to get a full picture of how different pixels or features affect the model's output, especially for the five most accurate predictions. SHAP, based on game theory principles, aims to elucidate the predictions made by machine learning models. It gives a unified framework for figuring out how important a trait is and gives overall reasons for how the model acts. To determine the feature significance values for the model's predictions on the validation and test sets, the Python SHAP module was utilized. These strategies for explainability enhance comprehension of the model's prediction process, both at the individual level (LIME) and across the entire dataset (SHAP).

3.4.3 Methods for visualization

Table comparisons are facilitated through the visual representation of the LIME explanations in the form of tables, which enable the examination and contrast of various iterations of the original image. This helps to deduce which parts of the picture are key to the model's forecasts. Visualization overlays are superimposed on the source images to generate saliency maps. These maps help with visual understanding by clearly showing places that the model considers noteworthy.

Applying a blurring mask to the photos hides irrelevant details, drawing attention to the model-identified regions of interest.

Visualizations are generated to demonstrate the influence of certain characteristics or pixels on the model's prediction for each image in the group. This aids academics and practitioners in understanding the internal mechanisms of the model.

3.4.4 Generating explanations

SHAP explanation class: This class is used to produce detailed explanations that include the model, the blurring masker, and the output class names, among other components. This comprehensive method helps in comprehending the model's decision-making process. Partition Explainer: The Partition Explainer instance is used to offer in-depth explanations for a set of images. This requires iterating through each image in the batch and calculating SHAP values to clarify the model's predictions. The lengths of each cycle are documented to gain insights into the time needed for generating explanations.

Data representation: Array Representation: SHAP results are stored in a five-dimensional array, containing important details such batch size, image dimensions, channels, and predicted classes. This organized depiction aids in the subsequent examination and understanding of the model's actions.

4. Result & discussions

4.1 Experimental setup

The base model is InceptionResNetV2, which has already been trained on ImageNet. A Batch Normalization layer receives the output from the underlying model. A 256-unit dense layer is constructed using kernel, bias, and L2 and L1 regularizers. Activation using ReLU is implemented. Normalization is done on the Dense layer by dropping out data at a rate of 0.45. The last layer of output is a dense layer that uses softmax to classify the data into a certain number of categories. Adamax optimizer with a learning rate of 0.001 was used to compile the model. Metrics for accuracy and categorical cross-entropy loss are employed. Investigation was conducted on Google Colab, a cloud-based machine learning platform equipped with powerful GPUs. Model employed a Tesla T4 GPU to train our deep learning model. The Tesla T4 GPU is a high-performance computing tool frequently utilized for deep learning applications. It provides substantial parallel processing capabilities, making it ideal for training extensive neural networks and speeding up computations in machine learning applications. Model apply explainable AI methods using LIME and SHAP. The complete execution, encompassing the deep learning model and the use of explainable AI, was carried out using Python. For LIME, the Keras and Lime packages were utilized, whereas the SHAP library was implemented for SHAP. This configuration supplied sufficient processing capacity for training and facilitated the smooth incorporation of model creation and analysis of interpretability in a Python environment.

4.2 Evaluation matrix

The assessment of the model's performance included using various metrics to evaluate its accuracy and capacity to categorize cases correctly. Accuracy (Acc), precision (Pre), recall

(Rec), F1-score metrics, and the confusion matrix are used to evaluate the proposed model.

4.2.1 Accuracy (ACC)
Accuracy decides the overall perfectness of the model's predictions.

$$Accuracy = \sum_c \frac{TP_c + TN_c}{TP_c + FP_c + TN_c + FN_c}, \quad c \in classes \tag{1}$$

4.2.2 Precision (Pre)
Precision quantifies a percentage of malignant lung pictures properly classified.

$$Precision \sum_c \frac{TP_c}{TP_c + FP_c}, \quad c \in classes \tag{2}$$

4.2.3 Recall (Rec)
Recall (responsiveness or real affirmative rate) measures the percentage of malignant lung pictures properly identified.

$$Recall = \sum_c \frac{TP_c}{TP_c + FN_c}, \quad c \in classes \tag{3}$$

4.2.4 F1-score
By integrating recall and accuracy, the F1 score balances both. It's the perfect balance between accuracy and recall.

$$F1_{score} = 2 \times \frac{precision \times sensitivity}{presicion + sensitivity} \tag{4}$$

4.2.5 Confusion matrix
Confusion matrix tables show categorization model efficacy. It counts true positive (TP), true negative (TN), false positive (FP), and false negative (FN) predictions (Table 2).

Table 2 Confusion matrix framework.

	Predicted: cancerous lung	Predicted: normal lung
Actual: Cancerous Lung	TP	FN
Actual: Normal Lung	FP	TN

TP: True Positive: Correctly identified images of a malignant lung.
TN: True Negative: Imaging of a normal lung is reliably identified as such.
FP: False Positive: Normal lung images are incorrectly classified as cancerous lung.
FN: False Negative: Images of cancerous lungs falsely identified as healthy lungs.

We can evaluate the model's accuracy, precision, recall, and balance between precision and recall (The score for F1) by calculating these metrics and analyzing the confusion matrix.

4.3 Result analysis

In this section, we describe the findings of our research project that aimed to improve the identification and classification of lung anomalies by utilizing the proposed model in conjunction with Explainable AI approaches. The proposed model, along with Explainable AI methods, has shown promise in finding and classifying lung cancer. The main points and results of your work are summed up here. The proposed model achieved a remarkable level of accuracy, reaching 99.92% when it came to identifying and categorizing lung cancer. The results indicate that the model excels on the test dataset, displaying minimal loss and remarkable accuracy. The high accuracy suggests that the model's predictions closely match the true labels in the test dataset. This strongly indicates the model's competence in appropriately identifying situations. Fig. 7 displays the accuracy and loss curves for detecting and classifying lung and colon cancer.

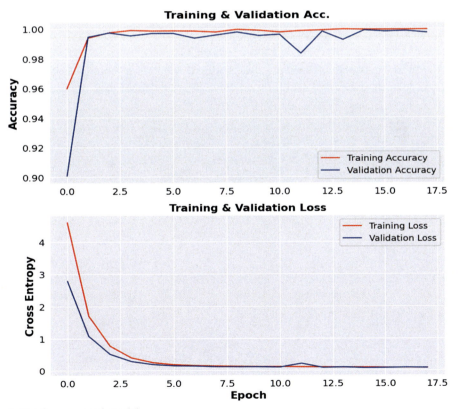

Fig. 7 Plot for proposed model.

The classification report in Fig. 8 tells, how well the model did in each class. Precision is the percentage of correct predictions out of all the cases that are put into a certain class. All of the precision values are 1, which means that the model properly finds all instances of each class and doesn't give any false positives. Recall, which is also called sensitivity, is the number of correct predictions out of all the cases that really do belong to a certain class. All of the recall values are 1, which means that the model properly finds all instances of each class without any false negatives. The harmonic mean of accuracy and recall is the F1-score. It gives a fair picture of how well a model does. All of the F1 scores are 1, which means that accuracy and recall are well balanced. Support is the number of examples of each type in the test set. There are 2500 instances of each class, and the support numbers are different for each one. The accuracy of a dataset is the percentage of properly classified instances out of all instances in the dataset. The average accuracy is 1, which means that the model does a great job of classifying the test data. The macro average finds the mean of the F1-score, precision, and memory scores for all classes, giving each class the same amount of weight. In this case, the overall average for accuracy, recall, and the F1-score is also 1, which means that performance was the same and fair across all groups. The weighted average finds the mean of the accuracy, recall, and F1-score for all classes, with the support of each class giving it a weight. The model did very well across all classes, even when taking into account differences between classes, as shown by the weighted average of 1.00 for accuracy, recall, and F1-score. Overall, the classification report shows that the model gets perfect accuracy and great performance across all classes. This shows that it is good at correctly putting instances in the test dataset into the right class.

```
⊡  50/50 [==============================] - 19s 295ms/step
                precision    recall  f1-score   support

           0        1.00      1.00      1.00       479
           1        1.00      1.00      1.00       519
           2        1.00      1.00      1.00       498
           3        1.00      1.00      1.00       516
           4        1.00      1.00      1.00       488

    accuracy                            1.00      2500
   macro avg        1.00      1.00      1.00      2500
weighted avg        1.00      1.00      1.00      2500
```

Fig. 8 Classification report for proposed model.

The confusion matrix in Fig. 9 for lung and colon cancer detection, with five labels for each class, offers a detailed summary of the model's ability to differentiate between various classes. The model's capability to accurately categorize instances into their respective classes and detect misclassifications can be evaluated through the examination of the confusion matrix. Confusion matrix indicates that the model has a high level of accuracy in making predictions. This accuracy is superior to that of models that are considered to be state-of-the-art. Experimental results showed that the proposed model did better than the transfer learning models.

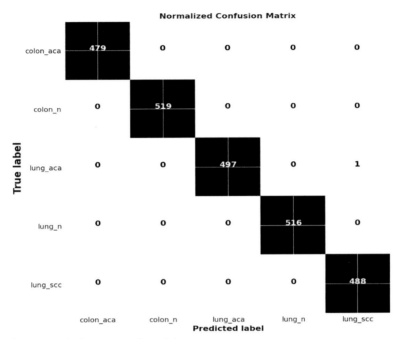

Fig. 9 Confusion matrix for proposed model.

4.3.1 Results of visualization using LIME and SHAP

The main objective of the study is to make it easier to understand what a deep learning model says about lung pictures. A number of explainable artificial intelligence techniques, especially LIME and SHAP, are utilized in order to accomplish this goal. Proposed research used two well-known explainable AI methods, LIME and SHAP, to decipher our deep learning model's predictions on the test and validation sets. Using LIME, picture segmentations were constructed, with an emphasis placed on most important regions for classification. In spite of the fact that LIME was able to successfully identify a number of characteristics in the lung images. An explanation of the model's prediction for a specific image is provided by the LIME explanation. The explanation places particular attention on the highest ranked predicted labels and the relevance of those labels. The LIME explanations are illustrated in Fig. 10 by comparing and contrasting two different versions of the initial image, with the emphasis being placed on crucial parts (Fig. 11). The

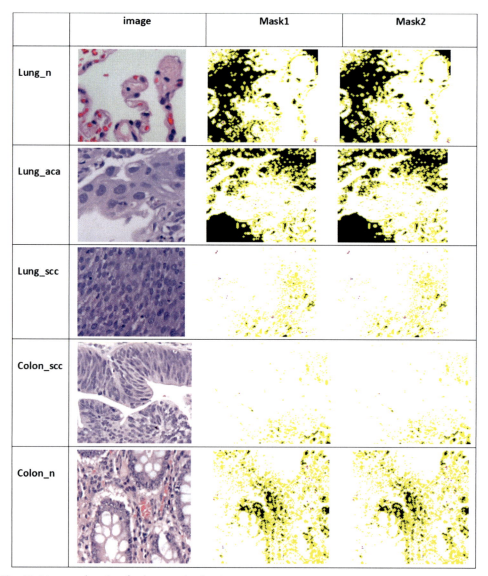

Fig. 10 Lime explanation for lung and colon images.

creation of a saliency map and its subsequent superimposition over the original image results in an improvement in visual perception by providing a more distinct depiction of areas that are noteworthy. It is possible to generate a blurring masker by applying the "blur (28,28)" command to the batch's photographs. For the purpose of generating an explanation that incorporates the model, masker, and output names (classes), the

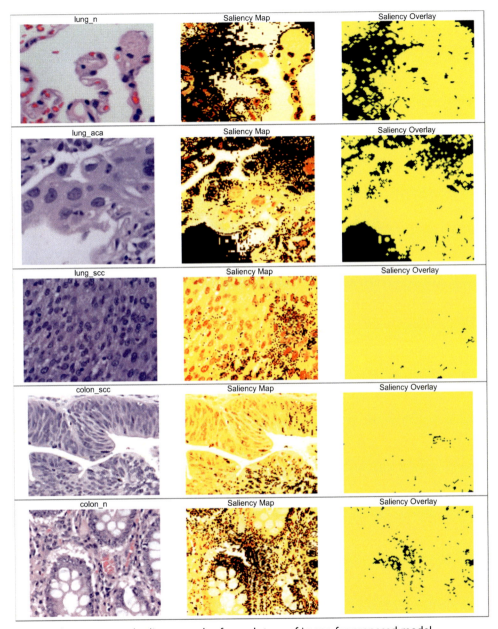

Fig. 11 Saliency map and saliency overlay for each type of image for proposed model.

SHAP explanation class should be utilized. A group of eight photographs are analyzed in order to ascertain their SHAP values by making use of the explanation that was developed earlier. Following the arrangement of the top five classes in descending order, the "outputs" argument is defined to prioritize the top five classes. How features, or pixels, contributed to the output of the model for the batch that was supplied is demonstrated by the Shap values that were generated as a result. A figure is presented that illustrates how various features (pixels) influence the output of the model for each individual picture in the batch. Furthermore, it provides information regarding the manner in which particular aspects of a picture either assist or hinder the model's predictions for the top five classes. In order to shed light on the model's predictions for a collection of photographs, a PartitionExplainer instance is now being utilized, as seen in Fig. 12. The progression is shown in the form of percentages, and each iteration corresponds to the explanation for a single image that is included in the batch. Additionally, the durations for each repeat are displayed, which also serves to demonstrate the amount of time that is necessary for the preparation of explanations. The array that was produced is a five-dimensional array with the dimensions [8, 128, 128, 3, 5] as its dimensions. The dimensions relate to the characteristics that are listed below: batch size, picture height, image width, number of channels, and number of classes.

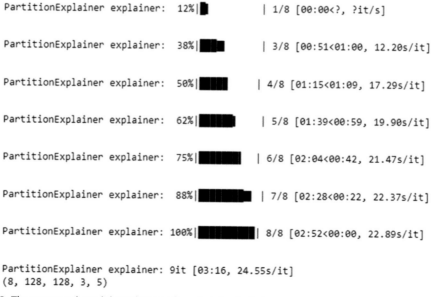

Fig. 12 The proposed model predictions for a batch of photos.

5. Conclusion

The study results demonstrate the effectiveness of the suggested methodology in identifying and categorizing lung and colon cancer. The suggested model shows superior accuracy compared to advanced transfer learning models, making it a promising tool for medical imaging interpretation. Precise diagnosis is crucial for improving patient results and determining suitable treatment strategies, highlighting the importance of the high accuracy attained by our model.

The model's exceptional success is due to its capacity to efficiently extract features from lung and colon pictures. We assessed multiple performance indicators such as Precision, Recall, F1 score, and Specificity to gain a thorough picture of the model's capabilities. Our model consistently beat transfer learning algorithms across various measures, demonstrating its proficiency in diagnosing lung and colon anomalies.

Furthermore, the dependability and consistency of our model in making predictions are essential for delivering precise and reliable diagnoses. The model shown outstanding performance on diverse datasets related to various diseases, demonstrating its ability to generalize effectively to new datasets and making it a versatile option for a range of medical uses. The incorporation of eXplainable Artificial Intelligence (XAI) methods like LIME and SHAP improved the transparency of our model, providing a deeper understanding of its decision-making mechanisms. This enhances trust and enables additional analysis by clinicians and academics. Moreover, our model's training time efficiency enhances its suitability for real-world applications, facilitating prompt and effective interpretation of medical imaging data, with potential to greatly influence patient care and clinical decision-making. Proposed model, utilizing eXplainable Artificial Intelligence (XAI) approaches, pinpointed crucial imaging characteristics linked to colon and lung cancer, facilitating early detection and tailored treatment plans.

We recognize the difficulties in implementing eXplainable Artificial Intelligence (XAI) using time series data, especially due to the temporal characteristics of this type of data.

Our study provides a thorough assessment of the suggested model for identifying lung anomalies, highlighting its exceptional performance and improved interpretability. Additional study and validation on other datasets are needed to confirm its wider application and efficacy in clinical settings, ultimately helping patients and healthcare professionals (Fig. 13).

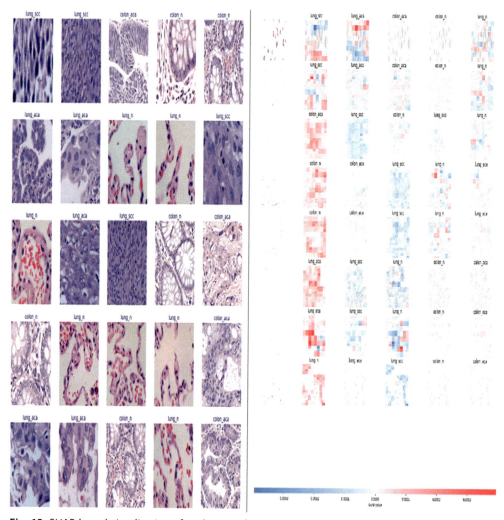

Fig. 13 SHAP-based visualization of each type of image.

6. Conclusion and future direction

6.1 Conclusion

Proposed model shows the effectiveness of deep learning models in precisely categorizing lung and colon tissues. The model demonstrated exceptional performance on the test set, achieving high accuracy, recall, precision, and F1 scores. The LIME and SHAP visualizations provided insightful local and global explanations, emphasizing crucial areas for categorization.

The LC25000 dataset comprises 25,000 histopathological pictures mostly centered on lung and colon tissues, making it highly helpful for classification tasks. Possible constraints include compression artifacts in JPEG format, an initial disparity in class distribution, restricted tissue representation beyond lung and colon, concerns regarding the quality and influence of augmentation, potential discrepancies in annotations, temporal fluctuations, inter-observer variability, and a dearth of comprehensive clinical data. Although there are limits, the dataset continues to be an important asset for histopathology image categorization and instructional purposes. Our work demonstrates a potential use of deep learning models to automate the diagnosis of lung and colon tissues.

We expect that our discoveries will stimulate more investigation in this field, perhaps expanding the model's range to encompass other ocular conditions and utilizing sophisticated methodologies to enhance performance and comprehensibility. Subsequent advancements may encompass the utilization of voluminous training data, sophisticated methods for augmenting data, the construction of multiple models, the incorporation of explainability, and the integration of supplementary patient data.

Human clinical experience remains essential in conjunction with AI forecasts, highlighting the importance of AI serving as a supplement rather than a replacement for clinicians. Inception-ResNet-v2 is a potent model that necessitates supplementary methods and data to improve its usefulness in healthcare and other practical situations. Overall, it shows what Inception-ResNet-v2 can do while also pointing out its flaws and the need for a more complete approach to AI solutions in areas like healthcare.

The study highlights the efficacy of the suggested methodology in enhancing the identification and categorization of lung and colon abnormalities. The findings significantly improve medical imaging analysis, resulting in enhanced diagnosis accuracy and decision support systems for respiratory and gastrointestinal diseases. We enhance the accuracy of detecting and categorizing lung and colon anomalies by using advanced machine learning techniques and using explainable artificial intelligence (XAI) methodologies. This also improves the clarity of the model's findings. Precision and comprehensibility are crucial for building trust in automated diagnostic systems and improving patient care and outcomes in the diagnosis and treatment of respiratory and gastrointestinal illnesses.

6.2 Future directions

As medical technology improves, it is very important to make sure that suggested models are fully tested with larger and more varied datasets. This makes sure that the model's results are reliable and can be used for a wide range of people and situations. Integrating strong security measures also protects private patient data, ensuring its integrity and secrecy. Analyzing possible cost savings from early identification and precise lung and colon abnormality classification is essential. A study can showcase the economic advantages of using the model in clinical settings, illustrating its usefulness to healthcare providers and administration. Cost reductions can result from decreased hospital stays, enhanced treatment strategies, and better distribution of resources. Focusing on early detection through the model's features can also improve patient results. Early detection of lung anomalies allows for immediate intervention and treatment, potentially halting disease advancement and associated issues. This proactive approach is in line with preventive medicine concepts, which aim to enhance general well-being and lessen healthcare expenses.

By taking these factors into consideration, the model can efficiently reduce the potential hazards linked to undetected or incorrectly classified pulmonary abnormalities. This helps healthcare services work better and last longer, which is good for patients as well. Overall, the successful application of such models in clinical practice requires a thorough strategy that takes into account patient-centered outcomes, cost analysis, security, and validation. This will eventually improve healthcare delivery and outcomes for all parties involved.

References

[1] S. Hadiyoso, S. Aulia, I.D. Irawati, Diagnosis of lung and colon cancer based on clinical pathology images using convolutional neural network and CLAHE framework, Int. J. Appl. Sci. Eng. 20 (1) (2023) 1–7.

[2] N. Yahia Ibrahim, A.S. Talaat, An enhancement technique to diagnose colon and lung cancer by using double CLAHE and deep learning, Int. J. Adv. Comput. Sci. Appl. 13 (8) (2022).

[3] S.U.K. Bukhari, A. Syed, S.K.A. Bokhari, S.S. Hussain, S.U. Armaghan, S.S.H. Shah, The Histological Diagnosis of Colonic Adenocarcinoma by Applying Partial Self-Supervised Learning, MedRxiv, 2020. 2020–08.

[4] F.M. Fatoki, E.K. Akinyemi, S.A. Phlips, Prediction of lungs cancer diseases datasets using machine learning algorithms, Curr. J. Appl. Sci. Technol. 42 (11) (2023) 15–23.

[5] M. Nishio, M. Nishio, N. Jimbo, K. Nakane, Homology-based image processing for automatic classification of histopathological images of lung tissue, Cancers 13 (2021) 1192.

[6] M. Masud, N. Sikder, A.A. Nahid, A.K. Bairagi, M.A. AlZain, A machine learning approach to diagnosing lung and colon cancer using a deep learning-based classification framework, Sensors 21 (3) (2021) 748.

[7] M.A. Abbas, S.U.K. Bukhari, A. Syed, S.S.H. Shah, The Histopathological Diagnosis of Adenocarcinoma & Squamous Cells Carcinoma of Lungs by Artificial Intelligence: A Comparative Study of Convolutional Neural Networks, MedRxiv, 2020. 2020–05.

[8] B.K. Hatuwal, H.C. Thapa, Lung cancer detection using convolutional neural network on histopathological images, Int. J. Comput. Trends Technol. 68 (10) (2020) 21–24.

[9] I. Dyah Irawati, S. Hadiyoso, A. Fahmi, Compressive sensing in lung cancer images for telemedicine application, in: The 4th International Conference on Electronics, Communications and Control Engineering, 2021, April, pp. 55–61.

[10] G. Zhang, L. Lin, J. Wang, Lung nodule classification in CT images using 3D densenet, J. Phys.: Conf. Ser. 1827 (1) (2021, March) 012155.

[11] H. Yu, Z. Zhou, Q. Wang, Deep learning assisted predict of lung cancer on computed tomography images using the adaptive hierarchical heuristic mathematical model, IEEE Access (2020) 8. 86400–86410.

[12] Q. Zhang, X. Kong, Design of automatic lung nodule detection system based on multi-scene deep learning framework, IEEE Access (2020) 8. 90380–90389.

[13] A. Messalas, Y. Kanellopoulos, C. Makris, Model-agnostic interpretability with shapley values, in: 2019 10th International Conference on Information, Intelligence, Systems and Applications (IISA), IEEE, 2019, July, pp. 1–7.

[14] D. Fryer, I. Strümke, H. Nguyen, Shapley values for feature selection: the good, the bad, and the axioms, IEEE Access 9 (2021) 144352–144360.

[15] C.J. Lin, S.Y. Jeng, M.K. Chen, Using 2D CNN with Taguchi parametric optimization for lung cancer recognition from CT images, Appl. Sci. 10 (7) (2020) 2591.

[16] L. Cai, T. Long, Y. Dai, Y. Huang, Mask R-CNN-based detection and segmentation for pulmonary nodule 3D visualization diagnosis, IEEE Access 8 (2020) 44400–44409.

[17] M. Tan, Q. Le, Efficientnet: rethinking model scaling for convolutional neural networks, in: International Conference on Machine Learning, PMLR, 2019, May, pp. 6105–6114.

[18] A. Masood, P. Yang, B. Sheng, H. Li, P. Li, J. Qin, et al., Cloud-based automated clinical decision support system for detection and diagnosis of lung cancer in chest CT, IEEE J. Transl. Eng. Health Med. 8 (2019) 1–13.

[19] Y. Nohara, K. Matsumoto, H. Soejima, N. Nakashima, Explanation of machine learning models using improved shapley additive explanation, in: Proceedings of the 10th ACM International Conference on Bioinformatics, Computational Biology and Health Informatics, 2019, September, p. 546.

[20] H. Guo, U. Kruger, G. Wang, M.K. Kalra, P. Yan, Knowledge-based analysis for mortality prediction from CT images, IEEE J. Biomed. Health Inform. 24 (2) (2019) 457–464.

[21] A.A. Borkowski, M.M. Bui, L.B. Thomas, C.P. Wilson, L.A. DeLand, S.M. Mastorides, Lung and Colon Cancer Histopathological Image Dataset (LC25000), 2019. arXiv preprint arXiv:1912.12142.

CHAPTER 10

Unveiling explainable artificial intelligence (XAI) in advancing precision medicine: An overview

Arpita Nayak[a,b,c], Ipseeta Satpathy[a,b,c], and Vishal Jain[a,b,c]
[a]KIIT School of Liberal Studies, KIIT DU, Bhubaneswar, India
[b]KIIT School of Management, KIIT DU, Bhubaneswar, India
[c]Department of Computer Science and Engineering, School of Engineering and Technology, Sharda University, Greater Noida, India

1. Introduction

A collection of procedures and techniques known as explainable artificial intelligence (XAI) enables human users to understand and have faith in the output and outcomes produced by machine learning algorithms. The term "explainable AI" refers to an AI model's predicted effects and possible biases. It contributes to defining model correctness, fairness, transparency, and decision-making results driven by AI. When implementing AI models into production, an organization needs to be able to explain AI in order to gain the confidence of its stakeholders. Explainability in AI aids in an organization's adoption of a responsible development strategy for AI. Humans are finding it more difficult to understand and follow the algorithm's path as AI advances [1]. The entire computation process becomes what's known as a "black box," meaning it's hard to decipher. The data is used directly to generate these black box models. Furthermore, neither the engineers nor the data scientists who developed the algorithm can fully comprehend or articulate the inner workings of it, let alone describe how the AI algorithm came at a particular conclusion [2]. An organization must not blindly trust AI; instead, it must fully comprehend the decision-making processes of AI, including model monitoring and accountability. Explainable AI can assist people in comprehending and elucidating neural networks, deep learning, and machine learning (ML) algorithms.

Explainable AI in Healthcare Imaging for Medical Diagnoses
https://doi.org/10.1016/B978-0-443-23979-3.00010-5

ML models are frequently perceived as unintelligible "black boxes."[a] Deep learning uses neural networks, which are among the most difficult for humans to comprehend. AI model training has always carried the possibility of bias, which is frequently based on racial, gender, age, or geographic factors. Furthermore, because training and production data are different, AI model performance may drift or deteriorate [3]. This means that in order to support AI explainability and gauge the commercial impact of deploying such algorithms, a firm must constantly monitor and manage its models. Additionally, explainable AI supports productive AI use, model auditability, and end user trust. Moreover, it lessens the reputational, legal, security, and compliance concerns associated with production AI [4]. Recent study in IBM stated that there are three primary methods for setting up XAI techniques. While decision understanding addresses human demands, prediction accuracy and traceability address technological requirements as shown in Fig. 1.

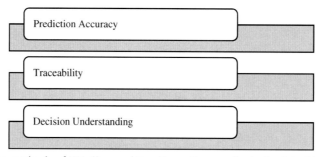

Fig. 1 Three main methods of XAI. *(Source: https://www.ibm.com/topics/explainable-ai.)*

Prediction accuracy: For artificial intelligence to be successfully used in daily operations, accuracy is a crucial factor. The prediction accuracy can be ascertained by executing simulations and contrasting the output of XAI with the outcomes in the training data set. Local Interpretable Model-Agnostic Explanations (LIME), which describes how classifiers are predicted by the ML algorithm, is the most widely used technique for this.

---

[a] The paragraph underpins the need for the explainable AI (XAI) to enhance the concerns and risks of AI model's training and deployment, especially prerequisites for precision medicine. It underlines the need to make AI systems more open to allow users to better trust, and be able to gain confidence in ethical standards and the law, as well as to detect and avoid security threats. These aspects align directly with the chapter title, "Unveiling Explainable Artificial Intelligence (XAI) in Advancing Precision Medicine: AI for Precision Medicine: An Overview", that points to the fact that XAI is one of the key components that enable closing the gap between the development of AI systems and their actual implementation in precision medicine. Per the principles of precision medicine, the contributions that XAI makes in terms of dealing with biases, improving model credibility, and promoting explicability fit into the framework of boosting predictive power and quality of decision making and ultimately, advancing patient personalized care. Once again, the importance of focusing on XAI for reaching these objectives while avoiding potential risks corresponds to the conceptual premise of the chapter.

Traceability: Another essential method for achieving XAI is traceability. This is accomplished, for instance, by restricting the options available for decision-making and by establishing a more constrained set of ML features and rules. DeepLIFT (Deep Learning Important FeaTures) is an example of a traceability XAI approach. It provides a traceable relationship between each active neurone and even reveals dependencies between them by comparing the activation of each neurone to its reference neurone.

Decision understanding: The human element is this. Although many people are skeptical of AI, they must come to trust it in order to work with it effectively. To do this, training the team utilizing the AI helps them comprehend how and why the AI makes judgments.

Precision medicine, which is often referred to as personalized medicine, is a customized approach to illness prevention and treatment that considers the individual genetics, lifestyle, and environment of each patient. Giving each patient the appropriate care at the appropriate time is the aim. Explainable AI may be described as the approaches and tools aimed at making the rationale behind the decisions of sophisticated AI systems more comprehensible to the end-users. In the context of high-staked settings, precision medicine for that matter, XAI stands as a key enabler to guarantee the revealed by ML models recommendations' interpretability as well as the consequent confidence by both clinicians and patients. The fact is that human's body functioning is extremely complicated while medical data are rather intricate as well, so it is crucial for clinicians to understand how an AI system came to such or such decision. Lacking this level of understanding, AI integration in clinical practice can turn into enabler of medical progress into enemy that hinders the process [5]. The complex nature of many machine learning models, particularly the deep learning networks, has also been identified as a major impediment to the use of AI in healthcare. These models are sometimes described as "black boxes," especially because the amateurs and even experts cannot understand how they work. In the case of precision medicine where Artificial Intelligence is employed in the analysis of biological data this kind of opacity leads to some disbelief among the doctors. In this case, the doctors themselves may not understand how a certain AI system has arrived at its conclusions, they may not go by what the system is putting out [6]. Explainable AI then goes ahead and addresses this challenge through giving information on how AI models arrive at their decisions. Other methods such as feature importance matrix and attention mechanisms show which features, for instance genetic profiles, patient history, or test results, have contributed a great deal in the judgment passing by the AI. In other words, XAI helps to create trust between clinicians and AI systems, means that clinicians are ready to follow recommendations suggested by AI models. This is very critical in areas such as oncology or genomics where precision medicine is advancing but nonetheless has to rely on a lot of trust on the data interpretations [7]. Failure in diagnosis is still a major problem in the delivery of health care with implications of death. Another well-explained aspect of the AI is its capability of enhancing diagnostic precision by recognizing patterns that can be unnoticed by medical practitioners and specialists. However, the technicality and opaqueness of the above-stated models can work to the detriment of their ability to perform [8]. By using XAI, the clinicians are able to have more insight on how a certain model came to a particular decision. For example, if ANNs have been developed to diagnose cancer risk based on patients' biomarkers and genetic mutations, XAI will reveal which of the biomarkers and mutations affected the decision. This way, the AI findings are checked against the knowledge possessed by the health care providers to have an assurance that they are correct. Moreover, with XAI, one may pinpoint the model's sources of bias to make sure that diagnostic tools are not only effective but also fairly applied across subpopulations of patients. Interpreting AI models from a fine-grained perspective leads to higher diagnostic quality and improvement on the clinical decision-making process [9]. The hope of high personalization in subsequent precision strategy—the main goal of the concept—is based on the idea of targeting primary treatments according to the actual molecular-end biological characteristics of the patient. This is made possible by using AI models in the identification of different treatment options by using genetic information, patient records, among other factors. But it would be essential to know more than that personalization is making intelligent use of available data; it has to do with understanding how recommendation engines are coming up with its recommendation results [10]. This understanding comes from explainable AI that gives a visualization of how certain inputs, whether it is genetic information, the patient's habits, or their medical history influences the results of the treatment. For instance, in the cancer therapy, the XAI models may help in understanding why a particular drug is preferred over the other depending on the patient's tumor profile. Such a high level of transparency can help the healthcare providers to deliver the therapies with more precision thus enhancing the treatment and the results that surround the patients. Further, the XAI assists clinicians in explaining out these recommendations to the patients, and thus more patient trust in their customized care plan [11]. AI models deployed in precision medicine are also used for risk factor estimations for instance, the probability of disease occurrence or the risk of having an allergic reaction to a particular medication or treatment. However, since the process of deriving these risk predictions is not well understood, the healthcare providers may not implement the same. Similarly, patients may be puzzled over being categorized as high risk or low risk and this may lead to doubt in the AI system. Regarding these issues, XAI comes up with an explanation of what goes into the construction of risk predictions. Through creating awareness of these factors, XAI enhances the accuracy of risk estimations which in turn helps healthcare providers to be wiser in their interventions. Furthermore, the follow-up risk-based recommendations are also better accepted by the patients if they know that the specific characteristics considered by the AI were used to make such a decision [12]. The implementation of AI particularly in the healthcare facilities is a well-regulated process to guarantee the safety and effectiveness of the technology as well as to assume liability for the same. This makes sense as many regulatory authorities like the FDA and the EMA have insisted that any artificial intelligence-driven medical devices as well as the algorithms powering these devices must be explainable and hence interpretable. In this way, XAI assists AI models to meet these requirements of regulations by presenting understandable explanations for the model's learned results. This also helps in determining approval of the systems and also check on the AI systems for fairness and transparency. In the field of Precision medicine, patients' involvement is one of the key factors to achieving the goals of individualized therapy. XAI gives the healthcare providers an opportunity to assist the patients understand the machine learning decisions made about them and come up with the right course of action to take. This helps the patients to be more active participants in their treatment thus improving compliance and other patient satisfaction parameters.

2. Research questions

- In what ways does Explainable AI (XAI) increase the level of trust in AI's decisional capacity in precision medicine within diagnostics and treatment decision making?
- In the context of precision medicine, how does XAI affect patients' clinical outcomes and engagement and how does its use shape the outcomes of targeted treatments for patients?

3. Research objectives

- To understand different ways XAI helps in AI-driven diagnostics and treatment recommendations more trusted by healthcare professionals.
- To understand how XAI promotes individualized treatment outcome and enables patients to understand and engage more effectively in their healthcare choices.

4. The role of explainable AI (XAI) in demystifying AI models: Building trust in precision medicine

In the sphere of precision medicine that implies the use of AI models for diagnosing diseases and estimating treatment efficacy, or ineffectiveness, the models' interpretability is a compelling issue. Most sophisticated AI systems, and particularly the algorithms that implement deep learning, can be therefore considered as "black boxes." This means that it can be challenging for healthcare workers to understand the process, through which these models come to the particular conclusions. Lack of transparency of some AI-based algorithms may slow the acceptance and proper implementation of the technology-recommended solutions and, accordingly, the use of these technologies in clinical practice. In order to resolve this problem and enhance the level of trust in AI among clinicians, Explainable AI (XAI) comes as a evolutional tool [13].

Indeed, Explainable AI can be defined as various approaches and methods which are aimed at the development of the interpretable machine learning algorithms. Through the provision of an understanding of how an AI system works and comprises, XAI makes it possible for clinicians to understand the rationale behind specific plans recommended to be enacted. This understanding is important for improving people's confidence in the use of AI models in precision medicine and their effectiveness [14]. An important aspect of how XAI leads to trust is through feature importance analysis. It will enable one to know some of the features or variables that contribute most to the performance of the AI model. For instance, when one is creating a model that aims at predicting the propensity of a certain patient to develop a certain disease, the importance assessment will identify the crucial factors that include genetic profile, past medical history or habits that formed the basis of the prediction. Due to the analyses of feature importance, clinicians are able to gain further understanding of which features the model operates to deliver its predictions to ensure that the AIED system generates reliable suggestions. This also helps the healthcare providers to have enhanced confidence in the AI system, as well as helps them to decide based on the outputs of the AI system [15]. Another one of XAI techniques is the Decision trees that also help to build trust as it provides clear and easily explainable decision making. While decision trees might have a look like complex neural networks, their operational mode is quite different—the structure of decision making is evident on the tree where each step is a result of certain criteria and certain threshold level being met. For example, if a decision tree is applied to diagnose a particular disease, then it will show how a patient's symptoms and a test outcome will result in a specific diagnosis. Interpretability of models at this third level can be beneficial for healthcare providers since it involves a more precise understanding of the AI's recommendations and/or-verifications of those recommendations with clinical knowledge. That is why decision trees contribute to enhancing the levels of trust in the AI system since the decision-making process is transparent and easily comprehensible [16]. There are extended XAI techniques like SHAP (SHapley Additive exPlanations) that explain how the certain feature predisposes to the model's decision. SHAP values break down a prediction into portions that are attributable to each feature thus giving a detailed analysis of which factor has an impact on the prediction. In the context of precision medicine, SHAP can give the clinician more explainable insights on how a certain genetic variant or biomarker affects the probability of a response to a specific treatment. This detailed explanation not only helps to prove the AI recommendations' accuracy in practice but also increases the understanding of their operation by the healthcare providers and makes the decisions themselves more transparent to the patients [17].

Currently, the process of training and implementing AI in precision medicine is highly opaque, thus by incorporating XAI into such models the healthcare field can improve trust among clinicians and patients in the models used. XAI techniques allow the health care professionals filter and verify the output generated by Artificial

Intelligence and as a result, make confident decisions by leveraging this kind of AI. In particular, the possibility of AI decision-making and explaining of its actions in terms of clinical knowledge means that it is easier to adopt this approach into the widespread practice, which in turn will increase its impact on the field of personalized healthcare [18]. The role of Explainable AI is crucial in increasing the confidence in the models implemented in precision medicine by ensuring that the decision making that went into the model is comprehensible. Tools like feature importance, decision trees, SHAP values enable the user to assess and explain the sources that drove AI models to come up with certain outcomes, which is a major advantage of AI systems in the healthcare sector. Therefore, in addition to enhancing clinicians' confidence in AI-based decisions, XAI also leads to the proper adoption of AI for precision medicine.

5. Improving diagnostic accuracy in precision medicine through explainable artificial intelligence (XAI)

Concerning precision diagnostics, the utilization of AI in diagnostic processes has gradually gained popularity in the field. Recent developments in machine disciplines have revealed impressive performance in identifying possible diseases based on a variety of individual records such as genomic data, life styles, everyday practices and medical histories. However, such AI models are not perfect and are sometimes wrong in their interpretation of these patterns or might include bias and may provide wrong diagnosis. This brings out the need for interpreting and explaining artificial intelligence, a need that is well met by Explainable AI (XAI). Explicable AI means the strategies and approaches developed to provide more clarity on Artificial Intelligence models and assist the user in understanding why this or that decision has been made by the model. In the diagnosis of accuracy, XAI offers a vital solution of showing exactly how the model of AI reaches a solution and therefore ensure that the findings of the model's predictions are accurate and there are no mistakes in the diagnosis [19]. The first and most direct method through which XAI enhances diagnostic accuracy is through feature importance analysis. It not only reveals and measures the role of particular characteristics (for example, genes, biomarkers, histories of patients) in AI's predictions. For instance, let there be a model that is intended to predict the susceptibility of certain disease with genotypic feature, the FI analysis will determine which of the genetic mutations or biomarkers played most crucial role from the model in making the prediction. When knowing which particular variables contributed most to the model's results, the practitioners can cross-check the findings of the AI against the human facts and expertise in medicine. This validation process enables the elimination of rein formation bias where the algorithms relied on accurate information to make the predictions [20]. Other XAI technique which helps in improving diagnostic accuracy by giving local explanations is Local interpretable model-agnostic explanations (LIME). LIME operates by approximating the AI model's behavior within the proximity of a certain sample, and hence, lifting the veil over how the model arrived at a certain prediction for a certain instance. For instance, if an AI model has deemed a patient to be at a high risk of a certain disease, LIME can indicate which feature or indeed which parts of the tests results or family history contributed most toward this decision. The advantage of such a localized explanation is that it enables a clinician to evaluate whether the AI's prediction is consistent with the patient's clinical picture and whether there might be a need for additional diagnostics or another diagnosis tested instead [21]. Decision trees are another useful instrument in XAI to enhance diagnostic accuracy. While deep learning methods are rather opaque and complicated structures, decision trees allow for a clear understanding of the decision-making process based on specific features and their thresholds. For instance, if the model being described was a decision tree which was used in a diagnostic model for a certain disease, the model illustrates how information regarding symptoms and tests which comprise of the disease would be interpreted in a sequence. Such interpretability level allows the healthcare providers to track down the flow of the model's decision-making process and ensure that its solution corresponds to the guidelines and diagnostic key points. Through decision trees, any discrepancies or errors committed by the model in making a diagnosis are made clear to the clinicians hence making the diagnoses accurate [22]. SHAP (SHapley Additive exPlanations) is another sophisticated XAI technique that could also increase diagnostic precision since it identifies the ways in which a specific feature has affected the AI model's outcome. SHAP values represent measures that break the prediction down to show feature importance and, in a way, provides a way of understanding the effect of a particular feature on the model output. With precision medicine, SHAP can assist experts in the identification of aspects of genetics differences, life style or other parameters that might lead to a diagnosis. Such detailed explanation goes hand in hand with validation of AI predictions explicitly to assist clinicians to detect the discrepancies or elaborate on them if needed [13]. Such XAI methods as feature importance analysis enable the healthcare providers to identify which features—the genetic markers, biomarkers, or the patient history—are the most critical for AI models to make predictions. For instance, if a high-risk of a genetic disorder has been estimated for a given patient through an AI model, the feature importance analysis can demonstrate which genetic features contributed to this risk estimation. This is helpful for clinicians in a way that they can see the results of an AI related to their knowledge and check what is related to those features. XAI enhances the rationale that underpins AI predictions since it reveals the process used to make the prediction hence minimizing chances of errors in diagnosis [23]. XAI brought into the precision medicine AI models that are used in diagnostics will also enhance the level of accuracy and reliability in the predictions made. Feature importance analysis, LIME, decision tree and SHAP are effective methods of XAI that can help the healthcare providers to understand the factors that underpin an AI's conclusions effectively. This results not only increases the accuracy of diagnoses but also minimizes accidents hence improving the health outcome of the patient and overall personalization of the care. One of the primary contributions that Explainable AI brings to the implementation of precision medicine is raising the effectiveness of diagnostic expertise by making the models used more understandable. It is possible to make use of feature importance analysis, LIME, decision trees, and SHAP, to help healthcare providers to verify and to better understand how the AI models arrived at these certain predictions. In this sense, through overcoming primary problems and drawbacks of AI models and taking into account bias within the diagnosis, XAI helps to work out a reliable, accurate, and knowledge-based diagnostic model to improve the quality of treatment and further development of precision medicine.

6. Explainable artificial intelligence (XAI) tailoring personalized treatment plans in precision medicine

The concept of precision medicine is to plan and organize medical management based on the person's specific characteristics such as his or her genetic profile, life experiences, and other aspects. They are meant to offer better and personal than mass approaches to the clients as compared to the generic practices. Thus, even though there are tremendous opportunities in PM for increasing the efficacy of cancer treatment, identification of individual therapy regimens remains a difficult task. The use of XAI into this field provides viable solutions to improve on the parameters that personalize therapies through the use of AI by making the decision-making process more comprehensible and explainable.

This works in the form of Explainable AI which is the set of methods that are aimed at providing the audiences with meaningful insights that can be easily understood by humans. When it comes to the topic of individualized therapy, XAI is the method that eliminates the distance between a sophisticated AI decision-making process and rational therapeutic decisions which physicians make [24]. Feature importance analysis is one of the methods XAI uses to improve the identification of customized patient treatments. It narrows down which aspects, for instance, genetic markers, bio-signatures or environmental factors have the greatest influence on the recommendation made by an AI. For instance, if an AI model recommends a certain drug for a patient, feature importance analysis can uncover which genetic mutations or biomarker was instrumental in making that recommendation. In realizing the part played by these parameters, Doctors are in a position to evaluate why a particular intervention is called for and whether it fits in the other characteristics possessed by the patient. Such transparency helps in decision making so that the healthcare provider either goes ahead with the recommended AI treatment or pursue another course.

Another example of XAI technique is Local interpretable model-agnostic explanations (LIME) which also aids in the process of each patient's individuality and his/her treatment plan. LIME gives a localized interpretation of an instance's prediction because it mimics the behavior of the AI model in the vicinity of a given data point. For instance, there is an AI model that works on the probability of a patient responding well to a particular type of therapy; it can be illustrated with features such as genetic markers or prior responses to treatment. This localized explanation helps the physicians to know the reason behind such a recommendation of an AI for a given patient and thus enables them to adjust their treatment regimen accordingly [25]. Decision trees provide a non-technical method of comprehending AI-predicted treatment plans. It is easy to visualize decision making as a number of if-then scenarios that offers a concise representation of how inputs affect recommendations to choose a treatment. Using a decision tree in personalized medicine, one may see how an individual's genetic characteristics, life style, and other medical-related facts result in a recommended course of action. The above make it easier for the physicians to understand the flow of the recommendation made by the AI thus providing grounds for establishing the validity of the recommendation made by the AI and make necessary changes on the recommended treatment plan [26]. SHAP or SHapley Additive exPlanations is another accurate form of XAI that offers an explanation of how each of the features shapes the AI model's advice. SHAP values are a disaggregate of a treatment recommendation by subcomponents of features, and show how each part influences the prescribed therapy. For example, with SHAP we can discover how specific genetic or environmental factors contribute to the decision made by the AI regarding which treatment should be given. By elaborating details, the physicians are in a position to understand the rationale for recommending such treatment hence enabling them come up with more accurate systems of managing the disease [27]. The integration of XAI in personalization also solves the problem of model validation and integration into the treatment plan as well. XAI enables the practitioners to verify the recommendations arrived by the AI against known clinical wisdom because the suggestions include an explanation component that is easily understandable. Such validation validates the AI treatment plans by checking their compliance with the standard medical treatment guidelines while at the same time verifying that the data fed into the model formulations is accurate. Further, XAI enhances the permeation of AI recommendations within clinical contexts as the recommendations become easier for the healthcare professionals to follow [28].

XAI in personalized medicine also helps in communication with the patient as a form of patient engagement. Thus, Explainable AI will enable physicians who recommend the kind of treatment to their patients to easily explain the rationale behind their decision. By explaining to the patients why a specific treatment is needed considering certain features, the patients will always follow their prescription since they understand the need for such solution. This enhanced communication leads to interprofessional partnership care, in which patient and provider are equals in the process of attaining the best result possible [29].

Some important aspects that need to be taken into consideration when developing IT solutions for personalized treatment planning are the ethical considerations and the patients' right to complete transparency. XAI, therefore, improves the levels of transparency since it reveals the mechanisms that the AI models use in decision making. This transparency also assists in solving some of the ethical issues associated with the use of AI technology in healthcare for instance, to eliminate bias and give equal health treatments to everyone. Thus, XAI supports the development of ethical and promotes trust among patients and healthcare workers [30]. Explainable AI improves the idea of precision medicine by increasing the model interpretability and applying personalization to treatment plans. It is for this reason that feature importance analysis, LIME, decision trees, and SHAP are peculiar techniques that can help physicians to gain insights on aspects that the artificial system has used to provide its recommendations and probably offer recommendations that match the patient's characteristics. The four primary benefits of applying XAI are accuracy and relevance of patients' treatment and increased effectiveness due to transparency, and better patient engagement; it helps to avoid cases when a patient is offered a treatment plan that does not suit them. Thus, the use of XAI in the further development of PM will be critical in the determination of the best personalized treatment plans and consequently the optimization of patients' care [31]. In the aim of achieving precision medicine, an essential aspect is the individualized treatment plans. They strive to give the correct therapies to the patients according to several factors that include genetics, food and other factors in the social setting. Here, there is an irony in the integration of AI as it is indispensable in processing and analyzing such and vast and complicated data, only the outcome is given, and the process leading to that outcome is often 'black boxed,' so the clinicians cannot interrogate it. Intriguingly, Explainable AI (XAI) solves this problem resolving the questions of how the treatment plans are refined and individualized, while taking into account patient needs [32].

There is a host of new and promising fields for the XAI to revolutionize and one of them is the ability to individualize combined therapies. Precision medicine is usually associated with multi-drug prescriptions primarily in chronic and difficult conditions such as cancer or autoimmune diseases. XAI can explain why some drugs and combination of drugs might be influential to a patent by showing how aspects such as; drug interaction, genetic and metabolic rate affect treatment combination. This has the potential to be most valuable in the creation of highly individualized polypharmacy strategies of treatment, where multiple medications are used simultaneously and their interaction can be better analyzed because of synergy [33]. XAI insights are also further used to support understanding of long-term patients outcomes. Compared to most current AI models that are more spot-oriented with regards to treatment prognosis, XAI can provide more light on the future effects of a certain treatment in the course of, for instance, 5 to 10 years. Such individualized predictive avails clinicians the opportunity to develop treatments that consider not only immediate outcomes but also the best probabilities for overall well-being in the long-term [34]. One more promising field for XAI in PM is Genomics, that is, interpretation of genomic data. Genomics also applies in the treatment especially in cancer and hereditary diseases, since most of the treatment plans are tailor-made depending on the genetic conditions of a patient. However, genomic data is well known to be highly diverse and it is a huge challenge to analyze it. It means that XAI can make it easier for doctors to understand how artificial intelligence models process large genomic data to make proper treatment suggestions based on impact of definite genes or definite pathways. This is especially true for such things as distinguishing between minor alleles frequency, which are not so well-documented in medical literature and are so significant for some individuals. With XAI the patient will be presented with a clear view of how these rare variants have been weighed in coming up with the treatment models by the AI models [35]. XAI is expected to benefit patients in a way that was not previously possible. The other area that seems to be missing in the literature of personalized treatment planning is the patients' preferences, concerns, and life. XAI can also help to shift to a more patient-oriented approach by explaining the models' decision-making in terms which are easier to understand not only for the clinicians, but for patients too. For instance, when a particular therapy is suggested, XAI can describe how the patient's mapped genes, their food preferences, and physical activity levels led to it. It also makes it easier for the patients to understand why a particular treatment was suggested to them so that they have a higher chance of fully cooperating with the treatment plan that has been chosen. This will bring the complicated processes of AI in simple ways that will make patients feel and be more involved and confident in their diseases management and care plans. As ethical factors gain significance in AI operational healthcare environment, XAI plays its part to a more ethical approach by promoting the balance and transparency in treatments prescriptions. The delivered individualized treatment plans are themselves insensitive to biases; this may be as a result of relying on AI-based models trained on non-diverse samples. XAI assists in finding out any bias in how the particular groups are treated by showing the impact of the variables on recommendations given to different groups. This is important so that more intelligent healthcare treatments do not merely preserve restrictive and unequal distributions of care throughout society [36]. XAI can be a game changer in enabling collaborative care approaches. In today's context of medicine especially in the management of complicated ailments, various physicians work on the same patient. For every oncologist, geneticist, or pharmacist, XAI allows to grasp how an AI model is recommending precise treatments based on the specialist's knowledge and experience. For instance, a pharmacogenetics can explain to the model to see how the patient's genetics affect drug processing and then explain how the same feature affects cardiac risks during treatment to a cardiologist. This must achieve greater alignment, and shared visual interpretation of AI for more integrated, multidisciplinary and highly coordinated personal care plans of clients [37]. Not only is it enhancing AI model interpretability to support precision medicine but also it is expanding new frontiers for enhancing patients' care mapping plans. Whether it is the ability to grasp the interplay of combination therapies, decisions made based on practical consequences, long-term outcomes predictions, or answering the ethical questions, XAI becomes the more and more crucial element of modern precise, patient-oriented, and ethically sound health care provision. Therefore, as machine learning matures in healthcare, sensitive and effective exposition of AI into clinical practice through XAI is the cornerstone in realizing the potential of personalized medicine.

7. Conclusion

Conclusively, Explainable Artificial Intelligence (XAI) is a progressive role in enhancing the Precision Medicine by providing Explanation, Accountability, and Trust in AI-mediated Health Care Systems. In this respect, considerations of XAI can help healthcare professionals that it is arguably possible to make better decisions when generating treatment recommendations from patient data analytics such as genetic markers, biomarkers as well as environmental factors when such models are explained in simpler terms. It not only helps to improve diagnostic reliability and individualization of treatment but also strengthens the interaction of clinicians and patients. Furthermore, XAI also solves the ethical issues as it ensures equality and fairness in the decision-making process made by the AI models. The advancement of AI in Precision Medicine demands the functionality of XAI to help enhance its reliability and performance hence improving the health of patients and the health care system.

References

[1] R. Confalonieri, L. Coba, B. Wagner, T.R. Besold, A historical perspective of explainable artificial intelligence, Wiley Interdiscip. Rev. Data Min. Knowl. Discov. 11 (1) (2021) e1391.

[2] P.P. Angelov, E.A. Soares, R. Jiang, N.I. Arnold, P.M. Atkinson, Explainable artificial intelligence: an analytical review, Wiley Interdiscip. Rev. Data Min. Knowl. Discov. 11 (5) (2021) e1424.

[3] L. Longo, R. Goebel, F. Lecue, P. Kieseberg, A. Holzinger, Explainable artificial intelligence: concepts, applications, research challenges and visions, in: International Cross-Domain Conference for Machine Learning and Knowledge Extraction, Springer International Publishing, Cham, 2020, August, pp. 1–16.

[4] A. Das, P. Rad, Opportunities and challenges in explainable artificial intelligence (XAI): a survey, 2020. arXiv preprint arXiv:2006.11371.

[5] A. Nayak, A. Patnaik, I. Satpathy, B.C.M. Patnaik, A. Khang, Incorporating artificial intelligence (AI) for precision medicine: a narrative analysis, in: AI and IoT-Based Technologies for Precision Medicine, IGI Global, 2023, pp. 16–35.

[6] A.G. Akpan, F.B. Nkubli, J.C. Mbazor, G. Luntsi, O. Udeme, XAI methods for precision medicine in medical decision support systems, in: Explainable Artificial Intelligence in Medical Decision Support Systems, vol. 50, 2022, p. 471.

[7] M. Gimeno, E. San José-Enériz, S. Villar, X. Agirre, F. Prosper, A. Rubio, F. Carazo, Explainable artificial intelligence for precision medicine in acute myeloid leukemia, Front. Immunol. 13 (2022) 977358.

[8] B. Allen, The promise of explainable AI in digital health for precision medicine: a systematic review, J. Pers. Med. 14 (3) (2024) 277.

[9] J.M. Durán, Dissecting scientific explanation in AI (sXAI): a case for medicine and healthcare, Artif. Intell. 297 (2021) 103498.

[10] F. Klauschen, J. Dippel, P. Keyl, P. Jurmeister, M. Bockmayr, A. Mock, et al., Toward explainable artificial intelligence for precision pathology, Annu. Rev. Pathol. 19 (1) (2024) 541–570.

[11] L. Saba, M. Maindarkar, N.N. Khanna, A.M. Johri, L. Mantella, J.R. Laird, J.S. Suri, A pharmaceutical paradigm for cardiovascular composite risk assessment using novel radiogenomics risk predictors in precision explainable artificial intelligence framework: clinical trial tool, Front. Biosci. 28 (10) (2023) 248.

[12] T. Hulsen, Explainable artificial intelligence (XAI): concepts and challenges in healthcare, AI 4 (3) (2023) 652–666.

[13] K. Sheriffdeen, S. Daniel, Explainable Artificial Intelligence for Interpreting and Understanding Diabetes Prediction Models, 2024 (Report [No. 2516-2314]).

[14] J. Stodt, C. Reich, M. Knahl, Demystifying XAI: requirements for understandable XAI explanations, Stud. Health Technol. Inform. 316 (2024) 565–569.

[15] V. Chamola, V. Hassija, A.R. Sulthana, D. Ghosh, D. Dhingra, B. Sikdar, A review of trustworthy and explainable artificial intelligence (XAI), IEEE Access 11 (2023) 78994–79015.

[16] S. Ali, T. Abuhmed, S. El-Sappagh, K. Muhammad, J.M. Alonso-Moral, R. Confalonieri, et al., Explainable artificial intelligence (XAI): what we know and what is left to attain trustworthy artificial intelligence, Inf. Fusion 99 (2023) 101805.

[17] V. Kumar, From black box to clarity: demystifying explainable AI in data engineering pipelines, Soc. Sci. Spectr. 3 (1) (2024) 157–171.

[18] K. Przystalski, R.M. Thanki, Explainable Machine Learning in Medicine, Springer, 2024.

[19] Y. Zhang, Y. Weng, J. Lund, Applications of explainable artificial intelligence in diagnosis and surgery, Diagnostics 12 (2) (2022) 237.

[20] S.M. Anwar, AIM and explainable methods in medical imaging and diagnostics, in: Artificial Intelligence in Medicine, Springer International Publishing, Cham, 2022, pp. 501–510.

[21] E. Patel, Into the Future: How Artificial Intelligence Is Shaping Precision Medicine, 2023.

[22] Á. Torres-Martos, A. Anguita-Ruiz, M. Bustos-Aibar, A. Ramírez-Mena, M. Arteaga, G. Bueno, et al., Multiomics and eXplainable artificial intelligence for decision support in insulin resistance early diagnosis: a pediatric population-based longitudinal study, Artif. Intell. Med. (2024) 102962.

[23] D. Croce, A. Smirnov, L. Tiburzi, S. Travaglini, R. Costa, A. Calabrese, et al., AI-driven transcriptomic encoders: from explainable models to accurate, sample-independent cancer diagnostics, Expert Syst. Appl. (2024) 125126.

[24] I. Rahman, AI-powered personalized treatment recommendation framework for improved healthcare outcomes, J. Comput. Soc. Sci. 8 (11) (2023) 42–51.

[25] A. Ayesha, N.N. Ahamed, Explainable artificial intelligence (EAI), in: Explainable Artificial Intelligence for Biomedical and Healthcare Applications, 2024, p. 162.

[26] A. Garg, A.K. Singh, A. Kumar, Mental disorders management using explainable artificial intelligence (XAI), in: Explainable Artificial Intelligence for Biomedical and Healthcare Applications, CRC Press, 2024, pp. 113–138.

[27] A. Ayesha, N.N. Ahamed, Explainable artificial intelligence (EAI): for healthcare applications and improvements, in: Explainable Artificial Intelligence for Biomedical and Healthcare Applications, CRC Press, 2024, pp. 162–196.

[28] İ. Uysal, U. Kose, Explainability and the role of digital twins in personalized medicine and healthcare optimization, in: Explainable Artificial Intelligence (XAI) in Healthcare, CRC Press, 2024, pp. 141–156.

[29] F.H. Yagin, S. Yasar, Y. Gormez, B. Yagin, A. Pinar, A. Alkhateeb, L.P. Ardigò, Explainable artificial intelligence paves the way in precision diagnostics and biomarker discovery for the subclass of diabetic retinopathy in type 2 diabetics, Metabolites 13 (12) (2023) 1204.

[30] M.A. Zahra, A. Al-Taher, M. Alquhaidan, T. Hussain, I. Ismail, I. Raya, M. Kandeel, The synergy of artificial intelligence and personalized medicine for the enhanced diagnosis, treatment, and prevention of disease, Drug Metabol. Personal. Ther. 39 (2) (2024) 47–58.

[31] D.S. Ross, A. Bala, J. Arun, Enhancing patient care and treatment through explainable AI: a gap analysis, in: Artificial Intelligence in Medicine, CRC Press, 2024, pp. 203–214.

[32] L. Gaur, D. Gaur, Explainable artificial intelligence (XAI) on neurogenerative diseases, in: AI and Neuro-Degenerative Diseases: Insights and Solutions, Springer Nature Switzerland, Cham, 2024, pp. 63–72.

[33] J. Gerlings, M.S. Jensen, A. Shollo, Explainable AI, but explainable to whom? An exploratory case study of xAI in healthcare, in: Handbook of Artificial Intelligence in Healthcare: Vol 2: Practicalities and Prospects, 2022, pp. 169–198.

[34] K. Shiva, P. Etikani, V.V.S.R. Bhaskar, S. Nuguri, A. Dave, Explainable AI for personalized learning: improving student outcomes, Int. J. Res. Innov. Multidiscip. Stud. 3 (2) (2024) 198–207. ISSN: 2960-2068.

[35] X. Gao, H. Li, C. You, L. Zhou, Q. Shi, Z. Yang, et al., Challenges and Advances on Explainable Artificial Intelligence (AI): Diagnosing and Treating Tumors of the Female Reproductive Systems, 2024 (preprints).

[36] U.T. Kumbhar, R. Phursule, V.C. Patil, R.K. Moje, O.R. Shete, M.A. Tayal, Explainable AI-powered IoT systems for predictive and preventive healthcare—a framework for personalized health management and wellness optimization, J. Electr. Syst. 19 (3) (2023).

[37] S.R. Sindiramutty, W.J. Tee, S. Balakrishnan, S. Kaur, R. Thangaveloo, H. Jazri, et al., Explainable AI in healthcare application, in: Advances in Explainable AI Applications for Smart Cities, IGI Global, 2024, pp. 123–176.

CHAPTER 11

Pneumonia and brain tumors diagnosis using machine learning algorithms

Srđan Filipović and Seifedine Kadry
Department of Applied Data Science, Noroff University College, Kristiansand, Norway

1. Introduction

Artificial intelligence (AI) is a broad field with numerous applications. The main three areas of research on AI are Natural Language Processing (NLP), Computer Vision (CV), and Robotics (RO). All these research areas are used in medicine, but CV is most widely applied for medical diagnoses. The idea of CV has been around since the mid-20th century, but the applications of CV increased after the development of AlexNet [1] in 2012. In the following years, different CNN architectures have been improved, and it is expected that they will keep on improving. AI computer systems are used substantially in modern medical research and medical practices. These computer systems help researchers advance research and help medical practitioners optimize their clinical decision-making. Some state-of-the-art systems are achieving comparable accuracy to experts in relevant medical fields.

There are many different types of AI computer systems, but all work to ease and improve medical practitioners' and researchers' jobs. Most newly developed medical software developed to help medical diagnosticians were built using Machine Learning (ML) and Deep Learning (DL). Some state-of-the-art software is software for the diagnosis of Dementia (specifically Alzheimer's) [2,3], Tumors (Brain and Brest) [4], and the newest for COVID-19 [5,6]. These diagnostic software works using image recognition built using different ML approaches and algorithms. Because of ethical reasons, this software cannot replace diagnosticians but only improve their diagnosis. The introduction of AI to medicinal diagnostics has had a significant impact on diagnosing rarer diseases and the reduction of false positive diagnoses. According to a chapter [7] dedicated to developing an AI system for diagnosing breast cancer, the usage of AI helped radiologists reduce false positive rates by 37.3% and biopsies by 27.8% while not reducing the sensitivity of diagnosis. Moreover, according to another chapter [8], the AI diagnosis scored 85% accuracy in detecting rare diseases while radiologists scored 47% accuracy. It is evident that AI has improved the total accuracy of diagnosis, and researchers continue to improve upon it. According to a review article [9], software for the broad diagnosis of lung X-rays has accuracy comparable to three to four radiologists. In comparison, a

Explainable AI in Healthcare Imaging for Medical Diagnoses
https://doi.org/10.1016/B978-0-443-23979-3.00011-7

narrower software for diagnosing cancerous pulmonary nodules (an abnormal growth that forms in a lung) exceeds the accuracy of 17 out of 18 radiologists. Moreover, similar findings can be seen in another research [8], where software brain MRI general diagnosis accuracy exceeded radiologists' accuracy by more than 20%. The highest difference can be seen with rarer diagnoses. Nevertheless, when radiologists work in conjunction with software, the accuracy of positive diagnosis increases while false positive diagnosis decreases even more. There are three main methods to achieve image classification, which are supervised, unsupervised, and object-based classifications. The principal algorithms used in unsupervised classification are K-means clustering and ISODATA (Iterative Self Organizing Data Analysis Technique). An example of K-means clustering can be seen from [10] analysis of combining CNN with K-Means clustering for lung cancer diagnosis. The most used algorithm used for supervised classification is SVM. An example of the usage of supervised classification can be seen in research for diagnosing of Alzheimer's disease using machine learning [2]. Both supervised and unsupervised classifications are pixel-based (each pixel belongs to a class). Object-based classification can be found in detail in [11] a review for remote sensing. This type of classification is rarely used in medicine, with few exceptions in some research [12,13]. Across various research chapters, there are many ways the images were augmented for training the model. Models with the highest Sensitivity and Specificity were trained using Deep Convolutional Neural Network (DCNN) alone or in combination with Support-Vector Machines (SVM). Nevertheless, in a recent research [2], the focus was primarily on feature selection while disregarding all DL techniques. In this research, researchers used many different supervised learning algorithms, while SVM was the most prevalent. The accuracy of the trained model, which was multi-diagnostic, was comparable to models trained using DL techniques. Most of the research chapters focusing on research of such models, which have been written in the last 4 years, use mostly DL techniques, while older articles extensively use SVM or hybrid methods using the SVM-KNN approach for classification. The following sections of the introduction will address this research's clear aims and objectives, what approach will be taken for achieving such objectives, and risks, if any, related to this research.

This chapter contribution can be divided into following points:
- State of the art pneumonia detection and differentiation accuracy
- State of the art Brain Tumor detection and differentiation accuracy
- Implementation of CNN-SVM for differentiation between Viral and Bacterial Pneumonia
- A simple and fast tool for automatic prediction

1.1 Problem statement

This chapter aims to develop software containing multiple Machine Learning models for accurately diagnosing Brain Tumors and Lung Pneumonia from radiology imaging. The software will incorporate these models into a single interface for efficient diagnosis. The study will explore the accuracy and effectiveness of the individual ML models and the integration of these models into a unified software platform with the possibility of expansion.

1.2 Aims & objectives

The primary objective of this research is to build multiple models for the diagnosis of Brain Tumors and Lung Pneumonia and incorporate those models within one software. To achieve this primary objective, the following secondary objectives must be achieved:

- Firstly, relevant data, in this case, medical images, must be acquired and processed for use.
- Secondly, it is important to identify different ML techniques which have been used in the last 5–10 years for similar research.
- Next objective is to build models using relevant techniques acquired in the research and train them separately for the prediction of Tumor and Pneumonia.
- Afterwards, ensemble techniques such as bagging, boosting, and stacking will be explored for combining the models and improving the overall accuracy of the software.
- Last objective will be to build an interface that will accept uploads of different formats of images which include PNG, JPG, and DCM (DICOM).

1.3 Scope and limits

The scope of this research is limited to the development of software that incorporates multiple ML models for the diagnosis of Primary Brain Tumors and Lung Pneumonia and the differentiation between its types. The Brain dataset only includes primary tumor types and does not include other brain pathologies, such as metastasis and brain cysts. As a result, models trained on this dataset are unable to distinguish between these different types of pathologies. These models were trained to classify individual images rather than full brain MRI scans, which limits their diagnostic accuracy. Additionally, due to time constraints, this work did not include other pathologies and parts of the body, such as breast cancer and lung cancer. Moreover, this work does not include PET scans or ultrasounds because of the lack of data and time limitations. It is essential to acknowledge these limitations in the interpretation of the models' evaluations.

1.4 Ethical considerations

As with any technology that involves human health, it is important to consider ethical implications when developing machine learning models for medical diagnosis. It is also essential to recognize that these models are meant to be a supportive tool for medical professionals and should not replace the expertise of trained physicians. The accuracy percentage displayed in the final GUI is based solely on the model's predictions and does not consider the accuracy of the underlying base models. Therefore, it should be interpreted with caution and not taken as an absolute indicator of accuracy. Privacy concerns were considered during the research, as no patient-identifying information was used during training. Only the pixel values of images were extracted from DICOM files.

2. Literature review

Radiologists often rely on X-rays, CT scans, or MRIs to determine if a patient has a disease. X-rays are mostly used to detect any problems in hard tissue such as bones and teeth, but sometimes they are used for diagnosing internal organs disease such as Pneumonia found in the lungs. To help medical professionals, researchers have developed ML models, which have, over time, outperformed most radiologists. Today, the ML models did not replace medical professionals in medical diagnosis, but they became a valuable tool which helps them reach more accurate and faster decisions. Even though the idea of Artificial Intelligence has existed since 1950 [14], its applications in medicine were almost non-existent until the introduction of DL in the year 2000 [15]. After its introduction, DL has greatly improved Computer Vision and its applications. In 2011 researchers [16] developed one of the first ML models for the detection of Pneumonia from CT scans using Cellular Neural Network architecture. Soon the Cellular NN and other image classification methods became obsolete with advancements in Convolutional Neural Network (CNN) architectures. The rapid advancement in DL techniques, particularly in the domain of CNNs, has led to significant improvements in the performance and capabilities models used for disease detection.

2.1 Convolutional neural networks

As mentioned earlier, advancements in CNN architectures have significantly improved image classification in many research areas and industries, including medical diagnostics, autonomous driving vehicles, intelligent video analytics, and more. CNN is a specialized NN that is designed to segment and filter pixels of an image as a feature extraction to reach a prediction. Its name is derived from an operation layer called convolution. A standard CNN architecture consists of three crucial layers which are:
- Convolution: A layer used for feature extraction that separates parts of the image. It applies a set of filters to an input image and creates a set of output feature maps that highlight different aspects of the input.

- Pooling: A layer that combines the output of the convolution layer, reducing its dimensionality and highlighting important extracted regions if Max Pooling is used. The most common type of pooling is Max Pooling, which selects the maximum value in a rectangular sub-region of the feature map.
- Fully connected: Layers used for final classification. These layers are fully connected to all the neurons in the previous and subsequent layers, allowing the model to learn complex relationships between extracted features and output classes.

AlexNet CNN architecture can be considered one of the most influential breakthroughs in Computer Vision as the original AlexNet chapter [1] was cited over 120,000 times, creating a way for the development of modern, more advanced CNN architectures in the future years.

2.1.1 AlexNet

One of the reasons why AlexNet was so influential was that it could be trained with multiple GPUs [1] which greatly reduced training time and allowed for the more complex architectures in the future. Using two GPUs allowed for parallel processing, with one GPU training on grayscale images and the other on the original-colored images. Furthermore, the use of multiple GPUs allowed AlexNet to handle large datasets complex tasks. Another key innovation of AlexNet architecture was the introduction of the ReLU activation function, which helped reduce vanishing gradient problem in the FCNN. AlexNet architecture has a simple and fast design which targets larger areas of the image at the first convolution with a large stride. With this, the model can assess important areas of the image and focus on them. Whole model consists of five convolution layers, two max pooling and three fully connected layers as can be seen Fig. 1. What cannot be seen in the figure is usage of Dropout layers which helped reduce overfitting by randomly setting portion (50% for AlexNet) of weights to 0.

2.1.2 VGG-16

In contrast to AlexNet, VGG-16 uses a small first convolutional layer that targets smaller areas and assesses more details from the images. VGG-16 was designed to capture smaller details through small filters and with that achieved state of the art accuracy with ImageNet dataset in 2014. Every convolution layer uses 3×3 size filters and the same Max Pooling. After each of the first three Max Pooling layers, the number of filters doubles, starting from 64 at the first convolution layer. The difference between VGG-16 and VGG-19 is that VGG-19 has three extra convolutional layers while other layers are identical. One drawback of the VGG architectures is their large number of parameters which result in very high computational complexity.

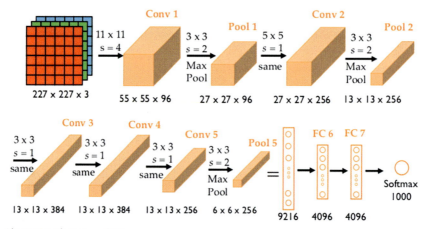

Fig. 1 AlexNet architecture [17].

2.1.3 DenseNet

The DenseNet architecture has four distinct variants that differ in block sizes and the total number of layers. DenseNet variants are DenseNet-121, DenseNet-169, DenseNet-201, and DenseNet-264. The number in the name of each variant represents the number of layers. Every variant trained on the ImageNet database introduced in the DenseNet chapter [18] consists of four Dense Blocks, three Transition Blocks, Classification Layer, and one simple convolution and Max Pooling, which can be seen in Fig. 2. Inside every Dense Block, each layer has a direct connection to each subsequent layer which allows for fewer parameters, and such architecture is less prone to overfitting.

Layers	Output Size	DenseNet-121	DenseNet-169	DenseNet-201	DenseNet-264
Convolution	112 × 112	7 × 7 conv, stride 2			
Pooling	56 × 56	3 × 3 max pool, stride 2			
Dense Block (1)	56 × 56	[1 × 1 conv; 3 × 3 conv] × 6	[1 × 1 conv; 3 × 3 conv] × 6	[1 × 1 conv; 3 × 3 conv] × 6	[1 × 1 conv; 3 × 3 conv] × 6
Transition Layer (1)	56 × 56	1 × 1 conv			
	28 × 28	2 × 2 average pool, stride 2			
Dense Block (2)	28 × 28	[1 × 1 conv; 3 × 3 conv] × 12	[1 × 1 conv; 3 × 3 conv] × 12	[1 × 1 conv; 3 × 3 conv] × 12	[1 × 1 conv; 3 × 3 conv] × 12
Transition Layer (2)	28 × 28	1 × 1 conv			
	14 × 14	2 × 2 average pool, stride 2			
Dense Block (3)	14 × 14	[1 × 1 conv; 3 × 3 conv] × 24	[1 × 1 conv; 3 × 3 conv] × 32	[1 × 1 conv; 3 × 3 conv] × 48	[1 × 1 conv; 3 × 3 conv] × 64
Transition Layer (3)	14 × 14	1 × 1 conv			
	7 × 7	2 × 2 average pool, stride 2			
Dense Block (4)	7 × 7	[1 × 1 conv; 3 × 3 conv] × 16	[1 × 1 conv; 3 × 3 conv] × 32	[1 × 1 conv; 3 × 3 conv] × 32	[1 × 1 conv; 3 × 3 conv] × 48
Classification	1 × 1	7 × 7 global average pool			
Layer		1000D fully-connected, softmax			

Fig. 2 DenseNet architecture [18].

2.1.4 NASNet

NAS stands for Neural Architecture Search, which is an ML algorithm used for automating the designs of NNs. Therefore, NASNet is not a CNN architecture. NASNet [19] uses a variant of reinforcement learning that uses policy gradient optimization to search for an optimal NN architecture. The idea behind this algorithm is to create an optimal CNN architecture for any given dataset. Modern architectures can be very complex, such as Inception architectures [GoogLeNet [20], Inception-v3 [21]], and creating them manually for each CV problem can be very time and resource-consuming. This is a problem that NASNet tries to address. In 2018 [22], NASNet achieved state-of-the-art accuracy on the ImageNet dataset and used nine billion fewer floating-point operations than the previously best-created architecture for the same dataset. In Fig. 3, the recurrent cell construction process can be seen depicted as three separate images.

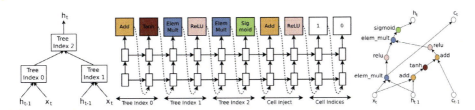

Fig. 3 Illustrations of the recurrent cell construction process in NASNet [19].

2.1.5 Classifiers

CNNs do not have to include Fully Connected layers, and it is possible to use another classification method in combination with an CNN. In recent years CNNs have been used for feature extraction, but the classifier differed from chapter to chapter. The most popular classifiers are Fully Connected Neural Networks (FCNN), K-Nearest Neighbor (KNN), Linear Regression (LR), and Support Vector Machine (SVM). Both the CNN architecture and the choice of classifier influence final model accuracy. One of the few chapters [23] that does not utilize CNN for feature extraction is from 2016. While they did not directly use CNNs, they combined a Learning Vector Quantization network with other non-CNN feature extraction methods to achieve 97.45% accuracy. However, this approach required a very complex feature extraction process and a custom ANN classifier for model training. This chapter highlights that it is possible to build models without using CNNs, but at the cost of time, resources, and potentially inferior performance compared to CNNs, especially for state-of-the-art architectures. The CNN architecture can be directly used for classification, without the need for Fully Connected layers, by using a variant called Fully Convoluted Networks (FCNs) [24]. FCNs rely entirely on convolutional layers for both feature extraction and classification. The Soft-Max function is applied to obtain class probabilities on the final convolutional layer. Because FCNs preserve the spatial information of the input image, it is well suited for image segmentation tasks.

2.1.6 Fully connected neural network

Fully Connected Neural Networks (FCNNs) are a type of neural network where every node in a layer is connected to every node in the subsequent layer, and these connections are represented by weights. FCNNs are used for various tasks, including classification, regression, and other ML problems. FC-NNs have shown to be exceptionally effective in modeling non-linear relations. Additionally, they are effective in a wide range of applications, including image and speech recognition, natural language processing, and robotics, among others. Furthermore, advances in hardware and software have made it possible to train deeper and larger FCNNs. After the CNN extracts image features, those features are typically flattened into a one-dimensional vector before being fed as input to the first layer of an FCNN. This flattening step effectively converts the spatial structure of the feature maps into a linear sequence of values. Conversion is necessary as the FCNN input layer expects one-dimensional input. The output of the last layer of the FCNN represents the network's predicted class probabilities, and the number of nodes in the output layer corresponds to the number of classes that the network is trained to predict. Example architecture of FCNN can be seen in Fig. 4. During training, FCNNs use forward propagation to compute the output of each layer, and they use backward propagation to update the weights based on the error between the predicted output and the true output. One of the biggest issues with FCNNs is the vanishing gradient problem that arises when the gradient of the loss function, with respect to the weights of earlier layers, becomes very small during backward propagation. This can result in slow or non-existent weight updates for these layers, leading to negligible improvements or even complete failure of the network to learn. Techniques such as random weight initialization, and layer activation functions like ReLU are used to mitigate this problem. The second biggest issue with FCNNs is overfitting, which occurs when the network has too many parameters relative to the amount of training data. This occurs because the network starts memorizing the training data instead of being able to generalize and perform well on new, unseen data.

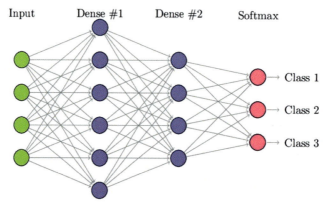

Fig. 4 Fully connected neural network architecture [25].

2.1.7 K-Nearest Neighbor

K-Nearest Neighbor (KNN) is an ML algorithm that can be used for both classification and regression. The KNN classification works in such a way that a new data points class is predicted based on the similarity of K neighbors where K represents a number. All data points are projected to a 2D plane, and the algorithm calculates a new data point's proximity to the other data points in the dataset, which it calculates based on the chosen distance metric. KNN has several advantages over other algorithms, and the biggest is that it is simple to understand and implement and can be used for both binary and multi-class classification problems. Additionally, KNN does not require any training time. Unfortunately, the computation complexity grows linearly with the size of the data. Moreover, the high dimensionality of data can result in a very high computation time, especially in image recognition tasks. To improve KNN's performance, feature space dimensionality reduction techniques, such as Principal Component Analysis (PCA) and t-distributed Stochastic Neighborhood Embedding (t-SNE), can improve KNNs' performance. Additionally, selecting the appropriate distance metric and K value is crucial for the effectiveness of KNN predictions. In Fig. 5, illustration of KNN data points can be seen, and how choosing between $K=5$ and $K=10$ can influence the model's final prediction. There are several methods to determine the optimal value of K, including trial-and-error, cross-validation, and grid search. Similarly, there are several distance metrics to choose from, but the most popular is Euclidean distance which represents the shortest line between two points. It is essential to experiment with different distance metrics and values of K to determine the optimal combination for the given problem.

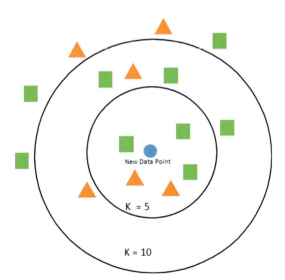

Fig. 5 Example of how K influences final prediction [26].

2.1.8 Support Vector Machine

Support Vector Machine (SVM) algorithm is a supervised learning method which is used for various tasks, including image classification. SVM can be a good choice for image classification tasks be-because it can handle high-dimensional data. In combination with CNNs (CNN-SVM method), SVM can be trained on CNN extracted features to perform image classification with high efficiency, especially for smaller datasets. This approach uses the strengths of both algorithms, with the CNN acting as an effective feature extractor and the SVM handling the final classification based on those features. Therefore, in the case when new images cannot be generated by augmentation and the dataset is small, CNN-SVM can be a good method to use. Another advantage of using SVM in combination with CNN is its ability to help regularize the model and prevent overfitting by optimizing the margin between classes. SVMs offer the ability to use different kernel functions, such as Linear (Fig. 6), Polynomial, or Radial Basis Function (RBF), which enable SVMs to handle both linear and non-linear classification problems. This versatility makes SVM a suitable choice for a wide range of image classification tasks, regardless of the underlying data distribution. There are however limitations and issues with using SVM. SVMs do not scale well with larger datasets, and choosing the best kernel and parameters can be a difficult and time-consuming process. Additionally, SVMs are only used for binary classification. While it is technically possible to use SVM for multiclass classification it can be done only by training multiple SVM models for each pair of the classes in the dataset which is impractical.

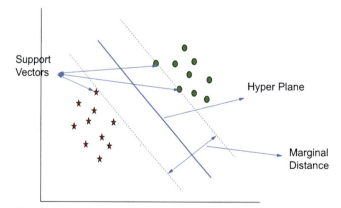

Fig. 6 Example of linear SVM [27].

2.2 Review of specific diseases

2.2.1 Pneumonia

Pneumonia is a highly infectious and deadly disease, and its early diagnosis is vital for treatment, especially for children. According to World Health (WHO) [28] there were 2.5 million people died from Pneumonia in 2019, and almost a third of these deaths were children under 5 years old. Pneumonia can be detected with different methods, which can be radiology imaging such as X-ray and CT. Even though it is easier to detect Pneumonia with CT, usage of CT is avoided because of both cost and increased radiation exposure. Therefore, most of the models built for its detection are trained on X-ray images. Pneumonia can be divided into three types depending on the type of infection which causes it. These are Bacterial, Viral, and Fungal. As Fungal Pneumonia is a very rare but very dangerous type of Pneumonia, it usually affects people with a weakened immune system. Nevertheless, because of limited radiology images, it will be disregarded in this research. The most common type is Bacterial Pneumonia, where the most common cause is Streptococcus bacteria [29]. Bacterial Pneumonia more commonly causes severe symptoms than Viral Pneumonia and is very likely to require hospital treatment, while Viral can resolve itself over time. Additionally, there is a much higher chance for Viral Pneumonia to spread from one person to another than other types of Pneumonia. Viral Pneumonia is most caused by flu, respiratory syncytial virus, or Coronaviruses. Even though COVID-19 is caused by the SARS-CoV-2 virus, we will not be grouping it with Viral Pneumonia as it is possible to distinguish COVID-19 in X-ray images from other Viral Pneumonia diseases such as Influenza and Respiratory Syncytial Virus and because of the recent pandemic. Pneumonia can also be categorized based on where and when the infection is acquired. Community-Acquired Pneumonia (CAP) occurs outside of hospitals, such as at home or in public spaces, while Hospital-Acquired Pneumonia (HAP) develops during their hospital stay. Even if a patient is diagnosed with Pneumonia inside of a hospital, it can be classified as CAP if Pneumonia develops within 48 h of the patient's stay in the hospital. HAP is often more severe and difficult to treat due to the potential of antibiotic-resistant bacteria. Identifying the source of the infection can play a crucial role in determining the appropriate treatment and management of the disease. Over the years, numerous studies have been focused on developing and improving techniques for Pneumonia detection, using a range of techniques and technologies to improve diagnostic accuracy and efficiency. In their chapter [30], researchers reported one of the best results for predicting if someone has Pneumonia and its type. They achieved 93.3% total accuracy while they used a pre-trained DenseNet201 CNN network. It is worth noting that the data they reported using does not match the test data they used. Even though the number of train data images is the same, the test data they reported for Normal, Bacterial, and Viral classes was 199, 197, and 201 images, respectively, while the dataset they reported using had 234, 242, and 148 images. During training, they also used image augmentation, such as rotation (+−45 deg), scaling (+−10%), and translation (+−10%). With this augmentation, they have had 4500 images for each of the classes. In the 2021 chapter [31], researchers reviewed AlexNet, SqueezeNet, VGG16, VGG19, and Inception V3 pre-trained CNN architectures in combination with different classifiers for the detection of Pneumonia without differentiating between different pneumonia types. The highest accuracy they achieved was 96.44% using AlexNet with a DNN classifier, while LR and KNN were not far behind. Similarly, in a study using pediatric X-rays [32], researchers reviewed AlexNet, DenseNet121, InceptionV3, GoogLeNet, and ResNet18 and ResNet18 architectures had the best results with 94.23% accuracy. When using Ensemble, the accuracy had risen to 96.39%. Since the COVID-19 pandemic began in 2020, a lot of chapters regarding the detection of COVID-19 from

X-rays came out. In one chapter [33], researchers achieved 95.79% accuracy using ResNet50 CNN architecture with an SVM classifier. In a similar research chapter [34], the researchers achieve an accuracy of 94.5% when detecting COVID-19. In other re-search chapters focused on classifying COVID-19 [35,36], researchers report an accuracy of over 98% using different combinations of CNN architectures in combination with different classifiers with best scores resulting from VGG19-RNN architecture with accuracy being 99.86%. In Table 1, it is possible to see a summary of the key findings from the various studies mentioned in this section, including their highest achieved accuracy, the methods they employed to reach final accuracy, and the specific classes each study focused on predicting.

Table 1 Chapters with focus on detection of Pneumonia.

Chapter	Dataset	Classes	Method	Accuracy
Ismael and Şengür [33]	Radiology assistant chest dataset (un-available)	COVID-19, Normal	ResNet50 +SVM	95.79
Haritha et al. [35]	Unavailable	COVID-19, Normal	GoogLeNet	98.52
Heidari et al. [34]	Combination of 5 datasets	COVID-19, Normal, Pneumonia	VGG16	94.50
Islam et al. [36]	Combination of 3 datasets	COVID-19, Normal, Pneumonia	VGG19-RNN	99.86
Chouhan et al. [32]	Guangzhou Woman and Children's Medical Center Dataset	Normal, Pneumonia	5 Model Ensemble	96.39
Muhammad et al. [31]	Kaggle Chest X-ray Images (Pneumonia)	Normal, Pneumonia	Inception-V3	97.19
Hashmi et al. [37]	Large Dataset of Labeled OCT and Chest X-ray Images	Normal, Pneumonia	5 Model Ensemble	98.43
Elshennawy and Ibrahim [38]	Kaggle Chest X-ray Images (Pneumonia)	Normal, Pneumonia	ResNet152V2	99.22
Rajaraman et al. [39]	Guangzhou Woman and Children's Medical Center Dataset	Normal, Pneumonia-Bacterial, Pneumonia-Viral	Custom VGG16	91.80
Rahman et al. [30]	Kaggle Chest X-ray Images (Pneumonia)	Normal, Pneumonia-Bacterial, Pneumonia-Viral	DenseNet201	93.30

2.2.2 Brain tumor

The brain is considered as one of the most important parts of our body. Therefore, an accurate and fast diagnosis of potential and current issues with it is vital. Brain tumors are abnormal growths in the brain that can be malignant (cancerous) or benign. They can be primary or secondary (metastasis). Primary brain tumors originate in the brain, while secondary spread to the brain from another part of the body, such as the lungs or breasts. Metastases are usually smaller than primary brain tumors, and there can be more of them which can result in a lower detection rate. When trying to detect a brain tumor using radiology imaging MRI is preferred over CT because of a couple of reasons. One of the main reasons is that MRI does not use ionizing radiation, whereas CT does. It also provides a better 3D image which allows an understanding of the exact location of the tumor. Additionally, MRI shows differences between various soft tissue, which CT does not. In this section, we will review three types of primary Brain Tumors relevant to this work: Gliomas, Meningiomas, and Pituitary Tumors. The percentages mentioned in the following paragraphs change depending on the source [40–43], therefore the average percentage will be mentioned. Example images of these tumors can be seen in Fig. 10.

Firstly, Gliomas are the most common malignant brain tumors, accounting for approximately 30% of all brain tumors and 80% of all malignant brain tumors. They originate in the glial cells (gliocytes), which are non-neuronal cells in the nervous system that provide support and protection for neurons. Gliomas can be found anywhere in the cerebral hemispheres of the brain, but they are most found in the frontal and temporal lobes. Secondly, Meningioma is the most common primary Brain Tumor, accounting for more than 33% of all Brain Tumors. Unlike Gliomas, Meningiomas are usually benign, meaning they do not spread or invade surrounding tissues. They originate in the meninges, which are the three layers that envelop the brain and spinal cord and separate them from the bone, providing protection and support for the central nervous system. Lastly, Pituitary Tumors are abnormal growths in the pituitary gland, also called hypophysis, which plays a critical role in hormone regulation and growth. Approximately 14% of all primary brain tumors are Pituitary Tumors. However, it is worth noting that most Pituitary Tumors do not cause any symptoms, which can result in the tumor not being diagnosed. Therefore, the actual percentage of Pituitary tumors could be much higher than reported. Improved brain tumor detection can significantly influence patients' treatment options and overall health. Early and accurate diagnosis of Brain Tumors allows for an early intervention, which can lead to better prognoses and reduced mortality rates. Furthermore, precise identification of tumor types and their locations can help specific treatment plans for individual patients, improving the effectiveness of therapies such as surgery, radiation, or chemotherapy. This personalized approach can also minimize the side effects and complications associated with more aggressive treatments. Additionally, advancements in Brain Tumor detection research can reduce healthcare costs by minimizing the need for multiple diagnostic tests and intrusive procedures. Given the significant benefits of improved Brain Tumor detection, a considerable amount of research chapters has focused on detecting Gliomas, Meningiomas, and Pituitary tumors. In one research chapter [44], the researchers claim a 97.1% accuracy using SVM classifier on GoogLeNet CNN extracted features while they achieved 92.3% using the standard CNN-ANN method. They have also tested the CNN-KNN method and achieved 98% test accuracy.

Even though they report that the CNN-KNN method has the highest accuracy, they propose using the CNN-SVM method. One of the highest achieving chapters regarding the detection of Gliomas, Meningiomas, and Pituitary tumors [45] reports 98.69% accuracy using modified VGG16 architecture. In this chapter, they reported using the "fine-tune and freeze" transfer learning technique. This technique represents freezing the pre-trained layers of the model and training only added layers, and after the training is finished, the earlier layers are unfrozen while the new layers are frozen. Afterward, the model is retrained on a new dataset. A particularly interesting study [46] proposed a three-model classification system for Brain Tumor detection. The first model determines the presence of any Brain Tumors, including primary tumors and metastases. The second model identifies the type of tumor (Gliomas, Meningiomas, Pituitary, or metastases), while the third model detects the tumor's grade. They used two Grid Search algorithms for the optimization of CNN architecture and hyper-parameters during training. With this CNN model, they achieved 99.33%, 92.66%, and 98.14% accuracy for the first, second, and third model, respectively.

Table 2 Chapters with focus on detection of Brain Tumors.

Chapter	Dataset	Classes	Method	Accuracy
Das et al. [47]	Figshare	Glioma, Meningioma, Pituitary	Custom CNN	93.33
Anjum et al. [48]	Figshare	Glioma, Meningioma, Pituitary	ResNet101	95.65
Sadoon and Ali [49]	Nanfang Hospital, China	Glioma, Meningioma, Pituitary	Custom CNN	96.1
Deepak and Ameer [44]	Figshare	Glioma, Meningioma, Pituitary	GoogLeNet-SVM	97.10
Rehman et al. [45]	Figshare	Glioma, Meningioma, Pituitary	Custom VGG16	98.69
Irmak [46]	Combination of 4 datasets	Normal, Glioma, Meningioma, Pituitary, Metastases	Custom Automatically Generated CNN	92.66

In conclusion, the Literature Review has highlighted the importance of early diagnosis and accurate classification of various diseases, such as Pneumonia, COVID-19, and Brain Tumors. This review has shown the increasing use of DL techniques, particularly CNNs, for detecting and classifying these diseases from medical images such as X-rays and MRIs. Different CNN architectures, such as AlexNet, DenseNet, VGG, ResNet, and Inception, have been employed with varying success rates in detecting and classifying these diseases. Comparing these architectures reveals the potential trade-offs between final accuracy, computational complexity, and total training time, which are important factors to consider before model development. The review has also discussed how combining CNN architectures with different classifiers, including FCNN, KNN, and SVM, has shown possibilities for improving detection accuracy. The findings presented in Tables 1 and 2 show that the optimal CNN architectures for Brain Tumor and Pneumonia detection differ. For Pneumonia detection, models with the highest accuracy typically have greater depth and a higher number of filters in their convolutional layers. In contrast, the most effective models for Brain Tumor detection generally have shallower architectures and fewer filters in their initial convolutional layers. It is also evident that using alternative classifiers, such as KNN and SVM, in combination with CNN can result in models having high accuracy. However, their performance tends to be outperformed by the more conventional CNN-FCNN approach. Numerous studies have addressed problems like those this work aims to tackle. For Pneumonia detection, the differentiation between Bacterial and Viral Pneumonia remains an area with potential for improvement. As for COVID-19, the recent pandemic has led to an increased number of research chapters, with high accuracy rates reported in detecting the virus. Regarding Brain Tumor detection, various studies have focused on detecting and classifying Gliomas, Meningiomas, and Pituitary Tumors with high accuracy. Even though some studies report high accuracy for both Brain Tumor and Pneumonia detection, further research is needed to address potential limitations and challenges. By exploring new and refining existing techniques, this work aims to contribute to the growing body of knowledge regarding medical diagnostics and improve the accuracy and clinical applicability of disease detection models.

3. Methodology/implementation

3.1 Datasets

In this work, multiple datasets were used for the models for detection of Pneumonia, one dataset for primary Brain Tumor detection and one dataset consisting of MRI images for the model designed for differentiating radiology images. Pre-processing of data was similar with each dataset for simplicity. Each dataset was divided in Train, Validation and Test folders with approximately 60%, 20%, 20% of total data, respectively. For each dataset each class had its own subfolder in which the images were stored. The data was stored this way because of usage of ImageDataGenerator Keras class and its flow from data frame method.

3.1.1 Pneumonia datasets

First dataset used was Darwin Labs COVID-19 Chest X-Ray Dataset [50] with the intention to use 4 classes (COVID-19, Normal, Viral, Bacterial). The downloaded dataset came in JSON format with multiple annotations including polygons like the GeoJSON coordinate polygon system. After further inspection of tags there were more than 200 unique tags and some of them were CT scans. After inspecting images, themselves I found that many of the COVID-19 images included Electrocardiogram electrodes and few included tubes such as chest and endotracheal tubes. Additionally, some of the images included black or white arrows pointing at specific areas of lungs. Moreover, the dataset included "COVID-19 confirmed" and "COVID-19" tags. Even though while using this dataset I have developed code which was used for training and testing of different models with other datasets as well, I have discarded this dataset and all the models trained with it because of many reasons of which some are mentioned above, and examples of these images can be seen in Fig. 7.

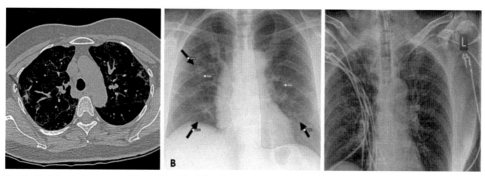

Fig. 7 Problematic images from Darwin labs dataset.

Second dataset [51] used for Pneumonia detection was a Kaggle Chest X-ray dataset, Version 2, which consisted of three classes which were Normal, Viral, and Bacterial and total of 5863 images. This dataset was organized in two folders with two subfolders, Normal and Pneumonia where each Pneumonia image name had "bacterial" or "viral" in it. This dataset was only used for the second testing to confirm accuracy of couple of models. The dataset [52] which was used for both training and testing of final models, contains 1281 COVID-19, 3270 Normal, 3001 Pneumonia-Bacterial, and 1656 Pneumonia-Viral X-ray images. The downloaded dataset came in the format of a single folder containing four subfolders representing the respective classes. Example images depicting these classes can be seen in Fig. 8, and the division of images in Train and Test sets in Table 3. An extensive dataset analysis reveals that some images have poor quality and other issues that could negatively impact model training and testing. Although not explicitly noted in the dataset description, several pediatric images are present, particularly in the Pneumonia-Bacterial and Pneumonia-Viral subfolders. Almost all images in the Normal subfolder adhere to the standard X-ray procedure, with the individual's arms raised above their head. However, the other subfolders feature a mix of images with varying arm positions and body orientations. Some images appear to include an X-ray image taken using a phone, while others display incomplete removal of DICOM information (excluding names and identifiable patient information) or exhibit color smudging. Some of these issues can be seen in Fig. 9. Most of these issues can be resolved in a real-world setting, and it would be optimal to develop the model without including images with the problems mentioned above. Nevertheless, no images will be removed from the dataset to maintain reproducibility, despite the potential impact on the final model's accuracy.

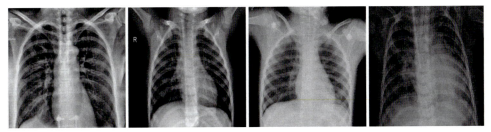

Fig. 8 Data samples from the final pneumonia dataset in order: COVID-19, normal, viral pneumonia, and bacterial pneumonia.

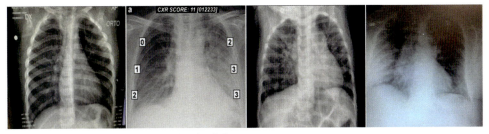

Fig. 9 Example images of issues in the final dataset.

Table 3 Number of images in training and testing sets in the pneumonia dataset.

Type	Training set	Test set
COVID-19	1025	256
Normal	2631	657
Pneumonia-Bacterial	1325	600
Pneumonia-Viral	2401	331

3.1.2 Brain dataset

The dataset [53] used for primary Brain Tumor detection was a Kaggle Brain Tumor MRI Dataset. Initially organized into two main folders, Training and Testing, it contains 7023 MRI images representing Glioma, Meningioma, No Tumor, and Pituitary classes. Examples of these images can be seen in Fig. 10, and the division of images in Train and Test sets in Table 4. This dataset does not contain many issues, aside from some orientation discrimination that could influence model accuracy and duplicate images. For instance, most Notumor images are taken in the axial plane (top-down view), and almost all Pituitary images are taken in either coronal or sagittal view (side views). This issue can potentially create a slight bias in the model training and testing, making the models less applicable in the real-world setting. The dataset description says that the image duplicates were removed, which is not the case. After testing the model there were identical misclassified images that can be seen in Fig. 11. After that it was evident that the dataset contained at least a couple of duplicates but because of the size of the dataset it is nearly impossible to find all of them.

Table 4 Number of images in training and testing sets in the brain tumor dataset.

Type	Training set	Test set
Glioma	1330	300
Meningioma	1339	306
Normal	1595	405
Pituitary	1457	300

In the "Note" section of the Kaggle website where the dataset was obtained, a link to a GitHub code is provided, which according to the Note, will improve the accuracy of the model. This code was used for experimentation, but the models using these pre-processed images had approximately 2% lower accuracy across multiple experiments. The process removes noise, removes extreme points on the images, and resizes images to 256×256 pixels. However, this or similar pre-processing was not incorporated, as the work primarily focuses on the development of model architectures and ensemble systems for the final program.

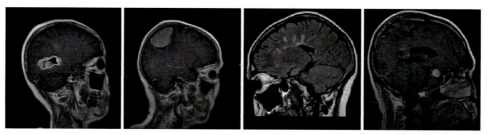

Fig. 10 Data samples from the brain dataset in order: Glioma, Meningioma, Notumor, and Pituitary.

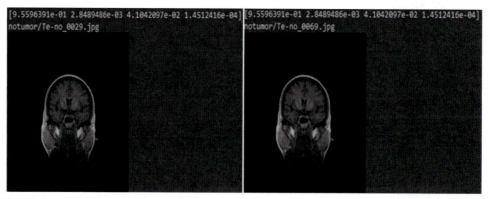

Fig. 11 Example of identical images in the dataset.

3.1.3 Other datasets

A combination of datasets were used for developing a model which differentiates between X-ray, CT, and MRI images. X-ray images were combination of Lung X-rays from datasets mentioned above and Body Parts Kaggle dataset [54]. CT and MRI images were taken from CHAOS dataset [55]. Examples of the X-ray, CT, and MRI images can be seen in Fig. 12.

3.2 Pre-processing

Some of the files used in this work came in DICOM (.dcm) format. DICOM files contain various types of information, including patient information, arrows, and anatomical labels. To prepare the data for the model training, the files were converted to PNG format with extracted pixel values from each image while discarding all excess information. Before training the models, the images were pre-processed using the ImageDataGenerator class from the Keras library. Images were scaled to 256×256 pixels and pixels were rescaled by a factor of $1/255$ to normalize pixel values between 0 and 1 to reduce computational complexity. Many trained models were tested on 512×512 pixels and while increasing time needed for model training this image size did not increase the general accuracy. Therefore, all models are trained on $256 \times 256 \times 3$ input, even the gray images were converted to RGB images. For the final Brain and Lung datasets used there was no need for future pre-processing. ImageDataProcessing image augmentation was used in couple of experiments, but it was discarded as most of the transformations greatly impacted model accuracy.

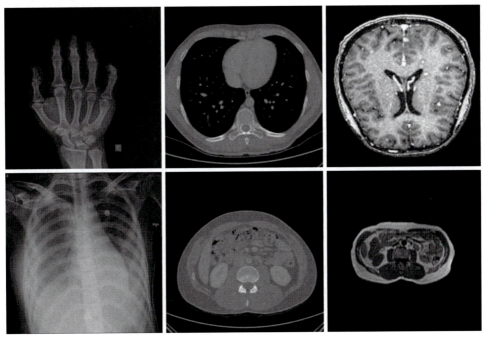

Fig. 12 Data samples of X-ray, CT, and MRI images on the left, middle and right respectively.

3.2.1 Model architectures

In this work, a variety of CNN architectures were tested. The chosen architectures included two custom sequential models, a custom AlexNet, DenseNet121, DenseNet169, Dense-Net201, VGG-16, VGG-19, NASNetMobile algorithm, and NASNetLarge algorithm. All non-custom architectures were directly taken from the TensorFlow Keras applications module, which provided pre-trained CNN models without classifiers such as ANN. These architectures were selected based on their proven effectiveness in previous research on automatic disease classification. To determine whether to use pre-trained weights, experiments were conducted with both random and "imagenet" pre-trained weights. For these experiments, a DenseNet201 CNN architecture with a three-layer deep ANN classifier was used. The training was stopped after 15 EPOCHs, with the results depicted in Fig. 13. The model initialized with random weights was overfitting heavily in the initial EPOCHs, with the difference between training and testing accuracy reducing to almost 10% in the 15th epoch. Additionally, using random initial weights, the first few epochs took slightly longer to complete, and the initial model accuracy was lower, with a slower increase com-pared to the pre-trained weights. In both instances, all layers were set to "trainable = True" to enable model training. After determining that using the "imagenet" pre-trained weights was better, an additional experiment was conducted, which included freezing the first half of the pre-trained model while training the second half and afterward doing the opposite. In the two experiments conducted on Brain Tumor and Pneumonia datasets, the results were almost identical to the models without frozen layers. As a re-sult, it was concluded that using the initial pre-trained "imagenet" weights without freezing was the better approach for all future models. This decision was based on the finding that using the freezing technique took twice as long to train models compared to training without freezing.

3.2.2 Classifier methodology

In this work, all CNN architectures were evaluated using fully connected output layers as classifiers. These fully connected output layers consisted of a global average pooling layer, followed by two dense layers with ReLU activation functions, batch normalization, and dropout regularization for the prevention of overfitting. After loading the pre-trained CNN models without the top layers, these output layers were appended to the model, as illustrated in Fig. 14. During the training process, these fully connected layers were used as primary classifiers. Subsequently, the extracted features from the best performing CNNs were combined with alternate classifiers, such as SVM and KNN, for further experimentation. To utilize the SVM classifier, the last six layers of the fully trained CNN model were removed, and the "model.predict" method was employed to extract image features. A portion of these features was used to determine the optimal hyperpara-meters using GridSearchCV. This process returned optimal values of "C" = 10, "gamma" = 0.01, and "kernel" = "rbf" for four-class experiments conducted on both

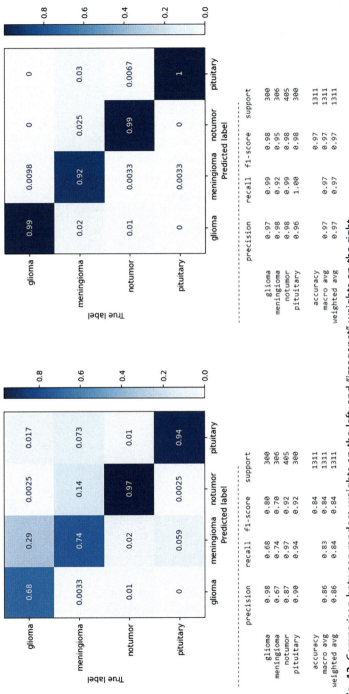

Fig. 13 Comparison between random weights on the left, and "imagenet" weights on the right.

Brain Tumor and Pneumonia datasets, and "C" = 0.1, "gamma" = 0.01, and "kernel" = "linear" for two-class experiments. Similarly, following the removal of the last six layers of a pre-trained CNN model, a KNN classifier was developed by combining it with the remaining layers used for feature extraction. The CNN model extracted features from the Train folder, which were then encoded into feature vectors. The KNN classifier computed the cosine similarity between a query image and the set of reference vectors, which were the encoded feature vectors of the training images. To prepare the reference vectors for the KNN classifier, the encoded feature vectors were flattened, normalized, and used as weights for the KNN dense layer with a linear activation function. The number of nearest neighbors used by the KNN classifier was experimented with various values, including 3, 5, 7, 10, 15, 20, 30, and 50, allowing for the identification of the most suitable configuration for the given classification task.

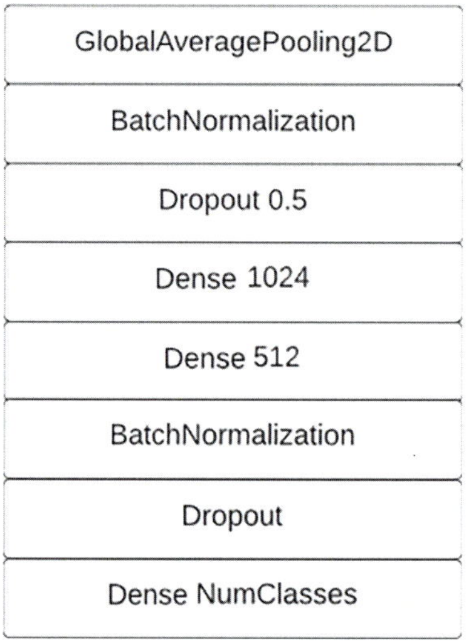

| GlobalAveragePooling2D |
| BatchNormalization |
| Dropout 0.5 |
| Dense 1024 |
| Dense 512 |
| BatchNormalization |
| Dropout |
| Dense NumClasses |

Fig. 14 Manually added layers to pre-built CNN architectures.

4. Evaluation

All experiments were performed in either the local Jupyter Notebook or in the Google Colaboratory in the personal Google Drive. Given the need for quick model training and limited time, multiple experiments and models were being trained at the same time to optimize efficiency. This approach allowed for the simultaneous execution of different processes, therefore significantly reducing the overall time required for the completion of the experimentation phase. Local Jupyter Notebook ran with i9 12,900 2.40 GHz CPU, 32GB 5600 MHz RAM, and NVIDIA GeForce RTX 3060 12GB GDDR6.

GPU. Google Colaboratory ran with P100 16GB HBM2 GPU, 12GB RAM, and 2 Intel Xeon 2.20 GHz CPUs. At the beginning of every script, np.random.seed(42) was called to ensure reproducibility. Because of the way random seed functions with both

Jupyter Notebook and Google Colaboratory, whole scripts need to be run in order from the beginning to ensure reproducibility. Different callbacks were tested, which can be used while training the model, and three, which were used in the final training, were ReduceLROnPlateau, EarlyStopping, and ModelCheckpoint. Mod-elCheckpoint monitored val accuracy and saved the model only when the validation accuracy in-creased. EarlyStopping monitored val accuracy as well with 15 patience. ReduceL-ROnPlateau monitored val accuracy as well and changed the learning rate by a factor of 0.5 every two times validation accuracy did not improve. Afterwards, the model was fitted with a train and validation data generator with 150 data samples per EPOCH. After the training was finished, the previously mentioned performance evaluation graphs were displayed. This process was automated with two functions which allowed for quick training and evaluation of each model. In this work, for both Brain Tumor and Pneumonia datasets, multiple models were trained, each tailored to handle a specific contiguous sublist of the list of classes. Specifically, every model was trained using a subset of the available data corresponding to a specific sublist. With this approach, it was possible to identify the specific classes that were challenging to differentiate and to develop a different approach to improve the accuracy of those classes. After training all models, the Cascading ensemble learning approach was used, which was specific to the dataset. The cascading ensemble learning approach will be discussed in more detail later in this work.

5. Results and discussion

The performance of all models was assessed using a Confusion Matrix, Classification Report, and Training and Validation accuracy and loss metrics. The final evaluation combined F1 score, Log Loss, and Micro ROC-AUC as key metrics. While Log Loss and Micro ROC-AUC are the most important metrics for tasks demanding high preciseness, such as the focus of this work, the F1 score was selected as the primary evaluation metric due to its widespread use in similar research studies, often as an alternative for accuracy. Given the imbalance in the testing data, testing accuracy was not considered. Nevertheless, Training and Validation accuracy and loss metrics were employed to ensure that the models did not overfit the training dataset. This was achieved by analyzing the differences between Training and Validation graph plots.

Explanation and equations of the metrics used as evaluation in this work:

- Precision: $\dfrac{\sum_{i=1}^{N} \mathrm{TP}_i}{\sum_{i=1}^{N} (\mathrm{TP}_i + \mathrm{FP}_i)}$

- Recall: $\dfrac{\sum_{i=1}^{N} \mathrm{TP}_i}{\sum_{i=1}^{N} (\mathrm{TP}_i + \mathrm{FN}_i)}$

- F1 score: $2 * \dfrac{\text{Precision*Recall}}{\text{Precision + Recall}}$

- Log loss: $-\dfrac{1}{N} \sum_{i=1}^{N} \sum_{j=1}^{M} y_{ij} \log\left(p_{ij}\right)$

- Micro ROC-AUC: Area Under the ROC curve with one-vs-rest approach, calculated by aggregating counts across all classes as if it were a binary classification problem. Inside the equations above, N represents the number of instances, M represents the number of classes, y_{ij} is a value that is either 0 or 1, indicating whether instance i belongs to class j, and p_{ij} is a value between 0 and 1, and it represents the predicted probability that instance i belongs to class j. Additionally, TP represents True Positive, TN True Negative, FP False Positive, and FN False Negative instances. To determine the best CNN architecture which will be used in the final experimentation, various architectures were tested on both the Brain Tumor and Pneumonia datasets. DenseNet201 achieved the best results for both datasets. The differences in accuracy between models trained on the Brain Tumor dataset were relatively small in comparison to models trained on Pneumonia dataset. The Recall and F1 scores were low, primarily when distinguishing between Glioma and Meningioma classes for all models except for DenseNet201 and DenseNet169. For the Pneumonia dataset, there were much greater differences in general performance from the best to the worst architecture. Table 5 displays all tested architectures for the Pneumonia dataset, along with their respective Log Loss, Micro ROC-AUC, Precision, Recall, and F1 scores.

Table 5 Performance evaluation of different CNN architectures on pneumonia dataset.

CNN architecture	Log Loss	Micro ROC-AUC	Precision	Recall	F1 score
VGG16	0.4211	0.9664	82.72	80.10	**75.92**
NASNet Mobile	0.6339	0.9455	77.25	78.93	**76.69**
VGG19	0.4193	0.9681	80.54	81.84	**80.65**
Custom AlexNet	–	–	82.34	82.36	**82.11**
NASNet Large	0.4203	0.9706	84.47	82.68	**83.12**
Custom1 CNN	–	–	84.25	85.05	**84.04**
DenseNet121	0.3222	0.9816	87.22	87.56	**87.12**
InceptionV3	0.3391	0.9807	87.89	87.84	**87.88**
DenseNet169	0.2860	0.9853	88.55	88.53	**88.02**
DenseNet201	0.2921	0.9850	88.95	89.07	**88.68**

DenseNet201 emerged as the top-performing model, with an F1 score of 88.68, a Log Loss of 0.2921, and a Micro ROC-AUC of 0.9850. The DenseNet169 model achieved similar results but had more significant differences between training and validation loss and accuracy, which suggests that the model was more prone to overfitting. When evaluating the computational complexity and training speed of the models, DenseNet201 was found to be one of the slowest models in the initial epochs, taking almost 3 s per batch. In comparison, the average time for the other models was 0.45 s per batch. However, the time it took to complete a single batch gradually decreased, reaching 0.35 s per batch after the tenth epoch. The fastest model was VGG16 which had an average of 0.3 s per batch. Considering the average time each model took to complete an epoch, it was deemed acceptable to use DenseNet201, as it was slow only in the initial epochs.

5.1 Pneumonia

In this section, a comprehensive evaluation was conducted on the Pneumonia dataset using a multi-class classification approach. Eleven distinct models were developed for each combination of classes, and their performance is illustrated in Fig. 15 using confusion matrices. It was observed that the models had difficulty distinguishing between Pneumonia-Bacterial and Pneumonia-Viral classes. To investigate this issue, misclassification percentages for the primary model, which predicts all four classes, and a sub-model, which only predicts Pneumonia-Bacterial and Pneumonia-Viral classes, were analyzed. Although many misclassifications varied, specific instances were identified where the primary model incorrectly predicted a class with nearly 50% probability, while the sub-model provided accurate predictions. To tackle this challenge, a two-stage prediction strategy was implemented. First, the primary model made a prediction for each image. If the predicted class were either Pneumonia-Bacterial or Pneumonia-Viral, the sub-model would then make the final prediction. This approach led to a slight improvement in the F1 score, reaching 89.92. An ensemble method was subsequently applied by averaging predictions from both models, which served as input for the calculation of Log Loss and ROC-AUC. This resulted in a Log Loss of 0.2785 and a ROC-AUC of 0.9865, demonstrating a slight increase in performance compared to using the base model alone. This experiment highlights the potential benefits of combining models with different prediction capabilities to improve classification performance in a challenging multi-class setting. The CNN-KNN method was also tested, using the number of neighbors as 3, 5, 7, 10, 15, 20, 25, 30, and 50. The best results were obtained with three neighbors, yielding an F1 score of 84.80, which is comparable to the Custom1 CNN but significantly lower than DenseNet201.

Finally, a CNN-SVM classifier was created using the methodology mentioned in Section 3. The results from the classifier trained on all four classes and the classifier trained on only Pneumonia-Bacterial and Pneumonia-Viral can be seen in Fig. 16. These models exhibited a similar issue to all other previously trained models, with the lowest accuracy between Pneumonia-Bacterial and Pneumonia-Viral classes. The four-class model achieved an F1 score of 89.84, surpassing all other standalone models. This outcome was due to the classifier predicting the two Pneumonia classes more accurately while over-predicting COVID-19. Therefore, another CNN-SVM model was trained which was predicting only the two Pneumonia classes. It achieved an F1 score of 85.36 which was superior to 82.06 when using only DenseNet201 as a classifier. To utilize the strengths of DenseNet201 and SVM, two cascading model ensembles were combined. The results of this ensemble can be seen in Fig. 17, achieving the highest F1 score of 91.44. A limitation of the two-class SVM classifier is that it sometimes predicts an equal chance for both classes (0.5, 0.5). In such cases, the primary prediction of the 4-class model was used. In most cases, when the SVM predicts an equal chance for both classes the true class is the opposite Pneumonia class primary model predicted. With that knowledge it would be possible to change the algorithm to predict the correct class, and increase final accuracy slightly over 95%, but that would be artificial increase and because it is dataset specific, it would represent manual overfitting.

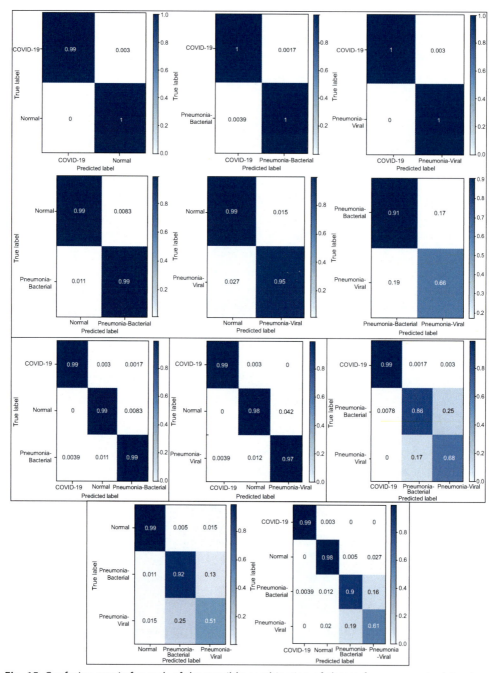

Fig. 15 Confusion matrix for each of the possible combination of classes for pneumonia dataset.

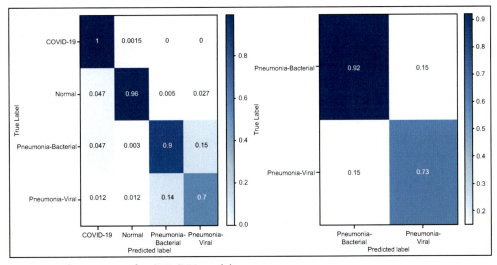

Fig. 16 Confusion matrix for CNN-SVM models.

Interestingly, when testing the models on different datasets, such as the Kaggle Chest X-ray Images dataset, the model accuracy improved further. The two-class SVM classifier achieved 89.23% accuracy and an F1 score of 90 when differentiating between Pneumonia-Bacterial and Pneumonia-Viral classes, which is substantially higher than the performance on the original dataset. This suggests that the model may have greater potential for generalization and could perform even better when applied to different data sources.

5.2 Brain Tumors

The evaluation of Brain Tumor models followed the same process as in the Pneumonia evaluation. Results from the first experiment can be seen in Fig. 18. The Log Loss of the four-class model was 0.0761, while the Micro ROC-AUC was 0.9987, and the F1 score was precisely 98, demonstrating exceptional performance. The models faced the most significant challenge in predicting the Meningioma class. Across all experiments, the models successfully predicted the presence of a tumor with a *Re*-call score of 100 and a Precision score of 97.78. In predicting the presence of a Brain Tumor, Recall is important, but Precision is even more critical. The models achieved similar prediction results for the Pituitary class, resulting in a total F1 score of 99.25. The misclassified Meningioma images were examined and can be seen in Fig. 19. A non-weighted ensemble was then tested with all 11 models, and the result can be seen in Fig. 20. This ensemble correctly classified eight additional images, increasing the ensemble's F1 score to 98.53. Unfortunately, Log Loss and ROC-AUC cannot be calculated for such an ensemble because each model predicts a different set of classes. The CNN-KNN method was also evaluated for the Brain Tumors dataset with the same number of neighbors as in the Pneumonia evaluation. The best result was obtained with 3 neighbors, yielding an F1 score of 94.13, which is significantly lower than the previous evaluation with DenseNet201. Every subsequent test with a higher number of neighbors resulted in a lower F1 score, with the lowest being 79.87.

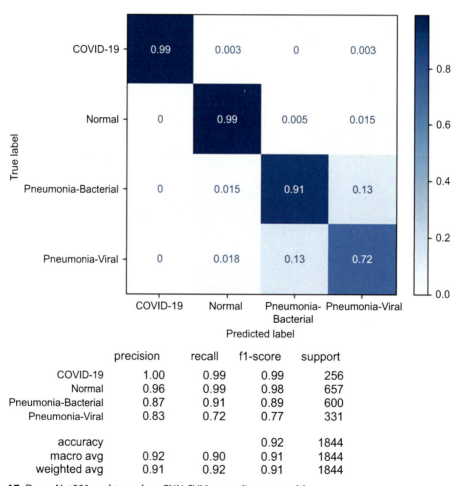

	precision	recall	f1-score	support
COVID-19	1.00	0.99	0.99	256
Normal	0.96	0.99	0.98	657
Pneumonia-Bacterial	0.87	0.91	0.89	600
Pneumonia-Viral	0.83	0.72	0.77	331
accuracy			0.92	1844
macro avg	0.92	0.90	0.91	1844
weighted avg	0.91	0.92	0.91	1844

Fig. 17 DenseNet201 and two-class CNN-SVM cascading ensemble.

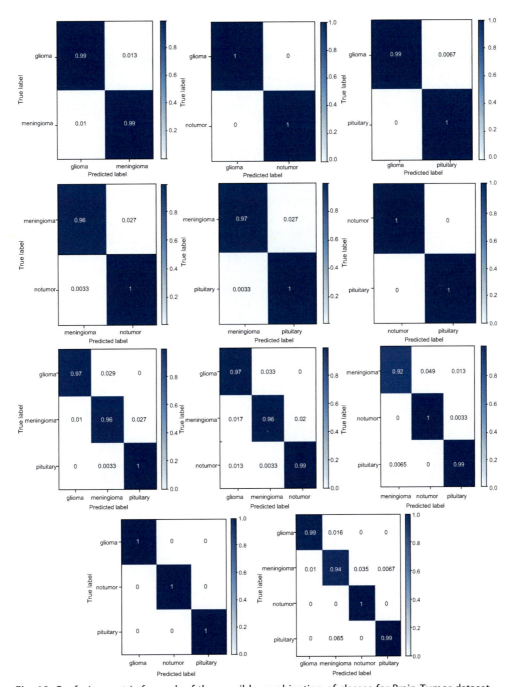

Fig. 18 Confusion matrix for each of the possible combination of classes for Brain Tumor dataset.

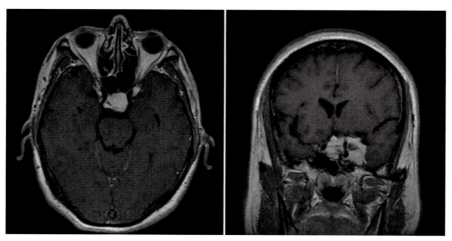

Fig. 19 Meningioma examples misclassified as pituitary.

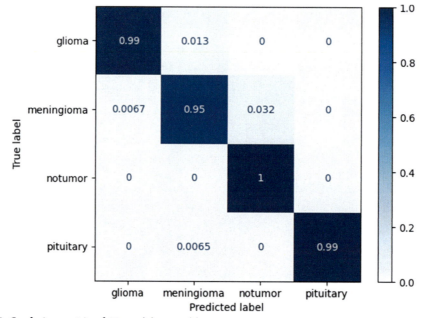

Fig. 20 Confusion matrix of 11 model ensemble.

Finally, the SVM classifier was tested. The evaluated SVM classifier achieved Micro ROC-AUC score of 0.9989, and Log Loss score of 0.0838 which was comparable to DenseNet201. This demonstrates that the SVM classifier is a viable alternative to the DenseNet201 model for the Brain Tumors dataset.

5.3 Radiology image type

In contrast to the Pneumonia and Tumor detection experiments, for radiological image type there were only three models trained with two being custom AlexNet CNN, and one was DenseNet201. All experiments achieved an accuracy of 100% with the dataset used, and that can be seen in the Confusion Matrix and Classification Report in Fig. 21. The difference between two AlexNet experi-ments was the image size, 512×512 and 256×256. In the end AlexNet model which was trained with 256×256 pixel images was used for the computational efficiency.

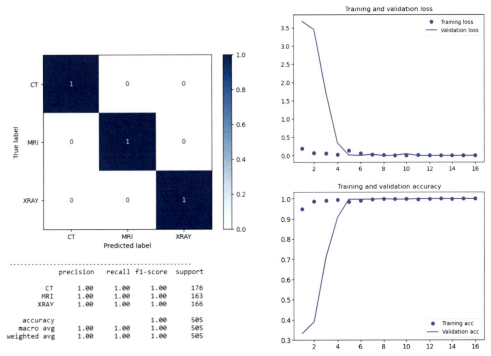

Fig. 21 Radiology image type final model graphs. Confusion matrix and classification report/training and validation accuracy and loss functions.

Following further experimentation with images outside of the dataset, the issues with the model was found. The model will correctly predict brain and abdomen MRI images as MRI, abdomen CT images as CT, and lung, head, and abdomen X-ray images as X-ray. As X-ray images of other parts of the body were also included in the training, such as body extremities model can predict them with high accuracy. However, it is evident that the model cannot accurately predict X-ray images containing two fists when the distance between fists is significant. This is likely due to the bias introduced by the vertical white line separating the fists in the training and testing dataset. Despite this issue, this model will be used in the final program is that it serves its purpose for differentiating between Brain MRI, Lungs X-ray, and CT images. Moreover, this model should not be perceived as proposed model for differentiation between different radiological image types as it requires more testing and more extensive datasets.

5.3.1 Comparison of results

The comparison of the results for the Brain Tumor models and the Pneumonia models are presented in Tables 6 and 7.

Table 6 Comparison of results for Brain Tumor models.

Chapter	Classes	Method	Accuracy	F1
Islam et al. [36]	COVID-19, Normal, Pneumonia	VGG19-RNN	99.86	99.78
Hashmi et al. [37]	Normal, Pneumonia	5 Model Ensemble	98.43	98.63
Elshennawy and Ibrahim [38]	Normal, Pneumonia	ResNet152V2	99.22	99.44
Rajaraman et al. [39]	Normal, Pneumonia-Bacterial, Pneumonia-Viral	Custom VGG16	91.80	91.00
Rahman et al. [30]	Normal, Pneumonia-Bacterial, Pneumonia-Viral	DenseNet201	93.33	–
This work	COVID-19, Normal, Pneumonia-Bacterial, Pneumonia-Viral	DenseNet201	89.09	88.68
This work	COVID-19, Normal, Pneumonia-Bacterial, Pneumonia-Viral	DenseNet201 +SVM Ensemble	91.97	91.44

Table 7 Comparison of results for Pneumonia models.

Chapter	Classes	Method	Accuracy	F1
Deepak and Ameer [44]	Glioma, Meningioma, Pituitary	GoogLeNet-SVM	97.10	97.00
Rehman et al. [45]	Glioma, Meningioma, Pituitary	Custom VGG16	98.69	–
Irmak [46]	Normal, Glioma, Meningioma, Pituitary, Metastases	Automatically Generated CNN	92.66	–
This work	Normal, Glioma, Meningioma, Pituitary	11 Model Ensemble	98.65	98.52
This work	Normal, Glioma, Meningioma, Pituitary	DenseNet201	97.96	98.00
This work	Normal, Glioma, Meningioma, Pituitary	2 Model Ensemble	98.09	98.26

6. Graphical user interface

The developed program's graphical user interface (GUI) is implemented using the Python library tkinter. The GUI features a simple design, including a menu that allows users to choose an individual image or an entire folder. Selected images are saved and can be navigated through the "Next" and "Previous" buttons. Upon clicking the "Predict" button, the program identifies the type of radiology image and then activates the appropriate disease prediction model based on the result. For instance, if the image is an MRI, the program will use the Brain Tumor model ensemble to classify the image. The same process applies to X-ray images for Pneumonia detection. The prediction result is displayed in the bottom-right text box. A visual representation of the GUI design can be seen in Fig. 22. It is worth noting that it may take approximately 10 s to open properly when launching the program, as it loads all the models before initializing the GUI. This ensures the GUI is ready to use once it appears on the screen. The program accepts a wide range of image formats, and prediction times may vary depending on the input image. They can range from under 1 s to up to 2 s if the loaded image is in DICOM format.

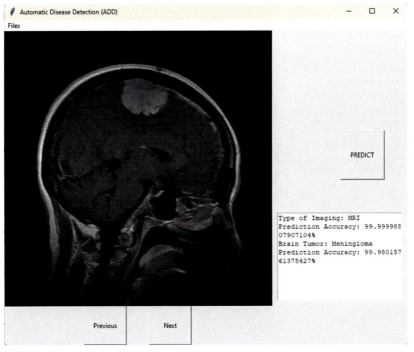

Fig. 22 GUI.

Fig. 23 shows a diagram of the program's algorithm, where round cells represent the text boxes in the GUI or output, and square cells represent internal algorithms. What is not shown is the processing of the input image. The program can read all image formats supported by the PIL library in addition to DICOM images. When a user chooses an image or a folder, the program first checks if the images are in DICOM format and removes any additional information from the image, if necessary. Regardless of the image format, the program normalizes the pixel values in the next step, saves the image, and finally displays it on the screen.

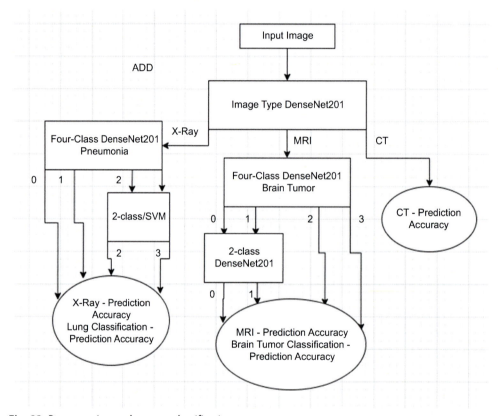

Fig. 23 Program -internal system classification.

7. Conclusion

The chapter has achieved its primary objective of building multiple models for diagnosing brain tumors and lung pneumonia and incorporating them into one software. We present several contributions to the field of automatic disease detection using DL techniques. The research successfully develops a novel two–part ensemble for Pneumonia detection, combining a DenseNet201 CNN and a DenseNet201-SVM to predict four classes with 91.97% accuracy. This approach highlights the potential for combining DL and traditional ML techniques to improve prediction performance. In addition, the work investigates multiple ensemble techniques for Brain Tumor detection, providing valuable insights into the performance and computational efficiency of different approaches, which could inform the design of future disease detection models. Furthermore, the research integrates automatic radiology image type detection and disease prediction within the developed software, simplifying the disease detection process and showing the potential for creating user-friendly and efficient automatic disease detection systems. These contributions can serve as a foundation for future research in developing more efficient and accurate disease detection models, which will be discussed in the following section.

7.1 Future work

One possible direction is to extend the scope of the research by including additional models that will predict the body part being imaged and detect a broader range of diseases, such as brain cysts, metastasis, breast cancer, and lung cancer. These models could be integrated into the final program, which would then automatically detect the type of body part and disease present in the image. Another direction for future work is adding another type of image analysis, such as analyzing images of tumor cells to determine their stage and aggressiveness. This would further enhance the diagnostic capabilities of the research. Another improvement could involve implementing a 3D reader that can process and analyze multiple MRI images of a patient and generate predictions based on the complete set of images. This could potentially improve the accuracy and reliability of the predictions, in addition to being more applicable to a real-world setting. Cooperation with a radiology department could further enhance the real-world applicability of the research and help address practical challenges in medical imaging and improve current classification accuracy. Additionally, a valuable feature that could be added to the program is the automatic outlining of detected diseases in the image displayed in the GUI. For example, if a tumor is detected, the program could automatically outline the tumor in the image. The Pneumonia detection model could be improved by tailoring it to specific age groups. By dividing the dataset into pediatric and adult X-rays and training separate models for each group, the accuracy of the predictions may be increased. Additionally, identifying and removing problematic images from the dataset, as well as eliminating duplicate images from the Brain Tumor dataset, could lead to more accurate and robust models. By exploring these ideas, future research can build upon the findings of this work and contribute to the development of more accurate and comprehensive diagnostic tools for medical imaging.

References

1. A. Krizhevsky, I. Sutskever, G.E. Hinton, Imagenet classification with deep convolutional neural net-works, Commun. ACM 60 (6) (2017) 84–90, https://doi.org/10.1145/3065386. https://dl.acm.org/doi/abs/10.1145/3065386.
2. V. Diogo, H.A. Ferreira, D. Prata, Early diagnosis of Alzheimer's disease using machine learning: a multi-diagnostic, generalizable approach, Alzheimers Res. Ther. 14 (1) (2022) 1–21, https://doi.org/10.1186/s13195-022-01047-y.
3. M. Tanveer, et al., Machine learning techniques for the diagnosis of Alzheimer's disease: a review, ACM Trans. Multimedia Comput. Commun. Applicat. 16 (1) (2020) 1551–6857, https://doi.org/10.1145/3344998.
4. J. Seetha, S. Selvakumar Raja, Brain tumor classification using convolutional neural net-works, Biomed. Pharmacol. J. 11 (3) (2018) 1457, https://doi.org/10.13005/bpj/1511. https://pdfs.semanticscholar.org/7946/1362af5899a45b420993af9ad9025dc6aedf.pdf.
5. S. Hassantabar, M. Ahmadi, A. Sharifi, Diagnosis and detection of infected tissue of COVID-19 patients based on lung X-ray image using convolutional neural network approaches, Chaos Solitons Fractals 140 (2020) 110170, https://doi.org/10.1016/j.chaos.2020.110170.
6. T. Ozturk, et al., Automated detection of COVID-19 cases using deep neural networks with X-ray images, Comput. Biol. Med. 121 (2020) 103792, https://doi.org/10.1016/j.compbiomed.2020.103792.
7. Y. Shen, et al., Artificial intelligence system reduces false-positive findings in the interpretation of breast ultrasound exams, Nat. Commun. 12 (1) (2021) 1–13, https://doi.org/10.1038/s41467-021-26023-2.
8. A.M. Rauschecker, et al., Artificial intelligence system approaching neuroradiologist-level differential diagnosis accuracy at brain MRI, Radiology 295 (3) (2020) 626, https://doi.org/10.1148/radiol.2020190283. https://www.ncbi.nlm.nih.gov/pmc/articles/PMC7263320/.
9. E.J. Topol, High-performance medicine: the convergence of human and artificial intelligence, Nat. Med. 25 (1) (2019) 44–56, https://doi.org/10.1038/s41591-018-0300-7.
10. Z. Rustam, et al., Analysis of architecture combining convolutional neural network (CNN) and kernel K-means clustering for lung cancer diagnosis, Int. J. Adv. Sci. Eng. Inf. Technol. 10 (3) (2020) 1200–1206, https://doi.org/10.18517/ijaseit.10.3.12113.
11. T. Blaschke, Object based image analysis for remote sensing, ISPRS J. Photogramm. Remote Sens. 65 (1) (2010) 2–16, https://doi.org/10.1016/j.isprsjprs.2009.06.004.
12. F. Ritter, et al., Medical image analysis, IEEE Pulse 2 (6) (2011) 60–70, https://doi.org/10.1109/MPUL.2011.942929.
13. M. Schwier, et al., Automated spine and vertebrae detection in CT images using object-based image analysis, Int. J. Numer. Methods Biomed. Eng. 29 (9) (2013) 938–963, https://doi.org/10.1002/cnm.2582.
14. A.N. Ramesh, et al., Artificial intelligence in medicine, Ann. R. Coll. Surg. Engl. 86 (5) (2004) 334. https://www.ncbi.nlm.nih.govn/pmc/articles/PMC1964229/.
15. V. Kaul, S. Enslin, S.A. Gross, History of artificial intelligence in medicine, Gastrointest. Endosc. 92 (4) (2020) 807–812, https://doi.org/10.1016/j.gie.2020.06.040. https://www.sciencedirect.com/science/article/pii/S0016510720344667.
16. A.A. Abdullah, N.M. Posdzi, Y. Nishio, Preliminary study of pneumonia symptoms detection method using Cellular Neural Network, in: International Conference on Electrical, Control and Computer Engineering 2011 (InECCE), IEEE, 2011, pp. 497–500. https://ieeexplore.ieee.org/abstract/document/5953933.
17. A.H. Reynolds, Large-Scale Image Recognition: AlexNet, October 2017. https://anhreynolds.com/blogs/alexnet.html.
18. G. Huang, et al., Densely connected convolutional networks, in: Proceedings of the IEEE Conference on Computer Vision and Pattern Recognition, 2017, pp. 4700–4708. https://openaccess.thecvf.com/content_cvpr_2017/html/Huang_Densely_Connected_Convolutional_CVPR_2017_chapter.html.
19. B. Zoph, Q.V. Le, Neural Architecture Search With Reinforcement Learning, 2016, arXiv preprint arXiv:1611.01578 https://arxiv.org/abs/1611.01578.

20. C. Szegedy, W. Liu, et al., Going deeper with convolutions, in: Proceedings of the IEEE Conference on Computer Vision and Pattern Recognition, 2015, pp. 1–9. https://www.cv-foundation.org/openaccess/content_cvpr_2015/html/Szegedy_Going_Deeper_With_2015_CVPR_chapter.html.

21. C. Szegedy, V. Vanhoucke, et al., Rethinking the inception architecture for computer vision, in: Proceedings of the IEEE Conference on Computer Vision and Pattern Recognition, 2016, pp. 2818–2826. https://www.cv-foundation.org/openaccess/content_cvpr_2016/html/Szegedy_Rethinking_the_Inception_CVPR_2016_chapter.html.

22. B. Zoph, V. Vasudevan, et al., Learning transferable architectures for scalable image recognition, in: Proceedings of the IEEE Conference on Computer Vision and Pattern Recognition, 2018, pp. 8697–8710. https://openaccess.thecvf.com/content_cvpr_2018/html/Zoph_Learning_Transferable_Architectures_CVPR_2018_chapter.html.

23. A. Moh'd Rasoul, D. Al-Hadidi, R.S. Razouq, Pneumonia Identification Using Organizing Map Algorithm, 2006. https://www.researchgate.net/profile/Osama-Dorgham-2/publication/301836004_Pneumonia_identification_using_organizing_mapalgorithm/links/5729fd4b08aef5d48d30bf5d/Pneumonia-identification-using-organizing-map-algorithm.pdf.

24. J. Long, E. Shelhamer, T. Darrell, Fully convolutional networks for se-mantic segmentation, in: Proceedings of the IEEE Conference on Computer Vision and Pattern Recognition, 2015, pp. 3431–3440. https://openaccess.thecvf.com/content_cvpr_2015/html/Long_Fully_Convolutional_Networks_2015_CVPR_chapter.html.

25. C. Pelletier, Example of Fully-Connected Neural Network Researchgate, 2023. https://www.researchgate.net/figure/Example-of-fully-connected-neural-network_fig2_331525817.

26. S.R. Rath, An Introduction to *K*-Nearest Neighbors in Machine Learning, January 2023. https://debuggercafe.com/an-introduction-to-k-nearest-neighbors-in-machine-learning/.

27. G. Chauhan, Sklearn SVM-Starter Guide, March 2021. https://machinelearninghd.com/sklearn-svm-starter-guide/.

28. World Health Organisation, WHO, The Top 10 Causes of Death, 2020. https://www.who.int/news-room/fact-sheets/detail/the-top-10-causes-of-death. (visited on 09/15/2022).

29. Bacterial Pneumonia: Symptoms, Causes, Treatment, Prevention (n.d.). https://www.webmd.com/lung/bacterial-pneumonia.

30. T. Rahman, et al., Transfer learning with deep convolutional neural network (CNN) for pneumonia detection using chest X-ray, Appl. Sci. 10 (9) (2020) 3233. https://www.mdpi.com/2076-3417/10/9/3233.

31. Y. Muhammad, et al., Identification of pneumonia disease applying an intelligent computational framework based on deep learning and machine learning techniques, Mob. Inf. Syst. 2021 (2021) 1–20. https://www.hindawi.com/journals/misy/2021/9989237/.

32. V. Chouhan, et al., A novel transfer learning based approach for pneumonia detection in chest X-ray images, Appl. Sci. 10 (2) (2020) 559. https://www.mdpi.com/2076-3417/10/2/559.

33. A.M. Ismael, A. Şengür, Deep learning approaches for COVID-19 detection based on chest X-ray images, Expert Syst. Appl. 164 (2021) 114054. https://www.sciencedirect.com/science/article/pii/S0957417420308198.

34. M. Heidari, et al., Improving the performance of CNN to predict the likelihood of COVID-19 using chest X-ray images with preprocessing algorithms, Int. J. Med. Inform. 144 (2020) 104284. https://www.sciencedirect.com/science/article/pii/S138650562030959X.

35. D. Haritha, N. Swaroop, M. Mounika, Prediction of COVID-19 cases using CNN with X-rays, in: 2020 5th International Conference on Computing, Communication and Security (ICCCS), IEEE, 2020, pp. 1–6. https://ieeexplore.ieee.org/abstract/document/9276753.

36. M.M. Islam, et al., Diagnosis of COVID-19 from X-rays using combined CNN-RNN architecture with transfer learning, BenchCouncil Trans. Benchmarks Stand. Eval. 2 (4) (2022) 100088. https://www.sciencedirect.com/science/article/pii/S2772485923000054.

37. M.F. Hashmi, et al., Efficient pneumonia detection in chest x-ray images using deep transfer learning, Diagnostics 10 (6) (2020) 417. https://www.mdpi.com/2075-4418/10/6/417.

38. N.M. Elshennawy, D.M. Ibrahim, Deep-pneumonia framework using deep learning models based on chest X-ray images, Diagnostics 10 (9) (2020) 649. https://www.mdpi.com/2075-4418/10/9/649.

39. S. Rajaraman, et al., Visualization and interpretation of convolutional neural network predictions in detecting pneumonia in pediatric chest radiographs, Appl. Sci. 8 (10) (2018) 1715. https://www.mdpi.com/2076-3417/8/10/1715.

40. M.D. Aaron Cohen-Gadol, Glioma: Expert Surgeon, January 2023. https://www.aaroncohen-gadol.com/patients/glioma/overview.

41. Brain Tumor Types, November 2021. https://www.hopkinsmedicine.org/health/conditions-and-diseases/brain-tumor/brain-tumor-types.

42. Brain Tumors (n.d.). https://www.aans.org/en/Patients/Neurosurgical-Conditions-and-Treatments/Brain-Tumors.

43. M. Weller, et al., Glioma, July 2015. https://www.nature.com/articles/nrdp201517.

44. S. Deepak, P.M. Ameer, Brain tumor classification using deep CNN features via transfer learning, Comput. Biol. Med. 111 (2019) 103345. https://www.sciencedirect.com/science/article/abs/pii/S0010482519302148.

45. A. Rehman, et al., A deep learning-based framework for automatic brain tumors classification using transfer learning, Circ. Syst. Signal Process. 39 (2020) 757–775. https://link.springer.com/article/10.1007/s00034-019-01246-3.

46. E. Irmak, Multi-classification of brain tumor MRI images using deep convolutional neural network with fully optimized framework, Iran. J. Sci. Technol. Trans. Electr. Eng. 45 (3) (2021) 1015–1036. https://link.springer.com/article/10.1007/s40998-021-00426-9.

47. S. Das, O.F.M. Riaz Rahman Aranya, N.N. Labiba, Brain tumor classification using convolutional neural network, in: 2019 1st International Conference on Advances in Science, Engineering and Robotics Technology (ICASERT), IEEE, 2019, pp. 1–5. https://ieeexplore.ieee.org/abstract/document/8934603.

48. S. Anjum, et al., Detecting brain tumors using deep learning convolutional neural network with transfer learning approach, Int. J. Imaging Syst. Technol. 32 (1) (2022) 307–323. https://onlinelibrary.wiley.com/doi/abs/10.1002/ima.22641.

49. T.A. Sadoon, M.H. Ali, Deep learning model for glioma, meningioma and pituitary classification, Int. J. Adv. Appl. Sci. 2252 (8814) (2021) 8814. https://ijaas.iaescore.com/index.php/IJAAS/article/view/20439.

50. COVID-19 Chest X-Ray Dataset, 2020. https://darwin.v7labs.com/v7-labs/covid-19-chest-x-ray-dataset.

51. P. Mooney, Chest X-Ray Images (Pneumonia), March 2018. https://www.kaggle.com/datasets/paultimothymooney/chest-xray-pneumonia.

52. U. Sait, Curated Dataset for Covid-19 Posterior-Anterior Chest Radiography Images (X-rays), May 2021. https://data.mendeley.com/datasets/9xkhgts2s6/3.

53. M. Nickparvar, Brain Tumor MRI Dataset, September 2021. https://www.kaggle.com/datasets/masoudnickparvar/brain-tumor-mri-dataset.

54. F. Kitamura, The UNIFESP X-ray Body Part Classification Dataset, April 2022. https://www.kaggle.com/datasets/felipekitamura/unifesp-xray-bodypart-classification?select=test.

55. ISBI, Chaos-Grand Challenge, April 2019. https://chaos.grand-challenge.org/.

CHAPTER 12

Explainable artificial intelligence in medical research: A synopsis for clinical practitioners—Comprehensive XAI methodologies

Fatma Hilal Yagin[a] and Abdulvahap Pinar[b]
[a]Department of Biostatistics and Medical Informatics, Faculty of Medicine, Inonu University, Malatya, Turkey
[b]Rectorate Unit, Adıyaman University, Adıyaman, Turkey

1. Introduction

Artificial intelligence (AI) has recently been successfully applied in healthcare thanks to the increasing availability of health data and the rapid development of big data analytics techniques. Powerful AI methods can reveal information hidden in large volumes of data that can aid clinical decision-making when guided by relevant clinical queries [1]. AI, the hardware and software systems used in image processing, computational comparison, speech and voice recognition and other fields are all included in the idea of AI. The discussion and application of AI on a global scale stems from the increasing use of intelligent machines in industry and service sectors. Various techniques, including deep learning, Machine Learning (ML) and artificial neural networks (ANN), are being applied in health sciences to utilize AI [2]. Artificial Intelligence is described by medical scientists as "Medical Technology" because it facilitates early and accurate diagnosis, reduces errors and problems, and shortens patients' hospital stays. Numerous medical specialities including cardiology, nephrology, neurology, neurology, gastroenterology, endocrinology, pulmonology, cancer and surgery have demonstrated the importance of AI in health sciences [3]. By offering previously unthinkable capabilities, AI has the potential to transform the healthcare industry. Applying AI to medical imaging is one well-known illustration of its transformational potential. AI-driven technologies may examine Magnetic Resonance Imaging (MRI), CT (computed tomography), and X-ray pictures in radiology to find irregularities or subtle patterns that the human eye could miss, including small malignant tumors. AI has proven to be as accurate as trained radiologists in detecting early-stage malignancies during breast cancer screening. This entails higher survival rates and more effective therapies due to early diagnosis for the patients [4–6]. Another area where AI is having a significant influence is predictive analytics. AI systems can anticipate the possibility of some diseases developing before symptoms appear by examining vast

Explainable AI in Healthcare Imaging for Medical Diagnoses
https://doi.org/10.1016/B978-0-443-23979-3.00012-9

amounts of patient data. To prevent major problems, healthcare practitioners can intervene in acute kidney injury 48 h in advance through AI models using an AI model created [7]. AI can be used in precision medicine, enabling early diagnosis and treatment. In oncology, AI is being used to optimize treatment regimens for diseases such as melanoma and lung cancer by analyzing the patient's genetic makeup to determine which treatments are most likely to succeed. These examples show how AI is advancing healthcare by not only aiding diagnosis but also customizing and elevating the overall standard of care [8,9]. AI is also significantly influencing the research and production of novel medications. AI technologies can screen possible medications faster and more effectively than ever before by analyzing billions of chemical molecules. Analyzing patient data in clinical decision support systems—which provide doctors treatment recommendations—further improves personalized medicine. AI's use in medicine enhances process efficiency and patient care quality, which helps the healthcare sector [10,11].

The explainability of the outputs of "Black Box" models using AI in healthcare allows the XAI technique to output more transparently. Factors such as patient safety, sensitivity and transparency for XAI can help physicians in terms of early diagnosis and treatment.

Explainable Artificial Intelligence (XAI) enables AI to bring the enormous advantages of AI to medical science, making it safer and more understandable. MRI and CT scans use AI algorithms to accurately identify diseases such as cancer [5]. But the processes used to make these choices are frequently unclear. XAI, clinicians may more easily comprehend and evaluate the AI's reasoning behind a given diagnosis by having transparent access to the AI's decision-making processes. XAI makes it understandable what data and how AI evaluates it to suggest a course of therapy in clinical decision support systems (CDSS). This strengthens the confidence that physicians and patients have in these technologies by enhancing the dependability of AI's potent diagnostic and treatment suggestions in the medical domain [8]. Doctors may diagnose, track, categorize, and offer treatments and solutions with the use of XAI applications in the healthcare industry. AI-based techniques can assist doctors in curing illnesses more quickly than they could with conventional methods, lowering the mistake rate in cancer detection, listening to patients' complaints and assisting them with therapy, diagnosing CT scans, and producing computerized medical reports. The adoption and use of new technology by physicians and other healthcare professionals is on the rise, as they bear the responsibility of providing healthcare [12].

This section addresses topics and methodologies pertaining to the interpretability of machine learning and artificial intelligence algorithms, categorizing them and discussing their applications in the medical domain. This section can offer additional justification for specific training in the medical field to harness the potential of algorithms.

2. Clinical decision support systems with XAI

To improve health and healthcare, clinicians, staff, patients or other individuals can have information and personalized knowledge intelligently filtered or presented at the right times through the use of computer systems called CDSS [13].

2.1 Explainability of intricate decision procedures

XAI is essential to make complicated healthcare decision-making processes comprehensible. The judgments made by the AI models used in CDSS must be comprehensible to healthcare professionals as they assist them with tasks like diagnosis, therapy, and patient management. Healthcare workers can better grasp how the model operates by having XAI clarify the reasoning behind these choices. When an AI model determines a patient has cancer it may illustrate and elucidate specific test data or medical pictures affected the diagnosis. Instead of viewing the model as a "black box," this enables medical professionals to make judgments that are more trustworthy and well-informed [14]. Patient safety is another benefit of XAI. Decisions made by AI systems must be clear and verifiable, as any error in the healthcare sector can have serious repercussions. The basis on which an AI model recommends a drug or treatment method is transparently displayed using XAI techniques. Thus, medical professionals can evaluate the treatment plan recommended by the model and provide substitute drugs if necessary. Patient safety is increased in this way and medical practitioners can better monitor the decisions of AI systems and produce results [14,15].

There are several phases involved in using AI, including gathering data, preparing it, configuring it, simulating it, evaluating it, and recalibration. Models must be transparent and comprehensible in each evaluation for outcomes to be considered trustworthy. However, in many scientific and technical domains, entirely opaque imaging of the model is desirable. The conversion of the convectional AI "black box" model into a "white box" model using XAI is seen in Fig. 1. XAI is utilized in health and medicine to provide more interpretability and transparency [16].

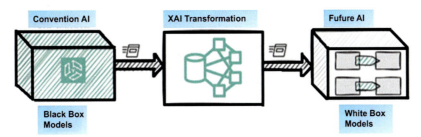

Fig. 1 The change of XAI from a black box to a white box model.

2.2 Patient safety and responsiveness

When it comes to patient safety and easier understanding of clinical decision processes, XAI is crucial for the healthcare industry. Thanks to XAI, physicians can better understand the complex and opaque processes of AI systems, also called "decision boxes," by making them more visible. Especially for serious diseases and high-risk patients, XAI transparency clarifies physicians' decisions. By using XAI techniques in the early diagnosis of life-threatening conditions such as sepsis, medical errors are reduced and patient safety is ensured [17].

A better explanation of the decisions made by the systems is another advantage of XAI for measuring patient response. Patients can thus better understand medical decisions and participate in treatment processes in a more informed way. As a result, patients have more confidence in medical procedures when the root causes of the diagnosis indicated by the AI are revealed. By revealing potential model biases, XAI enables more impartial and equitable decision-making. Due to these features, XAI is a vital tool for therapeutic precision and patient safety in the medical field [18,19].

2.3 Improving model performance

XAI is important for improving CDSS performance. Increasing the transparency and understandability of CDSS is one way to improve model performance. XAI gives physicians the confidence to use complex Machine Learning (ML) models and provides an explanation for these models and makes necessary changes. More specifically, by revealing what elements the model relies on to make decisions, methods such as Local Interpretable Model-agnostic Explanations (LIME) can reduce error rates by improving the precision of clinical judgments [20]. However, for XAI to work better, it needs to not only interpret the output of the model, but also improve the overall accuracy and reliability of the model. Shapley Additive Explanations (SHAP) increase model confidence to explore performance variations through a more balanced assessment of all features of the model. As a result, CDSSs improve in terms of responsiveness and reliability and increase their role in clinical decision-making [20,21].

3. Several biomarker types used in medicine using XAI

3.1 Cancer biomarkers

XAI makes the application of biomarkers, particularly gene expression profiles, to assess the probability of disease recurrence in cancer patients more visible and trustworthy. The purpose of XAI in this process is to make sense of the biological parameters that the AI model considers when predicting a patient's prognosis (i.e., the course of the disease and the likelihood of a recurrence) and to help physicians and researchers make intelligent decisions [22,23].

Moreover, biomarkers can be utilized to decide which therapy is most likely to be successful or as "predictive factors" or response modifiers for a certain program. Since somatic mutations in KRAS are linked to a poor response to therapy aimed against the epidermal growth factor receptor (EGFR), KRAS is a predictive biomarker for colorectal cancer [24].

The various stages of cancer biomarker identification, treatment and monitoring are illustrated in Fig. 2. Initially, biomarkers are used in disease diagnosis to determine the type and presence of cancer. Risk assessment determines a person's likelihood of developing cancer. Biomarkers are also essential for screening processes that detect cancer early. Individualized treatment of patients can be achieved by classifying their condition according to their genetic makeup and symptoms. During treatment, biomarkers can be used to monitor how the body responds to medication and thus improve its effectiveness. Biomarkers are also used to determine the stage of cancer and predict the course of the disease. Ultimately, cancer biomarkers guide the course of the disease and allow for individualized care and monitoring [25].

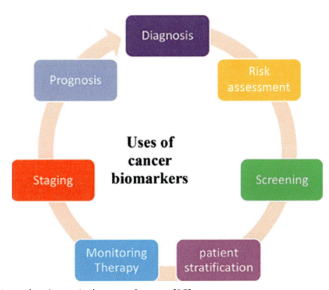

Fig. 2 Cancer biomarkers' practical use and usage [25].

3.2 XAI's function in neurological disease biomarker analysis

The application of XAI models in the diagnosis and treatment of neurological disorders is growing. Making AI model outputs clear and intelligible is the primary advantage of XAI. Researchers and physicians can now comprehend the biomarkers, clinical information, and other elements that the AI is considering while making judgments. By improving the transparency and dependability of AI models, the application of XAI in neurological disorders aids medical professionals in making decisions. Improved diagnostic and treatment outcomes are possible with a better understanding of the biomarker and clinical data evaluation process for illnesses including Parkinson's, Alzheimer's, multiple sclerosis, and epilepsy [26–28].

4. XAI methods

Concept Bottleneck Models (CBMs) are designed to improve interpretability by integrating human-understandable concepts into the model's prediction process. Rather than relying on raw input features, CBMs learn intermediate, high-level concepts that make the decision-making process transparent. This approach enables users to understand how specific concepts contribute to predictions and allows for human intervention if the model's concept-level reasoning is incorrect. CBMs are versatile, making them suitable for applications like medical diagnostics where understanding the reasoning behind predictions is crucial [29].

Explainable Graph Neural Networks (XGNNs) provide transparency for Graph Neural Networks by using methods such as subgraph extraction and counterfactual reasoning to highlight the nodes and edges that drive model predictions. These explanations are essential for complex, structured data like social networks, knowledge graphs, or protein interactions, where traditional feature importance methods may fall short. XGNNs offer both local explanations, focusing on individual predictions, and global explanations, which reveal general patterns within the data [30].

Counterfactual Generative Models use generative approaches like GANs or VAEs to create realistic counterfactual examples, offering insight into how slight changes in input data could lead to different model outcomes. By showing users what would need to be altered to achieve a desired result, these models provide actionable insights, especially in fields like finance, healthcare, and policy-making. The ability to generate realistic alternatives makes counterfactual generative models particularly useful for scenarios that require planning and decision-making under uncertainty [31].

Attention-based Explanation Models leverage attention mechanisms in models such as transformers, using attention weights as a tool for explaining which parts of the input data the model focuses on. By visualizing and interpreting these attention patterns, users can understand which features contributed most to the model's predictions. This approach is especially effective for sequence data, such as in natural language processing or time-series analysis, and has the added advantage of being scalable across a wide range of domains that employ attention mechanisms [32].

Prototype-based Explanations offer interpretability by identifying representative examples from the training data that are most similar to the current input, thereby providing a concrete reference for understanding the model's decision. This method is particularly useful in classification tasks, where it helps users grasp why a model assigned a certain label by comparing the input to known prototypes. By clarifying decision boundaries and linking new data to specific examples, prototype-based explanations make complex classification processes easier to interpret [33].

Multi-modal Explanations aim to explain model predictions by integrating insights from multiple types of data, such as images, text, and audio, and fusing these into a cohesive explanation. This approach is particularly valuable in domains where decisions depend on diverse sources of information, such as in medical diagnostics or autonomous systems. By providing explanations that consider multiple perspectives, multi-modal approaches offer a more comprehensive understanding of the model's reasoning than single-modality explanations [34].

Post-hoc Rule Extraction Models focus on generating human-readable rules that summarize the decision-making process of complex models, such as neural networks or ensemble methods. After the model has made a prediction, rule extraction techniques are applied to translate its internal logic into simple if-then rules that are easy for users to interpret. This method makes black-box models more transparent and accessible, and can be used across various applications to provide post-hoc rationales for decisions without altering the model's architecture [35].

4.1 SHAP (SHapley Additive exPlanations)

SHAP is an XAI technique grounded in cooperative game theory, which calculates the contribution of each feature to a model's prediction. By assigning Shapley values to features, SHAP ensures both local accuracy (explaining individual predictions) and global interpretability (explaining model behavior across the dataset). The method computes the marginal contribution of each feature by considering all possible feature subsets, providing a balanced way to allocate credit for the prediction. SHAP is widely used for its ability to consistently quantify feature importance and its flexibility across different model types, offering both local and global explanations [36].

4.2 LIME (Local Interpretable Model-agnostic Explanations)

LIME is a model-agnostic method that creates locally interpretable models around specific predictions of any black-box model. It perturbs the input data and observes how the predictions change, then uses the modified data to train a simple, interpretable model (like linear regression) that approximates the black-box model's behavior in the local vicinity of that prediction. This localized approach allows users to understand why a complex model made a particular prediction without requiring access to the model's inner workings. LIME's flexibility to be applied across various types of models makes it a versatile tool for interpretability [36].

4.3 CEM (Contrastive Explanation Method)

CEM provides contrastive explanations by highlighting both the features that led to a prediction (pertinent positives) and those whose presence would have changed the prediction (pertinent negatives). This dual approach offers a deeper understanding of model behavior by not only explaining why a certain prediction was made but also identifying what changes in the input would result in a different outcome. CEM is especially valuable in decision-making contexts where knowing both what caused the prediction and what could alter it is crucial. This method enhances traditional explanations by focusing on contrastive reasoning, providing richer insights into the model's decision boundaries [37,38].

4.4 DeepLIFT (Deep Learning Important Features)

DeepLIFT is an XAI technique designed for neural networks, especially deep learning models. It decomposes the output prediction by comparing the activation of each neuron to its reference activation, attributing changes in the output to specific inputs. DeepLIFT tracks how changes in input lead to changes in predictions, enabling a clear path of responsibility from input to output. This method is computationally efficient and is particularly valuable in deep learning models where standard methods like SHAP or LIME might struggle to provide fast, scalable explanations [39].

4.5 Integrated Gradients

Integrated Gradients is another explanation method designed for deep learning models. It computes feature attributions by integrating the gradients of the model's output with respect to the input, as the input varies from a baseline to the actual input. The method ensures that features contributing most to the prediction receive higher attribution scores. Unlike methods that only rely on gradients at a single point (which can miss the impact of certain features), Integrated Gradients accumulate information across the entire gradient path, providing a more robust explanation, especially for neural networks [40].

4.6 Grad-CAM (Gradient-weighted Class Activation Mapping)

Grad-CAM is a visualization technique for deep convolutional neural networks (CNNs), commonly used in computer vision tasks. It produces class-discriminative heatmaps that highlight the regions of an image most relevant to a model's prediction. Grad-CAM works by backpropagating the gradients of the target class through the network to generate importance scores for each spatial location in the final convolutional layer. This allows users to visually interpret which parts of the input image were critical for the classification, making it a highly intuitive and effective method for explaining image-based deep learning models [41].

4.7 Permutation Feature Importance

Permutation Feature Importance is a simple, model-agnostic method for explaining feature importance by randomly shuffling the values of individual features and measuring the change in the model's performance. If the model's accuracy significantly decreases when a feature is permuted, that feature is considered important. This method is widely applicable and easy to implement, making it a practical tool for obtaining global feature importance in any machine learning model. However, it provides only a global view of feature importance rather than localized or instance-based explanations [42].

4.8 Counterfactual explanations

Counterfactual explanations focus on generating alternative scenarios that would have resulted in a different prediction from the model. By identifying minimal changes to the input that could change the outcome, counterfactual explanations offer actionable insights into what must be altered to achieve a desired result. This method is useful in domains like finance or healthcare, where understanding how to influence outcomes is critical. Counterfactual explanations complement methods like SHAP or LIME by focusing on decision boundaries and providing insights into what could have caused a different prediction [43].

5. Conclusion

For biomarker discovery, disease prediction, and CDSS to be effectively implemented in medical practice, they must be dependable, comprehensible, and, most importantly, enhance human decision-making. Explainability is an essential element in attaining these objectives. Explainability enables engineers to detect shortcomings in a system and instills confidence in physicians regarding the decisions they make utilizing AI models. This section on XAI in Medicine examines the "where" and "how" of its application, allowing us to quantify certain gains achieved in this domain. Research concentrating on the phases of AI model development in Medicine is essential to accurately delineate the practical use of explainability in critical scenarios.

References

[1] F. Jiang, Y. Jiang, H. Zhi, Y. Dong, H. Li, S. Ma, Y. Wang, Q. Dong, H. Shen, Y. Wang, Artificial intelligence in healthcare: past, present and future, Stroke Vasc. Neurol. 2 (2017) 230–243.
[2] Y. Çilhoroz, O. Işık, Yapay zekâ: Sağlık hizmetlerinden uygulamalar, Ankara Hacı Bayram Veli Üniversitesi İktisadi ve İdari Bilimler Fakültesi Dergisi 23 (2021) 573–588.
[3] I. Salomon, S. Olivier, Artificial intelligence in medicine: advantages and disadvantages for today and the future, Int. J. Surg. Open 62 (2024) 471–473.
[4] G. Litjens, T. Kooi, B.E. Bejnordi, A.A.A. Setio, F. Ciompi, M. Ghafoorian, J.A. Van Der Laak, B. Van Ginneken, C.I. Sánchez, A survey on deep learning in medical image analysis, Med. Image Anal. 42 (2017) 60–88.
[5] S.M. McKinney, M. Sieniek, V. Godbole, J. Godwin, N. Antropova, H. Ashrafian, T. Back, M. Chesus, G.S. Corrado, A. Darzi, International evaluation of an AI system for breast cancer screening, Nature 577 (2020) 89–94.
[6] A. Esteva, B. Kuprel, R.A. Novoa, J. Ko, S.M. Swetter, H.M. Blau, S. Thrun, Dermatologist-level classification of skin cancer with deep neural networks, Nature 542 (2017) 115–118.
[7] N. Tomašev, X. Glorot, J.W. Rae, M. Zielinski, H. Askham, A. Saraiva, A. Mottram, C. Meyer, S. Ravuri, I. Protsyuk, A clinically applicable approach to continuous prediction of future acute kidney injury, Nature 572 (2019) 116–119.

[8] E.J. Topol, High-performance medicine: the convergence of human and artificial intelligence, Nat. Med. 25 (2019) 44–56.

[9] K. Kourou, T.P. Exarchos, K.P. Exarchos, M.V. Karamouzis, D.I. Fotiadis, Machine learning applications in cancer prognosis and prediction, Comput. Struct. Biotechnol. J. 13 (2015) 8–17.

[10] R. Qureshi, M. Irfan, T.M. Gondal, S. Khan, J. Wu, M.U. Hadi, J. Heymach, X. Le, H. Yan, T. Alam, AI in drug discovery and its clinical relevance, Heliyon 9 (2023) 1-23.

[11] K.-H. Yu, A.L. Beam, I.S. Kohane, Artificial intelligence in healthcare, Nat. Biomed. Eng. 2 (2018) 719–731, https://doi.org/10.1038/s41551-018-0305-z.

[12] G.Y. Tarcan, P.Y. Balçık, N.B. Sebik, Türkiye ve Dünyada Sağlık Hizmetlerinde Yapay Zekâ, Mersin Üniversitesi Tıp Fakültesi Lokman Hekim Tıp Tarihi ve Folklorik Tıp Dergisi 14 (2024) 50–60, https://doi.org/10.31020/mutftd.1278529.

[13] J.A. Osheroff, J.M. Teich, B. Middleton, E.B. Steen, A. Wright, D.E. Detmer, A roadmap for national action on clinical decision support, J. Am. Med. Inform. Assoc. 14 (2007) 141–145, https://doi.org/10.1197/jamia.M2334.

[14] O.N. Akarsu, Bölüm VI, BİDGE Yayınları, Kafkas University, Institute of Social Sciences, 2024.

[15] A. Alpkoçak, Sağlıkta açıklanabilir yapay zekâ, TOTBİD Dergisi S 23 (2024) 18–19.

[16] A. Lakhan, M.A. Mohammed, J. Nedoma, R. Martinek, P. Tiwari, A. Vidyarthi, A. Alkhayyat, W. Wang, Federated-learning based privacy preservation and fraud-enabled blockchain IoMT system for healthcare, IEEE J. Biomed. Health Inform. 27 (2022) 664–672.

[17] L. Wyatt, L. van Karnenbeek, M. Wijkhuizen, F. Geldof, B. Dashtbozorg, Explainable artificial intelligence (XAI) for oncological ultrasound image analysis: a systematic review 14 (18) (2024) 8108. https://doi.org/10.3390/app14188108.

[18] P.W. Koh, P. Liang, Understanding black-box predictions via influence functions, in: Proceedings of the 34th International Conference on Machine Learning, vol. 70, PMLR, 2017, pp. 1885–1894.

[19] K. Chaudhary, O.B. Poirion, L. Lu, L.X. Garmire, Deep learning-based multi-omics integration robustly predicts survival in liver cancer, Clin. Cancer Res. 24 (2018) 1248–1259.

[20] S.K. Mandala, XAI Renaissance: Redefining Interpretability in Medical Diagnostic Models, arXiv preprint arXiv:2306.01668, 2023.

[21] P. Korica, N.E. Gayar, W. Pang, Explainable artificial intelligence in healthcare: opportunities, gaps and challenges and a novel way to look at the problem space, in: 22nd International Conference on Intelligent Data Engineering and Automated Learning 2021, Manchester, United Kingdom, Lecture Notes in Computer Science, vol. 13113, Springer, 2021, pp. 333–342. https://doi.org/10.1007/978-3-030-91608-4_33.

[22] S. Paik, S. Shak, G. Tang, C. Kim, J. Baker, M. Cronin, F.L. Baehner, M.G. Walker, D. Watson, T. Park, A multigene assay to predict recurrence of tamoxifen-treated, node-negative breast cancer, N. Engl. J. Med. 351 (2004) 2817–2826.

[23] L.J. van't Veer, S. Paik, D.F. Hayes, Gene expression profiling of breast cancer: a new tumor marker, J. Clin. Oncol. 23 (2005) 1631–1635.

[24] C.J. Allegra, J.M. Jessup, M.R. Somerfield, S.R. Hamilton, E.H. Hammond, D.F. Hayes, P.K. McAllister, R.F. Morton, R.L. Schilsky, American Society of Clinical Oncology provisional clinical opinion: testing for KRAS gene mutations in patients with metastatic colorectal carcinoma to predict response to anti-epidermal growth factor receptor monoclonal antibody therapy, J. Clin. Oncol. 27 (2009) 2091–2096.

[25] M. Wang, F. Witzmann, Role of Biomarkers in Medicine, BoD–Books on Demand, 2016.

[26] A. Almohimeed, R.M. Saad, S. Mostafa, N. El-Rashidy, S. Farag, A. Gaballah, M. Abd Elaziz, S. El-Sappagh, H. Saleh, Explainable artificial intelligence of multi-level stacking ensemble for detection of Alzheimer's disease based on particle swarm optimization and the sub-scores of cognitive biomarkers, IEEE Access 11 (2023) 123173–123193, https://doi.org/10.1109/ACCESS.2023.3328331.

[27] F. Cavaliere, A. Della Cioppa, A. Marcelli, A. Parziale, R. Senatore, Parkinson's disease diagnosis: towards grammar-based explainable artificial intelligence, in: 2020 IEEE Symposium on Computers and Communications (ISCC), Rennes, France, IEEE, 2020, pp. 1–6. http://doi.org/10.1109/ISCC50000.2020.9219616.

[28] S. Jahan, F. Nowsheen, M.M. Antik, M.S. Rahman, M.S. Kaiser, A.S. Hosen, I.-H. Ra, AI-based epileptic seizure detection and prediction in internet of healthcare things: a systematic review, IEEE Access 11 (2023) 30690–30725.

[29] P.W. Koh, T. Nguyen, Y.S. Tang, S. Mussmann, E. Pierson, B. Kim, P. Liang, Concept bottleneck models, in: ICML'20: Proceedings of the 37th International Conference on Machine Learning, ACM, 2020, pp. 5338–5348. 495 https://dl.acm.org/doi/10.5555/3524938.3525433.

[30] H. Yuan, H. Yu, S. Gui, S. Ji, Explainability in graph neural networks: a taxonomic survey, IEEE Trans. Pattern Anal. Mach. Intell. 45 (2022) 5782–5799.

[31] S. Wu, W. Zhou, M. Chen, S. Zhu, Counterfactual generative models for time-varying treatments, in: Proceedings of the Proceedings of the 30th ACM SIGKDD Conference on Knowledge Discovery and Data Mining, ACM, 2024, pp. 3402–3413. https://doi.org/10.1145/3637528.367195.

[32] S. Liu, F. Le, S. Chakraborty, T. Abdelzaher, On exploring attention-based explanation for transformer models in text classification, in: Proceedings of the 2021 IEEE International Conference on Big Data (Big Data), IEEE, 2021, pp. 1193–1203. http://doi.org/10.1109/BigData52589.2021.9671639.

[33] Y.-M. Shin, S.-W. Kim, E.-B. Yoon, W.-Y. Shin, Prototype-based explanations for graph neural networks (student abstract), in: Proceedings of the AAAI Conference on Artificial Intelligence, 36 (No. 11: IAAI-22, EAAI-22, AAAI-22), AAAI, 2022, pp. 13047–13048. https://doi.org/10.1609/aaai.v36i11.21660.

[34] B. Finzel, D.E. Tafler, S. Scheele, U. Schmid, Explanation as a process: user-centric construction of multi-level and multi-modal explanations, in: Proceedings of the KI 2021: Advances in Artificial Intelligence: 44th German Conference on AI, Virtual Event, September 27–October 1, 2021, Proceedings 44, 2021, 2021, pp. 80–94.

[35] G. Vilone, L. Longo, A quantitative evaluation of global, rule-based explanations of post-hoc, model agnostic methods, Front. Artif. Intell. 4 (2021) 717899.

[36] V. Vimbi, N. Shaffi, M.J. Mahmud, Interpreting artificial intelligence models: a systematic review on the application of LIME and SHAP in Alzheimer's disease detection, Brain Inform. 11 (2024) 10.

[37] J. Labaien, E. Zugasti, X. De Carlos, Contrastive explanations for a deep learning model on time-series data, in: M. Song, I.Y. Song, G. Kotsis, A.M. Tjoa, I. Khalil (Eds.), Big Data Analytics and Knowledge Discovery. DaWaK 2020, Lecture Notes in Computer Science, vol. 12393, Springer, Cham, 2020, pp. 235–244. https://doi.org/10.1007/978-3-030-59065-9_19.

[38] P. Mishra, Extensions, Frameworks. Contrastive Explanations for Machine Learning, Apress, Berkeley, CA, 2022, pp. 279–298. https://doi.org/10.1007/978-1-4842-7158-2_11.

[39] A. Shrikumar, P. Greenside, A. Kundaje, Learning important features through propagating activation differences, in: ICML'17: Proceedings of the 34th International Conference on Machine Learning, vol. 70, JMLR.org, 2017, pp. 3145–3153. https://dl.acm.org/doi/10.5555/3305890.3306006.

[40] I. Čik, A.D. Rasamoelina, M. Mach, P. Sinčák, Explaining deep neural network using layer-wise relevance propagation and integrated gradients, in: 2021 IEEE 19th World Symposium on Applied Machine Intelligence and Informatics (SAMI), Herl'any, Slovakia, IEEE, 2021, pp. 000381–000386. https://doi.org/10.1109/SAMI50585.2021.9378686.

[41] R.R. Selvaraju, A. Das, R. Vedantam, M. Cogswell, D. Parikh, D.J.A. Batra, Grad-CAM: gradient-weighted class activation mapping, arXiv (2016). https://doi.org/10.48550/arXiv.1610.02391.

[42] H. Kaneko, Cross-validated permutation feature importance considering correlation between features, Anal. Sci. Adv. 3 (2022) 278–287.

[43] R. Guidotti, Counterfactual explanations and how to find them: literature review and benchmarking, Data Min. Knowl. Disc. 38 (2024) 2770–2824.

CHAPTER 13

Advancing explainable AI and deep learning in medical imaging for precision medicine and ethical healthcare

Tariq Mahmood[a], Yu Wang[b], Amjad R. Khan[a], and Noor Ayesha[c]
[a]Artificial Intelligence and Data Analytics (AIDA) Lab, CCIS Prince Sultan University, Riyadh, Kingdom of Saudi Arabia
[b]Shandong Research Institute of Industrial Technology, Jinan, China
[c]Center of Excellenec in Cyber Security (CYBEX), Prince Sultan University Riyadh, Saudi Arabia

1. Introduction

Artificial intelligence (AI) systems, which can recognize images, comprehend speech, make judgments, and translate languages, have applications across various sectors, including economics, biometrics, online commerce, and the automotive industry [1]. AI is poised to address the growing challenges caused by rising patient demands, increasing chronic diseases, and resource constraints. As digital health technologies become more widespread, they contribute to the exponential increase in healthcare data. Implementing AI could allow healthcare professionals to focus more on understanding disease origins and monitoring the effectiveness of treatments and preventive measures. Lawmakers, politicians, and decision-makers need to recognize this trend, as many experts argue that AI should be a central pillar of healthcare reform [2].

The global healthcare landscape faces rising costs, driven by factors such as the increasing prevalence of chronic illnesses, higher life expectancy, and the continual emergence of costly treatments. Researchers predict a challenging future for the sustainability of healthcare systems [3]. AI offers a potential solution by improving healthcare quality and cost-effectiveness. AI is already widely used in healthcare to support clinical decisions and diagnose chronic diseases. However, despite its promise, AI-based models trained on individual patient data do not provide a universal solution. These models come with various challenges, from technical to legal, medical, and ethical, requiring a multidisciplinary approach to overcome. One significant challenge is the explainability of AI systems, which is not only a technical issue but also raises critical medical, legal, ethical, and social concerns [4].

AI has proven to be a transformative force in medical imaging, particularly in automating complex tasks. One is image segmentation, a time-consuming process prone to errors when done manually [5]. Accurate automated medical image segmentation is critical for improving clinical workflows, providing healthcare professionals with reliable tools for extracting quantitative measurements and aiding decision-making. The surge in medical imaging across modalities such as radiography, endoscopy, computed tomography (CT), mammography (MG), ultrasound, magnetic resonance imaging (MRI), and positron emission tomography (PET) has further underscored the importance of AI in supporting radiologists. The shortage of radiologists worldwide compounds the need for AI-driven solutions to handle the growing volume of images efficiently [6].

Deep learning, particularly convolutional neural networks (CNNs), has revolutionized medical image analysis. While traditional machine learning techniques were dominant in the early stages, deep learning initiated a paradigm shift in the mid-2010s. The success of CNNs in tasks like the ImageNet Challenge demonstrated their potential in medical imaging, leading to a wave of research in AI applications for disease diagnosis, segmentation, and treatment planning [7]. These advancements are supported by improvements in computational infrastructure, such as graphical processing units (GPUs) and cloud computing, enabling more sophisticated AI models to process complex medical data [8].

This study provides a comprehensive review of deep learning techniques in medical imaging, covering methods such as segmentation, detection, classification, and enhancement. It explores various applications of AI in disease diagnosis and treatment, focusing on the role of deep learning in improving medical imaging processes [9]. Additionally, the study emphasizes the availability of publicly accessible datasets, particularly for brain cancer segmentation, and highlights recent advances in deep learning techniques for this application [10].

However, the present issues are still related to explanations in the case of AI as an applicable tool in medicine. It was also crucial to stress that Transparency in AI models is essential for building rapport and clarity with practitioners and AI systems. Thus, it supports the implementation of explainable AI (XAI) frameworks into medical imaging in healthcare [11]. Clinical requirements and responsibilities can be met by XAI to minimize the differences between current AI capabilities and the amount of supporting information that clinicians need to provide decisions that are based on the use of AI. Thus, it is crucial to involve technological, ethical, legal, and clinical stakeholders to discuss the social and ethical aspects of AI in health care [12].

Therefore, this paper aims to provide an exhaustive analysis of deep learning frameworks in medical imaging, specifically with a study of brain cancer segmentation and a discussion of XAI and its task to make AI work safer and more ethical in clinical practice [13]. Addressing the explainability challenges, this research will help fill the existing gap in understanding how AI should be further developed and used to enhance its benefits for patients and make healthcare more sustainable in the coming years.

1.1 The need for Explainable Artificial Intelligence in the medical field

In today's world, factors such as increased life expectancy, the rising incidence of chronic diseases, and the constant development of expensive new therapies [14] drive more research focus into the healthcare field. Artificial intelligence (AI) is expected to alleviate some of these challenges by improving healthcare and making it more cost-effective [15]. However, despite AI's undeniable potential, it is not a universal solution. Due to its direct connection with human health and life, the medical field often requires professionals to adopt a serious and cautious attitude. This uniqueness sets higher demands for AI's interpretability in medical applications [16].

In AI, when applied to lesion prediction or detection, interpretability explains how the model identifies itself or predicts the location of a lesion in a patient for diagnosis or prognosis [17]. When interpreting medical image analysis, doctors and clinical professionals must require the information generated by the AI model when arriving at the diagnosis or the positioning of a lesion to make the right decisions about the patient's state. For instance, in diagnosing and staging prostate cancer, MRI has been reported to enhance prostate cancer detection dramatically [3] and is more frequently recommended to direct a therapeutic strategy. MRI is the most sensitive non-invasive imaging modality for visualizing, detecting, and localizing cancer. Compared to ultrasound-guided biopsy alone, MRI–ultrasound fusion biopsy improves the detection of clinically significant prostate cancer. Histopathological analysis of prostate tissue obtained via biopsy helps determine the presence and grading of prostate cancer [18]. The location and extent of invasive cancer observed through non-invasive imaging can also guide treatment decisions, such as whether to pursue radical prostatectomy, local therapy, or active surveillance. However, the subtle visual differences between benign and malignant tissue on MRI often pose challenges for radiologists in image interpretation. At this point, an accurate and explainable method for lesion localization becomes crucial. Not only can it help doctors make more precise diagnoses and reduce medical errors, but it can also improve healthcare professionals' efficiency and free up medical resources.

In the medical field, doctors are constantly communicating with patients. From the patient's perspective, the critical question regarding interpretability is whether AI-driven decision support tools align with the intrinsic values of patient-centered care. Patient-centered care aims to respond to and respect individual patients' values and needs [19]. It views patients as active partners in care, emphasizing their choice and control over medical decisions. A key component of patient-centered care is shared decision-making, which seeks to identify the most appropriate treatment based on the patient's individual circumstances. This involves open dialogue between patients and clinicians, where clinicians inform patients about the potential risks and benefits of available options, and patients discuss their values and priorities [20]. Suppose clinicians cannot fully understand the internal workings and computations of the decision support tools. In that case, they cannot explain how specific outcomes or recommendations were derived to the patient. Interpretability can address this issue by providing personalized dialogue assistance to clinicians and patients based on individual patient characteristics and risk factors. By simulating the effects of different treatment or lifestyle interventions, explainable AI decision support can help increase patient awareness of their options and support clinicians in understanding patients' values and preferences.

1.2 Deficiencies in existing approaches

Although conventional diagnostic techniques are efficient, they may be time-consuming and lack the precision for personalized therapy. ML algorithms are a powerful instrument for early and precise cancer prediction, as they can analyze extensive data and identify nuanced patterns. Numerous intelligent healthcare systems are already capable of accurately predicting cancer lesions [21]. Moreover, medical professionals employ prediction systems to facilitate early diagnosis, significantly reducing the cost of diagnosis by reducing the number of cases requiring additional conventional diagnostic procedures. Many systems already possess a high degree of accuracy in predicting cancer abnormalities. Nevertheless, in critical applications of this nature, we consistently pursue the development of a more precise, resilient system that does not generate any false pessimistic predictions [22]. Consequently, recall is a critical determinant in evaluating the effectiveness of these systems. False pessimistic predictions are those that can jeopardize life, as the conventional diagnosis may or may not be pursued by the patient or physician, which can exacerbate cancer's progression. In the present day, the body places a naive faith in the accuracy of the predictions, and all individuals, including medical professionals and patients, seek to understand the rationale and justifications behind them. The significance of explanations is further emphasized if the cancer prediction is optimistic. Consequently, medical professionals and patients rely heavily on rationales and explanations before reaching a conventional cancer diagnosis. However, these methods set the mandate for the central idea of AI interpretability and plotted a course to that uncharted territory [23]. When complex models like deep learning are being used in various fields, the need for the interpretability of artificial intelligence is keenly felt. Deep learning has revolutionized medical image classification and segmentation despite numerous challenges and difficulties:

1. In medical image analysis, deep learning models depend on supervised data for accurate results, and small amounts of data can result in underfitting. Medical image data is limited, and it has limited availability of qualified datasets due to its specialized area of focus.

2. Among the factors defining the decision on which architecture is suitable for deep learning models, achieving high accuracy is paramount. It is essential to appreciate that the currently available literature might not always be conclusive and affine recommendations about the most appropriate architectural solution for some

image-processing medical tasks. This research helps to understand the changes in performance commonly resulting from extending the architecture, more specifically, additional layers and filters.

3. Class imbalance represented in medical image datasets hurts the reliability and performance of deep models, as disease samples are dominant over others in the majority.

4. The medical image analysis requires accurate prediction results; thus, deep learning models' interpretability is significantly appreciated.

5. Training deep models with several parameters entailing several hours of training and tuning. Current issues related to deep learning applied to the diagnosis and treatment of patients include a time-consuming training process, slow convergence for the models, and inefficiency.

6. The effectiveness of medical image functions of classification and segmentation is reduced by the differences between patients, morphological structure variations, and lesion severity quantization because there is poor contrast between different tissue types. In light of this, the model's accuracy has to be improved depending on the characteristics of the medical image.

7. In the present study, to compare the new model's performance with the other classification models, the computational complexity and execution time of the latest model were compared with those involved and needed by other advanced techniques.

The chapter discusses the potential of deep learning in medical image classification and segmentation tasks, particularly in functions. It highlights the limitations of medical image model accuracy and the challenges of training deep models, highlighting the research content and significant innovations in multi-medical images.

1.3 Contributions

The key contributions of this chapter are summarized as follows:

- This study explores the evolution of deep learning techniques for medical imaging, compares the properties and efficacy of contemporary cancer segmentation methods on benchmarks, and discusses deep learning-based image segmentation challenges and future directions.

- Employing a versatile XAI model, which combines the CNN framework with the ML technique, has significantly improved cancer detection accuracy and efficiency by leveraging CNN's feature extraction expertise and robust ensemble learning protocols.

- The hybrid approach utilizes XAI technology to provide healthcare practitioners with comprehensive explanations and valuable insights. This feature generates easily understandable forecasts, enhancing physicians' ability to make informed clinical choices by understanding various factors affecting outcomes.

- Highlighting specific contributions of different deep learning architectures in brain cancer segmentation.

1.4 Chapter organization

A review of recent literature on deep learning applications in medical imaging, with a particular concentration on cancer segmentation, is presented in Section 2. Section 3 discusses deep learning methodologies, frameworks, and evaluation metrics for models in biomedical imaging and analysis. It highlights the importance of various modalities in monitoring medical diagnosis and treatment. The study also discusses analytical methodologies used in medical image processing. It also explores recent developments in deep learning architectures for cancer segmentation, focusing on relevant medical image modalities and datasets. The primary challenges, issues associated, and findings with applying deep learning to medical imaging are addressed in Section 4. Lastly, Section 5 concludes with a summary, an outlook for further exploration in this field, and recommendations and conclusions for future research.

2. Literature review

2.1 History of Explainable Artificial Intelligence

Explainable Artificial Intelligence, abbreviated as XAI, originates from Explainable Artificial Intelligence. This concept was first introduced in 2016 by the U.S. Defense Advanced Research Projects Agency (DARPA) [24]. To avoid confusion, the abbreviation for Explainable Artificial Intelligence should be written as XAI, with the "X" for "Explainable." Since then, XAI has been widely accepted as the abbreviation for explainable AI. However, research on explainable AI can be traced back to 1982, when researchers proposed an artificial neural network called Neocognitron [25]. This network used a layered design that enabled computers to gradually "learn" to recognize visual patterns through multi-layer learning. Its performance improved incrementally by repeatedly training the network with a reinforcement strategy activated multiple times. Due to the relatively few layers and fixed learning content, the network had some degree of interpretability in its early stages. This can be seen as the starting point of deep learning visualization. The study also introduced a sensitivity analysis method based on statistical results that analyzed machine learning model outcomes to explore their interpretability [12]. These early systems provided rule sets that offered intuitive and easily understandable explanations. However, these systems showed limitations in handling complex problems and large-scale data, making them insufficient for increasingly complex demands in the future. However, these methods set the mandate for the central idea of AI interpretability and plotted a course to that uncharted territory. As deep learning and many other complex models are used in numerous application domains, there is a growing need for interpretable AI systems. Scientists have only started to look for ways to estimate the decision-making procedures for these models. Since the middle of the 2010s until now, several interpretability techniques have appeared, creating several subfields [26]. These methods can be divided into two main categories based on the scope of their explanations: local interpretability and global interpretability. Local interpretability techniques indicate the local usage of specific input features in light of the entire data set and assign the cause of the result to particular values [27]. On the other hand, the GIMs allow an understanding of the attribution of the decision-making of the whole model for various input data. XAI is an acronym for clarifying or explaining the decision-making and behavior of the learning models by AI systems. Implementing models' interpretability in practical use in machine learning and such spheres as individual rights, privacy, and ethical subjects is essential. Various successful applications of deep learning-based systems include image classification, speech recognition, and emotion analysis, etc., which mimic human performance; however, the use of these systems in the actual production environments is constrained by some issues, namely lack of interpretability of the final decision and lack of control over the output. The idea is to create AI systems that offer better explainability to make AI more useful and acceptable in the healthcare, finance, and law industries. Interpretability of the models used in AI is preferred, making concerns and doubts about AI systems less of an issue. Emphasizing interpretability is not only a technical decision but also a social and responsible issue in the development of AI [28].

1. Perceived reliability and robustness relate to how humans trust the model to perform soundly while making decisions. They also include the degree to which the model generates reasonable and convincing explanations.
2. From a societal perspective, AI's reasons for making decisions have to address equity issues. AI interpretability combines several ethical and social aspects, including impartiality of decisions, non-violation of privacy, and enough evidence and rationales in vital domains where the models are applied. This helps implement the models and prevent social implications that may arise in the future.
3. Sensitivity can be explained as feedback because the model can capture user responses and adapt. The benefits include enhanced performance, decreased errors, and improved user satisfaction.
4. Interpretability requires that machine learning models be created and worked on without hidden steps or elements. This is helpful since users and stakeholders can see how any model reaches a conclusion and make changes where needed based on an attitude of transparency.
5. Related to transparency, comprehensibility emphasizes that the model's output should be explained in a manner understandable to humans. This includes understanding the importance of features in the model, the factors influencing the decision, and how the model responds to different inputs.

2.2 Background knowledge

Understanding that explainable models are preferred in many decision-making processes because they increase reliability and user recognition and help acquire new knowledge. Consequently, the problem of obtaining explanations from ML models pervasively or designing inherently interpretable models has emerged [29]. With regard to the nature of the explanations, these methods can be local or global, and some can be both. Local interpretability techniques are centered on the model's behavior in the code's neighborhood. In contrast to global interpretability, local interpretability explains the model's decision-making for a particular instance or a small group of instances rather than for the entire dataset [30]. On the other hand, global interpretability methods provide insights into the model's decision-making process as a whole, deepening our understanding of the attributes of an array of input data. Moreover, based on the underlying algorithmic principles used in explainable methods, these approaches can be categorized into perturbation-based, gradient-based, counterfactual generation, and knowledge distillation-based methods. In this section, we will introduce and discuss these methods. Medical imaging has become widespread, leading to a dramatic increase in misdiagnoses—the given diagnosis is wrong because humans make mistakes when tired or simply inexperienced. These may result in identifying normal variants as diseases or failure to diagnose the pathology. This can lead to false negative affirmations, which worsens the patient's condition and leads to more tests and higher costs [31]. The need for computer contractors has emerged as evident from these arguments. Digital image analysis, putting data on an image into a computer, can be used to identify, divide, and diagnose organs, tissues, and lesion areas. This enhances the accuracy and reliability of medical diagnoses among patients. Computer-assisted instruments are beneficial in surgical planning and training due to the large amount of potential research and the broad applicability of these methods. The first studies on the applicability of computer technology to automatic medical image analysis started in the 1960s; however, due to the shortage of computational power and high-quality digitized images, the field did not receive much attention. However, with the advancement and development of artificial intelligence (AI) and machine learning, the initial machine-learning approaches for medical image analysis were inaccurate. Headers contained non-linear transformations, and multi-layered hierarchical structures were used for data processing and emerged as a new category of machine learning known as deep learning in 2006. The automated classification and segmentation workflow of medical images is a crucial step in computer-aided medical image interpretation since developments in computer vision have greatly improved medical image analysis. Another area of great importance is integrating deep learning with medical imaging and facilitating the automated understanding of complicated medical images. After learning hierarchical representations, convolutional neural networks (CNNs) have shown excellent results in several tasks, including image segmentation, classification, and diagnosis from raw data.

2.3 State-of-the-art schemes

Counterfactual generation methods use context-aware semantic-level features and generative adversarial networks, generate synthetic samples that respect data distribution, and outperform heuristic sample imputation. Recent methods use adversarial autoencoders to generate examples and counterexamples, mapping data into latent space, applying random perturbations, and decoding it to generate actual synthetic samples. Xie et al. [32] proposed a method that creates transmission paths in a learned low-dimensional manifold, allowing the transformation of samples into simulated ones and observing different classes while preserving their background features. This allows for the exploration and visualization of domain knowledge behind distinguishing rules. The image samples are encoded into a vector space, divided into two subspaces: class-relevant features (style subspace) and class-independent sample features (background subspace). By gradually modifying the sample, the class-related code can be shifted along a guided path, explaining each sample. Singla et al. [33] a backbone network based on StyleGANv2 to improve uncertainty quantification in pre-trained deep neural networks. Fine-tuning with augmented data and soft labels improves decision boundaries, effectively capturing uncertainty in ambiguous samples, unseen near-distribution samples, and label shift scenarios, addressing high accuracy but overconfident predictions. Singla et al. [34] developed a cGAN-based framework to audit a black-box classifier and assess its clinical relevance. The framework generates perturbed query images, changing negative and positive classification decisions. The authors added a specialized reconstruction loss to preserve anatomical shapes and foreign objects. The model explained cardiomegaly and pleural effusion decisions, confirmed by experienced radiology residents. The framework combines semantic segmentation and foreign object detection networks. Tao et al. [35] developed a lesion segmentation model using image-level labels, a faster and easier method than full annotations. The model uses an image classifier trained with these labels, followed by model visualization tools to generate an object heatmap for each training sample. The generated heatmaps, pseudo-annotations, and an adversarial learning framework are used to construct an image generator for edema region segmentation. This approach combines the benefits of supervised learning and adversarial training, achieving interpretability while addressing the challenge of fully annotated medical images.

Knowledge distillation-based explainability methods aim to enhance the interpretability of deep learning models by establishing a relationship between the predictions of complex models and corresponding interpretable models. The core idea is to transfer the knowledge from a complex black-box deep learning model to an interpretable model, making the original model's predictions easier to understand and explain. Knowledge distillation methods improve the interpretability of complex models by transferring their knowledge to simpler models, making their predictions easier to understand. This helps users better understand the model's behavior and decision-making process. High predictive accuracy can be maintained during distillation while computational resources for training and inference are reduced. However, compressing complex model knowledge into an interpretable model may result in information loss, reduce generalization ability, and negatively impact the quality of the final interpretable model. Overfitting to training data may also reduce generalization ability. Tan et al. [36] proposed a method for auditing proprietary or opaque black-box risk-scoring models. They treated a teacher model as a black-box model and trained a transparent student model to mimic its risk scores. The distilled student model was compared with a non-distilled transparent model trained on labeled outcomes. The difference between the two models provided insights into the black-box model. This method is useful for complex real-world data with unknown bias sources, as it doesn't require prior knowledge. Frosst and Hinton [37] introduced a method to create a more interpretable model from a trained neural network using soft decision trees trained using stochastic gradient descent. These trees use learned filters to make hierarchical decisions based on input examples, selecting specific static probability distributions as outputs. Although they generalize better than directly trained models, their performance is lower than the neural network used to train them. Che et al. [38] introduced Interpretable Mimic Learning. This knowledge distillation method uses gradient-boosted trees to learn interpretable features from deep learning models like stacked denoising autoencoders and LSTM networks. They demonstrated that this approach achieved similar or better performance than deep learning models in real-world clinical time-series datasets, providing interpretable phenotypes for clinical. The study aimed to transform deep neural networks into decision trees on multi-class datasets, addressing the challenge of understanding and reasoning about complex predictions. Liu et al. [39] applied knowledge distillation techniques to transform a deep neural network into a decision tree, achieving good performance and interpretability. Experiments showed that the student model's accuracy performance significantly exceeded that of regular decision trees at the same tree depth.

3. Methodology

Medical imaging provides anatomical and functional metabolic information about human tissues, critical in disease diagnosis, staging, and treatment evaluation. Preprocessing medical images is crucial to obtaining high-quality images, which are also necessary for AI, and AI has progressed in this field. The research summarizes recent work and development on using AI in image noise reduction, lesion segmentation, and quantitative analysis in CT, MRI, and PET/CT. Therefore, the objective of the paper is to improve clinical awareness of AI and encourage the integration of such systems into the practice of medical imaging.

3.1 Overview of medical imaging and XAI integration

Medical images used in diagnostics procedures include Mammography, Ultrasound, CT, and MRI [40]. However, the manual analysis of customer reviews is time-consuming and can result in many errors because of cognitive constraints and fatigue. Employing CNNs and parsing and classifying images automatically using AIMs, such as automatic image segmentation, enhances efficiency and precision. AI has revealed some of the best ways of identifying cancer lesions and cardiovascular imaging. However, it should be noted that these models are still partly opaque and, consequently, have not been implemented in practice settings. The solution to this problem is called explainable AI (XAI) [41]. XAI frameworks allow clinicians to understand how and why AI systems make certain decisions and build trust in the models by making them more understandable. They use Attention Mechanisms to associate the model attention with the observation made by the physician and Grad-CAM methods to display parts of the images that are relevant to the input image. This method overlaps the gap between complex human interpretability and fundamental machine learning algorithms.

3.2 Diverse medical imaging modalities

Screening modalities for cancer prediction include mammography, ultrasonography, MRI, and PET scans. These modalities identify architectural distortions of the parenchyma, such as masses that may indicate the presence of cancer abnormalities. These modalities may also be useful in diagnosing cancer early, thus helping the patient increase the survival rate. Further, its use might help to reduce the number of false-positive predictions, therefore avoiding excess biopsies [42]. This work considers the modalities involved and used by the CAD system for brain cancer prediction to aid physicians in screening and grading cancer lesions. It is critical to obtain high-quality data that will enable the learning of good computational models, and computer vision has led to improvement in this area. This is because as medical tools and faith progress, the volume and accuracy of acquired medical images are dramatically enhanced. Nevertheless, labeling the medical image data is complex, as medical data contains a patient's information, and annotating the medical images requires a high level of medical knowledge. Accessing large quantities of public, labeled medical image data sets is difficult. There is a huge reliance on medical images and computer-aided tools to solve a problem with public datasets, like OAI and MOST, for grading knee osteoarthritis. However, labeling is work-intensive and requires help from a high-level medical expert, a problem for deep learning-based medical image analysis technologies [43].

Clinical images are digital representations of a patient's body used to document injuries, burns, and lesions, aiding in tracking treatment efficacy and dermatological and cosmetic treatments. They play a crucial role in cancer detection, particularly melanoma. X-ray imaging detects fractures, bone dislocations, and illnesses like pneumonia or pneumothorax. Mammography detects changes in breast tissue using low-intensity X-ray photons. If dense tissue is not visible during screening, radiologists may refer to ultrasound, MR imaging, CT-SCane, or biopsy for more precise analysis [44]. Computed

Tomography (CT) is a sophisticated imaging technique that uses X-rays to generate detailed cross-sectional images, enhancing diagnostic precision and patient care. Magnetic resonance imaging (MRI) creates high-resolution images of tumors, often used when cancer is diagnosed to gather more information and may result in a biopsy if suspicious lesions are discovered. Ultrasound imaging (US) offers real-time visualization of internal organs and blood flow, facilitating safe clinical assessments.

3.3 Image acquisition and pre-processing

The acquisition of annotated images is crucial for creating an effective deep learning (DL) model, as they are manually labeled and segmented to avoid data corruption during mammography. However, images captured in the real world often contain flaws that degrade feature extraction quality, making it challenging to satisfy the need for multiple annotated images. Future research will focus on developing databases with millions of images and boosting segmentation performance to increase the efficacy of classifiers [45]. ML imaging technology has enabled the acquisition of more precise and detailed medical images, resulting in abundant medical images available for research and diagnosis. These images have allowed researchers to construct deep learning-based classification and prognosis algorithms for brain cancer. Before training DL methods, medical images must be pre-processed, which involves several procedures such as augmentation, region of interest (ROI), background separation, scaling, image upgrades, cropping, noise reduction, and muscle removal. Some of the chosen works contributed to pre-processing, while future work may reveal the need for more effective pre-processing procedures. Noise is a significant factor affecting image quality, and AI shows excellent potential in noise reduction, mainly through deep learning algorithms like residual learning, dense network learning, and batch normalization. These methods can significantly enhance the signal-to-noise ratio (SNR) of images. Noise reduction in CT imaging using AI is already relatively well-developed and has achieved remarkable results in vascular reconstruction. It can decrease noise in lower limb arterial CTA images and enhance image quality [46]. In MRI, high-quality diagrams can be obtained, and the body structures can be visualized better, thus enabling accurate diagnosis, which would otherwise have been detected late. Filtering and Interpolation reconstruction techniques based on AI and using elements of deep learning can enhance the signal-to-noise ratio and achieve an improved image in higher resolution &/or better quality—all without adding time. They proposed that these types of noise reduction based on artificial intelligence are more effective than the conventional filter techniques in abdominal MR imaging. In abdominal MR imaging, AI filtering and interpolation reconstruction improve diagnostic value and disease identification while increasing image quality without sacrificing time or signal-to-noise ratio. These deep-learning techniques to minimize noise levels are far better than filtering techniques.

3.4 Medical image analysis (image registration and object detection)

Medical image analysis involves several techniques and uses essential in making sense of image data. Image registration is often used in medical imaging to warp images to create the desired comparisons or combinations correctly [47]. This is particularly important for observing the illness's evolution and for radiation therapy or surgery. Precise determination of the spatial or temporal positions of Objects such as anatomical structures, organs, landmarks, or tasks in organs within the image is the objective of object localization, which is very important in medical imaging [48]. CNNs have become highly influential in object detection, where localized regions can be characterized accurately for abnormal images without human experts' supervision. The CNN algorithms can identify and locate pulmonary nodules in chest X-ray or CT scans, enhancing cancer diagnosis and therapy. The study aims to improve the quality of the X-ray images where DL algorithms are applied for cancer diagnosis. The pre-processing technique has involved the reduction of inherent image noise, correcting implicature pixel intensity differences, and improving the edges of anatomical structures. It helps in realizing improved accurate classification when implementing other subsequent deep-learning models. The mapped proposed study will be devoted to accurate cancer prediction based on the analysis of X-ray images, which involves input sample data containing discriminating features and corresponding output labeled data. Gray scaling is introduced as a predictive method for cancer detection using X-ray image analysis, converting images into grayscale, similar to black-and-white presentation. Data normalization promotes consistency across all variables and mitigates the impact of any one feature on machine learning algorithms. Augmentation processes, such as horizontal flipping, random rotation, and controlled randomness, are explored to improve the accuracy of cancer X-ray image forecasting.

3.5 Substructure segmentation

Multimodal imaging, such as CT, MR, and PET/CT, provides anatomical and metabolic information about lesions, and accurate segmentation of these lesions enables precise visualization of their morphology and relationship with surrounding structures. This, in turn, provides more accurate clinical diagnosis and treatment [49]. Traditional image segmentation largely relies on manual delineation, which is time-consuming and subjective. AI, intense learning-based CNN algorithms, can automatically identify and delineate lesions, significantly reducing subjectivity and improving segmentation efficiency and reproducibility. CT is a standard auxiliary diagnostic tool identifying lesions based on density differences. However, because CT produces grayscale images, extensive image review can lead to visual fatigue, resulting in missed or incorrect diagnoses. AI can rapidly and automatically annotate lesion location, size, quantity, and relationship with adjacent structures, reducing missed diagnoses and improving lesion identification accuracy. This is particularly mature in intracranial vessel segmentation [50]. MR images are particularly challenging due to their multiple sequences and complex signals. AI's ability to quickly and accurately process large volumes of MR data makes it highly promising for precisely segmenting various lesions. Tumor lesion segmentation and boundary localization are crucial for preoperative evaluation of resection margins and radiotherapy target delineation. PET/CT, a well-established multimodal imaging technique, provides anatomical and metabolic information about lesions, helping clinicians assess tumor boundaries more precisely. Compared to CT alone, PET/CT is better at delineating tumor contours and differentiating between normal and diseased tissue boundaries. However, manual segmentation is time-consuming, so there is an urgent need for AI-based automatic segmentation methods in clinical practice. Some researchers have developed a pulse coupled neural network (PCNN) model to segment PET/CT images automatically and compared the results with conventional manual segmentation. This conclusion shows that PET automatic segmentation was as accurate as manual target delineation and more precise than CT-based segmentation. Namely, PET automatic segmentation was less generalizing tumors with atelectasis than six Australian radiologists, while PET-based radiotherapy planning was significantly lower than CT-based [51].

U and V–Net models play an essential role in medical image segmentation, as well as in contextual and localization abilities as well. V–Net, introduced in 2016, generalizes this concept for 3D medical image segmentation and is used frequently in semantic segmentation [52]. A new model, SegNet, is inspired by the VGG16 model, and it uses an encoder similar to the first 13 convolutional layers of the VGG16 followed by a decoder that directly corresponds to the decoder layer in encoders and converts the encoded form into the image form. In more recent approaches, architectures called AttentionNet, Pyramid Scene Parsing Network (PSPNet), Mask R-CNN, and DeepLab rely on dilated convolutional layers to improve image segmentation. DeepLabV3, presented in 2017, uses dilated convolution to increase the receptive field while keeping the resolution high, which can be useful for semantic segmentation. The Unified Perceptual Parsing (UperNet) has a pyramid pooling module to capture multi-scale contextual information without losing spatial resolution. Others, such as Swin Transformer, ConvNeXt, and ClipSeg, extend the application of segmentation methods.

Case study of deep learning in cancer segmentation: Brain cancer care helps identify and determine malignant from benign tissue within the brain [53]. It also performs feature extraction to pick appropriate features, including the tumor's shape, size, texture, and margins, which are important in diagnosing and treating the disease. A proper separation is also required in radiation therapy to ensure adequate destruction of cancerous cells while doing minimal damage to other cells or tissues. It also compares segmented regions for monitoring and prognosis targeted at treatment effectiveness and recurrence detection. Brain cancer is the most common cancer type found in women across the world, with higher mortality than other types of cancer. Radiographers and other practitioners employ 584 segmentation methods to diagnose diseases, observe pathological changes, quantify tissue volume, study structures, and plan treatment. Fig. 1 shows the CAD system for the segmentation of medical images using the XAI methodology.

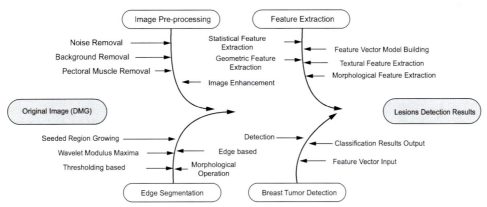

Fig. 1 CAD system for cancer lesion segmentation.

3.6 Feature explanation methods

The fundamental idea behind Feature-based explanation methods is to explain a model by iteratively probing the changes in output caused by different variations in the input [54]. These perturbations can be made at the feature level, where certain features are replaced with zero LIME (Local Interpretable Model-Agnostic Explanations), which is a method that generates human-interpretable representations by evaluating changes in model predictions within a local region near the input [55]. It uses superpixels to construct interpretable local models, developing a binary vector indicating which superpixels are most important for a particular class output in the explanation model. SHAP (Shapley Additive Explanations) is a method proposed by Lundberg and others that calculates Shapley values based on cooperative game theory [56]. This method treats data features as players in a game and calculates each feature's contribution to the final output prediction, enhancing interpretability. Zeiler et al. introduced a technique that visualizes neural activations of deep CNNs by occluding different parts of the input image and using a Deconvolutional Network (DeConvNet). DeConvNets creates an activation map that maps back to the input pixel space, visualizing neural activity. RISE (Randomized Input Sampling for Explanation) perturbs the input image by multiplying it with random masks [57]. The masked image is input into the model, and a saliency map corresponding to each image is captured. Confidence scores are then used to compute a weighted average of the masks, resulting in a final saliency map with positive heatmap values for individual predictions. Interpretability randomized test (IRT) and one-shot feature testing (OSFT) are methods introduced by Burns et al. to examine the importance of context in deep learning algorithms [58]. However, these methods require multiple perturbations and evaluations of the input, resulting in high time costs, and may not be suitable for time-sensitive applications or adversarial examples or by selecting one or a group of pixels (superpixels) to blur, shift, or mask.

3.7 Classification method

Feature extraction is crucial for early disease detection and diagnosis, enabling clinicians to identify pathological conditions and initiate timely interventions. Deep learning methods, particularly CNNs, have revolutionized classification tasks, enabling automated analysis with unprecedented accuracy and efficiency. CNN-based algorithms can classify suspicious lesions indicative of cancer, facilitating early diagnosis and treatment planning. This research report classifies cancer tissue, including pectoral muscle, fibro-glandular, and fatty tissue [59]. Machine Learning techniques are used to classify cancer tissue densities but are not precise enough. Classification models for cancer anomalies include supervised learning for feature categorization and a novel neuron structure for map representation. Methods for cancer lesion prognosis include incremental boosting convolutional networks, CNN networks, and a softmax activation function for grading issues. Deep learning-based XAI models like DCNN, U-Net variants, Transformers variants, and V-Net are widely used in medical image segmentation and classification, supporting tasks like tumor segmentation, organ detection, and disease classification [60]. CNN-based models have demonstrated remarkable performance in cancer imaging, identifying malignancies with high sensitivity and specificity. However, challenges remain in AI and medical imaging, such as data availability and quality, model generalization, and interpretability. Data availability and quality can be complex due to patient privacy concerns and expert annotation costs. Model generalization can be problematic, as many models exhibit overfitting, particularly in healthcare, where models must be robust across diverse populations. Additionally, traditional deep learning models are often opaque, making it difficult for healthcare providers to trust their decisions and reliant on AI systems.

Precision medicine is a field in which artificial intelligence treats patients, depending on their individuality. AI can sort through the enormous amounts of data that might exist and arrive at a favorable diagnosis or recommend the most effective treatment. However, it's only acceptable if it can be explained and integrated into our systems. XAI frameworks can enhance this by offering explicable reasons behind the AI explication, according to which the clinician can confirm the AI's prophecies with clinical data [61]. Clinicians can adapt interventions to address patients' situations if they comprehend how AI models develop individualized interventions. By applying XAI to segmentation models of cancer, decision-making involving AI systems is made more trusted. However, when applied to the healthcare sector, the following ethical and social implications arise privacy, responsibilities, prejudice, and objectivity. Thus, with the help of XAI, potential problems can be solved since clinicians can take full ownership of AI-supported diagnoses due to sophisticated, elaborated explanations of AI activities. Science and technology have made it very easy to collect medical data compared to the past. However, there are difficulties: The data exists in several distinct formats; the difference in quality arising from the use of different equipment; and, finally, the absence of standard disease profiles because of interpatient variability. X-ray Imaging, MRI, CT, Ultrasound, OCT, and PET have been widely used in early diagnosis, detection, and treatment of diseases. Following the structure and function of biological neural networks, deep learning techniques transformed the analysis and interpretation of medical images. Neural network models are normally implemented through deep learning frameworks with special features suitable for some applications [62]. TensorFlow is an open-source framework created by the Google Brain team in 2011, and, as of the most recent, it is highly flexible and efficient for the training of large-scale neural networks. It supports Python and C++ programming languages and is among the most popular solutions for research and industry uses. Caffe is a deep learning framework designed at the Berkeley Vision and Learning Center in 2014 that received great attention due to its speed and modularity, particularly for CNNs. Torch is an open-source ML library written in Lua scripting language used in research and development. Many of these can be traced to Theano, a high-level numerics computation library written in Python by the Montreal Institute for Learning Algorithms. Keras is the neural network API that can be run on top of TensorFlow, Theano, or Microsoft Cognitive Toolkit. MatConvNet is another deep learning framework written in MATLAB, and it is primarily used for constructing convolutional neural networks because it is easily integrated with the MATLAB environment, which is common in academia. Deeplearning4j is the one created by Java for business purposes, and it has a lot of facilities for a distributed environment and is fully optimized. Fig. 2 depicts the stepwise prediction disease prediction using explainable AI approaches.

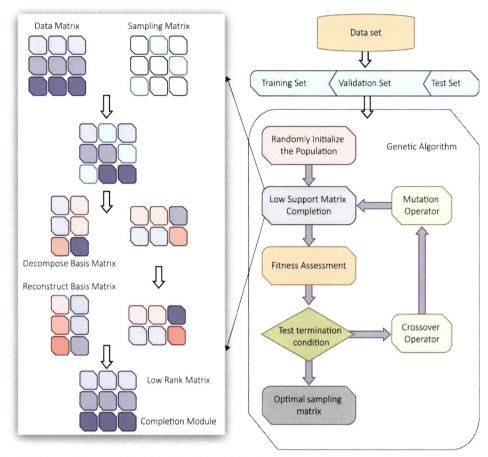

Fig. 2 Flow architecture of explainable AI model for disease prediction.

3.7.1 Convolutional neural networks

CNNs is a kind of deep learning model commonly applied in medical imaging for image categorization, detection, identification, partition, and correspondence. They maintain local image relations while performing the dimensionality reduction that results in a spatial feature map of the input data and replacing the large feature map with the feature map, which is smaller in size, through the pooling layer. CNNs have improved diagnostic accuracy and early cancer detection, especially cancer detection. They can identify particular characteristics intrinsic to medial images and hence make far more reliable predictions [63]. CNNs have been used in medical image analysis, with pre-trained models through transfer learning used to enhance results in small data sets. However, areas like data imbalance and model over-specialization need further research. CNNs have a shorter training time than large explicit pre-trained models, making them suitable for advanced medical centers and raw environments. Data augmentation and class weighting improve the model's capacity to classify unseen data and avoid overfitting. Fig. 3 explained the layered architecture of CNN-based model.

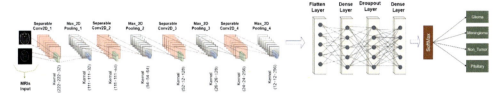

Fig. 3 CNN-based architectures.

3.7.2 Recurrent neural networks (RNNs)

RNNs are DL-based networks that identify patterns in data sequences, such as time series or natural language. They have connections that form directed cycles, allowing information to persist. Standard RNNs are neural network architectures with hidden states and feedback loops that recognize sequential patterns, making them valuable for tasks involving language models and time-series data analysis [63]. RNNs have gained attention due to their flexibility in capturing non-linear relationships. RNNs are similar to traditional neural networks but have a modified hidden layer incorporating self-recurrence to continuously record historical information in the secret state vector. This structure holds data from each time step, constructs a relationship between data from different time steps, and gradually inputs the influence into the last stage of the model. Techniques in applying RNNs show that they fit time series data much more effectively than other types of neural networks. However, they also suffer a disadvantage when considering the long input sequences because of the well-known issues of vanishing or exploding gradients, which complicate the model training. The activation functions in RNNs are typically the tanh functions, which only respond to inputs that are in a set bound. If the sequence length increases, the gradient either goes to infinity or zero, making network training impossible and the model non-convergent.

3.7.3 Long-short-term memory networks (LSTMs)

LSTMs are a type of RNN that mitigates the vanishing gradient problem with a more complex cell structure that allows for long-term dependencies in data. They are highly effective for tasks like language modeling and speech recognition. The LSTM cell processes the information flow using the current input, the previous hidden state, and the previous cell state to update the current hidden and cell states via sigmoid and tanh activation functions. This structure allows LSTMs to capture the long-term dependency of sequential data efficiently.

3.7.4 Gated recurrent units (GRUs)

GRUs are similar to LSTMs but have a more straightforward structure, often making them faster to train while still maintaining the ability to capture long-term dependencies. The GRU cell uses these gates to control the flow of information. The reset gate modulates how much of the previous hidden state should be forgotten. In contrast, the update gate determines how much of the new information should be incorporated into the hidden state. The candidate hidden state is updated using a combination of the current input and the past hidden state via non-linear activation functions such as sigmoid (σ) and tanh. This structure allows GRUs to maintain relevant information over longer sequences without the complexity of LSTMs.

3.7.5 Generative models

Generative models are a class of models that can generate new data samples from the learned distribution of the training data. Restricted Boltzmann machines (RBMs), structured with visible and hidden layers but no connections within layers, have gained attention for their role in building robust neural networks. Particularly useful in unsupervised learning, RBMs contribute to various applications. Deep belief networks (DBNs) feature multiple layers containing an RBM and offer promise in semi-supervised learning scenarios. DBNs are efficiently trained using labeled and unlabeled data, which aid in feature extraction and representation learning across different levels. GANs consist of two neural networks, a generator and a discriminator, trained simultaneously through adversarial learning. GANs have created realistic synthetic data, including images and audio.

3.7.6 Autoencoders

Autoencoders are a neural network used to learn efficient data representations, typically for dimensionality reduction or feature learning. Standard Autoencoders, another unsupervised representation learning model, encode input data into lower-dimensional representations. Particularly beneficial for datasets with abundant unlabeled data, autoencoders facilitate feature extraction and dimensionality reduction in medical image analysis. Variational autoencoders (VAEs) introduce a probabilistic approach to the latent space representation, enabling the generation of new data samples similar to the training data. They are widely used in generative modeling. Attention Mechanisms and Transformers: Attention mechanisms have revolutionized deep learning, particularly in natural language processing. They improve neural network performance by enabling the model to focus on the most relevant parts of the input data, enhancing the handling of sequences and spatial information. Transformers utilize self-attention mechanisms and effectively process medical images by capturing dependencies across spatial and contextual dimensions. These models excel in tasks such as image classification, segmentation, and anomaly detection, providing state-of-the-art performance in medical image analysis.

Fig. 4 illustration of an encoder-decoder architecture often used in burst-sequence models. The encoder processes the input sequence through layers of RNNs, often with skip layers, to prevent overfitting and transforms the input into a fixed-length context vector. The decoder receives the context vector from the encoder. It produces the output sequence with different RNN layers, with dropout and linear layers that generate the final predictions. This architecture is often used in applications such as machine translation, where the input sequence might be a sentence in one language, and the output sequence is translated into another.

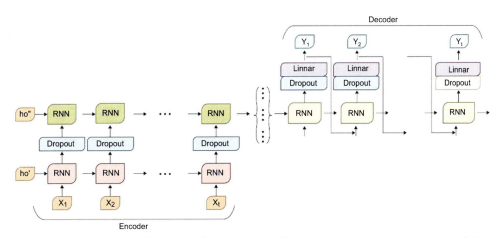

Fig. 4 Depicts the encoder-decoder architecture typically used in sequence-to-sequence models.

3.7.7 Graph neural networks (GNNs)

GNNs are deep learning models that operate on graph structures, making them suitable for tasks where data is naturally represented as a graph, such as social networks and molecular biology. GCNs extend the principles of convolutional networks to graph-structured data, enabling effective feature learning and representation of nodes within a graph. GANs are deep learning models that sample directly from a desired data distribution without explicitly modeling the underlying probability density function. GANs consist of a generator and a discriminator, trained to compete against each other.

In the past decade, there is a notable development in the application of deep learning for image classification, As AlexNet, GoogleLeNet, VGG-16, ResNet50, Inception-v3, DenseNet, EfficientNet, Detection Transformer(DETR), Vision Transformer (ViT), Data–efficient Image Transformer (DeiT), Pyramid Vision Transformer (PVT), and Deformable Attention Transformer (DAT). AlexNet from year 2012 again brought out a novel architecture of CNNs with deeper architecture and ReLU activation along with dropout. GoogleLeNet also known as Inception v1 which consists of a new "Inception module" which help the model to learn features at multiple resolutions. As for the networks, let me talk about VGG-16: it is rather simple but highly effective. The ResNet50 mobilized a minor revolution in deep learning since it sought to solve the vanishing gradients issue, adopting residual connections. Inception-v3, introduced in 2015, refined the inception module architecture and 316 incorporated batch normalization for improved training stability. DenseNet, introduced in 2017, promoted feature reuse and enhanced gradient flow throughout the network. EfficientNet, developed in 2019, uses a compound scaling method to achieve state-of-the-art accuracy with fewer parameters and computational resources. Detection transformer (DETR) is a transformer-based model designed for object detection, while vision transformer (ViT) adapts the transformer architecture from natural language processing to image classification tasks. Deep learning models use backpropagation algorithms to calculate the gradient of the loss function concerning model parameters, indicating changes in the loss function as the model moves through the parameter space. These methods produce a heatmap or saliency map, indicating which input parts contribute most to the model's decision. Gradient class activation mapping (CAM) is a method that uses global average pooling for all pooling operations instead of max pooling. Grad-CAM and Grad-CAM+ + have improved CAM operations for deeper CNNs but still have high noise and lack of fine-grained interpretation issues. DeConvNet is a gradient-based method that generates saliency maps for convolutional networks using backpropagation for activation visualization. DeConvNets give relative importance to gradient values during backpropagation, allowing for accurate visualization. The GuidedBP method also improves on this work.

3.8 Proposed model

This chapter addresses the shortcomings of current brain tumor MRI image segmentation methods, which tend to overlook the differences and connections between the feature information of multiple modalities of brain tumors. Additionally, these methods often consider a single dimension when modeling global contextual information. This chapter proposes a cross-modal brain tumor MRI image segmentation network model called MCAT-Net, based on a multi-view coupled Transformer model to solve these issues. This network model effectively extracts the feature space information between multiple modalities and leverages the local and global features in each modality's spatial and depth dimensions. As a result, it improves the accuracy of brain tumor lesion segmentation. It enhances the performance of the segmentation network in general, thus helping doctors diagnose brain tumors and similar diseases. Fig. 5 presents the overall architecture of the proposed model, showcasing its various components, including the input layer, feature extraction modules, multi-scale connections, and output segmentation layer, designed to enhance the accuracy of brain tumor segmentation in medical imaging.

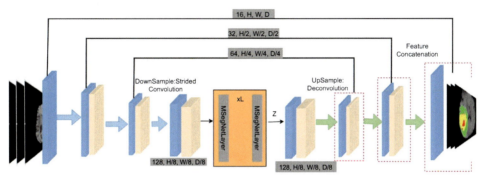

Fig. 5 The proposed model architecture uses straggled convolutions to reduce input image spatial dimensions and enhance depth, passing residual layers for feature extraction. The intermediate bottleneck layer handles the smallest, most abstract image. After deconvolutions restore spatial dimensions, fine-grained information is preserved in upsampling by feature concatenation with downsampling layers.

To address the problem of extracting and fusing the feature information of the four modalities in the brain tumor MRI image data, this paper introduces a cross–modal attention mechanism between the four types of data and a cross–modal attention fusion module for the multimodal brain tumor MRI image segmentation. This module extracts information mapping of the local features and models the global relations with multiple modalities. For the current problem of extracting local contextual features and modeling global features, this chapter presents a multi-view coupled cross–modal attention brain tumor magnetic resonance imaging (MRI) image segmentation network derived from the Transformer model called MCAT-Net. The MCAT-Net consists of three major components: Three MCAT-Net3D CNN encoders, the MCCA Transformer network layer encoder, and the MCAT-Net3D CNN decoder. To acquire contextual feature information of the brain tumor MRI images, first, the encoder utilizes the MCAT-Net3-D CNN to learn the 3D spatial features of an echo. Subsequently, the extracted volumetric blocks are passed into the transformer layer of our MCCA Transformer network to perform cross–modal feature extraction and global feature mapping in the in-depth and spatial domains. Finally, the decoder embeds the feature information obtained from the MCAT-Net network model into the volumetric blocks and progressively upsamples them to predict detailed full-resolution subregion maps for brain tumor segmentation.

To accurately segment the three nested subregions (ET, TC, and WT) in multimodal brain tumor MRI images, this paper proposes a multi-view coupled cross-modal attention brain tumor MRI image segmentation network based on the Transformer model. Overall, the segmentation network follows a classic encoder-decoder U-shaped structure. The architecture of MCAT-Netis is illustrated in Fig. 5. In MCAT-Net, to capture the local contextual feature information of brain tumor MRI images, the encoder first utilizes MCAT-Net3D CNN to extract the 3D spatial features of the brain tumor. At the same time, the input multimodal MRI images are downsampled, and their feature information is mapped. The processed feature volumetric blocks are fed into the subsequent network for cross-modal feature extraction and global feature modeling in in-depth and spatial dimensions. In Fig. 5, the input brain tumor MRI image is represented as $X \in H \times W \times D \times C$, where $H \times W$ denotes the spatial resolution of the brain tumor MRI image, D indicates the depth dimension (i.e., the number of slices), and C represents the number of channels (i.e., the number of modalities). In this process, black arrows represent the downsampling of feature maps, while red arrows represent the upsampling. Each time the feature maps are downsampled, their size is halved. To effectively map the spatial relationships between modalities, the spatial dimensions of feature maps, and the depth dimension of brain tumor MRI images into the segmentation network, we employ the MCAT-Net3D CNN encoder to generate compact feature maps that capture spatial and depth information—additionally, the MCCA Transformer encoder models long-range dependencies in global space. The MCAT-Net model also employs skip connections, which link shallow and deep features, reducing the loss of spatial feature information caused by multiple downsampling operations. The detailed processing flow of the MCAT-Net model. It consists of the MCAT-Net3D CNN encoder, the MCCA Transformer encoder, and the MCAT-Net3D CNN decoder. The decoder progressively upsamples the embedded feature information obtained from the MCAT-Net model into volumetric blocks, ultimately predicting detailed full-resolution brain tumor subregion segmentation maps. Furthermore, this paper proposes a cross-modal attention fusion module to associate the feature information extracted from multimodal input images while processing multimodal brain tumor MRI images. This relationship is then mapped to the segmented output image. After completing cross-modal attention fusion, the MRI images of each modality are divided into three views and input into the network to facilitate feature extraction from specific perspectives.

This paper aims to show how the feature information belonging to the respective modalities can be linked to the image segmentation module to solve the noted problem in current models where the spatial data from multiple modalities is partially utilized in the brain tumor MRI image segmentation segmentation. In particular, we propose a cross-modal attention approach to optimize the brain tumor MRI image segmentation task. This research presents an innovative four-modal attention mechanism that uses a cross-modal attention mechanism. The mechanism outputs the targeted feature information from the multichannel inputs for the multimodal MRI image inputs and fuses the feature information with correlation. The fused information is then transferred into the MRI image segmentation of brain tumors. This enables the cross-modal attention mechanism to model the spatial relationship between the different modalities globally, enhancing segmentation results.

3.9 Evaluation metrics for deep learning models

This chapter focuses on the concept of explainability within deep learning algorithms, which currently have no standard methodology for defining explanation properties. However, much work remains to be done, and the overall focus of the chosen area is, to a significant extent, in its exploration stage. The paper urges the community to design reasonable algorithms for assessing the explanations tools. Carrying out evaluation is very important when it comes to defining the capability of deep learning models in medical imaging tasks. These metrics are defined from the confusion matrix for binary segmentation tasks: TP = True Positive, FP = False Positive, TN = True Negative, FN = False Negative, and predicted AUC. The selection of the evaluation metrics is an important step when assessing the performance of deep learning models in medical imaging tasks since each is important for making correct conclusions regarding the task result.

3.9.1 Accuracy

A measure of the general accuracy of a model's predictions is accuracy. Out of all the occurrences in the dataset, the percentage of accurately predicted cases is calculated. This can be mathematically represented below equation.

$$Accuracy = \frac{TP + TN}{TP + TN + FP + FN}$$

Herein, TP, TN, FP, and FN denote true positive, true negative, false positive, and false negative, respectively.

3.9.2 Loss

The **modification** among the model's predictions and the ground truth labels in the training dataset is represented by loss, sometimes referred to as the cost or objective function. The deep learning model's training process aims to minimize this loss function. For various tasks, typical loss functions include:

Mean Squared Error (MSE) for tasks involving regression

Cross-entropy loss (categorical cross-entropy or binary cross-entropy) is used for classification jobs involving many classes or binary classification.

3.9.3 Precision

Precision is a metric that assesses how well the model predicts the future. It establishes the ratio of correctly foreseen positive outcomes to all positive outcomes, including both true and false positives. Precision is crucial when false positives have serious repercussions. The recipe for accuracy is:

$$Precion = \frac{TP}{TP + FP}$$

4. Result analysis and discussion

In this work, a significant concern is identifying which explainability method to use for image classification with BraTS2020, a medical imaging dataset consisting of 3D brain MRI scans and their correspondences of ground truth labels. The dataset was divided into two classes: diseased and non-diseased. The researchers also discovered that not all explainability methods can be used for saliency and binary segmentation map creation. He and his team chose 10 appropriate explainability methods for this experiment. As a confirmation, the researchers carried out comparative learning on a similar task, specifically on the dataset using the ResNet50 model. They applied different explainability techniques for identifying lesion areas and obtaining flamingos' segmentation maps. To objectively measure the performance of the segmentation, the IOU and DICE measures between the obtained segmentation and the ground truth were used. While IOU is used for more accurate localization between the ground truth bounding boxes or regions and predicted ones, DICE is used for size and shape variations. These measures serve as an improvement in the performance evaluation of the algorithm and guarantee that the algorithm works well.

4.1 Dataset

The brain tumor MRI imaging training set is obtained from MRI scans from different medical centers called BraTS2020. Still, it does not include images from the TCIA cancer immunology atlas database. It omits images from the cancer immunology atlas database of TCIA and consists of 369 glioma cases. Each case has four modalities and four labels, which make diagnosis between tumor and edema possible, although the latter is not sensitive to internal heterogeneity. In this paper, the BRATS benchmark, known to provide accurate results when used in brain tumor MRI image segmentation tasks, is utilized. This standard, developed from the MICCAI 2012 and MICCAI 2013 challenges, compares ground truth labels from the BraTS series datasets with predicted segmentation results from different models. Evaluation metrics for the three tumor sub-structures are WT (Whole Tumor), TC (Tumor Core), and ET (Enhancing Tumor).

4.2 Performance analysis

The BraTS2020 dataset was also used to train the MCAT-Net brain tumor MRI image segmentation model on an NVIDIA Tesla 3040Ti GPU (32GB). During training, the model randomly sampled image patches of size $128 \times 128 \times 128$, with a batch size set to 16 and a learning rate 0.0005. After training, we obtained the MCAT-Net loss curve, as shown in Fig. 6A and B. Observing the overall trend of the curve, it is clear that as the number of training iterations increases, the accuracy gradually increases, and the Loss value gradually decreases. When the number of epochs exceeds 07, the fluctuation in the Loss value becomes smaller, and after 12 epochs, the curve begins to stabilize, indicating that the function has essentially converged. Therefore, we set the number of epochs to 30 to reduce training time while ensuring adequate training. In the BraTS2020 dataset, the proposed MCAT-Net segmentation model achieved average Dice scores of 79.13%, 93.56%, and 97.19% for ET, TC, and WT, respectively, through five-fold cross-validation experiments. We also tested the BraTS2020 validation set and compared the MCAT-Net segmentation model with the SOTA 3D models. The specific experimental results are shown in Table 1.

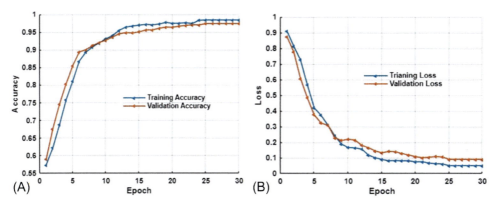

Fig. 6 Performance comparison, loss, and accuracy curves during training, and validation during testing. (A) Model accuracy curve. (B) Loss curve of the model.

Table 1 Table shows the segmentation performance was better than previous SOTA 3D methods.

Model	Dice score			Hausdorff distance		
	WT	TC	ET	WT	TC	ET
3D UNet	83.89	82.91	73.92	14.681	14.023	53.567
V-Net	88.23	79.85	65.02	19.901	15.523	48.256
TransBTS	93.56	86.23	79.36	5.895	12.562	19.964
MCAT-Net	97.19	93.56	80.39	12.125	12.982	19.860

Overall, the segmentation performance was better than previous SOTA 3D methods, especially for the tumor region (WT) and the tumor core region (TC), with significant improvements. Specifically, the model achieved up to a 19.69% improvement for the tumor core region and a 9.85% improvement for the whole tumor region. Although the segmentation results for the enhancing tumor region (ET) were not the best, the Dice score still outperformed mainstream brain tumor MRI segmentation models such as 3D UNet and V-Net.

Regarding Hausdorff distance, the MCAT-Net segmentation model also demonstrated an advantage, indicating that the segmentation results obtained by MCAT-Net closely matched the ground truth images manually annotated by expert physicians. The model performed exceptionally well in segmenting the tumor core region and the whole tumor. While the edge handling for the enhancing tumor region may be somewhat coarse, the model could still delineate and segment the lesion areas in brain tumors. The MCAT-Net effectively supports clinical diagnosis and treatment of brain tumor-related diseases by assisting professional physicians.

Similarly, to provide a more intuitive comparative analysis of the MCAT-Net model's segmentation performance in the WT, TC, and ET regions, we used the segmentation results of a sample case to compare brain tumor MRI image segmentation visually, as shown in Fig. 7. The detailed comparison is shown in Fig. 8. In this figure, we compare the segmentation results from the MCAT-Net model with the original images and the manually segmented results from three perspectives: axial, coronal, and sagittal views. Overall, the differences between the MCAT-Net model and manual segmentation are minimal. The model successfully separates the tumor from the surrounding normal tissue, with precise boundary segmentation for the WT, TC, and ET regions. Specifically, in the axial view, the MCAT-Net model demonstrates more accurate and precise boundary segmentation of the tumor lesion area than the manual segmentation. In the coronal view, the model distinguishes the enhancing tumor region from the surrounding edema area, with finer boundary segmentation than manual segmentation. In the sagittal view, the MCAT-Net model effectively segments the tumor core region,

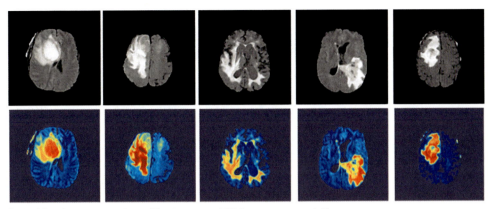

Fig. 7 Visual representation of brain tumor MRI image segmentation.

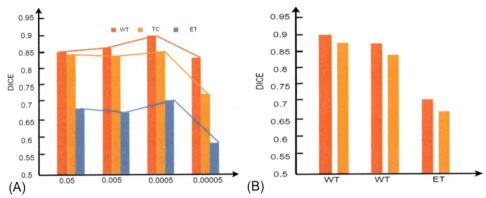

Fig. 8 Performance comparison (A) using different learning rates and (B) performance based on the dice coefficient.

which is morphologically irregular and varies significantly in size. Therefore, the proposed MCAT-Net model can assist medical professionals by reducing the workload and error rate associated with manual segmentation. Whether in terms of segmentation accuracy, precision, or efficiency, the MCAT-Net model significantly supports clinicians in diagnosing and treating brain tumor patients.

4.3 Ablation experiments

Ablation experiments examine the impact of core components in a model or theoretical system. The MCAT-Net brain tumor MRI image segmentation model improves segmentation accuracy by mapping spatial correspondence across modalities and extracting features from spatial and depth dimensions. The cross-modal fusion module is a critical component in the MCAT-Net model, embedding spatial relationships between modalities (T1, T2, FLAIR, and T1ce) into feature maps. The cross-modal fusion module was removed in this experiment, and the model's performance was evaluated on the BraTS2019 dataset with the same hyperparameters. Fig. 8 shows the performance reduction without the cross-modal fusion module. The Dice scores for the WT, TC, and ET regions decreased significantly, with ET being the most affected. This indicates that the cross-modal fusion module is crucial for accurately segmenting complex tumor regions.

4.4 Challenges and future directions

The objective of the investigation is to create a deep-learning model for the diagnosis of cancer abnormalities through the use of classification algorithms. Nevertheless, practical execution is impeded by the time-consuming availability of medical images, computational resources, and the limitations of machine learning techniques. To enhance the effectiveness of DL schemes, researchers must consider factors such as the type of data, the type of deep learning model, and the training parameters. Using the OTSU filter, designating ROI using the symmetric approach, and utilizing a multimodal imaging database with patient information and context are all recommended. The scarcity of annotated data, large labeled datasets, and ethical and regulatory considerations such as patient privacy, data security, and algorithmic bias are among the challenges encountered by deep learning in medical imaging, particularly for cancer segmentation. The ethical development and deployment of medical imaging systems based on deep learning necessitate collaborative efforts among researchers, clinicians, policymakers, and regulatory bodies. An excellent AI system should possess strong, intelligent performance and remain transparent, interpretable, reliable, fair, privacy-conscious, sustainable, and capable of interacting with humans in a friendly manner, all while upholding social responsibility and ensuring alignment with ethical and societal values. Blindly trusting the results of predictive classifiers is inadvisable under today's standards due to inherent data biases, credibility issues, and the influence of adversarial examples in machine learning.

In this chapter, we explored the critical importance of explainable AI (XAI), covering several aspects of XAI and classifying them based on their underlying algorithmic principles to explain deep neural networks. Considering the unique challenges of medical imaging data, our evaluation experiments observed that existing explainability methods applied to medical imaging datasets suffer from low accuracy and vague semantic directionality. In light of these limitations, we proposed several improvement suggestions for future researchers to explore further. Overall, AI explainability remains a highly focused area of research, with ongoing efforts to enhance model interpretability and make the decision-making processes of deep learning models more understandable. As explainability tools continue to evolve, they are expected to be widely applied in healthcare and finance. Future research in AI explainability could combine global and local methods to gain a more comprehensive understanding of AI systems. Deductive global methods can correspond to the model's behavior, and the actions' activation in diverse circumstances to determine how and which decision rules derived from them are converted into coherent local explanations. One possible area for future work could be guided counterfactual generation, which can be achieved by mapping the dataset into the latent space where there is content or feature information. Another exciting line of work for interactive explainability could be investigating how humans interact with the model and the options it comes up with instead of how the model makes decisions. Perhaps bidirectional value learning is where the system and humans always learn from the other's perspective to update the value function towards human values. External knowledge expansion in deep learning models, like knowledge graphs cooperating with deep learning methods, could be interesting to research. This would mean that the learning of the model will be done based on the knowledge and rules that people have gathered over time, hence the usability and explication of the outcomes. These directions point to possible ways to address the current issues in AI interpretability so that AI matches people's expectations and ethical norms.

5. Summary/conclusion

This chapter focuses on reviewing deep learning approaches for disease diagnosis, especially in diagnosing disorders from medical images. Reference is made to the more recent studies on the application of deep learning for cancer diagnosis to show how this can transform management and care. The paper also describes some public datasets in the medical domain containing cancer images and highlights the need for Rational Medical Image Analysis using DEEP LEARNING MODELS. It expands the capacities for reforming disease diagnosis, presents inspiring solutions, and enhances patients' lives. Even as interest in applying deep learning to medical imaging grows, more work must be done to maximize this technology's potential in medicine. Incorporating AI in clinical practice requires explainability to ensure a smooth revolution in the healthcare imaging industry. By generating models that can be interpreted by clinicians to match clinical goals, XAI contributes to the most significant utilization of AI techniques in precision medicine. AI's digital revolution in healthcare innovates diagnostics and treatment and is still aligned with the ethics of healthcare professionals and service consumers.

Acknowledgment

The authors would like to acknowledge the support of AIDA Lab CCIS & CYBEX Prince Sultan University, Riyadh Saudi Arabia.

Availability of data and materials: The authors confirm that the data supporting the findings of this study are available within the Section 4.1.

Conflicts of interest: Authors declare no conflicts of interest for this research.

References

[1] D. Göndöcs, V. Dörfler, AI in medical diagnosis: AI prediction & human judgment, Artif. Intell. Med. 149 (2024) 102769.

[2] K. Meethongjan, M. Dzulkifli, A. Rehman, A. Altameem, T. Saba, An intelligent fused approach for face recognition, J. Intell. Syst. 22 (2) (2013) 197–212.

[3] S. Aminizadeh, A. Heidari, M. Dehghan, S. Toumaj, M. Rezaei, N.J. Navimipour, et al., Opportunities and challenges of artificial intelligence and distributed systems to improve the quality of healthcare service, Artif. Intell. Med. 149 (2024) 102779.

[4] Z. Li, Ethical frontiers in artificial intelligence: navigating the complexities of bias, privacy, and accountability, Int. J. Eng. Manage. Res. 14 (3) (2024) 109–116.

[5] Z. Liao, S. Hu, Y. Xie, Y. Xia, Modeling annotator preference and stochastic annotation error for medical image segmentation, Med. Image Anal. 92 (2024) 103028.

[6] M.N. Yeasmin, M. Al Amin, T.J. Joti, Z. Aung, M.A. Azim, Advances of AI in image-based computer-aided diagnosis: a review, Array (2024) 100357.

[7] A. Nazir, A. Hussain, M. Singh, A. Assad, Deep learning in medicine: advancing healthcare with intelligent solutions and the future of holography imaging in early diagnosis, Multimed. Tools Appl. (2024) 1–64.

[8] U. Iqbal, T. Davies, P. Perez, A review of recent hardware and software advances in GPU-accelerated edge-computing single-board computers (SBCs) for computer vision, Sensors 24 (15) (2024) 4830.

[9] B. Abhisheka, S.K. Biswas, B. Purkayastha, D. Das, A. Escargueil, Recent trend in medical imaging modalities and their applications in disease diagnosis: a review, Multimed. Tools Appl. 83 (14) (2024) 43035–43070.

[10] R. Sajjanar, U.D. Dixit, V.K. Vagga, Advancements in hybrid approaches for brain tumor segmentation in MRI: a comprehensive review of machine learning and deep learning techniques, Multimed. Tools Appl. 83 (10) (2024) 30505–30539.

[11] S.R. Sindiramutty, W.J. Tee, S. Balakrishnan, S. Kaur, R. Thangaveloo, H. Jazri, et al., Explainable AI in healthcare application, in: Advances in Explainable AI Applications for Smart Cities, IGI Global, 2024, pp. 123–176.

[12] M. Fontes, J.D.S. De Almeida, A. Cunha, Application of example-based explainable artificial intelligence (XAI) for analysis and interpretation of medical imaging: a systematic review, IEEE Access 12 (2024) 26419–26427.

[13] N.A. Wani, R. Kumar, J. Bedi, I. Rida, Explainable AI-driven IoMT fusion: unravelling techniques, opportunities, and challenges with explainable AI in healthcare, Inform. Fusion (2024) 102472.

[14] J. Amann, A. Blasimme, E. Vayena, D. Frey, V.I. Madai, Precise4Q Consortium, Explainability for artificial intelligence in healthcare: a multidisciplinary perspective, BMC Med. Inform. Decision Making 20 (2020) 1–9.

[15] D. Higgins, V.I. Madai, From bit to bedside: a practical framework for artificial intelligence product development in healthcare, Adv. Intell. Syst. 2 (10) (2020) 2000052.

[16] A. Chaddad, J. Peng, J. Xu, A. Bouridane, Survey of explainable AI techniques in healthcare, Sensors 23 (2) (2023) 634.

[17] I. Bhattacharya, Y.S. Khandwala, S. Vesal, W. Shao, Q. Yang, S.J. Soerensen, et al., A review of artificial intelligence in prostate cancer detection on imaging, Ther. Adv. Urol. 14 (2022) 17562872221128791.

[18] M. van der Leest, E. Cornel, B. Israel, R. Hendriks, A.R. Padhani, M. Hoogenboom, et al., Head-to-head comparison of transrectal ultrasound-guided prostate biopsy versus multiparametric prostate resonance imaging with subsequent magnetic resonance-guided biopsy in biopsy-naïve men with elevated prostate-specific antigen: a large prospective multicenter clinical study, Eur. Urol. 75 (4) (2019) 570–578.

[19] P.A. Noseworthy, J.P. Brito, M. Kunneman, I.G. Hargraves, C. Zeballos-Palacios, V.M. Montori, H.-H. Ting, Shared decision-making in atrial fibrillation: navigating complex issues in partnership with the patient, J. Interv. Card. Electrophysiol. 56 (2019) 159–163.

[20] J.C. Bjerring, J. Busch, Artificial intelligence and patient-centered decision-making, Philos. Technol. 34 (2021) 349–371.

[21] S. Nazir, D.M. Dickson, M.U. Akram, Survey of explainable artificial intelligence techniques for biomedical imaging with deep neural networks, Comput. Biol. Med. 156 (2023) 106668.

[22] T. Hulsen, Explainable artificial intelligence (XAI): concepts and challenges in healthcare, AI 4 (3) (2023) 652–666.

[23] B.M. de Vries, G.J. Zwezerijnen, G.L. Burchell, F.H. van Velden, M.-v. der Houven, C.W. van Oordt, R. Boellaard, Explainable artificial intelligence (XAI) in radiology and nuclear medicine: a literature review, Front. Med. 10 (2023) 1180773.

[24] D. Gunning, D. Aha, DARPA's explainable artificial intelligence (XAI) program, AI Mag. 40 (2) (2019) 44–58.

[25] K. Fukushima, Neocognitron: a self-organizing neural network model for a mechanism of pattern recognition unaffected by shift in position, Biol. Cybern. 36 (4) (1980) 193–202.

[26] S.S. Band, A. Yarahmadi, C.C. Hsu, M. Biyari, M. Sookhak, R. Ameri, et al., Application of explainable artificial intelligence in medical health: a systematic review of interpretability methods, Inform. Med. Unlocked 40 (2023) 101286.

[27] E. Pintelas, I.E. Livieris, P. Pintelas, A grey-box ensemble model exploiting black-box accuracy and white-box intrinsic interpretability, Algorithms 13 (1) (2020) 17.

[28] W. Liu, F. Zhao, A. Shankar, C. Maple, J.D. Peter, B.G. Kim, et al., Explainable AI for medical image analysis in medical cyber-physical systems: enhancing transparency and trustworthiness of IoMT, IEEE J. Biomed. Health Inform. (2023) 1–25.

[29] S. Ali, F. Akhlaq, A.S. Imran, Z. Kastrati, S.M. Daudpota, M. Moosa, The enlightening role of explainable artificial intelligence in medical & healthcare domains: a systematic literature review, Comput. Biol. Med. (2023) 107555.

[30] S. Bharati, M.R.H. Mondal, P. Podder, U. Kose, Explainable artificial intelligence (XAI) with IoHT for smart healthcare: a review, in: Interpretable Cognitive Internet of Things for Healthcare, 2023, pp. 1–24.

[31] M.N. Alam, M. Kaur, M.S. Kabir, Explainable AI in healthcare: enhancing transparency and trust upon legal and ethical consideration, Int. Res. J. Eng. Technol. 10 (6) (2023) 1–9.

[32] R. Xie, J. Chen, L. Jiang, R. Xiao, Y. Pan, Y. Cai, Accurate Explanation Model for Image Classifiers using Class Association Embedding, 2024. arXiv preprint arXiv:2406.07961.

[33] S. Singla, N. Murali, F. Arabshahi, S. Triantafyllou, K. Batmanghelich, Augmentation by counterfactual explanation-fixing an overconfident classifier, in: Proceedings of the IEEE/CVF Winter Conference on Applications of Computer Vision, 2023, pp. 4720–4730.

[34] S. Singla, M. Eslami, B. Pollack, S. Wallace, K. Batmanghelich, Explaining the black-box smoothly—a counterfactual approach, Med. Image Anal. 84 (2023) 102721.

[35] Y. Tao, X. Ma, Y. Zhang, K. Huang, Z. Ji, W. Fan, et al., LAGAN: lesion-aware generative adversarial networks for edema area segmentation in SD-OCT images, IEEE J. Biomed. Health Inform. 27 (5) (2023) 2432–2443.

[36] S. Tan, R. Caruana, G. Hooker, Y. Lou, Distill-and-compare: auditing black-box models using transparent model distillation, in: Proceedings of the 2018 AAAI/ACM Conference on AI, Ethics, and Society, 2018, December, pp. 303–310.

[37] N. Frosst, G. Hinton, Distilling a Neural Network Into a Soft Decision Tree, 2017. arXiv preprint arXiv:1711.09784.

[38] Z. Che, S. Purushotham, R. Khemani, Y. Liu, Distilling Knowledge From Deep Networks With Applications to Healthcare Domain, 2015. arXiv preprint arXiv:1512.03542.

[39] X. Liu, X. Wang, S. Matwin, Improving the interpretability of deep neural networks with knowledge distillation, in: 2018 IEEE International Conference on Data Mining Workshops (ICDMW), IEEE, 2018, November, pp. 905–912.

[40] H.M. Fahad, M.U. Ghani Khan, T. Saba, A. Rehman, S. Iqbal, Microscopic abnormality classification of cardiac murmurs using ANFIS and HMM, Microsc. Res. Tech. 81 (5) (2018) 449–457.

[41] K. Raghavan, S. Balasubramanian, K. Veezhinathan, Explainable artificial intelligence for medical imaging: review and experiments with infrared breast images, Comput. Intell. 40 (3) (2024) e12660.

[42] A. Husham, M. Hazim Alkawaz, T. Saba, A. Rehman, J. Saleh Alghamdi, Automated nuclei segmentation of malignant using level sets, Microsc. Res. Tech. 79 (10) (2016) 993–997.

[43] N. Hussain, M.A. Khan, M. Sharif, S.A. Khan, A.A. Albesher, T. Saba, A. Armaghan, A deep neural network and classical features based scheme for objects recognition: an application for machine inspection, Multimed. Tools Appl. (2024) 1–23.

[44] S. Iftikhar, K. Fatima, A. Rehman, A.S. Almazyad, T. Saba, An evolution based hybrid approach for heart diseases classification and associated risk factors identification, Biomed. Res. 28 (8) (2017) 3451–3455.

[45] S. Jabeen, Z. Mehmood, T. Mahmood, T. Saba, A. Rehman, M.T. Mahmood, An effective content-based image retrieval technique for image visuals representation based on the bag-of-visual-words model, PLoS One 13 (4) (2018) e0194526.

[46] A. Jamal, M. Hazim Alkawaz, A. Rehman, T. Saba, Retinal imaging analysis based on vessel detection, Microsc. Res. Tech. 80 (7) (2017) 799–811.

[47] R. Javed, M.S.M. Rahim, T. Saba, A. Rehman, A comparative study of features selection for skin lesion detection from dermoscopic images, Netw. Model. Anal. Health Inform. Bioinform. 9 (1) (2020) 4.

[48] M.A. Khan, M.I. Sharif, M. Raza, A. Anjum, T. Saba, S.A. Shad, Skin lesion segmentation and classification: a unified framework of deep neural network features fusion and selection, Expert. Syst. 39 (7) (2022) e12497.

[49] S. Larabi-Marie-Sainte, L. Aburahmah, R. Almohaini, T. Saba, Current techniques for diabetes prediction: review and case study, Appl. Sci. 9 (21) (2019) 4604.

[50] B. Mughal, N. Muhammad, M. Sharif, A. Rehman, T. Saba, Removal of pectoral muscle based on topographic map and shape-shifting silhouette, BMC Cancer 18 (2018) 1–14.

[51] B. Mughal, M. Sharif, N. Muhammad, T. Saba, A novel classification scheme to decline the mortality rate among women due to breast tumor, Microsc. Res. Tech. 81 (2) (2018) 171–180.

[52] Z.F. Muhsin, A. Rehman, A. Altameem, T. Saba, M. Uddin, Improved quadtree image segmentation approach to region information, Imaging Sci. J. 62 (1) (2014) 56–62.

[53] A. Rehman, Brain stroke prediction through deep learning techniques with ADASYN strategy, in: 2023 16th International Conference on Developments in eSystems Engineering (DeSE), IEEE, 2023, December, pp. 679–684.

[54] T. Saba, A. Rehman, Effects of artificially intelligent tools on pattern recognition, Int. J. Mach. Learn. Cybern. 4 (2013) 155–162.

[55] L. Tan, C. Huang, X. Yao, A concept-based local interpretable model-agnostic explanation approach for deep neural networks in image classification, in: International Conference on Intelligent Information Processing, Springer Nature Switzerland, Cham, 2024, April, pp. 119–133.

[56] Y. Liu, Y. Fu, Y. Peng, J. Ming, Clinical decision support tool for breast cancer recurrence prediction using SHAP value in cooperative game theory, Heliyon 10 (2) (2024).

[57] H. Naveed, S. Anwar, M. Hayat, K. Javed, A. Mian, Survey: image mixing and deleting for data augmentation, Eng. Appl. Artif. Intell. 131 (2024) 107791.

[58] M. Beitollahi, A. Bie, S. Hemati, L.M. Brunswic, X. Li, X. Chen, G. Zhang, Parametric Feature Transfer: One-shot Federated Learning With Foundation Models, 2024. arXiv preprint arXiv:2402.01862.

[59] B. Tahir, S. Iqbal, M. Usman Ghani Khan, T. Saba, Z. Mehmood, A. Anjum, T. Mahmood, Feature enhancement framework for brain tumor segmentation and classification, Microsc. Res. Tech. 82 (6) (2019) 803–811.

[60] T. Saba, S. Al-Zahrani, A. Rehman, Expert system for offline clinical guidelines and treatment, Life Sci. J. 9 (4) (2012) 2639–2658.

[61] T. Saba, S.T.F. Bokhari, M. Sharif, M. Yasmin, M. Raza, Fundus image classification methods for the detection of glaucoma: a review, Microsc. Res. Tech. 81 (10) (2018) 1105–1121.

[62] T. Saba, A. Rehman, Z. Mehmood, H. Kolivand, M. Sharif, Image enhancement and segmentation techniques for detection of knee joint diseases: a survey, Curr. Med. Imaging 14 (5) (2018) 704–715.

[63] K. Yousaf, Z. Mehmood, T. Saba, A. Rehman, A.M. Munshi, R. Alharbey, M. Rashid, Mobile-health applications for the efficient delivery of health care facility to people with dementia (PwD) and support to their carers: a survey, Biomed. Res. Int. 2019 (1) (2019) 7151475.

CHAPTER 14

Leveraging explainable AI in deep learning for brain tumor detection

Deep Kothadiya, Aayushi Chaudhari, Prisha Faldu, and Dulari Gajjar
U & P U Patel Department of Computer Engineering, Faculty of Technology (FTE), Chandubhai S. Patel Institute of Technology (CSPIT), Charotar University of Science and Technology (CHARUSAT), Changa, India

1. Introduction

Brain tumors are one of the most aggressive and life-threatening conditions, requiring precise diagnosis, treatment planning, and monitoring. Magnetic resonance imaging (MRI) is the standard imaging modality for brain tumor assessment due to its superior soft tissue contrast and the ability to provide detailed structural information. Accurate segmentation of brain tumors from MRI scans plays a vital role in clinical workflows, including tumor diagnosis, surgical planning, radiation therapy, and prognosis evaluation [1].

Manual segmentation of brain tumors is a labor-intensive and time-consuming process, often leading to variability in results due to human error and subjectivity. Moreover, the manual approach becomes increasingly challenging as the size and complexity of medical image datasets grow. To overcome these limitations, automated brain tumor segmentation techniques have been developed to provide accurate, fast, and reproducible results. Such automated methods can assist clinicians in making more informed decisions, improving patient outcomes [1,2]. Brain tumor segmentation is a critical task in medical image analysis, aimed at accurately identifying and delineating tumor regions in brain scans, primarily from magnetic resonance imaging (MRI) [2]. Precise segmentation is essential for diagnosis, treatment planning, and monitoring the progression of brain tumors. Traditional manual segmentation by radiologists is time-consuming and prone to inter- and intra-observer variability, leading to the growing interest in automated segmentation methods.

Over the past few decades, advancements in machine learning, particularly deep learning, have significantly improved segmentation performance. Convolutional neural networks (CNNs) have become the cornerstone of automated medical image segmentation due to their ability to capture spatial hierarchies in imaging data, achieving state-of-the-art results across various segmentation challenges. Methods such as U-Net and its

Explainable AI in Healthcare Imaging for Medical Diagnoses
https://doi.org/10.1016/B978-0-443-23979-3.00014-2

variants have shown great success in brain tumor segmentation tasks, providing high accuracy and robustness even in complex cases [3]. Brain tumors are heterogeneous in nature, exhibiting significant variations in size, shape, texture, and intensity across patients and tumor types. Typically, a brain tumor consists of multiple subregions, includings: (1) Necrotic core: The dead tissue within the tumor. (2) Enhancing tumor core: The actively proliferating tumor region. (3) Peritumoral edema: Swelling caused by fluid accumulation around the tumor.

Each of these subregions may have different appearances on MRI images, making accurate segmentation more challenging. The anatomical variability between individuals, noise and artifacts in MRI scans, and the inherent difficulty of distinguishing tumor boundaries from healthy tissues are additional factors that complicate the segmentation task [4]. Despite these advancements, brain tumor segmentation remains challenging due to the heterogeneous nature of brain tumors, which can vary widely in size, shape, and appearance across patients. Moreover, the presence of different tumor subregions, such as necrosis, edema, and enhancing core, further complicates the task [5]. Therefore, robust segmentation techniques are required to deal with these complexities, ensuring reliable and reproducible results. Initial approaches to brain tumor segmentation involved traditional machine learning techniques, which relied on handcrafted features extracted from the images [6]. These techniques used features such as intensity, texture, and shape to classify pixels as tumor or non-tumor. While these methods improved over manual approaches, their effectiveness was limited due to the need for extensive feature engineering and the lack of ability to generalize across different datasets.

2. Deep learning in brain tumor recognition

The rise of deep learning, particularly convolutional neural networks (CNNs), has revolutionized the field of medical image analysis, including brain tumor segmentation. CNNs automatically learn hierarchical features from data, eliminating the need for manual feature engineering. Architectures such as U-Net [7] and its extensions have become the backbone of many state-of-the-art brain tumor segmentation models. U-Net, with its encoder-decoder structure, is particularly well-suited for pixel-wise segmentation tasks as it captures both low-level and high-level image features [8].

In recent years, more advanced architectures have been proposed, including variants of U-Net that incorporate attention mechanisms, dilated convolutions, and 3D CNNs, which further enhance the model's ability to capture fine-grained tumor details while preserving spatial context [9]. These networks are often trained on large, annotated datasets such as the Multimodal Brain Tumor Image Segmentation (BraTS) challenge datasets, which provide labeled MRI scans of brain tumors [10].

In the late 1990s, researchers started to use Hopfield Neural Networks coupled with active contours to state the tumor boundaries and enlarge the tumor region. However, at that time, neural network training was affected significantly since constraints on computational resources and technical support were imposed. From the late 1990s to the early 2000s, most brain tumor segmentation techniques were heavily relying on the classical machine learning algorithms that used hand-crafted features [11], such as expert systems retaining multi-spectral histograms, template-based segmentation, graphical models based on intensity histograms [12], and boundary detection from latent atlases [13]. These early approaches pioneered the use of machine learning for brain tumor segmentation but came with prominent drawbacks.

First, these early approaches generally focused on segmenting the whole tumor region and, therefore, delivered segmentation results with a single category. In comparison to current algorithms, earlier theory developed with restricted assumptions and conditions. Then, manually handcrafted feature engineering is limited because of prior knowledge, leading to less effective generalization. Lastly, early works failed to address issues like appearance variability and data imbalance [14]. Several unanswered problems still remain in brain tumor segmentation. In which one of area is brain tissue segmentation or anatomical brain segmentation, which targets to categorize each unit into a predefined brain tissue type, assuming the brain image is tumor free or does not contain other anatomical anomalies [15]. Another task, white matter lesion segmentation, focuses on separating white matter lesions from normal tissue. This approach typically doesn't involve sub-regions like tumor cores and is usually tackled by binary classification.

Tumor detection task consists of identifying abnormal tumors or lesions and predicting a class of label for each of the tissue types. Generally, this process delivers a bounding box as the detection output and a label as the classification result [16]. Also, it is important to mention that some classification methods of brain tumors return either a single-label segmentation mask or the center point of the tumor core, without providing detailed sub-region segmentation. Classification of disorders is another suitable task in this field. This task involves first extracting the predefined features from brain scans and subsequently classifying the feature representations into specific disorder categories like high-grade gliomas (HGGs) and low-grade gliomas (LGGs) or mild cognitive impairment (MCI), Alzheimer's disease (AD), and Schizophrenia [17–19].

Survival prediction attempts to determine patterns and behaviors of the tumor in order to provide an estimate of the patient's survival. This is yet another clinical diagnosis aid [20,21]. Both disorder classification, and survival prediction, can be classified as downstream tasks, depending upon the segmentation results of the tumor as shown in Table 1.

Table 1 Challenges in brain tumor detection [22–24].

Challenges in tumor detection	Possible solution
Data imbalance challenge: One of the most critical challenges in brain tumor classification is the issue of data imbalance. Certain tumor types, such as high-grade gliomas (HGG), are more prevalent in medical datasets compared to rare types like low-grade gliomas (LGG). This imbalance results in models biased towards predicting the more common classes, which leads to poor performance in identifying less frequent tumor types. The classifier often becomes insensitive to minority classes, reducing its overall accuracy and generalizability	To address data imbalance, techniques such as oversampling and undersampling can be applied. Oversampling increases the number of minority class samples by duplicating them or generating synthetic data through methods like SMOTE (synthetic minority over-sampling technique). On the other hand, undersampling reduces the number of majority class samples to balance the dataset. Additionally, cost-sensitive learning can be employed, where misclassifications of the minority class are penalized more heavily during training to encourage the model to pay more attention to underrepresented tumor types. Loss functions like focal loss can also help by down-weighting the loss of well-classified examples and focusing on harder, misclassified samples
Limited annotated data: The availability of labeled medical imaging data for brain tumor classification is another major challenge. Manually annotating MRI scans requires expertise and is highly time-consuming, especially when sub-regions of tumors are involved. The lack of large-scale annotated datasets makes it difficult to train deep learning models effectively, leading to overfitting and poor generalization to new data	Transfer learning can be a powerful solution to this challenge. By leveraging pre-trained models, typically trained on large datasets like ImageNet, one can fine-tune the models on the limited available medical data. Data augmentation techniques such as rotation, flipping, zooming, and elastic deformation can artificially increase the size of the dataset and introduce variability, which helps the model generalize better. Moreover, semi-supervised learning approaches can be used, where a small labeled dataset is complemented with a larger set of unlabeled images, with models trained to predict labels for the latter
Variability in MRI scans: Medical imaging data, especially MRI scans, show high variability across different acquisition machines, scanning protocols, and image modalities (T1, T2, FLAIR, etc.). This variability makes it difficult for a model trained on one type of data to perform well on another. Additionally, tumors can vary greatly in shape, size, and location, which complicates the classification process	A multi-modality approach, which integrates data from different MRI modalities (e.g., T1, T2, and FLAIR), can help capture complementary information about tumor structure and appearance. Domain adaptation techniques can also be employed to address the variability across different MRI acquisition settings. In domain adaptation, a model trained on one domain (e.g., a specific hospital's MRI data) is adapted to generalize better to another domain with different data characteristics. Additionally, normalization techniques like histogram matching or intensity normalization can be applied to standardize the input images and reduce variability

Table 1 Challenges in brain tumor detection—cont'd

Challenges in tumor detection	Possible solution
Class overlap and ambiguity: Brain tumor types often have overlapping visual features, especially in cases like low-grade gliomas (LGG) and high-grade gliomas (HGG), making classification challenging. Tumors with similar appearances may belong to different classes, leading to confusion in distinguishing between these classes. This challenge is exacerbated when tumor regions do not have clear boundaries or when there are low-contrast regions in the MRI scans	One way to address this issue is to adopt multi-task learning where the model is trained to perform both segmentation and classification simultaneously. Segmentation helps by identifying tumor sub-regions, which can provide more detailed spatial information that aids in classification. Moreover, using attention mechanisms in neural networks can help the model focus on the most relevant regions of the MRI scan, improving its ability to differentiate between similar-looking tumors. Hybrid models, which combine CNNs with recurrent neural networks (RNNs) or transformers, can also improve feature extraction by focusing on the spatial and contextual relationships between tumor regions
High computational cost: Deep learning models for brain tumor classification, especially multi-modal and multi-model approaches, can be computationally expensive. Training large models on high-resolution MRI data requires significant GPU resources and time. This is particularly challenging in clinical settings where real-time or near-real-time diagnosis is crucial	Model compression techniques like pruning, quantization, and knowledge distillation can be applied to reduce the size and complexity of the models without significant loss in performance. Pruning removes less significant weights, while quantization reduces the precision of model weights and activations. Knowledge distillation transfers the knowledge of a large, complex model (teacher) to a smaller, more efficient model (student), which can achieve similar performance with fewer computational resources. Additionally, cloud-based solutions and edge computing can be used to handle large computational loads remotely, thereby enabling faster processing in clinical environments
Interpretability and explainability: The "black box" nature of deep learning models poses a significant challenge in the medical field, where model decisions need to be interpretable to ensure trustworthiness. For brain tumor classification, clinicians require an understanding of how and why a model makes a particular prediction, especially when it comes to sensitive diagnoses like cancer	To enhance interpretability, techniques such as Grad-CAM (gradient-weighted class activation mapping) can be employed, which generates heatmaps to highlight the areas of the image that the model focused on when making its predictions. Saliency maps and SHAP (SHapley Additive exPlanations) values are other popular methods to interpret the model's decisions by quantifying the contribution of each feature (or region of the

Continued

Table 1 Challenges in brain tumor detection—cont'd

Challenges in tumor detection	Possible solution
	image) to the final prediction. Furthermore, using simpler, more interpretable models like decision trees or rule-based systems as a post hoc analysis on top of complex models can provide additional insight into the model's behavior
Tumor sub-region segmentation: Brain tumors often consist of different sub-regions (e.g., the core, the enhancing tumor, and the edema), each requiring separate identification and classification. Accurately segmenting these sub-regions is critical for clinical diagnosis and treatment planning, but it remains a challenge due to the complexity of the tumor's structure and the high variance in appearance between patients	U-Net-like architectures with an encoder-decoder structure can be highly effective in tackling this challenge. 3D CNNs can also be used to capture volumetric information from MRI scans, which helps in better sub-region segmentation. Additionally, incorporating hierarchical classification that first segments the tumor into sub-regions and then classifies the type of tumor can help improve accuracy. Using custom loss functions like the Dice loss, combined with cross-entropy, helps deal with imbalanced sub-region segmentation by focusing more on the difficult-to-segment areas
Uncertainty and ambiguity in diagnosis: Medical images, particularly brain scans, often exhibit uncertainty due to noise, artifacts, or low-quality images, which can lead to ambiguous classifications. The variability in tumor appearance across different patients further complicates the classification process	Uncertainty quantification techniques can be integrated into deep learning models to give probabilistic outputs rather than deterministic ones. Bayesian deep learning models and Monte Carlo dropout can help estimate the uncertainty in predictions. By assessing the model's confidence in its predictions, clinicians can be alerted when further investigation is needed for ambiguous cases. Additionally, incorporating ensemble learning with multiple models can provide more robust predictions by combining the strengths of different models to minimize uncertainty
Cross-institutional variability: Differences in MRI scan acquisition protocols between institutions lead to variations in data quality, making it difficult for models trained on one institution's data to generalize well to data from another. This variability can cause significant drops in model performance when deployed in new clinical settings	To combat cross-institutional variability, domain generalization techniques can be employed. These techniques allow a model to learn representations that are invariant to changes in data distribution between institutions. Transfer learning can also be applied, where a pre-trained model is fine-tuned on a small dataset from the target institution to adapt to the new data distribution. In addition, collecting and training on multi-institutional datasets can make the model more robust to variations in data acquisition and processing

2.1 Interpretability in brain tumor detection

2.1.1 Clinical trust and deciding results

Physicians and radiologists are dependent on segmentation models to support their decisions. A deep learning model may effectively segment a brain tumor; however, if the model's reasoning cannot be explained or understood, the clinician may lack trust in the model. Conversely, an interpretable model that is capable of showing why some regions within an image are annotated tumor tissue increases the clinician's trust in the fidelity of the model, allowing for more actionable results for the clinical settings in medical diagnoses and treatment [25].

2.1.2 Accountability in high-stakes conclusions

Brain tumor segmentation is subsequently utilized to determine treatment alternatives (i.e., surgery, radiotherapy, or chemotherapy). An inaccurate segmentation leads to the risk of interpretation that may lead to grave risks, such as unnecessary treatment, or to misdiagnosis. An accountable approach allows the clinician to backtrack and contextualize the model's opinion, increasing their assurances that decisions are made from informed interpretation [26].

2.1.3 Corresponding deviations of the model

Interpretability aids in identifying the weaknesses of the segmentation model. For example, because of the lack of data on certain types, the model may misclassify certain tumor regions even if they are understandable and recognizable by the model. It is explained how one may be able to see the decision process of the model, and this makes it possible to identify some of these problems and fix them by updating the model or enhancing the data [27].

2.1.4 Legal and ethical considerations

Particularly in the case of medical usage, authorities such as a regulatory agency like the FDA place high criteria on the level of explanation required for AI systems. There is no path breaking or risk to parameter definition because the brain tumor segmentation model has a good level of interpretability as it enables clinicians and their regulators to understand the reasoning behind its predictions and assess the implications of its outputs whether they are valid, ethical and clinically appropriate [28].

2.1.5 Effective integration of AI and medical professionals

Interpretability augments cooperation between AI tools and clinicians. By assisting in the task of providing insight into the model's rationale for segmenting various tumor locations, clinicians can inject their own knowledge, refuting or confirming the model's outputs almost instantaneously. This situation maintains the notion of AI as an adjunct to human decision making, not a replacement [29].

2.1.6 Enhancing generalizability

Interpretable models allow researchers to identify the nascent reasons for error or outliers in predictions. This positions them to adjust the model, or even identify what parts of the data should be improved (such as imaging artifacts or variations in tumor appearance), improving overall generalizability and perhaps improving model performance across a patient population [29].

2.1.7 Accommodating data variability and uncertainty

Brain tumor imaging varies significantly between patients with respect to factors like MRI modalities, MRI scanners, or appearances of tumors. Interpretability accommodates this variability in the evaluation of uncertainty in the model's predictions. The explicit explanations provide reasons to the clinician when and why a model may be uncertain about its prediction leading to a more accurate and practical clinical decision [26].

2.1.8 Understanding sub-region analysis

Most advanced brain tumor segmentation models will require distinguishing between various tumor sub-regions (e.g., tumor core, enhancing, necrotic, etc.) being assessed. Interpretation will provide the practitioner explanations and determine how the model obtained certain sub-regions. Understanding these factors enable us to make more precise tumor characterization and aiding in personalized treatment plans [27].

3. Deep learning models for Brain tumor recognition

3.1 Convolution neural networks

Convolution neural network simply known as CNN [30] is part of deep neural network (DNN) learning module, while ANN is an advanced DNN learning model. Three different kinds of layers like the pooling layer (PL), the convolutional layer (CL), and the fully connected layer (FCL) are used to construct CNNs, illustrated in Fig. 1. The CL multiplies the input features in the perceptual field by matrix element multiplication and uses the deviation to extract features from the input data or upper layer characteristics. The extraction of local spatial correlation features from the input data is controlled by the size of the convolution kernel in the CL, which may enhance certain aspects of the original signal while minimizing the effect of noise. The PL is responsible of reducing the spatial size of the Convolved feature. To lower the amount of processing resources required to examine the data, it employs dimensionality reduction techniques. Furthermore, it helps sustain the process of successfully training the model by extracting pertinent characteristics that are rotationally and positionally invariant.

Learning nonlinear combinations of the high-level characteristics, as represented by the output of the CL, is effectively accomplished by adding an FCL. In the space [31], the

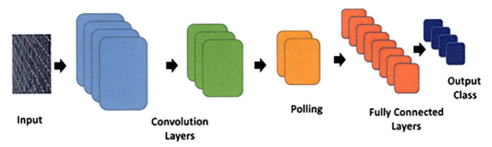

Fig. 1 Functional architecture of CNN for PV cell fault detection.

FCL is now learning a function that may not be linear. The diagnostic procedure is more reliable and the generalization is enhanced due to the translation invariance of the properties that CNN extracted [31]; (iii) massive real-time data often engulfs the data that might identify brain tumor. To create samples, the developed countermeasure networks may learn the probability distribution of actual data [30], which is appropriate for small sample sizes.

Compared to other neural networks, CNNs are more sophisticated. Fig. 2 shows the construction of a CNN, which consists of input, convolution, and pooling, fully linked output. A CNN may be utilized for image categorization since it has more data processing power. As an example, by performing convolutional computations on the source feature map and the convolutional kernel as shown in Eq. (1), the convolutional layer is able to extract local features from the image using LeNet-5 [32].

$$Z^{l+1} = \left[Z^l * W^{l+1} \right](m, n) + b \tag{1}$$

where (m, n) 2 $f0, 1, \dots L^{l+1}$ g; L^{l+1} is the scale of Z^{l+1}; $Z(m, n)$ is the pixel; K is the number of passes in the eigenmap; f is the convolution kernel size; and s_0 is the convolution step size [33]. The pooling layer's role is to select key features on the outcomes of the previous layer, as Eq. (2),

$$A_k^l(m, n) = g\left(Z_k^l(m, n)\right) \tag{2}$$

where the ReLU is often used and g is the excitation function [34]. When used with the softmax classifier in the final layer, the results of the classification are provided by the fully connected layer, which scales the input features into a one-dimensional matrix.

3.2 Recurrent neural network (RNN)

Recurrent neural networks (RNNs) are network architectures where all nodes are linked together in a chain and the inputs are time-series data [35]. Unlike multi-layer perceptron (MLP), RNN can learn sequences that change over time because they have a sense of time and a recollection of previous network states. Long short-term memory (LSTM) [36] and

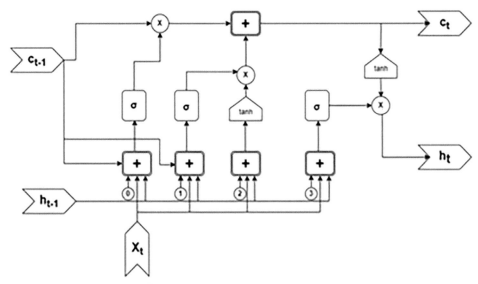

Fig. 2 LSTM gate architecture for PV cell fault detection.

gated recurrent unit (GRU) [37] networks are the two RNN types that are now most often employed. Conceptual architecture of gate learning models like LSTM and GRU illustrated in Fig. 2. To prevent long-term dependency, each recurrent unit may adaptively capture the dependence of various time scales by incorporating gates [36]. However, there is a length restriction when using the RNN because of gradient bursting and disappearing. In order to address lengthy sequence prediction issues, RNN versions such LSTM [36] and GRU [37] neural networks have been created. The hidden state vector, which is different for each input sequence and always starts with a zero-element vector for $t = 0$, introduces memory into the network. Eq. (3) may be used to compute Reset Gate. The information that will decide what will be eliminated from the earlier concealed time steps is represented by the resultant reset vector, or r. The forget operation is applied via element-wise multiplication, indicated by the Hadamard product operator [38], much as in the forget gate. Eq. (4) with the input $X(t)$ is used to compute the reset vector as a linear combination of the input vectors for the most recent concealed state and the current timestamp. Matrix manipulation was used to accomplish the Bothe gate computation. The timestamp vector value ranges from 0 to 1, with 0 or close to 0 indicating that feedback is no longer required and 1 or close to 1 aiding in backpropagation timestamp memory.

$$r_t = \sigma(W_i x_t + b_i + W_h h_{t-1} + b_h) \tag{3}$$

$$z_t = \sigma(W_{iz} x_t + b_{iz} + W_{hz} h_{t-1} + b_{hz}) \tag{4}$$

With various trainable matrices and biases this time, we compute another representation of the input vector x and the prior hidden state. Z stands for the update vector, while W

and b for the weights and bias, respectively. The second component is the reset vector r, which is used in the previous hidden state. Be mindful that the forget/reset vector is not applied to the hidden state of an LSTM cell, but rather to the intermediate representation of the cell vector c. Eq. (5a) may be used to determine the final output of an LSTM cell, where $n(t)$ is an output gate with a $\tanh$ activation function [39].

$$h_t = (1 - z_t)\bigodot n_t + z_t\bigodot h_{t-1} \tag{5a}$$

The benefits of the RNN-based tumor detection include: (1) Time-series data are the RNN's inputs, and its depth varies with the length of the input sequence, making it suitable for brain tumor detection. (2) Recurrent networks are Turing complete, and the feedback connection mode helps extract and represent nonlinear dynamic feature of tumor images. (3) The RNN is stable when the lengths of the learning and testing sequence differ. (4) RNN are Turing complete. For FDD with numerous flaws and high noise, the authors in Ref. [40] created three RNN-based models (vanilla RNN, LSTM, and GRU), and they demonstrated that the LSTM and GRU models performed better than the vanilla RNN. To address the gradient explosion or disappearance issue, the authors in Ref. [41] employed GRU in the RNN model, which decreases the parameters by managing the gate mechanism. The suggested approach in Ref. [42] uses LSTM networks to extract features, and the chosen features are then input into a softmax regression classifier for tumor identification. In addition to having a geographical dependency in the measurement space, tumor detection system data are distinguished by their temporal relevance.

Recurrent networks are struggling with short term memory problem. If a sequence is too long, they will find it difficult to transfer knowledge from earlier time steps to later ones. RNN's may exclude crucial information from the start if you're attempting to analyze a text content to make predictions. Recurrent neural networks experience the vanishing gradient issue during back propagation. The weights of a neural network are updated via gradients. The vanishing gradient problem refers to the fact that the gradient decreases as time passes. If a gradient value is really little, it doesn't add much learning. Therefore, learning in recurrent neural networks stops for layers that get a little gradient update. Typically, they are the older layers. Since these layers don't learn, RNNs may as a result forget what they have seen in longer sequences, leading to a short-term memory.

3.3 Attention model

For the identification of tumor cell detection using MRI image data, SOTA Deep Learning models like CNN and RNN, or maybe a variation of CNN and RNN, have shown encouraging results in the past. Transformer architecture [43] has shown substantially more accurate prediction than SOTA deep learning models. Originally intended for language processing tasks, the vision transformer (ViT) [44] architecture is now utilized for various computer vision applications, including detection, identification, and segmentation. Transformer design places much emphasis on feature extraction using an attention model. General attention and multi-head attention are the two divisions of the attention mechanism.

As it computes the dot product for every query (q) with each of its corresponding keys (k), the general attention approach, also known as the scaled dot-product attention model, then divides each result by (d_k) before applying the softmax function. By doing this, it receives the weights, V [45], which are then used to scale the data. Fig. 3 shows the dot product attention model conceptual design.

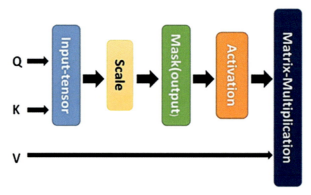

Fig. 3 Conceptual architecture of attention learning models.

The complete collection of questions may be processed effectively at once using the scaled dot-product attention calculations. The matrices Q, K, and V are given as inputs to the attention function to do this as Eq. (5b), F_A, represents the feature matrix from attention learning

$$F_A = softmax\left(\frac{QK^T}{\sqrt{d_q}}\right) * V \tag{5b}$$

The goal of multi-head attention is to make it feasible for the attention function to extract data from several representational subspaces, which is hard to achieve using a single attention head. They use a newly learnt projection each time as part of their multi-head attention method, which linearly projects the queries, keys, and values h_n (number of heads) times. Then, each of these h_n projections is subjected to the single attention mechanism in parallel to create h_n outputs, which are then concatenated and projected one more to produce the final result. The multi-head self-attention functional architecture is shown in Fig. 4 [46].

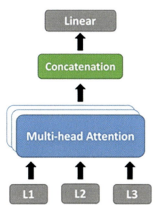

Fig. 4 Conceptual diagram of multi-head self-attention transformer architecture.

The multi-head attention model multiplies the corresponding weight matrices, W_i^Q, W_i^K, and W_i^V, one for each h_i, to calculate the linearly projected versions of the queries, keys, and values. By multiplying the queries and keys matrices, the single attention function is applied for each head [46]. After that, weighting the values matrix and using the scaling and softmax procedures to create an output for each head. The next step is concatenating all heads vectors into h_n, representing the final concatenated feature matrix. The final output is then produced by applying a linear projection on the concatenated output and multiplying it by the weight matrix, *Wf*.

Authors [45] developed a U–Net–based attention model to identify tumor cells. The authors also used convolutional attention model combinations like LinkNet and FPNet to compare the proposed attention–based model's performance assessment. The authors of [46] used an ensemble learning strategy to identify brain tumor. Authors have developed a coordinate attention model to improve feature extraction from the proposed ResNet152 and Xception deep learning models. In addition to transformer–based attention, writers [47] have improved feature extraction from images by combining channel and spatial attention models. Still, additional research is required to develop the process of attention for PV cell identification.

3.4 Autoencoder

Neural networks of the autoencoder type are used in deep learning. Multiple-hidden neural networks called stacked auto encoder (SAE) [48] are created by stacking several Autoencoder networks. The following auto-encoder network layer receives input from the previous layer's output. Referring to Fig. 5, an encoder and a decoder make up each auto-encoder network [49]. The encoder converts the network input into the hidden layer representation, while the decoder translates the hidden layer information back to the original input. An autoencoder's main components are encoded, code, and decoder. The encoder layer compresses the input image into a latent space representation. It creates

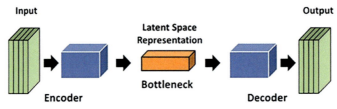

Fig. 5 Deep learning autoencoder architecture for PV cell fault detection.

a compressed version of the supplied image in a smaller dimension. The code layer represents the decoder layer's compressed input. The decoder layer restores the image's original dimensions after decoding it. Latent space representation is used to recreate the decoded image [50]. Table 2 demonstrates the different types of autoencoders that can be used for brain tumor detection.

Table 2 Types of autoencoder [50].

Autoencoder type	Discerption
Under complete autoencoders	You may utilize an unsupervised neural network under full autoencoders to produce a shortened version of the input data. These types of autoencoders are mostly used to create latent spaces, also known as bottlenecks, which serve as compressed substitutes for the input data and are rapidly decompressible with the aid of a network
Sparse autoencoders	By adjusting the number of nodes at each buried layer, sparse autoencoders may be managed. The activation of certain neurons in hidden layers is penalized by sparse autoencoders, which make it difficult to create neural networks with a variable number of nodes at their hidden layers. The neural network is regularized using the sparse autoencoder using the L1-loss and KL-divergence methods
Contractive autoencoders	A regularization term is also included in the contractive autoencoder to stop the network from discovering the identity function and translating input into output. We must make sure that the hidden layer activation derivatives are minimal relative to the input in order to train a model that complies with this condition
Denoising autoencoders	Regular autoencoders and denoising autoencoders both accept an input and output, respectively. The fact that they don't utilize the input image as their ground truth, however, makes them different. In its place, they use a loud variant. The reason for this is because dealing with images makes it challenging to remove image noise. L1 and L2 loss functions are often utilized
Variational	Models known as variational autoencoders (VAEs) aim to solve a particular issue with ordinary autoencoders. When an autoencoder is trained, it learns to only represent the input in what is known as the latent space or bottleneck. However, this latent space formed after training

AEs map input vectors to the hidden representation $y \in Rd$ The deterministic mapping $y = s_n (W_x + b)$, uses weights and bias. Each input x_i is mapped to a corresponding y_j, which is then mapped to a reconstruction z_i, such that $z_i \approx x_i$, the set of parameters θ, which minimizes the loss function, is shown in Eq. (6) [51]. L can be chosen as the traditional mean squared error (MSE), stated in Eq. (7).

$$\theta = arg_{min} \sum L(m, n) \tag{6}$$

$$L(m, n) = \frac{1}{2} \sum_{i=0}^{t} (x_i - z_i)^2 \tag{7}$$

The cross-entropy (CE), as described in Eq. (8) may be employed [49] if the input x is understood as either a series of bits or a sequence of bit probabilities (i.e., they are Bernoulli probability vectors).

$$L(m, n) = \sum_{i=0}^{n} x_i \log z_i + (1 - x_i) \log(1 - z_i) \tag{8}$$

3.5 Generative adversarial networks

Generative adversarial networks (GAN) are a method for generative modeling that uses deep learning tools like convolutional neural networks [52]. Generative models as a machine learning technique that automatically recognizes and learns the regularities or patterns in incoming data in order to enable the model to generate new examples that might have been properly deduced from the first dataset. When the two models are trained simultaneously in an adversarial, zero-sum game, the discriminator model is misled around half the time, showing that the generator model is providing plausible examples. The promise of generative models is fulfilled by GANs in their capacity to provide realistic examples across various problem domains, such as defect detection, object identification, and many more. The produced samples are distinguished from ordinary or original ones by the second network acting as a discriminator model. The actual ones are usually provided as samples from a training data set, referred to as x below (Eq. 9) [52].

$$f(x) = arg_{min}(\theta) - \frac{1}{2} \log D(x) - \frac{1}{2} \log(1 - D(G(\gamma))) \tag{9}$$

The discriminator is trained to optimize loss by providing an output of 1 for the actual samples and 0 for the generated ones, where x stands for real samples and y represents for want output of given $D(x)$. Employing the discriminator and creating samples that the discriminator recognizes as genuine is the key to reducing generator loss, as shown in Eq. (10) [53].

$$arg_{min}(\theta) - \frac{1}{2} \log D(G(\gamma)) \tag{10}$$

Fig. 6 presents the hypothetical GAN diagram. The Generator's principal goal is to force the discriminator to categorize its output. A discriminator network classifies the created data using a generator network, which turns the random input into data instances. Generator loss, which punishes the Generator for not fooling the discriminator. By assessing the weight's influence on the output, the backpropagation technique is utilized to move each weight in the proper direction. The generator weights may be changed using these gradients, which are also obtained using it [54].

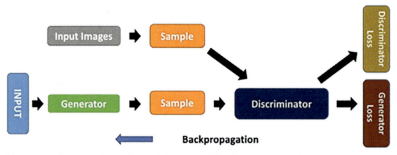

Fig. 6 Architecture of generative adversarial network for brain tumor detection.

4. Transfer learning approach

Transfer learning is a natural and powerful approach in deep learning, especially for learning medical tasks like brain tumor segmentation, where annotated medical data is usually limited. Rather than start from scratch by training a model, transfer learning allows one to leverage a previously trained model that has been trained on a large, general dataset (i.e., ImageNet) [55]. This pre-trained model captures valuable features like edges, textures, and shapes, and can be tuned to task of brain tumor segmentation. Transfer learning allows for quicker convergence on the task, higher performance, and more efficient use of available data, and is especially appropriate for medical applications, where it is often difficult or impossible to assemble a large dataset.

4.1 Choosing a pre-trained model

In the domain of brain tumor segmentation, common backbones include VGG16, VGG19, ResNet, DenseNet, or U-Net architecture [56]. These models are initially pre-trained on larger datasets with various image representations, ultimately containing layers that are sensitive to certain features in images. However, since these models were originally designed for classification tasks, the final layers (i.e., fully connected layers) needed to be removed or revised to encompass segmentation models that required a pixel-wise classification stage. Choice of pre-trained model can influence results due to some remaining models performing analysis on fine details, while others are better capable of working through classic structures such as tumors [57].

4.2 Modifying the architecture

After choosing a pre-trained model, one would modify the model's architecture for segmentation. In general, the fully connected layers (which are utilized for classification) are replaced by an encoder-decoder architecture. The encoder is usually the pre-trained model while the decoder is made up of upsampling layers, which increase the spatial resolution of feature maps for making pixel-wise predictions. For example, in a U-Net architecture, the encoder could be replaced by a pre-trained network, such as VGG or ResNet, while the decoder reconstructs segmented brain tumor regions using transposed convolutions or other upsampling methods.

4.3 Fine-tuning the model

After modifying the pre-trained model, it needs to be fine-tuned to the task of brain tumor segmentation. Fine-tuning involved training the model on the brain tumor dataset, benefits from training, but the learning rate is lowered in order to not disturb the learned pre-trained features. In general, the earlier layers, which encode general image features (e.g., edges, textures) are frozen for training. The earlier refers to updates of their weights during training (each iteration) is prevented. Typically, the new layers (e.g., the segmentation head) are the only parts of the model trained at first. Once effective training has occurred, layers of the pre-trained backbone are unfrozen, and fine-tuning training (in this case, the brain tumor medical dataset) makes new updates, then the entire model is fine-tuned to the medical dataset to improve accuracy of the task at hand [58].

Moreover, data preprocessing is vital in brain tumor segmentation, as medical images (especially MRIs) are not the same as natural images. MRI images must be resized first so they comply with the input staggering of the pre trained model, or model architecture (e.g., 224×224 pixels). Then, normalization or standardization of pixel intensity is generally applied. Using normalization or standardization will adjust the pixel values into the range which the pre-trained model expects. Brain tumor segmentation usually involves multiple modalities for MRIs (i.e., T1, T2, FLAIR). Those differing modality images may be treated as multi-channel input to the model, similar to RGB images of natural images for classification tasks. These preprocessing steps ensure the model is interpreting images appropriately, as instructed by the pre-trained model [58]. Data augmentation: Since brain tumor datasets are often small, data augmentation serves an important purpose to avoid overfitting and enhance model generalizability. Data augmentation has many technique options available for implementation, such as rotating, flipping, scaling images, intensifying transformations, etc. Artificially increasing the size of the dataset is critical to early-stage model training. There are several effective ways to simulate natural variability of the brain structure, such as elastic deformities and random cropping to exhibit some variations for training [59]. These augmentation strategies educate the AI model to recognize tumors approaching from different angles and treating akin imaging conditions, therefore generalizing the model application to unseen data better.

4.4 Loss function

The selection of the loss function is critical for brain tumor segmentation, as the task involves classification at the pixel level. The most commonly applied loss function is categorical cross entropy, which is suitable for segmenting sub-regions of tumors or performing multi-class segmentation. However, categorical cross-entropy scores tend to be skewed values due to the high class imbalance between the tumor region and the background area. Therefore, Dice Loss is frequently used, as it uses the overlap between the predicted segmentation mask and the ground truth to inform the segmentation model, maximizing the intersection and reduces variation in loss scores between each iteration [59]. Some models apply Dice Loss and cross-entropy loss together in order to further enhance segmentation scores, particularly in the presence of imbalanced datasets.

4.5 Evaluation metrics

The effectiveness of the transfer learning model for brain tumor segmentation can be evaluated with multiple evaluation metrics. The Dice Coefficient is a favored evaluation metric, as it expresses the overlap between the predicted segmentation mask and true segmentation mask. Similarly, another popular evaluation metric is Intersection over Union (IoU) [60], or percentages of similarity between the predicted segmentation and true segmentation. Precision and recall can be calculated to evaluate the model's effectiveness in identifying the tumor region, while not creating excessive false positives or false negatives. Additionally, evaluation metrics like Hausdorff Distance can be used to calculate accuracy of predicted tumor boundaries.

4.6 Post-processing

Post-processing techniques can improve the output of the segmentation model. Generally, morphological operations, including erosion and dilation, are applied to enhance the results. These techniques can clean up the segmentation masks by eliminating small isolated regions that are not relevant for the tumor. Another step is, using connected component analysis to preserve the largest tumor region and eliminate small disconnected regions which are more likely noise. These post-processing steps ensure the final segmentation output is clinically useful and smooth [58].

Transfer learning has several appealing features for segmentation of brain tumors. First, we can use kinds of pre-trained models to partially generate segments to make the learning faster and less data-dependent to guarantee better results. Second, we can use deep learning frameworks pre-trained on huge datasets like ImageNet, so that the medical image segmentation model can also recognize the smaller details in medical images that the model had hard time learning due to the small sample size from scratch. Finally, transfer learning reduces the need for large amounts of annotated medical data, which can be challenging and time-consuming to obtain in the healthcare domain.

5. Multimodal approaches

In brain tumor classification, a multi-model approach combines the strengths of different models to achieve better classification performance. This is particularly important in medical imaging, where various types of brain tumors, such as gliomas, meningiomas, and pituitary tumors, exhibit complex and diverse characteristics. A single model may not effectively capture all the features needed for accurate classification. The multi-model approach uses multiple classifiers or architectures, such as convolutional neural networks (CNNs), support vector machines (SVMs), and ensemble techniques to improve accuracy, robustness, and generalizability [61].

The motivation for employing a multi-model approach arises from the limitations of using a single model in brain tumor classification. Different machine learning models have distinct strengths: CNNs are powerful in feature extraction from medical images, SVMs are effective in separating classes when there is a clear margin between them, and ensemble methods like Random Forest or Gradient Boosting enhance robustness by combining multiple weak learners. By integrating these models, we can mitigate individual weaknesses, ensuring that features learned from the data are captured from multiple perspectives. This approach also addresses data imbalance, where certain tumor classes may be underrepresented in the dataset. One common multi-model approach involves combining CNNs with traditional machine learning models. CNNs can be used to extract high-level features from MRI scans, which are then passed to models like SVMs or Random Forest for classification [62].

Ensemble learning is another multi-model strategy where predictions from multiple models are combined to improve classification performance. Bagging (Bootstrap Aggregating) and Boosting are common ensemble techniques used in brain tumor classification. In Bagging, multiple classifiers (such as decision trees or CNNs) are trained independently on different subsets of the training data, and their predictions are averaged. The equation for the final prediction in Bagging is calculated as Eq. (11) [62].

$$\widehat{Y} = \frac{1}{N} \sum_1^N h_i(x) \tag{11}$$

where $h_i(x)$ is the prediction of the ith model on input. In Boosting, models are trained sequentially, and each new model focuses on correcting the mistakes of the previous ones. The final prediction is a weighted sum of all models as Eq. (12) [63].

$$\widehat{Y} = \sin \sum_1^N a_i h_i(x) \tag{12}$$

Transfer learning can be an integral part of a multi-model approach, especially when dealing with limited medical imaging data. Pre-trained CNNs like VGG16, ResNet, or DenseNet can be fine-tuned on a brain tumor dataset and combined with other models for classification. The multi-model approach can leverage transfer learning by using pre-trained networks for feature extraction and combining these features with classifiers like SVM or ensemble methods for improved classification.

Voting-based classifier ensembles another technique used in multi-model brain tumor classification is voting-based classifier ensembles. In this method, different classifiers are trained on the same dataset, and the final classification decision is made by voting. There are two common voting strategies: hard voting and soft voting. Evaluating the performance of multi-model approaches for brain tumor classification involves using metrics such as accuracy, precision, recall, F1-score, and AUC (area under the curve) [64]. Due to the medical significance of brain tumor classification, metrics like sensitivity (true positive rate) and specificity (true negative rate) are critical to ensure the model identifies tumor types correctly while minimizing false positives and negatives. Cross-validation or k-fold validation is often employed to validate the performance of the multi-model approach on small datasets, ensuring that the model generalizes well across different patient samples.

6. Importance of interpretability in tumor detection using deep learning models

6.1 Importance of interpretability in clinical settings

In medical applications such as brain tumor detection, the interpretability of AI models is essential. Deep learning models, especially complex architectures like CNNs, are often considered "black boxes" due to their lack of transparency in how they arrive at decisions. In the context of brain tumor detection, clinicians need to understand why the model made a certain diagnosis or prediction. Without interpretability, it becomes difficult to trust the model's outputs, especially in critical cases like cancer diagnosis, where treatment decisions can be life-altering. Clinicians need interpretable models to verify predictions, ensure reliability, and gain insights into the biological reasoning behind classifications [65].

6.2 Challenge of black box models in tumor detection

Deep learning models, such as those used for brain tumor detection, often achieve high accuracy, but their decision-making processes are opaque. These models learn complex patterns in the data, but the learned features are typically high-dimensional and abstract, making it hard to map them back to human-understandable concepts. In medical imaging, particularly with MRI scans, tumors can have subtle features, and without interpretability, it is difficult to know which parts of the image influenced the model's prediction. This "black box" nature becomes a major concern when a model misclassifies a tumor, as there is no clear explanation for the error, making it harder to correct or improve the model [66].

6.3 Impact on trust and adoption in healthcare

The lack of explainability in tumor detection models hinders the broader adoption of AI tools in healthcare. Clinicians and patients alike are hesitant to trust a system that provides no rationale for its decisions. In high-stakes medical environments, any wrong prediction or misclassification can lead to serious consequences. For example, an incorrect tumor classification could result in a patient receiving inappropriate treatment. Hence, clinicians are more likely to use interpretable systems where they can understand the model's reasoning, validate the outputs, and apply their domain knowledge when necessary. Without explainability, AI models are less likely to be integrated into clinical workflows [67].

6.4 Legal and ethical implications

There are also legal and ethical considerations tied to interpretability in medical AI applications. As AI models start playing a more significant role in clinical decision-making, there is growing pressure to provide explanations for predictions, especially when it comes to patient care. In regions where regulatory frameworks like the GDPR (general data protection regulation) apply, there is a legal requirement for "the right to explanation," meaning that patients should be able to understand how an AI system arrived at its decision regarding their health. This challenge becomes even more crucial when AI is used in brain tumor detection, as misdiagnosis can have life-threatening implications [68].

6.5 Tools for improving interpretability

To address this challenge, a variety of interpretability tools have been developed. One popular method is Grad-CAM (gradient-weighted class activation mapping), which visualizes the regions of the MRI scan that most strongly influenced the model's decision. By producing a heatmap that highlights these regions, Grad-CAM helps clinicians understand what parts of the brain the model was "looking at" when making predictions. Other techniques like SHAP (SHapley Additive exPlanations) provide feature importance scores, showing how much each input feature contributed to the model's output. These tools enhance interpretability by offering explanations that are easier for clinicians to interpret and validate [67].

6.6 Balancing accuracy and interpretability

A critical trade-off often arises between accuracy and interpretability in brain tumor detection models. Highly accurate models, such as deep neural networks, are typically less interpretable due to their complexity. On the other hand, simpler models like decision trees or logistic regression are more interpretable but often lack the accuracy needed for complex medical tasks like brain tumor classification. The challenge is to strike a balance between the two, ensuring that models remain transparent and interpretable without sacrificing too much accuracy. One approach is to use hybrid models that combine the high accuracy of deep learning with interpretable post hoc analysis tools [69].

6.7 Role of explainability in improving model performance

Explainability not only builds trust but also helps improve the model itself. By understanding how a model makes decisions, developers can identify biases, weaknesses, or areas where the model is overfitting. For instance, if an interpretability tool like Grad-CAM reveals that the model is focusing on irrelevant regions of the brain scan, it signals that the model may not be learning the correct features for tumor detection. This feedback allows researchers to adjust the model architecture, loss function, or training data to improve performance. Interpretability thus plays a vital role in both model refinement and clinical validation [68].

6.8 Explainability in tumor sub-region detection

Brain tumors often consist of multiple sub-regions, each with distinct characteristics (e.g., enhancing tumor, necrotic core, and edema). Segmenting and classifying these sub-regions accurately is critical for diagnosis and treatment planning. Explainability becomes even more important in such scenarios because clinicians need to understand why a model predicts certain sub-regions. Without this transparency, there may be ambiguity in interpreting the tumor's boundaries or identifying the exact sub-region affected, which could lead to incorrect treatment approaches. Interpretability tools that highlight specific sub-regions can help clinicians verify if the model is detecting the correct regions [69].

6.9 Addressing uncertainty with explainability

Another challenge in brain tumor detection is the inherent uncertainty in medical imaging. Tumor boundaries are often unclear, and MRI scans may be noisy or contain artifacts. Explainability can help in managing this uncertainty by providing confidence scores for the model's predictions. For instance, probabilistic models or ensemble methods can generate uncertainty estimates, which can be visualized to show how certain or uncertain the model is about a particular classification. This allows clinicians to make more informed decisions, knowing when the model's prediction is uncertain and might require further investigation or alternative diagnostic approaches [70].

6.10 Future directions and research opportunities

As the field of AI-driven brain tumor detection advances, the focus on interpretability and explainability will only grow. Future research will likely explore more sophisticated methods for making models transparent without losing accuracy. For instance, new architectures could be developed that are inherently interpretable, combining the high performance of deep learning with easily understandable decision-making processes. Additionally, more user-friendly interfaces for explainability tools will be essential for enabling clinicians to interact with AI models effectively. The ultimate goal is to create a symbiotic relationship between AI and human expertise, where models are both accurate and explainable, leading to better healthcare outcomes [71].

7. Conclusion

Brain tumor is very critical medical condition and needs to be handled with appropriate care, in this chapter we have discussed about various challenges that can prove as an obstacle in predicting the tumor within the brain and further we have discussed upon the deep learning models like CNN, RNN, LSTM, attention models, GANs, Autoencoders, transfer learning approaches that can be involved in predicting the tumors from images. Also we have enclosed in various multimodal approaches to detect the tumors. Further to guide the readers towards the research in brain tumor detection, future directions are included. This chapter highlights various obstacles in accurately predicting tumors, such as variations in imaging quality, differences in tumor presentation, and the limitations of traditional diagnostic methods. We then delve into advanced deep learning models, including convolutional neural networks (CNNs) for image classification, recurrent neural networks (RNNs) and long short-term memory (LSTM) networks for temporal data, and attention mechanisms that enhance model focus on critical image features. Generative adversarial networks (GANs) and Autoencoders are discussed for their roles in data augmentation and feature extraction, while transfer learning approaches leverage pretrained models for improved accuracy with limited data. Additionally, we explore multimodal strategies that integrate various data types—such as imaging and genomic information—to enhance diagnostic precision. Finally, we outline future research directions, encouraging further exploration of hybrid models and real-time applications to ultimately improve brain tumor detection and treatment outcomes.

Availability of data and materials

The authors confirm that the data supporting the findings of this study are available within the chapter sections.

Conflicts of interest

Authors declare no conflicts of interest for this research.

References

[1] G. Mohan, M.M. Subashini, MRI based medical image analysis: survey on brain tumor grade classification, Biomed. Signal Process. Control 39 (2018) 139–161.

[2] S. Bakas, et al., Identifying the Best Machine Learning Algorithms for Brain Tumor Segmentation, Progression Assessment, and Overall Survival Prediction in the BRATS Challenge, 2018. arXiv preprint arXiv:1811.02629.

[3] B.H. Menze, et al., The multimodal brain tumor image segmentation benchmark (BRATS), IEEE Trans. Med. Imaging 34 (10) (2015) 1993–2024, https://doi.org/10.1109/TMI.2014.2377694.

[4] O. Ronneberger, P. Fischer, T. Brox, U-Net: convolutional networks for biomedical image segmentation, in: Medical Image Computing and Computer-Assisted Intervention—MICCAI, 2015, 2015, pp. 234–241, https://doi.org/10.1007/978-3-319-24574-4_28.

[5] Y. Zhang, et al., Segmentation of brain MR images through a hidden Markov random field model and the expectation-maximization algorithm, IEEE Trans. Med. Imaging 20 (1) (2001) 45–57, https://doi.org/10.1109/42.906424.

[6] T. Saba, A. Rehman, Effects of artificially intelligent tools on pattern recognition, Int. J. Mach. Learn. Cybern. 4 (2013) 155–162.

[7] M. Futrega, A. Milesi, M. Marcinkiewicz, P. Ribalta, Optimized U-Net for brain tumor segmentation, in: International MICCAI Brainlesion Workshop, Springer International Publishing, Cham, 2021, September, pp. 15–29.

[8] A. Jamal, M. Hazim Alkawaz, A. Rehman, T. Saba, Retinal imaging analysis based on vessel detection, Microsc. Res. Tech. 80 (7) (2017) 799–811.

[9] F. Isensee, et al., nnU-Net for brain tumor segmentation, in: A. Crimi (Ed.), Lecture Notes in Computer Science, Springer, Cham, 2021, pp. 118–132. https://doi.org/10.1007/978-3-030-72087-2_11.

[10] M.K. Balwant, A review on convolutional neural networks for brain tumor segmentation: methods, datasets, libraries, and future directions, Irbm 43 (6) (2022) 521–537.

[11] K. Meethongjan, M. Dzulkifli, A. Rehman, A. Altameem, T. Saba, An intelligent fused approach for face recognition, J. Intell. Syst. 22 (2) (2013) 197–212.

[12] S. Bauer, R. Wiest, L.P. Nolte, M. Reyes, A survey of MRI-based medical image analysis for brain tumor studies, Phys. Med. Biol. 58 (13) (2013) R97.

[13] A. Rehman, Brain stroke prediction through deep learning techniques with ADASYN strategy, in: 2023 16th International Conference on Developments in eSystems Engineering (DeSE), Istanbul, Turkiye, 2023, pp. 679–684.

[14] M.S. Mahaley, C. Mettlin, N. Natarajan, E.R. Laws, B.B. Peace, National survey of patterns of care for brain-tumor patients, J. Neurosurg. 71 (6) (1989) 826–836.

[15] J. Liu, M. Li, J. Wang, F. Wu, T. Liu, Y. Pan, A survey of MRI-based brain tumor segmentation methods, Tsinghua Sci. Technol. 19 (6) (2014) 578–595.

[16] D.J. Sharp, C.F. Beckmann, R. Greenwood, K.M. Kinnunen, V. Bonnelle, X. De Boissezon, J.H. Powell, S.J. Counsell, M.C. Patel, R. Leech, Default mode network functional and structural connectivity after traumatic brain injury, Brain 134 (8) (2011) 2233–2247.

[17] H.M. Fahad, M.U. Ghani Khan, T. Saba, A. Rehman, S. Iqbal, Microscopic abnormality classification of cardiac murmurs using ANFIS and HMM, Microsc. Res. Tech. 81 (5) (2018) 449–457.

[18] A. Husham, M. Hazim Alkawaz, T. Saba, A. Rehman, J. Saleh Alghamdi, Automated nuclei segmentation of malignant using level sets, Microsc. Res. Tech. 79 (10) (2016) 993–997.

[19] N. Hussain, M.A. Khan, M. Sharif, S.A. Khan, A.A. Albesher, T. Saba, A. Armaghan, A deep neural network and classical features based scheme for objects recognition: an application for machine inspection, Multimed. Tools Applic. (2020) 1–23.

[20] M.A. Balafar, A.R. Ramli, M.I. Saripan, S. Mashohor, Review of brain MRI image segmentation methods, Artif. Intell. Rev. 33 (2010) 261–274.

[21] S. Iftikhar, K. Fatima, A. Rehman, A.S. Almazyad, T. Saba, An evolution based hybrid approach for heart diseases classification and associated risk factors identification, Biomed. Res. 28 (8) (2017) 3451–3455.

[22] N. Birkbeck, D. Cobzas, M. Jagersand, A. Murtha, T. Kesztyues, An interactive graph cut method for brain tumor segmentation, in: Workshop on Applications of Computer Vision, 2009, pp. 1–7.

[23] M. Cabezas, A. Oliver, X. Lladó, J. Freixenet, M. Bach Cuadra, A review of atlas-based segmentation for magnetic resonance brain images, Comput. Methods Programs Biomed. 104 (2011) e158–e177.

[24] M.A. Deeley, et al., Comparison of manual and automatic segmentation methods for brain structures in the presence of space-occupying lesions: a multi-expert study, Phys. Med. Biol. 56 (2011) 4557–4577.

[25] R.C. Gonzalez, Digital Image Processing, Pearson Education India, 2009.

[26] I. Bankman, Handbook of Medical Imaging: Processing and Analysis Management, Academic Press, 2000.

[27] M.A. Khan, M.I. Sharif, M. Raza, A. Anjum, T. Saba, S.A. Shad, Skin lesion segmentation and classification: a unified framework of deep neural network features fusion and selection, Expert. Syst. 39 (7) (2022) e12497.

[28] N. Zhang, S. Ruan, S. Lebonvallet, Q. Liao, Y. Zhu, Kernel feature selection to fuse multi-spectral MRI images for brain tumor segmentation, Comput. Vis. Image Underst. 115 (2) (2011) 256–269.

[29] T. Magadza, S. Viriri, Deep learning for brain tumor segmentation: a survey of state-of-the-art, J. Imaging 7 (2) (2021) 19.

[30] W. Ayadi, W. Elhamzi, I. Charfi, M. Atri, Deep CNN for brain tumor classification, Neural. Process. Lett. 53 (2021) 671–700.

[31] S. Kumar, R. Dhir, N. Chaurasia, Brain tumor detection analysis using CNN: a review, in: 2021 International Conference on Artificial Intelligence and Smart Systems (ICAIS), IEEE, 2021, March, pp. 1061–1067.

[32] A.M. Al-Zoghby, E.M.K. Al-Awadly, A. Moawad, N. Yehia, A.I. Ebada, Dual deep CNN for tumor brain classification, Diagnostics 13 (12) (2023) 2050.

[33] P. Tiwari, B. Pant, M.M. Elarabawy, M. Abd-Elnaby, N. Mohd, G. Dhiman, S. Sharma, Cnn based multiclass brain tumor detection using medical imaging, Comput. Intell. Neurosci. 2022 (1) (2022) 1830010.

[34] M.S.I. Khan, A. Rahman, T. Debnath, M.R. Karim, M.K. Nasir, S.S. Band, A. Mosavi, I. Dehzangi, Accurate brain tumor detection using deep convolutional neural network, Comput. Struct. Biotechnol. J. 20 (2022) 4733–4745.

[35] R. Vankdothu, M.A. Hameed, H. Fatima, A brain tumor identification and classification using deep learning based on CNN-LSTM method, Comput. Electr. Eng. 101 (2022) 107960.

[36] J. Amin, M. Sharif, M. Raza, T. Saba, R. Sial, S.A. Shad, Brain tumor detection: a long short-term memory (LSTM)-based learning model, Neural Comput. Applic. 32 (2020) 15965–15973.

[37] E. Dandıl, S. Karaca, Detection of pseudo brain tumors via stacked LSTM neural networks using MR spectroscopy signals, Biocybernet. Biomed. Eng. 41 (1) (2021) 173–195.

[38] A. Rajendran, S. Ganesan, T.K.S. Rathis Babu, Hybrid ResGRU: effective brain tumour classification approach using of abnormal images, J. Intell. Fuzzy Syst. (2023) 1–15 (preprint).

[39] A.M. Gab Allah, A.M. Sarhan, N.M. Elshennawy, Classification of brain MRI tumor images based on deep learning PGGAN augmentation, Diagnostics 11 (12) (2021) 2343.

[40] S. Alsubai, H.U. Khan, A. Alqahtani, M. Sha, S. Abbas, U.G. Mohammad, Ensemble deep learning for brain tumor detection, Front. Comput. Neurosci. 16 (2022) 1005617.

[41] S. Shanthi, S. Saradha, J.A. Smitha, N. Prasath, H. Anandakumar, An efficient automatic brain tumor classification using optimized hybrid deep neural network, Int. J. Intell. Netw. 3 (2022) 188–196.

[42] K. Shah, K. Shah, A. Chaudhari, D. Kothadiya, Comprehensive analysis of deep learning models for brain tumor detection from medical imaging, in: International Conference on Data Science and Applications, Springer Nature Singapore, Singapore, 2023, July, pp. 339–351.

[43] Z.K. Abbas, Z.A. Alsarray, A.H.H. Al-obeidi, M.R. Mutashar, Enhancing brain tumor detection: integrating CNN-LSTM and CNN-BiLSTM models for efficient classification in MRI images, Int. J. Adv. Technol. Eng. Explorat. 11 (115) (2024) 888.

[44] N.S. Shaik, T.K. Cherukuri, Multi-level attention network: application to brain tumor classification, SIViP 16 (3) (2022) 817–824.

[45] Y. Lin, C.W., Hong, Y. and Liu, J., Aggregation-and-attention network for brain tumor segmentation, BMC Med. Imaging 21 (1) (2021) 109.

[46] W. Jun, Z. Liyuan, Brain tumor classification based on attention guided deep learning model, Int. J. Comput. Intell. Syst. 15 (1) (2022) 35.

[47] B.C. Mohanty, P.K. Subudhi, R. Dash, B. Mohanty, Feature-enhanced deep learning technique with soft attention for MRI-based brain tumor classification, Int. J. Inf. Technol. 16 (3) (2024) 1617–1626.

[48] D.R. Nayak, N. Padhy, P.K. Mallick, A. Singh, A deep autoencoder approach for detection of brain tumor images, Comput. Electr. Eng. 102 (2022) 108238.

[49] F. Bashir-Gonbadi, H. Khotanlou, Brain tumor classification using deep convolutional autoencoder-based neural network: multi-task approach, Multimed. Tools Appl. 80 (13) (2021) 19909–19929.

[50] J. Amin, M. Sharif, N. Gul, M. Raza, M.A. Anjum, M.W. Nisar, S.A.C. Bukhari, Brain tumor detection by using stacked autoencoders in deep learning, J. Med. Syst. 44 (2020) 1–12.

[51] B. Ahmad, J. Sun, Q. You, V. Palade, Z. Mao, Brain tumor classification using a combination of variational autoencoders and generative adversarial networks, Biomedicines 10 (2) (2022) 223.

[52] G. Neelima, D.R. Chigurukota, B. Maram, B. Girirajan, Optimal DeepMRSeg based tumor segmentation with GAN for brain tumor classification, Biomed. Signal Process. Control 74 (2022) 103537.

[53] B. Ahmad, J. Sun, Q. You, V. Palade, Z. Mao, Brain tumor classification using a combination of variational autoencoders and generative adversarial networks, Biomedicines 10 (2) (2022) 223.

[54] R.K. Gupta, S. Bharti, N. Kunhare, Y. Sahu, N. Pathik, Brain tumor detection and classification using cycle generative adversarial networks, Interdiscip. Sci.: Comput. Life Sci. 14 (2) (2022) 485–502.

[55] R. Mehrotra, M.A. Ansari, R. Agrawal, R.S. Anand, A transfer learning approach for AI-based classification of brain tumors, Mach. Learn. Applicat. 2 (2020) 100003.

[56] R. Chelghoum, A. Ikhlef, A. Hameurlaine, S. Jacquir, Transfer learning using convolutional neural network architectures for brain tumor classification from MRI images, in: IFIP International Conference on Artificial Intelligence Applications and Innovations, Springer International Publishing, Cham, 2020, May, pp. 189–200.

[57] T. Sadad, A. Rehman, A. Munir, T. Saba, U. Tariq, N. Ayesha, R. Abbasi, Brain tumor detection and multi-classification using advanced deep learning techniques, Microsc. Res. Tech. 84 (6) (2021) 1296–1308.

[58] B. Tahir, S. Iqbal, M. Usman Ghani Khan, T. Saba, Z. Mehmood, A. Anjum, T. Mahmood, Feature enhancement framework for brain tumor segmentation and classification, Microsc. Res. Tech. 82 (6) (2019) 803–811.

[59] S. Kumar, S. Choudhary, A. Jain, K. Singh, A. Ahmadian, M.Y. Bajuri, Brain tumor classification using deep neural network and transfer learning, Brain Topogr. 36 (3) (2023) 305–318.

[60] K. Yousaf, Z. Mehmood, T. Saba, A. Rehman, A.M. Munshi, R. Alharbey, M. Rashid, Mobile-Health Applications for the Efficient Delivery of Health Care Facility to People With Dementia (PwD) and Support to Their Carers: A Survey, BioMed Research International, 2019.

[61] M.I. Sharif, M.A. Khan, M. Alhussein, K. Aurangzeb, M. Raza, A decision support system for multimodal brain tumor classification using deep learning, Complex Intell. Syst. (2021) 1–14.

[62] I.B. Santoso, S.N. Utama, Multi-model of convolutional neural networks for brain tumor classification in magnetic resonance imaging images, Int. J. Intell. Eng. Syst. 17 (5) (2024).

[63] B.H. Menze, A. Jakab, S. Bauer, J. Kalpathy-Cramer, K. Farahani, J. Kirby, Y. Burren, N. Porz, J. Slotboom, R. Wiest, L. Lanczi, The multimodal brain tumor image segmentation benchmark (BRATS), IEEE Trans. Med. Imaging 34 (10) (2014) 1993–2024.

[64] P.A. Dino, S. Lakshmanan, A multimodel accurate classification of brain tumor based on deep learning techniques, in: 2024 2nd International Conference on Networking and Communications (ICNWC), IEEE, 2024, April, pp. 1–8.

[65] C. Rudin, Stop explaining black box machine learning models for high stakes decisions and use interpretable models instead, Nat. Mach. Intell. 1 (2019) 206–215.

[66] A. Das, P. Rad, Opportunities and Challenges in Explainable Artificial Intelligence (XAI): A Survey, 2020. arXiv preprint arXiv:2006.11371.

[67] A. Chaddad, J. Peng, J. Xu, A. Bouridane, Survey of explainable AI techniques in healthcare, Sensors 23 (2) (2023) 634.

[68] F.K. Došilović, M. Brčić, N. Hlupić, Explainable artificial intelligence: a survey, in: 2018 41st International Convention on Information and Communication Technology, Electronics and Microelectronics (MIPRO), IEEE, 2018, May, pp. 0210–0215.

[69] G. Schwalbe, B. Finzel, A comprehensive taxonomy for explainable artificial intelligence: a systematic survey of surveys on methods and concepts, Data Min. Knowl. Disc. (2023) 1–59.

[70] E. Tjoa, C. Guan, A survey on explainable artificial intelligence (xai): toward medical xai, IEEE Trans. Neural Networks Learn. Syst. 32 (11) (2020) 4793–4813.

[71] I. Shahzadi, T.B. Tang, F. Meriadeau, A. Quyyum, CNN-LSTM: cascaded framework for brain tumour classification, in: 2018 IEEE-EMBS Conference on Biomedical Engineering and Sciences (IECBES), IEEE, 2018, December, pp. 633–637.

CHAPTER 15

Unveiling the root causes of diabetes using explainable AI

P.K. Nizar Banu[a], Ahmad Taher Azar[b,c,d], and Nashwa Ahmad Kamal[e]
[a]Department of Computer Science, CHRIST (Deemed to be University) Central Campus, Bangalore, Karnataka, India
[b]College of Computer and Information Sciences, Prince Sultan University, Riyadh, Saudi Arabia
[c]Automated Systems and Soft Computing Lab (ASSCL), Prince Sultan University, Riyadh, Saudi Arabia
[d]Faculty of Computers and Artificial Intelligence, Benha University, Benha, Egypt
[e]Faculty of Engineering, Cairo University, Giza, Egypt

1. Introduction

It is commonly known that 422 million individuals globally suffer from diabetes. Blood sugar levels that are either too high or too low are known as diabetes. Since it is a chronic illness, a person will have to live with it for the remainder of their lives after receiving a diagnosis. It is a non-communicable disease, requires advanced medical technologies for early diagnosis. Diabetes affects more women than males and can be fatal or significantly worsen quality of life [1].

Diabetes is risky for pregnant women, and unborn children are likely to be affected by this disease. To stop the rise, it is imperative to diagnose the condition early. Diabetes is a lifelong ailment for which there is no known cure; the only thing that can slow its progression is early diagnosis, the illness can be controlled with the appropriate care, consistent diet, and medication [2,3]. According to certain clinical data and physical data, a medical professional performs the illness diagnosis. Artificial intelligence (AI) tools have demonstrated exceptional performance in interpreting available data. In the medical profession, in particular AI-based techniques are utilized for the diagnosis or treatment of various ailments due to their rapid and effective outcomes. Cancer diagnostic studies are a few examples of these are [4], hear diseases [5], Alzheimer's [6], brain tumors [7], diabetes [8], COVID-19 [9], etc.

Big data based on AI is typically interpreted using machine learning algorithms, On the basis of the data observations and samples; the relationship between them is discovered. ML techniques such as artificial neural networks, support vector machines, k-nearest neighbors, decision trees, and Naïve Bayes are commonly used. These techniques discover the relationship between the target and input data [10–14]. There has been increasing advancements in artificial intelligence and computer processors, leading to the further development of ANN and the rise of deep learning, which combines feature extraction and categorization. Convolutional neural networks are the model most frequently utilized in applications based on deep learning for medical diagnostics. CNN have become more popular because of its deep architecture and high-level feature representation [15–17].

Machine learning is crucial in many fields because it can enhance creativity, efficiency and decision-making [18–22]. It enables precise data sorting and categorization in classification tasks, which is necessary for applications like as spam detection, image recognition and medical diagnosis [23–27]. Machine learning models use past data to predict future events, which help with demand planning, weather forecasting and financial market analysis. Machine learning is enabling robots to learn, make decision, and adapt more effectively. This allows robots to be used in manufacturing healthcare and autonomous vehicles, transforming these industries with increased productivity, safety and precision [28–31].

Currently, machine learning appears to be a viable tool for improving diabetic patients' quality of life. Machine learning algorithms and deep learning algorithms have been developed, but their accuracy and interpretability are not always reliable. Explainable AI (XAI) is a collection of techniques that helps people understand and trust the output of machine learning algorithms, enhance model performance, produce feature attributions, visualize behavior, create systems that are easy to use, and comply with legal requirements. It supports ethical norms, validates predictions, and helps healthcare workers enhance patient care techniques.

Because AI systems rely significantly on data, biased data may result in different choices. Thus, XAI algorithms can detect and rectify skewed data. This chapter covers an interpretable diabetes prediction model. Diabetes typically manifests as weight loss, obesity, acetone breath, frequent urination, impaired vision, vomiting, nausea and intense fatigue. High blood sugar levels are a symptom of diabetes mellitus, which is caused by insufficient or lacking pancreatic synthesis of insulin. Type-1, type-2 and gestational diabetes are the three main forms of the disease [32]. Type-2 diabetes mellitus, often known as adult-onset diabetes, is the most prevalent form of the disease in people with diabetes [32]. Since type-2 diabetes affects a significant proportion of the worlds' population, its sufferers typically exhibit resistance to insulin [33].

Diabetes, a prevalent and potentially devastating disease, can significantly impact healthcare systems due to numerous risk factors that increase the likelihood of developing the condition. A few predictors have been found by researchers that should be taken into account when evaluating individuals who might develop diabetes in the future. Among those predicators are an elevated body mass index, a family history of diabetes and the smoker's current smoking status [34]. Advancements in artificial intelligence and machine learning enable early disease detection, allowing researchers to predict conditions and prevent disease progression. These technologies help forecast diseases by turning clinical and genomic data into insightful knowledge [35].

In Ref. [36] the study employed the KNN machine learning classifier to predict diabetes illness. To authenticate the study, the findings were compared with another machine learning classifier called SVM. The scientists employed a convolution neural network (CNN) to diagnose diabetes type-2 disease and compared their results to the linear regression (LR) model and multilayer perceptron (MLP). The neural network was employed to diagnose diabetes type-2 conditions in this study. For the authenticity of the work, two machine learning classifiers were employed to compare the results. The feature selection technique was also used in Ref. [37]. We can minimize the processing capabilities of the system and improve accuracy by using the feature selection technique. For comparison and credibility of results, multiple machine learning models were used. Computational technologies are applied with bloodless procedures for diabetic illness prediction. As per the research, the accuracy gained with this method was 91.67%. The deep (DNN) approach detected diabetic retinopathy. CNN achieved a 74.4% accuracy rate. Artificial intelligence was used to detect multiclass retinal illness. The CNN classifier was utilized, and it had a 92% accuracy rate. The following models were applied to the two different diabetes datasets [37]: Linear regression classification: 74.6% and 77.7%, Naive Bayes: 72% and 76.5%, random forest: 74.8% and 81.2%, K nearest neighbor: 73.3% and 77.7%, decision tree: 74% and 84.3% and support vector machine: 74.6% and 84.3%. In Ref. [37] a new model: SHRS-M3DP was proposed which employs deep machine learning. This algorithm performed better than all the above-mentioned models which give 99.6% accuracy.

In Ref. [38], a benchmark PIMA Indian Diabetes dataset is used for experimental evaluation using machine learning-based approach for diabetes classification, early-stage identification, and prediction has been proposed. It also shows a possible IoT-based diabetes monitoring device for healthy and affected people to track their blood glucose (BG) levels. Three distinct classifiers were used to classify diabetes: logistic regression (LR), multilayer perceptron (MLP), and random forest (RF). They used long short-term memory (LSTM), moving averages (MA), and linear regression for predictive analysis (LR). First, Ref. [38] proposed an MLP-based diabetes classification system as well as a deep learning-based LSTM for diabetes prediction. They also presented a hypothetical real-time diabetic monitoring system based on the Internet of Things. The suggested approach will assist users in determining their diabetes risk at an early stage and get future projections of their blood glucose levels. MLP and LSTM have been fine-tuned for diabetic categorization and prediction. On the PIMA Indian Diabetes dataset, the recommended methodologies are tested. Both approaches outperformed state-of-the-art ways with an accuracy of 86.083% and 87.26%, respectively. In Ref. [39], the attributes that were used for the analysis to detect diabetes are gender, age, urea, creatinine ratio, body mass index, LDL, cholesterol, VLDL, and triglycerides (TG), HDL, HBA1C, and class. The procedure used in this analysis is to first prepare the data by various tests like ANOVA test, residual analysis, correlation test, and removal of insignificant attributes.

Following this, the obtained data is separated into training and test data. Performed tests on the trained model and assess its accuracy. Finally, the fitted model was evaluated, and as a result of which we can find the people who are potentially at the risk of being diabetic.

Predictive analysis uses historical and present data to extract knowledge and predict future events by integrating a variety of machine learning algorithms, statistical approaches and data mining methodologies [40]. Machine learning and predictive models are widely utilized in healthcare for strategic objectives, but their results often lack clarity for non-specialists. Various machine learning models applied for diabetes dataset is presented in Table 1.

Table 1 Machine learning models for diabetes prediction.

S. no.	Machine learning/Deep learning techniques	Accuracy (%)	References
1	Naïve Bayes	72.0	[36]
2	KNN	73.3	[36]
3	Decision tree	74.0	[36]
4	Linear regression	74.6	[36]
5	SVM	74.6	[36]
6	Random forest	74.8	[36]
7	mRMR-RF	77.21	[39]
8	Feature selection and SVM	77.37	[40]
9	LR	77.6	[41]
10	Stacking of MLP, SVM, LR	78.2	[42]
11	Ensemble of SVM and NN	80.04	[43]
12	Support vector machine	83.1	[44]
13	Stacked autoencoders-DNN	86.26	[45]
14	DCNN/SMOTE/Outlier detection	86.29	[43]
15	UTA-NN and GA	87.46	[46]
16	Deep MLP	88.41	[47]
17	LSTM-AR	89	[48]
18	Conv-LSM	91.38	[49]
19	Artificial neural network	92.00	[50]
20	ResNet18 and ResNet50-Relief	92.19	[51]
21	SHRS-M3DP	99.6	[36]

Explainable Artificial Intelligence (XAI) is a technique bridging the gap between machine learning experts and non-experts, advancing algorithms and predictive models for high-quality, interpretable, trustable, and accountable AI systems [52] Machine Learning algorithms often create "black boxes" where decisions are not explicitly explained, making it difficult for end-users to understand. XAI is often implemented to address these issues and improve understandability [53]. This chapter discuses machine learning model that uses an XAI approach for predicting if a person has diabetes or not and to generate reliable, interpretable, and explainable predictions for diabetes.

2. Explainable artificial intelligence

AI has revolutionized healthcare by diagnosing diseases, predicting outcomes, and improving patient care, particularly in diabetes prediction models. However, their complexity often leads to ambiguous decision-making processes. XAI aims to improve interpretability of AI systems by providing insights into their predictions and decisions, enhancing trust, facilitating collaboration between humans and machines and promoting informed decision-making by explaining their inner workings. XAI is crucial for healthcare professionals in diabetes prediction models to understand factors driving predictions, identify biases and improve treatment plans. It also empowers patients to make informed lifestyle choices and participate in healthcare management.

Enhancing AI system comprehension can enhance its utilization and address ethical concerns related to AI. Conventional machine learning/deep learning methods enhance model output performance by improving evaluation metrics, but they lack interpretability and often lead to trust issues. GBDT-based XGBoost for good performance and XAI technique for interpretability is presented in Ref. [54]. The responsible implementation of XAI in healthcare requires avoiding biases, maintain impartiality and resolving any potential ethical issues highlighting the importance of XAI in enhancing the transparency and interpretability of diabetes prediction models. This chapter delves into the use of XAI techniques like feature importance analysis, SHAP values and LIME in interpreting and understanding diabetes prediction models. The goal is to enhance trust and facilitate collaboration between healthcare professionals and AI systems. The development of artificially intelligent models which provide clear, understandable explanations for their judgements, predictions or recommendations is known as Explainable AI (XAI). It seeks to bridge the knowledge gap that exists between advanced AI systems and humans.

2.1 Explainable AI in healthcare

The healthcare industry is one of the most promising to adopt Explainable AI (XAI). With the advancement of complex machine learning models, XAI is crucial for addressing significant issues and opportunities in the healthcare sector. Improving the precision of diagnosis is one such opportunity. XAI is a disruptive force in the healthcare sector since it has the ability to drastically change medical diagnosis. The examination of medical imaging data from modalities like MRIs, CT scans, and X-rays is one of the most fascinating uses of XAI in medicine. A vast range of ailments, such as diabetes, cancer, fractures, heart problems, and neurological disorders, can only be found with the use of diagnostic equipment. Even though AI models have shown an incredible ability to absorb and interpret these images, many stakeholders rely on these models to make judgments that make sense.

Interpretability is crucial for radiologists and physicians to trust and validate AI models. They need to understand the model's reasoning to integrate AI into clinical procedures. By closely inspecting areas of interest and understanding patterns, they can diagnose patients more confidently, combining medical expertise with AI's analytical skills.

Patients can also gain many benefits from the interpretability that XAI in medical imaging provides. People's lives may be altered by the outcomes of diagnostic tests. A medical image may indicate that a patient has a dangerous illness that requires immediate attention or that there is a less serious issue that should just be closely monitored. Patients should be informed of the reasons behind their unique diagnoses in both cases. XAI empowers patients by giving them insight into the reasoning behind the AI's decisions. This enhanced transparency has led to patients feeling more secure in the recommended treatment strategy. Patients are more inclined to trust medical advice when they can understand it. XAI promotes collaboration between patients, doctors, and AI systems for precise medical diagnoses. It enhances diagnostic accuracy in various clinical scenarios, including laboratory interpretation, predictive analytics, and drug interaction identification.

2.2 Approaches of XAI

XAI approaches can be broadly categorized into several techniques as shown in Fig. 1.

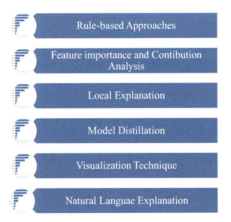

Fig. 1 Approaches of XAI.

Rule-based approaches: The AI model's decision-making process is illustrated using explicit rules or decision trees, offering explanations depending on particular requirements and standards.

Feature importance and contribution analysis: The methods emphasize key factors in AI model predictions, enabling explanations of specific forecasts by understanding the significance and contribution of various features.

Local explanations: LIME (Local Interpretable Model-Agnostic Explanations) is a technique that simplifies and explains models, resembling the behavior of sophisticated AI models, for specific cases.

Model distillation: Model distillation is a process that simplifies complex AI models by retaining their primary decision-making features, thereby enhancing comprehension and elucidation.

Visualization techniques: Visualization techniques like heatmaps, saliency maps, and activation maps display the internal workings of AI models, highlighting the areas of input data that significantly influence decision-making.

Natural language explanations: The aim of explanations in natural language is to give easily comprehensible explanations in a language that is readable by humans. The AI model itself may produce these explanations or specific algorithms may supply them as post-hoc explanations.

XAI enhances human comprehension, auditing, ethical compliance, identifying biases, errors, and limitations in AI models, enhancing trust in sectors like healthcare, finance, and autonomous vehicles, and providing clear, responsible justifications for AI decisions.

2.3 Local interpretable model-agnostic explanations (LIME)

LIME is an important method in explainable AI, aiming to provide clear, understandable justifications for predictions from complex machine learning models. Its model independence makes it a flexible tool for interpretability across various models as shown in Fig. 2.

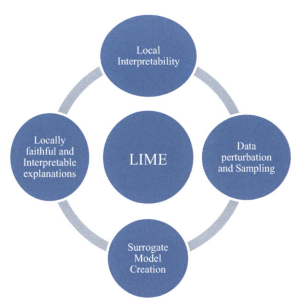

Fig. 2 Features of lime.

Local interpretability: When an explanation is required for a specific forecast, LIME focuses on the local area of the relevant data point. This local viewpoint ensures that the explanation is unique to the individual characteristics of each instance rather than attempting to describe the complete behavior of the model.

Data perturbation and sampling: LIME intentionally manipulates data point properties to cause perturbations and generate explanations, resulting in a new dataset reflecting local fluctuations. Randomness is added to assess feature combinations' impact on predictions, and sampling techniques are used to effectively explore the altered data space.

Surrogate model creation: LIME creates a surrogate model that resembles AI behavior using perturbed data space, typically a linear model, and is model-agnostic, allowing for the use of various machine-learning models.

Locally faithful and Interpretable explanations: Interpretable and locally accurate explanations are the end result of the LIME procedure. These explanations, which highlight the contributions of distinct features to the specific prediction, offer an in-depth understanding of the reasoning behind the AI model's decision-making for the given data point query.

Various XAI models used for highlighting the results of machine learning models that are applied to diabetes dataset are listed in Table 2.

Table 2 XAI methods for interpreting ML/DL results.

S. no.	Machine learning/deep learning techniques	XAI method	References
1	Random forest	SHAP/LIME/ELI5	[55]
2	Ensemble of XGBoost and random forest	LIME/SHAP	[56]
3	SVM	SHAP	[57]
4	Bayesian-optimized TabNet	SHAP/LIME/ELI5	[58]
5	DNN	SHAP	[59]
6	XGBoost	LIME	[60]
7	XGBoost with ADASYN	LIME/SHAP	[61]
8	XGBoost	SHAP	[62]
9	Ensemble stack	SHAP/LIME/ELI5/Qlattice/Anchor	[63]

2.4 Importance of interpretability in AI models

Numerous uses of AI have been found in healthcare industry, including clinical decision-making, predictive medicine, health services management, and patient diagnostics [64–76]. Explainable artificial intelligence (XAI) is the development of AI models that provide concise, intelligible justifications for their decisions, forecasts, or suggestions. It aims to bridge the knowledge gap between humans and sophisticated AI systems, allowing people to understand and trust their reasoning abilities. For many reasons as presented in Fig. 3, interpretability is a crucial component in the development and application of AI models.

Openness and Confidence

- Interpretable AI algorithms provide transparent explanations for their predictions and forecasts, fostering confidence and acceptance among users, including patients and medical professionals.

Accountability and Ethical Considerations

- In high-stakes fields like healthcare, interpretability in AI models guarantees accountability by enabling stakeholders to closely examine decision-making procedures.
- It ensures that ethical standards be observed by assisting in the identification of potential biases and unfair treatments.

Domain Knowledge Integration

- AI models enhance precision and applicability by incorporating domain knowledge from healthcare workers.
- Interpretability facilitates collaboration between AI systems and human specialists, resulting in more accurate and trustworthy forecasts.

Training and Validation:

- Diabetes prediction models undergo training and validation phases using machine learning methods to learn data patterns and relationships.
- Understanding these procedures helps evaluate model validity and generalizability.

Performance Evaluation Metrics:

- Various performane indicators are used to evaluate diabetes prediction models in order to determine their accuracy and dependability.
- These measurements include AUC-ROC, positive predictive value, negative predictive value, accuracy, sensitivity (true positive rate) and specificity (true negative rate).
- Gaining an understanding of these indicators and hoe to interpret them will help you understand teh predictive power and constraints of the model.

Feature Importance and Contribution:

- Diabetes prediction methods help healthcare practitioners identify critical risk factors and prioritize interventions.
- Understanding the limitations and uncertainties of these models is crucial. They may not fully represent all variables and should be regarded cautiously due to their probabilistic nature. Understanding these elements can inform clinical decision-making.

Regulatory Compliance:

- In order to guarantee accountability, equity, and transparency in the application of AI technology, interpreted AI models are essential for regulatory compliance in the industries including healthcare, banking and legal systems.

Error Detection and Model Improvement:

- Interpretability may be used to identify biases in the data, unexplained correlations, or inadequacies in the model's training procedure. Through repeated refinement and model improvement, prediction become more accurate and dependable over time.

Fig. 3 Importance of interpretability in AI models.

3. Understanding diabetes prediction models

Diabetes prediction models are artificially intelligent systems that detect a person's risk of getting diabetes using a variety of machine learning techniques. These models produce risk scores or probabilities of diabetes onset by analyzing a variety of patient-specific variables, including demographics, biomarkers, lifestyle choices and medical history. Healthcare practitioners must comprehend these prediction models in order to make well-informed decisions about patient care and intervention techniques. Here are certain key aspects to consider when looking for understanding diabetes prediction models:

Data types: For diabetes prediction models to be accurate, a variety of data sources are needed. These include biomarkers (glucose levels, insulin sensitivity) lifestyle factors (diet, physical activity, smoking), clinical data (blood pressure, cholesterol levels, body mass index and demographic data (age, gender, and ethnicity). Comprehending the relevance and significance of these data kinds is crucial to understanding the predictions made by the model.

Machine learning algorithms: Diabetes prediction models can make use of a variety of machine learning techniques, including random forests, decision trees, logistic regression, support vector machines, and neural networks. Learning the key concepts and assumptions of each algorithm is essential to understanding how the model generates its predictions acknowledging each algorithms' advantages and disadvantages.

3.1 Challenges with black box models

Black box models can introduce biases, making it difficult to identify and mitigate them, potentially leading to unfair outcomes in healthcare, compromising ethical considerations. Major challenges encountered by black box models are shown in Fig. 4.

Fig. 4 Challenges with black box models.

Lack of explainability: Black box models, particularly in crucial fields requiring transparency, are challenging to interpret and maintain trust.

Limited insights into decision-making: Because black box models lack insight into underlying factors, they are unable to determine precise reasons behind predictions, which limits their ability to be validated or refined according to domain expertise.

Ethical concerns and bias: Black box models have the potential to be biased or to discriminate against particular qualities or groups. It is more difficult to identify and reduce these biases when there is no interpretability, which could have discriminatory consequences.

Trust and acceptance: Black box models in healthcare often face distrust and suspicion from users, potentially undermining trust and preventing their adoption and use in the industry.

Regulatory compliance and legal requirements: Within regulated sectors like healthcare, transparent decision-making procedures are crucial, but black box models may struggle to meet these requirements, making integration into real-world systems challenging.

Safety and accountability: Safety issues with black box models in medical diagnosis and treatment suggestions due to difficulty in understanding model logic in the event of errors or unfavorable results.

Addressing obstacles in healthcare AI application is crucial. Explainable AI tools aim to provide interpretability, transparency, and accountability for black box models, promoting trust and acceptance in healthcare AI. Various factors of XAI when applied on machine learning, deep learning are shown in Fig. 5.

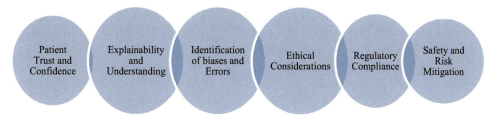

Fig. 5 Factors of XAI in ML/DL models.

Patient trust and confidence: AI systems enhance patient confidence and trust by providing accountability and transparency, promoting patient-centered strategies, and allowing active engagement in healthcare choices, ultimately strengthening trust and faith in the care they receive.

Explainability and understanding: Transparency in AI model results aids healthcare workers in understanding and analyzing AI models, enabling informed decision-making and improved patient care through collaboration between AI systems and human professionals.

Identification of biases and errors: Transparent AI systems help in identifying and reducing biases and mistakes, ensuring accurate and reliable healthcare services. Encouraging transparency helps identify biases, understand their origins, and take corrective action, thereby preventing negative effects on patient health.

Ethical considerations: Healthcare AI adheres to ethical standards like fairness, privacy, and autonomy, ensuring ethical practices through transparent platforms for evaluation and verification by healthcare professionals.

Regulatory compliance: Healthcare regulatory agencies demand accountability and transparency in AI-based decisions, requiring explanations and arguments for crucial conclusions like diagnosis and treatment suggestions. Transparent AI systems help meet these standards.

Safety and risk mitigation: Transparent AI technologies reduce risks and safeguard patients by allowing healthcare practitioners and patients to understand AI algorithms' judgments, enabling continuous evaluation, enhancement, and detection of safety issues.

The ethical use of AI in healthcare relies on accountability and openness, enhancing patient safety, trust, and detecting biases. It can improve outcomes and services by prioritizing accountability and openness.

3.2 Interpreting and understanding diabetes prediction models

Analyzing the model's features, looking at its performance characteristics and taking into account its limits are all necessary when interpreting and comprehending diabetes prediction models. To assist in interpreting and comprehending diabetes prediction models, follow the following steps:

Feature importance: These days research aims to identify key factors influencing the model's diabetes prediction, using coefficients, feature weights, or significance scores to identify variables that significantly increase diabetes risk.

Domain knowledge: Examine the model's results against established risk variables like age, blood pressure, and family history to verify predictions and identify inconsistencies.

3.2.1 Dataset details

The work carried out in this chapter studies the Explainable AI for publicly available PIMA Indian diabetes dataset [58]. The details of the dataset are presented in Table 3. The participants involved in this dataset are only women.

Table 3 PIMA Indian dataset.

Attribute	Description	Range
Pregnancies	Number of times a women is pregnant	0–17
Glucose	Glucose levels during oral glucose tolerance test	0–199
BloodPressure	Diastolic blood pressure (mmHg)	0–122
SkinThickness	Triceps skinfold thickness (mm)	0–99
Insulin	2-hour serum insulin levels	0–846
BMI	Body mass index obtained from height and weight	0–67.1
DiabetesPedigreeFunction	This is a function which describes diabetes family history	0.078–2.42
Age	Age of participants	21–81
Outcome	Has diabetes/No diabetes	0/1

3.3 Machine learning algorithms

Few machine learning algorithms that are applied on the diabetes dataset are discussed below.

3.3.1 K-nearest neighbors

K-nearest neighbors (KNN) is a simple, easy-to-implement supervised machine learning technique that may be used to address both classification and regression issues. In order to forecast the values of new data points, the KNN algorithm employs feature similarity, which further implies that the new data point will be assigned a value depending on how closely it matches the points in the training set. KNN algorithm is implemented using Python. The KNN algorithm assumes that similar things exist in close proximity. In other words, similar things are near to each other.

3.3.2 Decision tree

Decision trees are designed to mirror human thinking abilities when making decisions, making them simple to comprehend. Because the decision tree has a tree-like form, the rationale behind it is simple to comprehend. Decision trees have internal nodes that reflect dataset properties, branches that represent decision rules and leaf nodes that deliver the tree structured classifier's decision. A decision tree consists of two nodes: the decision node and the leaf node. In contrast, decision nodes which consists of multiple branches are utilized to create any kind of decision. Leaf nodes provide the output of these decisions and do not have any more branches. The procedure for determining the class of a given dataset in a decision tree starts at the root node of the tree. This algorithm checks the values of the root attribute with the values of the record (actual dataset) attribute and then follows the branch and jumps to the next node based on the comparison. The algorithm compares the attribute value with the other sub-nodes and moves on to the next node. It repeats the process until it reaches the tree's leaf node.

3.3.3 Random forest

Random forest is a well-known machine learning algorithm that uses the supervised learning method. In machine learning, it can be utilized for both classification and regression issues. It is based on ensemble learning, which is a method of integrating several classifiers to solve a complex problem and increase the model's performance. The random forest predicts the outcome based on the majority votes of forecasts, gathering forecasts from multiple decision trees rather than depending only on one. Table 4 presents the accuracy of various machine learning algorithms applied.

Table 4 Accuracy of the algorithms.

S. no.	Algorithm	Accuracy
1.	KNN	0.92
2.	Decision tree	0.94
3.	Random forest	0.95

3.3.4 Use of SHAP in diabetes prediction

SHAP is one of the most popular XAI technique used for explaining the importance of features in a dataset. It uses a distinct approach based on game theory to recognize the importance of every feature in the dataset, and explains how well they influence the model's predictions [59]. Shapley value defines the importance of a feature to the models decision by observing its influence across all possible combinations of other features. The influence of the features present in the dataset discussed in this study is given in Fig. 6.

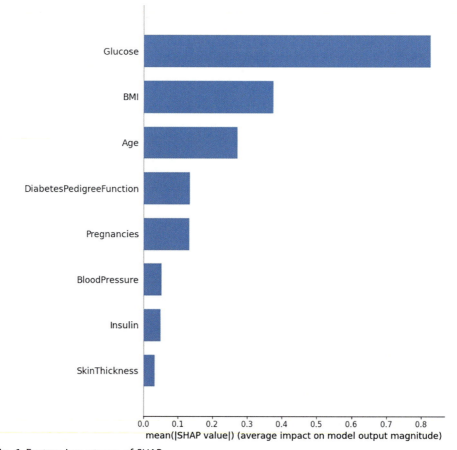

Fig. 6 Feature importance of SHAP.

From Fig. 6, we can observe that Glucose, BMI, Age are the primary features with more predictive power, whereas Pregnancies, Insulin, BloodPressure, SkinThickness, are contributing less.

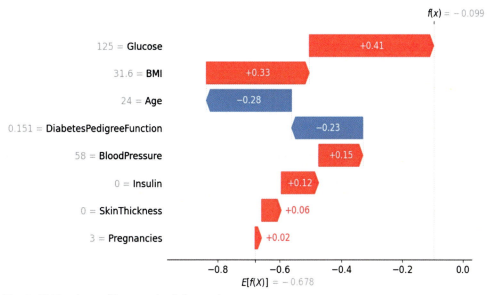

Fig. 7 SHAP values of features in diabetes dataset.

Fig. 7 displays the contribution of each feature to the final SHAP value for a specific prediction. Bars are layered on top of each other to represent the impact of each characteristic. The contributions of each feature are added or subtracted, beginning from a base value and then adding or subtracting contributions from individual features. We can able to comprehend how features affect the predictions made by models.

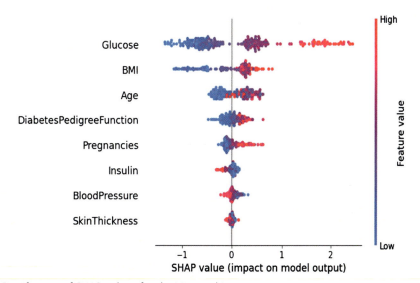

Fig. 8 Significance of SHAP values for decision making.

In Fig. 8, the features are ranked by their average SHAP values and represented vertically, horizontal values represent SHAP values. Positive values for a given feature push the model's prediction closer to the label being examined, In contrast, negative values push towards the other class in case of binary classifier.

3.3.5 Relationship between features and predicted outcome

Fig. 9 shows the dependency between glucose and pregnancies that contributes for model prediction, dependencies between age and glucose is presented in Fig. 10. Fig. 11 shows the dependencies between BMI and glucose.

According to the prediction of decision tree, Fig. 12 shows the combination of all the features with certain proportions that contribute to Class 0—which has diabetes. Fig. 13 shows for Class 1 considering all the factors contributing to Class 1—that has no diabetes.

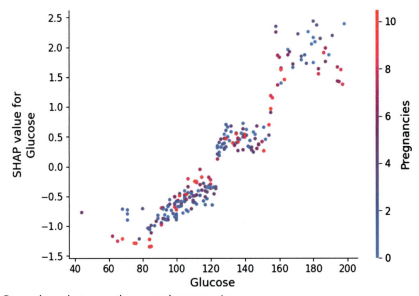

Fig. 9 Dependency between glucose and pregnancies.

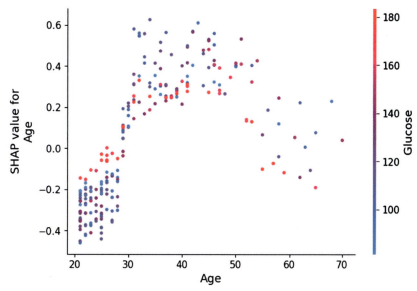

Fig. 10 Dependency between age and glucose.

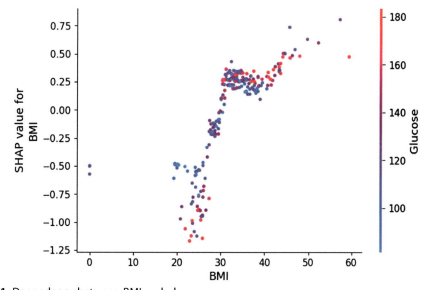

Fig. 11 Dependency between BMI and glucose.

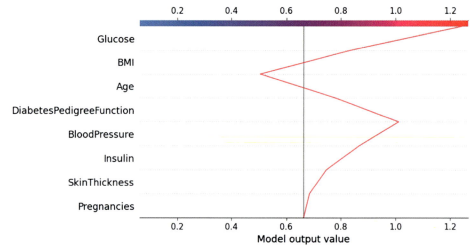

Fig. 12 Proportion of factors influencing for class—"Has Diabetes."

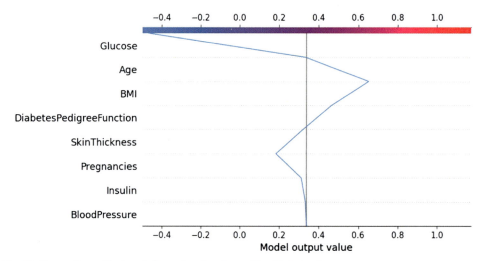

Fig. 13 Proportion of factors influencing for class—"No Diabetes."

4. Comprehending LIME and SHAP methods

Understanding the internal structure of a model is advantageous for multiple purposes, including ensuring model safety, complying with regulatory standards, troubleshooting models, and fostering confidence in the models' predictions. Both SHAP and LIME are helpful tools for explaining models. Machine learning models such as Naïve Bayes, linear regression, logistic regression, support vector machines, decision trees, random forest, gradient boosted tree, neural network may benefit from LIME and SHAP. To provide customized explanations for each prediction, LIME approximates any black box machine learning model using a local, interpretable model [60]. In XAI space, LIME can use any sort of data such as tables, photos, videos, text and provides reasoning for any supervised machine learning model. It provides local optimal explanations by calculating relevant features close to a given exceptional case. Essentially LIME is applicable to all significant machine-learning fields and can be expanded upon. XAI research for diabetes has advanced from simple ML models to complex DL models. These models use methods like LIME and SHAP to provide more meaningful and interesting information about how they make decisions.

5. Conclusion and future directions

This chapter explored the promising direction of XAI research in diabetes diagnosis. One of the XAI method called SHAP is experimented on PIMA, a very popular dataset. This experiment helps to learn various risk factors for diabetes, and other possible disease a patient may be prone. If diabetes is not identified and treated from the very beginning, it can be the underlying cause of many serious illnesses that could be fatal. There are many machine learning/deep learning algorithms are applied. However, in contrast to other healthcare applications there is still opportunity for improvement in predicting diabetes. Through the integration of different forms of data and the application of explainable AI methods, researchers can create more robust interpretable models. Finally, this may result in more accurate diagnosis, better treatment choices and better patient outcomes.

Acknowledgments

This work is derived from a research grant funded by the Research, Development, and Innovation Authority (RDIA)—Kingdom of Saudi Arabia—with grant number (13382-psu-2023-PSNU-R-3-1-EI-). The authors would like to thank Prince Sultan University, Riyadh, Saudi Arabia, for support with the article processing charges (APC) of this publication. The authors specially acknowledge the Automated Systems and Soft Computing Lab (ASSCL) at Prince Sultan University, Riyadh, Saudi Arabia.

References

[1] S.K. Bhoi, Prediction of diabetes in females of PIMA Indian heritage: a complete supervised learning approach, Turk. J. Comput. Math. Educ. (TURCOMAT) 12 (2021) 3074–3084.
[2] A. Azrar, Y. Ali, M. Awais, K. Zaheer, Data mining models comparison for diabetes prediction, Int. J. Adv. Comput. Sci. Appl. 9 (2018) 320–323.
[3] S. Larabi-Marie-Sainte, L. Aburahmah, R. Almohaini, T. Saba, Current techniques for diabetes prediction: review and case study, Appl. Sci. 9 (2019) 4604.
[4] M.F. Aslan, Y. Celik, K. Sabanci, A. Durdu, Breast cancer diagnosis by different machine learning methods using blood analysis data, Int. J. Intell. Syst. Appl. Eng. 6 (2018) 289–293.
[5] Y. Baashar, G. Alkawsi, H. Alhussian, L.F. Capretz, A. Alwadain, A.A. Alkahtani, M. Almomani, Effectiveness of artificial intelligence models for cardiovascular disease prediction: network meta-analysis, Comput. Intell. Neurosci. 2022 (2022) 5849995.

[6] T. Abuhmed, S. El-Sappagh, J.M. Alonso, Robust hybrid deep learning models for Alzheimer's progression detection, Knowl. Based Syst. 213 (2021) 106688.

[7] T. Saba, A.S. Mohamed, M. El-Affendi, J. Amin, M. Sharif, Brain tumor detection using fusion of hand crafted and deep learning features, Cogn. Syst. Res. 59 (2020) 221–230.

[8] N. Asiri, M. Hussain, F. Al Adel, N. Alzaidi, Deep learning based computer-aided diagnosis systems for diabetic retinopathy: a survey, Artif. Intell. Med. 99 (2019) 101701.

[9] M.F. Aslan, K. Sabanci, A. Durdu, M.F. Unlersen, COVID-19 diagnosis using state-of-the-art CNN architecture features and Bayesian Optimization, Comput. Biol. Med. 142 (2022) 105244.

[10] A. Koubaa, A. Ammar, M. Alahdab, et al., DeepBrain: experimental evaluation of cloud-based computation offloading and edge computing in the internet-of-drones for deep learning applications, Sensors 20 (18) (2020) 1–25, https://doi.org/10.3390/s20185240.

[11] H.A. Ibrahim, et al., A hybrid deep learning based autonomous vehicle navigation and obstacles avoidance, in: Proceedings of the International Conference on Artificial Intelligence and Computer Vision (AICV2020). AICV 2020. Advances in Intelligent Systems and Computing, vol. 1153, Springer, Cham, 2020, pp. 296–307.

[12] S.M. Fati, et al., Hybrid and deep learning approach for early diagnosis of lower gastrointestinal diseases, Sensors 22 (11) (2022) 4079, https://doi.org/10.3390/s22114079.

[13] A.S. Sayed, et al., Deep learning based kinematic modeling of 3-RRR parallel manipulator, in: Proceedings of the International Conference on Artificial Intelligence and Computer Vision (AICV2020). AICV 2020. Advances in Intelligent Systems and Computing, vol. 1153, Springer, Cham, 2020, pp. 308–321.

[14] S. Aslam, et al., Towards electric price and load forecasting using CNN-based ensembler in smart grid, Sustain. For. 13 (22) (2021) 12653, https://doi.org/10.3390/su132212653.

[15] J. Ganesan, et al., Deep learning reader for visually impaired, Electronics 11 (20) (2022) 3335, https://doi.org/10.3390/electronics11203335.

[16] N.A. Mohamed, et al., Experimental kinematic modeling of 6-DOF serial manipulator using hybrid deep learning, in: Proceedings of the International Conference on Artificial Intelligence and Computer Vision (AICV2020). AICV 2020. Advances in Intelligent Systems and Computing, vol. 1153, Springer, Cham, 2020, pp. 283–295.

[17] H.A. Elkholy, et al., Classifying upper limb activities using deep neural networks, in: Proceedings of the International Conference on Artificial Intelligence and Computer Vision (AICV2020). AICV 2020. Advances in Intelligent Systems and Computing, vol. 1153, Springer, Cham, 2020, pp. 268–282.

[18] T. Ashfaq, et al., A machine learning and blockchain based efficient fraud detection mechanism, Sensors 22 (19) (2022) 7162, https://doi.org/10.3390/s22197162.

[19] A.S.A. Aziz, et al., Genetic algorithm with different feature selection techniques for anomaly detectors generation, in: 2013 Federated Conference on Computer Science and Information Systems (FedCSIS), Kraków, Poland, September 8–11, 2013, p. 2013.

[20] H.H. Inbarani, et al., Leukemia image segmentation using a hybrid histogram-based soft covering rough K-means clustering algorithm, Electronics 9 (1) (2020) 188, https://doi.org/10.3390/electronics9010188.

[21] E. Emary, et al., Retinal blood vessel segmentation using bee colony optimisation and pattern search, in: IEEE 2014 International Joint Conference on Neural Networks (IJCNN 2014), July 6–11, Beijing International Convention Center, Beijing, China, 2014.

[22] E. Emary, et al., Retinal vessel segmentation based on possibilistic fuzzy c-means clustering optimised with cuckoo search, in: IEEE 2014 International Joint Conference on Neural Networks (IJCNN 2014), July 6–11, Beijing International Convention Center, Beijing, China, 2014.

[23] P.K.N. Banu, et al., Fuzzy firefly clustering for tumor and cancer analysis, Int. J. Model. Identif. Control (IJMIC) 27 (2) (2017) 92–103.

[24] H.I. Elshazly, et al., Ensemble classifiers for biomedical data: performance evaluation, in: IEEE 8th International Conference on Computer Engineering & Systems (ICCES), 26 Nov–28 Nov 2013, Ain Shams University, 2013, pp. 184–189, ISBN: 978-1-4799-0078-7, https://doi.org/10.1109/ICCES.2013.6707198.

[25] H.H. Inbarani, et al., Feature selection using swarm-based relative reduct technique for fetal heart rate, Neural Comput. Applicat. 25 (3–4) (2014) 793–806, https://doi.org/10.1007/s00521-014-1552-x.

[26] G. Jothi, et al., Hybrid tolerance rough set: PSO based supervised feature selection for digital mammogram images, Int. J. Fuzzy Syst. Applicat. (IJFSA) 3 (4) (2013) 15–30.

[27] A.M. Anter, et al., Automatic computer aided segmentation for liver and hepatic lesions using hybrid segmentations techniques, in: 2013 Federated Conference on Computer Science and Information Systems (FedCSIS), Kraków, Poland, September 8–11, 2013, p. 2013.

[28] F. Ajeil, et al., Autonomous navigation and obstacle avoidance of an omnidirectional mobile robot using swarm optimization and sensors deployment, Int. J. Adv. Robot. Syst. 17 (3) (2020) 1–15, https://doi.org/10.1177/1729881420929498.

[29] A.A. Najm, et al., Genetic optimization-based consensus control of multi-agent 6-DoF UAV system, Sensors 20 (12) (2020) 3576, https://doi.org/10.3390/s20123576.

[30] A.A. Al-Qassar, et al., Grey-Wolf optimization better enhances the dynamic performance of roll motion for tail-sitter VTOL aircraft guided and controlled by STSMC, J. Eng. Sci. Technol. 16 (3) (2021) 1932–1950.

[31] A.J. Humaidi, et al., Social spider optimization algorithm for tuning parameters in PD-like Interval Type-2 Fuzzy Logic Controller applied to a parallel robot, Meas. Control 54 (3–4) (2021) 303–323, https://doi.org/10.1177/0020294021997483.

[32] N. Singh, R. Kesherwani, A.K. Tiwari, D.K. Patel, A review on diabetes mellitus, Pharma Innov. (2016) 36–40.

[33] N. Chaudhary, N. Tyagi, Diabetes mellitus: an overview, Int. J. Res. Dev. Pharm. Life Sci. 7 (4) (2018) 3030–3033, https://doi.org/10.21276/ijrdpl.2278-0238.2018.7(4).3030-3033.

[34] V. Lyssenko, A. Jonsson, P. Almgren, N. Pulizzi, B. Isomaa, T. Tuomi, G. Berglund, D. Altshuler, P. Nilsson, L. Groop, Clinical risk factors, DNA variants, and the development of type 2 diabetes, N. Engl. J. Med. 359 (21) (2008) 2220–2232, https://doi.org/10.1056/nejmoa0801869.

[35] J. Li, J. Huang, L. Zheng, X. Li, Application of artificial intelligence in diabetes education and management: present status and promising prospect, Front. Public Health 8 (2020), https://doi.org/10.3389/fpubh.2020.00173.

[36] I. Tasin, T.U. Nabil, S. Islam, R. Khan, Diabetes prediction using machine learning and explainable AI techniques, Healthcare Technol. Lett. 10 (1–2) (2022) 1–10, https://doi.org/10.1049/htl2.12039. PMID: 37077883; PMCID: PMC10107388.

[37] K. Iizuka, D. Yabe, Dietary and nutritional guidelines for people with diabetes, Nutrients 15 (20) (2023) 4314, https://doi.org/10.3390/nu15204314. PMID: 37892390; PMCID: PMC10610139.

[38] U.M. Butt, S. Letchmunan, M. Ali, F. Hassan, A. Baqir, H.H.R. Sherazi, Machine learning based diabetes classification and prediction for healthcare applications, J. Healthcare Eng. (2021) 9930985, https://doi.org/10.1155/2021/9930985.

[39] Q. Zou, K. Qu, Y. Luo, D. Yin, Y. Ju, H. Tang, Predicting diabetes mellitus with machine learning techniques, Front. Genet. 9 (2018) 515.

[40] N. Sneha, T. Gangil, Analysis of diabetes mellitus for early prediction using optimal features selection, J. Big Data 6 (2019) 13, https://doi.org/10.1186/s40537-019-0175-6.

[41] A. Jakka, J. Vakula Rani, Performance evaluation of machine learning models for diabetes prediction, Int. J. Innov. Technol. Explor. Eng. (IJITEE) 8 (2019) 1976–1980.

[42] S.K. Kalagotla, S.V. Gangashetty, K. Giridhar, A novel stacking technique for prediction of diabetes, Comput. Biol. Med. 135 (2021) 104554, https://doi.org/10.1016/j.compbiomed.2021.104554. Epub 2021 Jun 8. PMID: 34139440.

[43] R. Zolfaghari, Diagnosis of diabetes in female population of PIMA Indian heritage with ensemble of bp neural network and SVM, Int. J. Comput. Eng. Manage. 15 (2012) 2230–7893.

[44] M.O. Edeh, O.I. Khalaf, C.A. Tavera, S. Tayeb, S. Ghouali, G.M. Abdulsahib, N.E. Richard-Nnabu, A. Louni, A classification algorithm-based hybrid diabetes prediction model, Front. Public Health 10 (2022) 829519.

[45] K. Kannadasan, D.R. Edla, V. Kuppili, Type 2 diabetes data classification using stacked autoencoders in deep neural networks, Clin. Epidemiol. Glob. Health 7 (2019) 530–535.

[46] S.M.H. Dadgar, M. Kaardaan, A hybrid method of feature selection and neural network with genetic algorithm to predict diabetes, Int. J. Mechatron. Electr. Comput. Technol. (IJMEC) 7 (2017) 3397–3404.

[47] A. Ashiquzzaman, A.K. Tushar, M. Islam, D. Shon, K. Im, J.-H. Park, D.-S. Lim, J. Kim, Reduction of overfitting in diabetes prediction using deep learning neural network, in: IT Convergence and Security 2017, Springer, Singapore, 2018, pp. 35–43.

[48] A. Massaro, V. Maritati, D. Giannone, D. Convertini, A. Galiano, LSTM DSS automatism and dataset optimization for diabetes prediction, Appl. Sci. 9 (2019) 3532.

[49] M. Rahman, D. Islam, R.J. Mukti, I. Saha, A deep learning approach based on convolutional LSTM for detecting diabetes, Comput. Biol. Chem. 88 (2020) 107329.

[50] S. Srivastava, L. Sharma, V. Sharma, A. Kumar, H. Darbari, Prediction of Diabetes Using Artificial Neural Network Approach, Springer, Singapore, 2019, pp. 679–687.

[51] M.F. Aslan, K. Sabanci, A novel proposal for deep learning-based diabetes prediction: converting clinical data to image data, Diagnostics 13 (2023) 796, https://doi.org/10.3390/diagnostics13040796.

[52] A. Das, P. Rad, Opportunities and challenges in explainable artificial intelligence (xai): a survey, 2020. Preprint arXiv:2006.11371.

[53] R. Calegari, G. Ciatto, J. Dellaluce, A. Omicini, Interpretable narrative explanation for ML predictors with LP: a case study for XAI, in: WOA, 2019, pp. 105–112.

[54] O. Sagi, L. Rokach, Approximating XGBoost with an interpretable decision tree, Inf. Sci. 572 (2021) 522–542, https://doi.org/10.1016/j.ins.2021.05.055.

[55] A. Guha, Building explainable and interpretable model for diabetes risk prediction, Int. J. Eng. Res. Technol. 9 (9) (2020) 1037–1042.

[56] H.B. Kibria, et al., An ensemble approach for the prediction of diabetes mellitus using a soft voting classifier with an explainable AI, Sensors 22 (19) (2022) 1–37.

[57] Y. Du, et al., An explainable machine learning-based clinical decision support system for prediction of gestational diabetes mellitus, Sci. Rep. 12 (1) (2022) 1–14.

[58] V. Vishwarupe, et al., Explainable AI and interpretable machine learning: a case study in perspective, Procedia Comput. Sci. 204 (2022) 869–876.

[59] N. El-Rashidy, et al., Utilizing fog computing and explainable deep learning techniques for gestational diabetes prediction, Neural Comput. Applic. 35 (10) (2023) 7423–7442.

[60] F. Curia, Explainable and transparency machine learning approach to predict diabetes development, Heal. Technol. 13 (5) (2023) 769–780.

[61] I. Tasin, et al., Diabetes prediction using machine learning and explainable AI techniques, Healthcare Technol. Lett. 10 (1–2) (2023) 1–10.

[62] G. Dharmarathne, et al., A novel machine learning approach for diagnosing diabetes with a self-explainable interface, Healthcare Anal. 5 (2024) 1–13.

[63] V.V. Khanna, et al., Explainable artificial intelligence-driven gestational diabetes mellitus prediction using clinical and laboratory markers, Cogent Eng. 11 (1) (2024) 1–19.

[64] G. Atteia, et al., Adaptive dynamic dipper throated optimization for feature selection in medical data, Comput. Mater. Continua 75 (1) (2023) 1883–1900.

[65] S. Khan, et al., Towards interoperable blockchains: a survey on the role of smart contracts in blockchain interoperability, IEEE Access 9 (2021) 116672–116691, https://doi.org/10.1109/ACCESS.2021.3106384.

[66] A.M. Anter, et al., Automatic liver parenchyma segmentation system from abdominal CT scans using hybrid techniques, Int. J. Biomed. Eng. Technol. 17 (2) (2015) 148–168.

[67] G.A.R. Ibraheem, et al., A novel design of a neural network-based fractional PID controller for mobile robots using hybridized fruit fly and particle swarm optimization, Complexity 2020 (2020) 1–18, https://doi.org/10.1155/2020/3067024. Article ID 3067024.

[68] H.H. Inbarani, et al., Hybrid rough-bijective soft set classification system, Neural Comput. Applicat. 29 (8) (2018) 67–78, https://doi.org/10.1007/s00521-016-2711-z.

[69] M.A. Cheema, N. Ashraf, A. Aftab, H.K. Qureshi, M. Kazim, A.T. Azar, Machine learning with blockchain for secure E-voting system, in: The First International Conference of Smart Systems and

Emerging Technologies (SMART TECH 2020), November 3–5, 2020, Riyadh, Saudi Arabia, 2020, pp. 177–182, https://doi.org/10.1109/SMART-TECH49988.2020.00050.

[70] S.S. Kumar, et al., Rough set-based meta-heuristic clustering approach for social e-learning systems, Int. J. Intell. Eng. Inform. 3 (1) (2015) 23–41.

[71] H.H. Ammar, et al., Metaheuristic optimization of fractional order incremental conductance (FO-INC) maximum power point tracking (MPPT), Complexity 2019 (2019) 1–13, https://doi.org/10.1155/2019/7687891. Article ID 7687891.

[72] A.T. Azar, et al., Expert system based on neural-fuzzy rules for thyroid diseases diagnosis, in: International Conference on Bio-Science and Bio-Technology (BSBT 2012), December 16–19, 2012, Korea. Vol. 353 of the Communications in Computer and Information Science Series, Springer, 2012, pp. 94–105, ISBN: 978-3-642-35520-2, https://doi.org/10.1007/978-3-642-35521-9_13.

[73] T.S. Gorripotu, et al., TLBO algorithm optimized fractional-order PID controller for AGC of interconnected power system, in: Soft Computing in Data Analytics. Advances in Intelligent Systems and Computing, vol. 758, Springer, Singapore, 2019.

[74] Pima Indians Diabetes Database, Kaggle, 2016. Available at: https://www.kaggle.com/datasets/uciml/pima-indians-diabetes-database.

[75] J.A. Robert, S. Hart, Handbook of Game Theory With Economic Applications, 2, Elsevier Science, 1992, pp. 1–818.

[76] C. Rodriguez-León, et al., Mobile and wearable technology for the monitoring of diabetes-related parameters: systematic review, JMIR mHealth uHealth 9 (6) (2021).

CHAPTER 16

Explainable AI for melanoma diagnosis through dermosopic images: Recent findings and future directions

Khadija Safdar[a], Shahzad Akbar[a], Usama Shahzore[a], Sajid Iqbal[b], Noor Ayesha[c], and Sahar Gull[a]

[a]Riphah College of Computing, Riphah International University, Faisalabad, Pakistan
[b]Department of Information Systems, College of Computer Sciences and Information Technology, King Faisal University, Al Ahsa, Saudi Arabia
[c]School of Clinical Medicine, Zhengzhou University, Zhengzhou, Henan, China

1. Introduction

Melanoma is among the extreme lethal types of skin cancer. Normally, the cells of skin develop in an orderly and controlled manner, healthy new cells develop and old cells eventually die. Any abnormality in melanocytes (melanin producing cell) or DNA damage causes melanoma to occur. The known causes of melanoma are greater exposure to UV (ultra-violet) radiations, tanning beds or lamps etc. Melanoma can occur anywhere on the body, especially in the areas more exposed to sun for instance face, arms, legs, back. It can also develop on areas which are less likely to receive sun light such as palms, fingernail bed, soles of feet. The hidden melanoma is more likely to develop in darker skin people. The initial indicator of melanoma include any change in existing moles, occurrence of unusual looking growth or any new pigment [1].

Melanoma may not always necessarily begin like a mole, it may appear to be a normal skin mole as well. Generally, normal moles have a stable color such as brown, black or tan. They have a visible borderline which separates the mole from its surrounding skin. Usually they are round or oval shaped with diameter smaller than 6 mm or ¼ inches. Contrary to this, unusual moles which may indicate melanoma or any other skin cancer have ABCDE characteristics. ABCDE characteristics represent asymmetrical shape, irregular borders, color changes, diameter growth and evolution over time such as bleeding or itching. Malignant melanoma may vary greatly in their appearance, some may or may not depict all of the above stated characteristics. Atypical nevi indicates the borderline cases including the lesions which are not melanoma, however, they are visually alike to melanoma. The similarity between normal moles and malignant melanoma makes their visual inspection very difficult and risky. The risk quotient that may increase the

Explainable AI in Healthcare Imaging for Medical Diagnoses
https://doi.org/10.1016/B978-0-443-23979-3.00016-6

likelihood of melanoma involve fair skin (less melanin/pigment), light colored eyes, red or blond hair, severe sunburn history, solar keratoses due to excessive UV exposure, weak immune system, greater number of moles (more than 50) or unusually larger, irregular and mixed color moles termed as "Dysplastic nevi" [2].

Malignant melanoma accounts for only 1% (approx.) of all the skin cancers but it is the leading cause of deaths due to skin cancer. According to statistics, the cases of melanoma have significantly increased since early 1970s, an average of 4% annually in United States. Fig. 1 depicts the melanoma affected human skin [3].

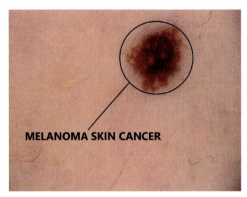

Fig. 1 Melanoma affected dermoscopic skin.

Statistics show that melanoma is 20 times more common in Whites as compared to African Americans. The life time risk of melanoma occurrence is 0.1% for Blacks, 0.6% for Hispanics and 2.6% for White people. Overall, melanoma is more likely to arise in men over 50, and higher in women below 50 years of age. Melanoma risk increases with age and on an average, melanoma is diagnosed at an age of 65. Melanoma is on the rise and according to American Cancer Society, the estimate of melanoma occurrence in United States for the year 2021 will be 106,110 new melanoma cases (about 43,850 in women and 62,260 in men) and 7180 expected fatalities from melanoma (about 2580 women and 4600 men) [4].

The massive progress in automated clinical support systems help the doctors in reliable disease detection and treatment plan of the patients [5,6]. Medical imaging is a practice to acquire images of various bodily structures and organs for medical use: to accurately study and diagnose a particular disease [7,8]. These procedures are rapidly developing due to frequent advancement in methods of image processing such as image recognition, image analysis and image enhancement [9]. The medical images can be expertly processed, accurately measured and can be easily and securely shared via appropriate communication protocols like PACS (Picture Achieving & Communication System) and DICOM (Digital imaging & communication in medicine) [10].

In digital images, a pixel is a luminous point of an image which is captured via electronic sensors. Digital images are produced by the ordered arrangements of pixels with varied intensities of colors. Each and every pixel depicts one intensity of green, red and blue (RGB system). The combination of RGB system resultantly forms one color of light spectrum, varying from white (as maximum intensity of RGB) and black (as absence of colors). All registered pixels are codified on a photographic image and the codification permits the researchers to do measurements like distance calculations, color intensities, areas and pattern recognition in the images. Resolution of a digital photograph is its overall number of pixels, higher the number of overall pixels, greater is the image resolution. Therefore, an elementary structure which forms a digital image is a pixel [11].

For appropriate digital dermatological photography, several crucial factors must be taken into consideration such as crystal lenses which outperform acrylic lenses. Use of macrophotography by implementing large sized amplification (75–110 mm) macro objectives in clinical practice are adequate to use. Compact cameras allow focus distance smaller than 50–80 cm, which is mostly used in dermatological photos. Flash can be used to enhance the Flash Photographic quality in the field of dermatology. Many compact cameras with built-in flash provide an option of quality control and auxiliary flash that provides better flexibility in representing lesions. Additionally, several other key factors which make dermatological photography reliable include proper approximation to the images using optic zoom (30–40%) in-order to smooth the luminosity of incorporated flashes and to reduce the deformities of the prominent areas. Higher resolution (approximately 3-Mpx) images resultantly lead to processing over-head and may slow the presentation process. The morphologic features of skin lesions are very directly related to their accurate diagnosis, therefore, dermatology is regarded as a specialty with a notable visual component that has led to the progress of significant number of iconographic representation procedures. Fig. 2 represents melanoma dermoscopic images gathered from Melanoma Skin Cancer Dataset of 10000 Images [12].

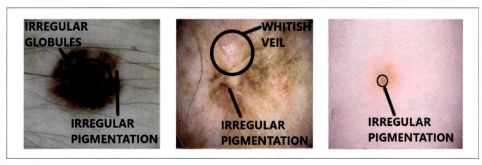

Fig. 2 Melanoma dermoscopic images [12].

Melanoma images have irregular pigmentation, whitish veil, irregular streaks and multiple holes/globules. Asymmetrical shaped lesions can also be seen.

Computer aided diagnosis system (CAD) can be termed as a cutting edge intelligence and expert system in the Medical and Computer science domain. CAD based frameworks use several diagnostic rules in-order to imitate the approaches used by a skilled man in his diagnostic purpose. Many latest CAD techniques possess the ability to analyze the clinical data and later deduce new and useful piece of knowledge [13]. The new knowledge improvises the present diagnostic rules through training and facilitate the expert system to escalate their performance accuracy in the course of time. CAD system adopts artificial intelligence, data mining, machine and deep learning based approaches for analysis of structured and unstructured clinical data. The generic expert system has a user who interacts with the knowledge base of system through the Inference engine. A knowledge engineer builds a knowledge base which consists of a set of diagnostic rules and facts. The Inference engine applies a backward or forward chaining technique on knowledge base to make a decision or give recommendations. The existing melanoma diagnosis CAD systems are capable of performing efficient image processing, lesion segmentation, useful feature extraction and lesion classification. A general pipeline procedure of melanoma detection using skin lesion images is demonstrated in Fig. 3 [14].

1.1 Image analysis techniques of skin lesion

Medical Image Processing comprises of five major steps including image formation, visualization, image analysis, image management and enhancement. Image analysis is one of the key steps of Medical Image processing. It comprises of various pre-processing and segmentation techniques. Accurate segmentation of ROI can be achieved using advanced deep learning models stated below.

1.1.1 Pre-processing techniques

The pre-processing technique prepares images for accurate detection of features and provides enhanced processing. It includes image acquisition as input, conversion to gray scale image, filtering noise and generation of binary image. Binarization is a process of transforming gray scale image to black and white that reduces the information within an image from 256 different shades of gray to black and white binary image [15]. Contrast adjustment, intensity adjustment, histogram equalization are the methods included in pre-processing of lesion images. Contrast adjustment improves the visibility of an image via histogram extension [16]. Intensity adjustment provides a high display quality by increasing the values of image intensity. Histogram equalization is carried out to provide an enhancement in global contrast of lesion images.

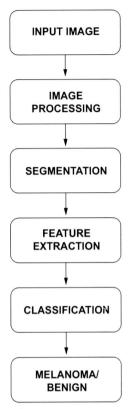

Fig. 3 Pipeline approach of melanoma diagnosis.

1.1.2 Segmentation techniques

Image segmentation is a prime step in skin lesion image analysis process. This process aims to highlight the region of interest (ROI) by separating diseased or lesion area from healthier skin area. Segmentation techniques can be applied in different ways including Handcrafted-features based method, Region and Edge based methods and supervised segmentation based on intelligence [17].

• Conventional intelligence based approaches

The conventional system is based on Artificial Intelligence (AI) which performs image analysis via its ability to learn, reason and perceive by using large image repositories/datasets. Supervised segmentation approach includes deep-learning based Artificial Neural Network (ANN). Note that both un-supervised and supervised approaches are used for skin lesion segmentation [18].

• Deep learning based segmentation approaches

Deep learning-based network is a large family of classifiers comprising of hierarchical probabilistic model, neural network, various unsupervised and supervised feature learning techniques. The segmentation task in image analysis process is very complex and requires great care. Convolutional Neural Network (CNN), U-Net, Fully Convolutional Network (FCN) are some of the best image segmentation approaches implemented by researchers. Deep learning models allow several processing layers that learn and present data with multiple abstraction stages. Hierarchal feature representation (pixel-high level) semantic features are performed in deep learning models. It captures intricate structure of given images very precisely as compared to conventional intelligence systems. Training of deep learning models is more complex and has a cost over-head. It also requires larger storage memory and more time to train and process smaller training datasets [19].

1.2 Challenges in skin lesion detection

A large variation exists in skin lesion images which make the detection of melanoma very complicated. Presence of noise and artifacts, multi-shape and multi-size lesions, fuzzy borders, low color illumination and low-contrast lesion images offer great difficulty and require efficient pre-processing techniques for precise analysis.

Artifacts and noise: The compromising signals which may affect the image interpretation via manual or automated means. Blood vessels, hair, scales or bubbles are the artifacts which serve as a noise in images.

Varied size and shape: The lesion size, location and shape varies from person to person. Advance pre-processing techniques are required for successful and accurate image examination.

Lower contrast and color illumination: Image segmentation is the most crucial phase of melanoma detection. A lower contrast between the two (normal skin and lesion) leads to poor segmentation. Also, the illumination of dermoscopic or non-dermoscopic images vary due to light rays, reflections or color texture and causes multi-resolution in images.

Fuzzy boundaries: Irregular borders lead to faulty lesion contour and boundary localization. The prediction of symmetric or asymmetric type of skin lesion is troublesome due to the presence of irregular and fuzzy borders [20]. Challenges in skin lesion identification are depicted in Fig. 4.

1.3 Publically available datasets

The automated CAD system for correct melanoma diagnosis need a huge data repository. To counter this issue, there are various publically available datasets present in-order to conduct a reliable diagnosis of skin cancer. The publically accessible datasets

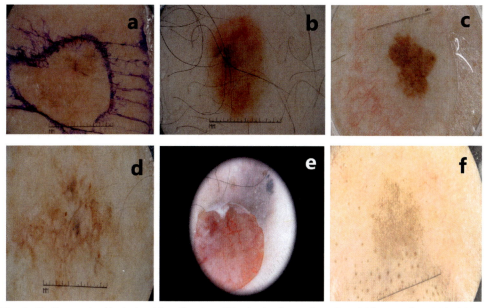

Fig. 4 Challenges in skin lesion identification: (A) ruler/scale mark artifact, (B) hair artifact, (C) bubbles, (D) fuzzy boundaries, (E) frame, (F) low contrast [20].

comprise of Med-Node, DermIS, PH^2, DermQuest, Dermnet, Interactive Dermoscopy Atlas, ISBI challenge 2016, 2017 and onwards. Others include image datasets gathered from internet sources. A brief description of some of the publically available datasets is stated below.

1.3.1 PH^2

This dataset has been established as a fine standard for the purpose of research and bench-marking, which ultimately facilitate the classification and segmentation of dermoscopic images. It is a database of dermoscopic images gathered at the Dermatology Service of Pedro Hispano Hospital, Matosinhos, Portugal [21]. This image dataset has a collection of 200 dermoscopic melanocytic lesion images, comprising 80 atypical-nevi, 40 melanoma, and 80 common-nevi. PH^2 dataset also comprises of various medical annotations of dermoscopic images specifically medical lesion segmentation, histological and clinical diagnosis and assessment of various dermoscopic criteria (blue-whitish veil, colors, streaks, dots/globules, regression areas, pigment network).

1.3.2 DermIS

It is known as the wide-ranging service of dermatology information accessible online. It comprises of 397 dermoscopic images. It provides elaborated image atlases (PeDOIA and DOIA) along with diagnoses, differential diagnoses, variety of case reports and other useful guidelines on a number of skin diseases [22].

1.3.3 Dermnet

Dermnet is an atlas of skin disease with a website support which contains around 23,000 skin images separated into 23 different classes. It mostly contains photographic images captured with smart phones [23]. The image size and quality of Dermnet data set varies widely. The 23 sub-classes of skin diseases are taxonomically categorized into 642-sub-classes. Although, the dataset contains some empty, duplicate and irrelevant sub classes as well.

1.3.4 Med-Node

It is a dermatology database that is available online. It is used in Med-Node systems for detection of skin cancer using macroscopic image dataset. It contains 170 images (70 melanoma and 100 nevus) from digital image catalog of UMGC (Dermatology Department) [24].

1.3.5 ISIC

The ISIC (The International Skin Imaging Collaboration) catalog comprises of the world's largest collection of non-polarized, quality controlled, publicly available dermoscopic images of the skin lesions [25]. Their current database comprises of more than 20,000 dermoscopic images and new images are added to the ISIC archive in an on-going fashion [26]. DICOM is the format of images present in this dataset. The ISIC archive can be easily accessed via commonly available library such as Pydicom. It comprises of both metadata and images. TFRecord and JPEG format is also supported via TFRecord and JPEG directories. TFRecord format images are uniformly resized to a 1024×1024. Metadata is also available in CSV file format. Each image has a binary target (0.0 or 1.0) against it. 0 represents benign and 1 represents malignant [27].

1.3.6 HAM1000

This dataset comprises of 10,015 dermoscopy images which are provided as a purpose of training data set for the academic machine learning approaches in the field of dermoscopy [28]. This dataset is publicly accessible through ISIC archive. In this dataset, more than 50% of the lesions have been verified by pathology. Below Table 1 presents an over-view of several benchmark databases. These datasets have been excessively adopted by researchers in their melanoma detection study.

Table 1 Benchmark skin lesion datasets in the domain of automated melanoma diagnosis.

Dataset	Total dermoscopic images	Melanoma images
PH^2 [21]	200	40
DermIS [22]	397	146
Dermnet [23]	~23,000	517
Med–Node [24]	170	100
ISIC [26]	~23,906	21,659
HAM1000 [28]	10,015	~1101

1.4 Performance metrics

The performance analysis of both machine and deep learning approaches for melanoma detection is based on the following metrics: sensitivity, specificity, precision, accuracy, positive predictive value, negative predictive value, Dice coefficient, ROC (receiver operating characteristic curve) AUC (area under ROC curve) and Jaccard index [29].

Sensitivity: Defined as the ability of a classifier to accurately detect the diseased (melanoma) state. True positive rate.

$$SE = \frac{TP}{TP + FN}$$

Specificity: Defined as the ability of a classifier to accurately detect the benign (normal, not-melanoma) case termed as true negative rate.

$$SP = \frac{TN}{TN + FP}$$

Accuracy: It is the number of accurate predictions divided by total number of predictions. It is the ratio of truly diagnosed lesion images to all cases.

$$ACC = \frac{TP + TN}{TP + FP + FN + TN}$$

Precision: It is the fraction of all relevant instances among the retrieved instances. It can be said as equal to the positive predictive values.

$$PREC = \frac{TP}{TP + FP}$$

PPV (Positive Predictive Value): It is a likelihood that whether a subject with the positive test, actually has the disease (melanoma).

$$PPV = \frac{TP}{TP + FP}$$

NPV (Negative Predictive Value): It is a likelihood that whether a subject with a negative screening test, truly does not have a disease (melanoma).

$$NPV = \frac{TN}{TN + FN}$$

ROC AUC: It is equivalent to the probability that a classifier will rank any randomly selected positive case higher than any randomly selected negative case. It is defined as a graph between false positive rates vs. true positive rates.

Dice Coefficient: It is an overlap index that analyze the overlap measure between a ground truth and automatic segmentation of skin lesion.

$$DC = \frac{2.TP}{2.TP + FN + FP}$$

Jaccard Index: It is a measure of the same aspects as Dice coefficient.

$$JA = \frac{TP}{TP + FN + FP}$$

This review paper is divided into four sections. Section 1 gives a comprehensive introduction of malignant melanoma. It also informs about latest Image analysis techniques used in CAD systems. This section also provides information about publically available benchmark skin lesion datasets. The performance metrics to analyze the performance of melanoma detection system is stated above. Section 2 provides a detailed over-view of existing machine and deep learning based melanoma detection approaches. A comparative study is done in order to highlight the best performing models. Section 3 provides a critical discussion on the existing melanoma diagnosis frameworks. Finally, the last section concludes the overall performance of CAD approaches in melanoma detection and highlights deep learning based models like Inception V3, ResNet and their ensemble as one of the best performing models. Future directions intends to provide an idea of a robust and ultra-modern melanoma detection system in near future.

2. Literature review

The exponential increase in melanoma cases has largely promoted the need and adoption of CAD based practices for its effective diagnosis and identification.

2.1 Melanoma detection through conventional machine learning

Machine learning, an application of Artificial Intelligence, aspires to provide machines with expert-level abilities of collecting data through cognition, similar to human abilities and senses, later processing the collected data using various machine learning techniques and computational intelligence in-order to make decisions and conduct predictions. Four general methods of machine learning exist which are termed as supervised, semi-supervised, un-supervised and reinforcement learning methods [30]. A computer program is expected to learn from (E) experience according to the class of (T) tasks and (P) performance measure, if the performance at tasks (T), as measured by the performance (P), is improved with experience.

The information processors in ML perform tasks like assembling, sorting, assimilating and classifying the information [31]. Supervised learning is defined as the approach in which the machine infers or performs reasoning on a function using a set of training data. Two categories namely Classification and Regression lies under supervised machine learning approach. Decision trees, Naïve Bayes, Random Forest are few examples based on supervised learning approach. Unsupervised learning is defined as an approach in which a machine aims to discover hidden structures in an unlabeled data set. Clustering algorithms, Apriori algorithm and Principal Component Analysis (PCA) are few un-supervised learning algorithms. In semi-supervised learning, the data is a blend of unclassified and classified datasets. It is targeted to perform well in classifying the classes as compared to supervised learning. Reinforcement learning is based on agents (intelligent programs) which make observations that are collected from the environment through continuous interaction in order to make decisions or take actions which minimize the risk factors and maximize the rewards. Various advance applications of ML are centered upon NLP (Natural languages processing), computer vision, cognitive computing, pattern recognition, medical image diagnosis, disease detection, outbreak prediction, defense and security, text mining etc. The basic architecture of automated melanoma detection using dermoscopic image is set forth in Fig. 5 [32].

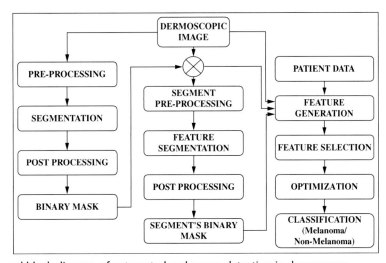

Fig. 5 General block diagram of automated melanoma detection in dermoscopy.

The general overview of automated melanoma detection architecture using dermoscopic images is illustrated. The key steps of the block diagram include lesion segmentation, feature extraction and lesion classification.

Li et al. [33] proposed an automated melanoma diagnosis approach (desktop application) based on a spectroscopic system. Three classifiers were adopted in this system namely ANN, K-nearest neighbor, Naïve Bayes. Noise reduction was performed using Median Filter. Min-max normalization was performed to normalize pixel intensities. The pixel intensities varied for malignant melanoma (0–1) and benign skin (0.2–0.4) using V scan and P scan. After the feature extraction phase, classification was performed and Naïve Bayes outperformed with an accuracy of 89%, sensitivity 89% and specificity 89% on a dataset of 187 images gathered from a clinic using a spectroscopic device.

Jafari et al. [34] proposed an automated Melanoma diagnosis system based on Support Vector Machine (SVM) classifier for the early diagnosis of lesion as benign (nevus) and cancerous (melanoma). In the pre-processing phase, the noise of the input image was reduced by "Guided filters" and high gradients in HSV color space were smoothed to reduce the illumination effects. An initial segmentation in HSV color space was performed by K-means. In the post-processing phase, the Dermatological Analysis and classification based on ABCD rule of dermatology were performed using SVM classifier based on SMO (sequential minimal optimization) and Linear kernel functions. The proposed method showed an accuracy of 0.79, sensitivity 0.90, specificity 0.72, NPV (Negative predicted value) 0.91 and PPV (positive predicted value) 0.70 on a data set of 170 non-dermoscopic images (100 nevus, 70 melanoma) from digital image archive of UMGC. Moussa et al. [35] proposed a computerized Melanoma detection technique based on well-known K-Nearest Neighbor (KNN), applied to geometric extracted features. In the initial phase, mole segmentation was performed using thresholding. In second phase, feature extraction was performed and three ABD features (Asymmetric index, Border irregularity, Diameter) were extracted. For border irregularity, two indices namely Circularity Index (CRC) and Adjusted Irregularity indexes (AIrA) were extracted from segmented lesion images. K-Nearest Neighbor classified the data as 0 (non-cancerous) and 1 (melanoma) with an accuracy of 89% on 15 standard digital images of DermIS and DermQuest datasets.

Waheed et al. [36] put forward a method based on SVM classifier to efficiently identify melanoma lesion using Dermoscopic images. Feature extraction was conducted to extract the color and texture properties. HSV color space was used to extract nine distinct statistical features. Texture analysis using Grey Level Co-occurrence Matrix (GLCM) which catered dark and light skin lesion of varied intensities was performed. In post-processing phase, classification was performed on 13-D features vector using SVM classifier. The results of this proposed system were evaluated using WEKA tool having sensitivity 96%, specificity 83%, accuracy 95%, AUC 0.8906 and using MATLAB having specificity 84%, sensitivity 97%, accuracy 96% and AUC 0.90 on 200 dermoscopic images of PH^2 data set.

Munia et al. [37] proposed an automated Melanoma diagnostic system in-order to predict mole as malignant or benign based on SVM supported by Linear Kernel. In the first stage, a union of K-means clustering and Otsu thresholding technique was applied for segmentation. Feature extraction was performed based on ABCD rule of Dermatology and Non-Linear features (Sample entropy, Hurst component and approximate entropy). The experimental results displayed combined segmentation accuracy of 89.07%, sensitivity 87.14%, specificity 91%, AUC 0.91on the dataset obtained from the archive of Dermatology Department of UMC, Groningen. Ozkan and Koklu [38] proposed an automated melanoma diagnostic approach based on ANN, K-NN, SVM and Decision Tree which classified skin lesion into abnormal, normal and melanoma category. Feature extraction was done based on ACD rules of dermatology (Asymmetry, color, dotted beaded structures). Ten layer cross validation was applied to the dataset for performance analysis. MATLAB statistics, MATLAB NN Tool box and ML Tool Box functions were applied. The performance rates of various classifiers were compared and ANN proved to be most efficient based on 92.50% accuracy, 90.86% sensitivity, 93.49% balance accuracy, 96.11% specificity, 92.38% precision, and 90.45% F1-score on PH^2 dataset containing 200 dermoscopic images.

Roslin [39] proposed a computer aided Melanoma classification method based on multiple supervised machine learning techniques. The pre-processing phase included image enhancement using un-sharp masking and anisotropic diffusion. K-means clustering algorithm was applied for melanoma segmentation. Feature extraction was performed based on texture features and intensity of the segmented image. Five different classifiers KNN, SVM, Decision tree, Multi-layer perceptron, Random Forest were trained on extracted features and 10-fold cross validation was applied to improve classifier's performance. Random forest out-performed other classifiers with an accuracy of 93% on a dataset of 900 images. Hameed et al. [40] proposed a futuristic CAD tool for the diagnosis of multiple types of skin lesion. In the pre-processing stage, Dull Razor algorithm was applied for effective hair removal. Feature extraction from the ROI was performed using Otsu's thresholding technique. Texture features were extracted using GLCM and NGTDM. The classification phase was based on two strategies, 3-label classification i.e. classify into three classes given as healthy, non-inflammatory disease, inflammatory disease and 6-label classification i.e. classify into six classes (acne, healthy, eczema, psoriasis, malignant and benign). Multiple classifiers (K-NN, SVM, Ensemble classifier, Decision tree) were applied in both strategies and non-linear SVM (Quadratic) out-performed other classifiers with 92.3% accuracy in three label classification and 83% classification in six label classification on a dataset of 1800 images collected from various sources.

Ramezani et al. [41] presented a novel technique to recognize malignant melanoma from pigmented, benign skin lesion using macroscopic images with high spatial resolution. The effect of noise and non-uniform illumination was reduced by using median filter in the pre-processing phase. Single channel image was applied for the segmentation of healthy skin and lesion. Bottom hat morphological transformations were applied to remove thick hair effect. Five descriptors including ABCD criteria and Texture feature were applied in the feature extraction phase. The features were extracted using unsupervised linear feature extraction method "PCA." SVM with a radial basis kernel function was applied for classification which achieved an accuracy 82.2%, specificity 86.93% and sensitivity 77% on the dataset obtained from DermIS, DermQuest, and Dermnet atlases. Chakkaravarthy and Chandrasekar [42] proposed a CAD framework for the diagnosis of malignant melanoma using Fuzzy C means technique. In the pre-processing phase, Gaussian filter was used to remove noise and unwanted pixel from dermoscopic images and RGB image was transformed to Grey scale image. Segmentation was performed using Fuzzy C means. In the post-processing phase, new cluster points were detected for each of the segmented image. Performance measure of the segmented image was performed to acquire the correct detection of lesion with an accuracy of 96.05% on DermIS dataset, 99.21% on DermQuest dataset and 96.48% on ISIC (2016) dataset. Vocaturo et al. [43] proposed an application of Multiple Instance-learning for the classification of melanoma by dysplastic nevi and dysplastic nevi by common nevi. For both tasks of classification, 5-fold and 10-fold cross validation were performed. This MIL optimization model results were compared to that of linear SVM and SVM (RBF) approaches. Results indicated that MIL overcame both SVM approaches with respect to sensitivity (92.56%) and correctness (87.50%) in classification of melanoma vs. dysplastic nevi. The CPU time and F-score (88.31%) of MIL also performed well as compared to classic SVM. However, all three classifiers showed unsatisfactory results in classifying dysplastic nevi and common nevi. The classification was performed using PH^2 dataset and two online databases namely Dermins and DermQuest. A comparative analysis of ML approach is given in Table 2.

Table 2 Comparative analysis of machine learning approaches in automated melanoma diagnosis.

Sr. no.	Author/s	Year	Techniques/model/approach	Dataset	Result
1	Li et al. [33]	2014	ANN K-nearest neighbor Naïve Bayes Min Max normalization	Local data set (187 images)	**Naïve Bayes outperformed** Accuracy 89% Sensitivity 89% Specificity 89%
2	Jafari et al. [34]	2016	SVM classifier K-means → initial segmentation Moore–Neighbor trace algorithm	Digital image archive of UMGC 170 non-dermoscopic images	Accuracy 79% Sensitivity 90% Specificity 72% NPV (negative predicted value) 91% PPV (positive predicted value) 70%
3	Moussa et al. [35]	2016	K-Nearest Neighbor Classifier Feature extraction → ABD features Circularity Index & AIrA → border irregularity	DermIS & DermQuest dataset (15 standard digital images)	Accuracy 89%
4	Waheed et al. [36]	2017	SVM classifier GLCM → Texture analysis WEKA tool & MATLAB for evaluation	PH^2 (200 dermoscopic images)	WEKA TOOL Sensitivity 96% Accuracy 95% Specificity 83% AUC 0.8906 MATLAB Sensitivity 97% Specificity 84% Accuracy 96% AUC 0.90
5	Munia et al. [37]	2017	Support Vector Machine Segmentation → K-means clustering & Otsu thresholding Guided Filter and Morphological operations 5-Fold cross-validation	Archive of Dermatology Department of UMC, Groningen	Accuracy 89.07% Sensitivity 87.14% Specificity 91% AUC 0.91
6	Ozkan and Koklu [38]	2017	ANN, K-NN, SVM and Decision Tree Feature Extraction → ACD rules 10 Layers cross validation MATLAB Statistics, MATLAB NN Tool box and ML Tool Box	PH^2 dataset (200 dermoscopic images)	**ANN outperformed** Accuracy 92.50% Sensitivity 90.86% Balance–accuracy 93.49% Specificity 96.11% Precision 92.38% F1–score 90.45%

Continued

Table 2 Comparative analysis of machine learning approaches in automated melanoma diagnosis—cont'd

Sr. no.	Author/s	Year	Techniques/model/approach	Dataset	Result
7	Roslin [39]	2018	Segmentation → K-means clustering Classification → KNN, SVM, Decision tree, Multi-layer perceptron, Random Forest 10-Fold cross validation	Dataset of 900 images	**Random Forest outperformed** Accuracy 93%
8	Hameed et al. [40]	2018	K-NN, SVM, Ensemble classifier, Decision tree Smoothing → Gaussian Filter, Dull Razor algorithm Texture feature extraction → GLCM and NGTDM	1800 images (various sources)	**SVM (Quadratic) out-performed** Accuracy 92.3% (3-label classification) 83% (6-label classification)
9	Ramezani et al. [41]	2019	SVM (radial basis kernel) Feature Extraction → ABCD Criteria and Texture feature → PCA	DermIS, DermQuest, Dermnet atlases	Accuracy 82.2% Specificity 86.93% Sensitivity 77%
10	Chakkaravarthy and Chandrasekar [42]	2019	Segmentation and Classification → Fuzzy C means technique Gaussian filter	ISIC (2016) dataset DermIS & DermQuest Dataset	Accuracy DermIS dataset (96.05%) DermQuest dataset (99.21%) ISIC (2016) dataset (96.48%)
11	Vocaturo et al. [43]	2020	Classification → Multiple Instance-learning approach 5-fold and 10-fold cross validation	PH^2 DermIS, DermQuest dataset	Classification of melanoma vs. dysplastic nevi Sensitivity (92.56%) Correctness (87.50%) F-score (88.31%) Classification of dysplastic nevi and common nevi Unsatisfactory result

2.2 Melanoma detection through deep learning techniques

Deep learning is an advancement of Machine learning which is based on ANN that aims to mimic the multi-layered cognition system of a human. Deep learning is rapidly excelling and gaining immense acceptance in analyzing and processing big data especially in health-care sector. In the practice of Radiology, structural abnormalities are detected and classified into respective disease categories. Since 1980s, several ML techniques with different mathematical bases and implementations have performed these classification tasks very efficiently. Radiology imaging (X-ray, MRI, and CT), pathology imaging, genomic sequences have been a great source of gathering large-scale beneficial data for the training and testing of ML algorithms. In early 2000s, CAD system generated more false positives as compared to human experts which resultantly led to additional biopsies and greater assessment time and cost. Eventually, advanced deep learning-based technologies were developed to overcome ML limitations in previous CAD system. The new DL based artificial intelligence approaches have a greater potential to conduct automated lesion detection, recommend differential diagnosis, compose data-rich radiology reports etc. IBM, the largest globally integrated company, is developing one of the largest "Dr. Watson" radiology applications. Several open-source DL libraries like Microsoft Cognitive ToolKit (CNTK), Caffe, Theano, Torch, Tensor flow are currently available [44].

Convolutional Neural Net (CNN) is one of the most extensively used deep learning based neural network. It comprises of multiple layers of neural connections. CNN architecture has pooling and fully connected convolutional layers. Convolutional layer detects distinctive local edges (motif-like), lines and several visual elements. Convolution are learned which are the parameters of the specialized filter operator. After learning the meaningful kernel, the operation mimics to extract several visual features, for instance colors, edges, similar as that of visual cortex.

2.2.1 Significance of deep learning in efficient melanoma detection

Deep learning models have the following key benefits over the conventional machine learning approaches which make them more reliable, user-friendly and robust in automated melanoma diagnosis.

- Deep learning networks omit the need of complex image pre-processing, feature extraction and lesion segmentation processes.
- Robust end-to-end training of deep networks is carried out using raw pixels and image labels, using both dermoscopic and clinical lesion images.
- A number of publically available labeled image datasets now aid researchers to counter the problem of restricted amount of annotated data present in the domain of medical imaging.
- The deep neural networks are now capable of having multiple hidden layers due to the use of fast Graphical Processing Units (GPUs).

- CNN has the ability to deal with limited amount of annotated data by employing several pre-trained models and data augmentation methods.
- The automated identification of melanoma has become troublesome due to the presence of unwanted artifacts such as hair, ruler marks, fuzzy boundaries and lower contrast. The higher inter-class and intra-class variations also serve as a hindrance in melanoma diagnosis. This issue is effectively encountered by deep learning models which carry out automated deep feature learning and produce useful feature maps [45].

A simple architecture of deep learning is presented below in Fig. 6.

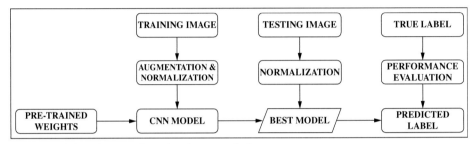

Fig. 6 General deep learning based architecture [45].

Several studies have been conducted in automated melanoma detection based on advanced deep learning approaches. Sheha et al. [46] presented an automated diagnosis system to classify malignant melanoma and Melanocytic Navi using Multi-layer perceptron (MLP) classifier. Feature extraction was carried out using Gray-level Co-occurrence matrix (GLCM) based on texture analysis. Twelve high score features were picked in the feature selection phase to improve the detection accuracy using the Fisher score ranking technique. Two approaches were used for the MLP classifier called as Automatic MLP and Traditional MLP. Training, validation and testing were performed in both approaches in which Traditional MLP outperformed with 92% and 100% of testing and training accuracy on dermoscopy image dataset.

Codella et al. [47] proposed a melanoma recognition system based on two tasks: Classification of images into melanoma or atypical and benign group, and melanoma or atypical group. Four modeling approaches called Ensemble, Caffe CNN, Sparse coding and Fusion of all models were applied in each task. Deep learning techniques used two parallel approaches i.e. CNN features were learned from natural photographs and unsupervised learning of features was carried out using dermoscopy images through sparse coding. Non-Linear SVM based on sigmoid feature normalization and histogram intersection kernels was trained for each approach and models were then combined for score averaging. The Fusion approach outperformed with a sensitivity of 94.9% and 73.8%, accuracy of 93.1% and 73.9%, specificity 92.8% and 74.3% in both tasks (Melanoma vs. Benign vs. Atypical and Atypical vs. Melanoma) respectively on ISIC dataset.

Premaladha and Ravichandran [48] proposed a futuristic CAD approach based on Hybrid-Ada Boost and DLNN algorithms to classify skin lesion as healthy or malignant. CLAHE preprocessing technique was applied to obtain contrast enhanced images. Median Filter was applied to remove noise and NOS technique was used for image segmentation. Statistical and shape features were extracted by GLCM Matrix and Geometrical based feature. Lastly, classification was done using ensemble based Hybrid-Ada-Boost and Deep Learning Neural Network (DLNN) architecture. The proposed CAD system showed an accuracy of 92.89% (DLNN), 91.73% (Hybrid-Ada Boost), sensitivity 94.83% (DLNN), 94.08% (Hybrid-Ada Boost), specificity 90.45% (DLNN), 88.71% (Hybrid-Ada Boost), Kappa value 0.856 (DLNN), 0.831 (Hybrid-Ada Boost) on 992 dermoscopy images obtained from PH^2 and Med-Node databases.

Nasr-Esfahani et al. [49] presented a highly computational, complex approach based on CNN for the detection of melanoma lesions using GPU equipped servers. Noise and illumination effects were discarded in the pre-processing step. K-mean classifier was applied on the pre-processed image to produce segmentation mask. Gaussian filter was applied to smooth the healthy skin region. Automatic feature extraction was done to obtain discriminative features. In the final stage, classification was done by CNN with two (5×5 kernel) convolving layers. The last two-layer network of CNN, with linear transfer functions, classified the lesion into malignant or benign with a sensitivity of 0.81, specificity 0.80, PPV 0.75, NPV 0.86, and accuracy 0.8170 on non-dermoscopic images obtained from UMGC digital images archive.

Jafari et al. [50] introduced a deep learning based automated melanoma detection system which aided in correct lesion region's extraction. Guided filter was applied in the pre-processing phase to remove unwanted artifacts. Local and Global patches were defined which showed the pixel as a lesion or normal region. These patches were fed to CNN which labeled the central pixel of patch. A segmentation mask for the image was produced in the post-processing phase and morphological operations of holes filling were applied. Segmentation performance analysis of the proposed system was done that showed an accuracy 98.5%, sensitivity 95%, and specificity 98.9% on 126 digital image of DermQuest database.

Sabouri and Gholam Hosseini [51] put forward a computerized Melanoma recognition model based on CNN. Lesion segmentation was performed using CNN applied to (50×50) image patches that classified each image patch into two classes' i.e. lesion and normal skin (background). ReLU activation function was applied which helped in faster training of the system. For smoothing of lesion border, various post-processing algorithms were applied and morphological operation was the best method. Testing accuracy was computed by using the 52 most challenging images from the dataset. The proposed system showed an accuracy 86.67% on a large dataset collected through online sources.

Shoieb et al. [52] proposed an advance and robust computer aided system for early melanoma diagnosis using SVM classifier. In the pre-processing phase, morphological operations were applied to a hair mask and each hair pixel was replaced by neighboring skin's pixel. Next, ROI segmentation was performed using pixel based classification. The enhanced segmentation composed of three steps, RGB to HSV color conversion, texture feature extraction of ROI and region clustering through K-means. Pre-trained model of CNN was used for lesion features extraction. In the last step, classification was performed using linear SVM which accurately classified the input image with a specificity 88%, sensitivity 100% and accuracy 94% on DermQuest and DermIS dataset that composed of 337 skin images. The system out-performed Robert Amelard system which proposed an enhanced texture based segmentation for melanoma diagnosis.

Sabbaghi et al. [53] put forward a stacked sparse-auto-encoder model using BOF (Bag of Features) technique for melanoma diagnosis and classification using a novel deep learning application "Stacked sparse auto-encoder." In BOF model, the descriptors were a blend of color features and SIFT properties. The extracted features were then clustered using k-means algorithm. Images were mapped to binary codes using sparse auto encoder. For the classification of skin lesion, training of both raw input data and BOF into SSA were performed and performance of both methods was evaluated. Deep BOF successfully improved the classification results and obtained specificity score 94.9%, sensitivity score 95.4% and best ROC on 8-bit RGB images gathered from National Institute of Health, United States.

Yang et al. [54] proposed a new multi-purpose melanoma detection deep learning approach which performed lesion segmentation and classification at the same time. The system provided a robust analysis of image dataset based on GoogleNet and U-Net architecture. Lesion segmentation, localization and detection of dermoscopic feature and melanoma classification were performed. In classification stage, the first task was to distinguish between (seborrheic keratosis, nevus) and melanoma and second was to classify between (melanoma, nevus) and seborrheic keratosis. This multi-task DCNN performed better than other conventional DCNN and showed an AUC value 92.6%, Jaccard index 72.4% on ISIC 2017 challenge dataset.

Zhang [55] proposed an enhanced deep neural network framework to distinguish between normal skin lesions and cancerous lesions. The applied neural network comprised of multiple convolution layers, filters, drop-out layers, SoftMax layer, and activation function. The gradient disappearance problems was mitigated using non-linear activation function (ELU and ReLU). As the training data was small, parameterized regulation was performed to solve the problem of over-fitting. The Loss algorithm was optimized using RMSprop/Adam algorithms. The proposed architecture performed well and showed an accuracy of 91%, specificity 94% and sensitivity 95% on 2017 ISBI dataset.

Ali and Al-Marzouqi [56] presented a simple melanoma recognition approach based on standard (CNN) framework "LightNet" which classified the lesion images into malignant or benign. 7-layers based CNN network was implemented and classification was performed using SoftMax layer. SoftMax classifier used SoftMax function to calculate the prediction probability of melanoma as 0 or 1. Classification outcome of the proposed system were compared to existing (CUMED) and this system outperformed in terms of specificity 98%. However, the proposed system achieved an accuracy of 81.6% and sensitivity 1.49% on ISIC dataset (1279 dermoscopic images). The overall accuracy and sensitivity could be improved through segmentation and useful feature input based on ABCD rule.

Kaymak et al. [57] proposed an automated two-step dermoscopic procedure for skin lesion detection based on Alex Net architecture. In the first step, a deep learning algorithm was applied to classify non-melanocytic and melanocytic lesions. Melanocytic category was further classified as melanocytic nevus and melanoma by using a second deep learning model. Non-melanocytic class was then categorized to malignant and benign classes using a third deep learning model of Alex Net architecture. The classification performance of the second deep learning approach outperformed the first and third approach with an accuracy of 84%, sensitivity 84.7%, specificity 83.8% on 10,015 dermoscopy images of ISIC challenge data set (2018).

Maia et al. [58] presented a CNN based architecture to detect the presence of Melanoma. Several CNN techniques were performed to test the feature vectors that were extracted using ResNet50, InceptionV3, VGG16 and VGG19 architectures. CNN architectures were applied to extract the image characteristics through Transfer learning technique using the Keras library. Next, four set of feature vectors were tested using six different classifiers. The best classification metrics was shown by Logistic regression coupled with VGG19 architecture that attained an accuracy of 92.5%, precision of 85.71%, recall 75%, specificity 96.88% and F1-score 80% on PH^2 database with 200 dermoscopic images.

Salido and Ruiz [59] put forth a pre-trained, Image-net based deep learning technique "AlexNet" for accurate diagnosis of melanoma using dermoscopy images. Harmonic in-painting and morphological operations were performed in pre-processing phase. AlexNet was used as a pre-trained model in Transfer learning and it was further refined to build a network particularly for dermoscopy image classification. Network training was performed using a SGD (stochastic gradient descent) algorithm with a momentum solver. Firstly, the classification of lesion as Melanoma (with and without hair) was performed with specificity 96%, accuracy 96% and sensitivity 100%. Next, a three- category model was applied to classify lesion as melanoma, atypical nevus and common nevus. The combined classification performance (with and without hair) showed an average accuracy of 90% on dermoscopy images of PH^2 dataset.

Jadhav et al. [60] proposed an automatic melanoma diagnosis system based on the concept of deep learning. A 3-layer CNN was implemented for the process of feature extraction and segmentation of the testing skin images. Cubic Support Vector machine (SVM) was applied for the sub-region classification of the lesion. For improved results, partial post-processing was done to remove noise and for smoothing unwanted artifacts.

A binary mask was generated as a lesion border that helped in effective diagnosis. The performance measure of the proposed system was calculated which showed a precision of 97.8%, accuracy 98.7%, and Jaccard index 95.8% on 12 most tough and challenging (15 × 15 input layers architecture) images chosen from a local dataset.

Hosny et al. [61] presented a computerized skin lesion diagnosis and classification system based on pre-trained DCNN (Deep CNN) "AlexNet." Augmentation of data was done to increment the number of colored skin images for training and testing purpose. The classification of images was done into three categories namely melanoma, atypical nevus and common nevus. In Transfer learning, the final layer of AlexNet was substituted with Softmax layer for classification into three classes. The proposed system was evaluated in two experiments i.e. with and without augmented images of the dataset. Segmentation was performed on the ROI to minimize image size as suitable for AlexNet. The proposed model performed very well on the augmented images with an accuracy of 98.61%, sensitivity 98.33%, specificity 98.93% and precision 97.73% using PH^2 dataset.

Ünlü and Çınar et al. [62] performed binary classification using "AlexNet" with ReLU activation function to categorize the image as melanoma or non-melanoma. SoftMax was used as the last layer of the model. In the next phase, second classification was performed to categorize the image into three; melanoma, nevus, seborrheic keratosis. The performance evaluation of the proposed system was done that showed an accuracy of 86% and an average accuracy 81% on 2000 colored images of ISIC dataset. Moreover, the running time of the proposed classification method was decreased as pre-processing and segmentation steps were not performed.

Jaworek-Korjakowska [63] presented an automatic detection system for diagnosis of atypical vascular structures (linear, comma, dotted, polymorphous etc.) which aided in differentiation between melanoma and normal skin lesion using deep learning-based U-Net encoder-decoder architecture. ReLu activation function was firstly applied to input image and then max pooling operation was performed. Classification into two classes (present or absent) was performed using Softmax classifier. The performance evaluation of the proposed vascular segmentation system was performed that showed sensitivity 85%, specificity 81% and DSC 84% on the colored dermoscopic image collection taken from Interactive atlas of Dermoscopy and PH^2 database.

Dorj et al. [64] put forth an intelligent skin lesion classification approach using ECOC SVM for efficient melanoma classification and deep neural network for feature extraction. The proposed architecture was designed to classify lesion into four categories namely Actinic Keratoses, Basal cell carcinoma, Melanoma and Squamous cell carcinoma. A pre-trained AlexNet model was employed for feature extraction. In deep neural network architecture, Max Pooling operation was performed to reduce the size of input neuron. For skin lesion classification, ECOC SVM classifier was used that efficiently classified skin lesions with an accuracy of 95.1% for Squamous cell carcinoma, 92.3% for Actinic keratosis, 91.8% for Basal cell carcinoma and 94.2% for Melanoma.

Devassy et al. [65] proposed an automatic melanoma detection system with a novel integration of hand crafted and deep features to categorize a skin lesion as melanoma and non-melanoma. The proposed work aimed to test whether using standard hand crafted features instead of complex features would aid in better classification or not. Gray world approach was used as a color consistency algorithm in the pre-processing phase. ResNet was applied for extraction of deep features and SIFT algorithm was applied for extraction of hand crafted features. Training and evaluation of the model was performed by employing only deep features, only hand crafted features and a blend of both extracted features on SVM and RUSBoot classifier. The results of deep only features with the SVM classifier outperformed with an accuracy 83.24%, specificity 95.68%, sensitivity 30.63%, recall 30.63% and precision 60.22% on ISIC 2017 challenge dataset.

Nie et al. [66] proposed a Melanoma diagnosis system using supervised deep learning YOLO techniques based on Deep CNNs. The YOLO technique comprised of three networks YoloV1, YoloV2 and YoloV3. Three different frameworks namely Darknet-19, Darknet53 and 24 convolutional layer were applied in all three Yolo networks. Data augmentation with random scaling was done to expand training data size and to enhance generalization ability of the model. Performance analysis of all three YOLO networks was done and YOLO-V2 outperformed with a mAP of 0.83 using Daknet-19 features extraction technique with larger sizes of grid on ISIC dataset of 200 skin images.

Adegun and Viriri [67] proposed an enhanced melanoma detection CNN architecture with an improved supervised Encoder Decoder network for strong feature extraction and learning of the model. Firstly, Softmax classifier was implemented for pixel- wise classification of the lesion. A novel approach "Lesion classifier" performed the classification of images as non-melanoma or melanoma. The Softmax module aided in less costly and successful feature extraction, segmentation and classification. The system was evaluated and compared with other state-of-the-art architectures on the basis of different variables and it showed 95% accuracy, 92% dice score, 97% sensitivity, 96% specificity on 2017 ISIC dataset and 95% accuracy, 93% dice score, 93% sensitivity and 95% specificity on PH^2 dataset respectively.

Hagerty et al. [68] put forth a CAD system for effective diagnosis of skin cancer mainly melanoma. Lesion feature extraction was done by three image processing modules and clinical module was used to extract patient's other information. The probability of Melanoma classification was done using ResNet 50 network. The overall probability of melanoma was calculated through an ensemble of all modules using Logic regression (LR). The fusion technique of ResNet 50 network and classical hand-crafted image processing module outperformed the individual performances done by ResNet 50 classifier and conventional classifier of image processing. The ensemble technique reached a classification accuracy of 94% on NIH dermoscopy studies.

Gavrilov et al. [69] presented a deep ANN for the early identification of skin diseases using Inception V3 architecture. Images from ImageNet archive were obtained for model training. The performance was enhanced by using an ensemble of five networks which made decisions based on decision voting principle. The specificity and sensitivity of the best models among these ensemble models was 92% and 85%. The accuracy of skin lesion classification was obtained as 91% with binary classification, AUC-ROC as 96% applied on ImageNet dataset.

Hekler et al. [70] proposed a novel deep learning approach for a histopathologic melanoma diagnosis using CNN algorithm. 695 lesion images were firstly classified by an expert histopathologist who labeled 345 melanoma and 350 nevi images. Hematoxylin eosin (H&E) stained images were digitalized through Nano-Zoomer slide scanners and were randomly cropped with 10-fold magnification. Transfer learning of ResNet50 was performed using cropped 595 image sections from 595 histopathologic slides. The misclassification rate of the CNN was evaluated as 20% (nevi), 18% (melanoma) and 19% (all set of images). 95% Confidence Interval (CI) was evaluated for all the rates by means of normal distribution approximations. The image slides dataset was acquired from the wide-reaching regional institute of dermatohistopathologic institute.

Hirano et al. [71] proposed a melanoma detection architecture based on a CNNs "GoogleNet" and Hyperspectral data (HSD). HSD is represented as 3-dimensional data (2-D position information and 1-D wavelength information). HSD image size was adjusted to provide a suitable input to GoogleNet using a novel network "Mini Network." The results indicated 77.2% accuracy, 72.3% sensitivity and 81.2% specificity on augmented dataset. The system could not provide the accuracy as achieved by previously existing systems. Resultantly, Mini Network should be improved and number of lesion images must be increased.

Razzak et al. [72] proposed a fully automated lesion classification method using three stage deep neural network to classify seven types of skin lesions namely Basal cell carcinoma, Actinic Keratosis, Benign Keratosis, Dermatofibroma, Melanoma nevus, Melanoma, Vascular lesion. FcResNet architecture contained numerous dense stages with fully connected layers. Various number of residual blocks and mini ResNet were used in each block. The classification performance was further improved by using both the imaging and clinical data. The proposed FcResNet (with transition layer) system performed superior than other state of the art architecture and outperformed ISIC 2018–19 challenge dataset with significant performance of 88.6%. In addition to this, other parameters of evaluation were higher with F-score 96.1%, precision 96%, specificity 96%, sensitivity 96.4%, accuracy 96.1% and AUC 98.4%.

Polat and Koc [73] proposed an efficient skin disease detection method based on CNN and OVA (One versus All) approach. The proposed architecture successfully classified seven different skin diseases namely Basal cell-carcinoma, benign keratosis,

Melanocytic type, actinic keratoses and intraepithelial carcinoma, Melanoma, Vascular lesions and Dermatofibroma. Two models (alone CNN) and (CNN and OVA) were proposed for multi-classification. In the later approach, all the classes were pair wise classified and average of pair-wise results was calculated. The combination of OVA and CNN outperformed the first model and increased the performance accuracy from 77% to 92.90%. The proposed method showed remarkable results without performing extensive pre-processing and feature extraction on HAM10000 dataset of 10,015 dermoscopic images.

Bama et al. [74] proposed a systematic approach for automatic segmentation of malignant melanoma using super-pixel region growing technique. The system was designed for better extraction of ROI from the images. Circular mask was applied to remove the shades from the corner of the image. GMM (Gaussian mixture model) super-pixels were implemented to segment the image into homogeneous regions. Specific colors (red, maroon, white, olive, black) were assigned to the super-pixel regions based on super pixel mean colors in segmentation phase. The performance metrics for image segmentation showed best accuracy of 86.83%, sensitivity of 86.45%, specificity 97.32%, precision 97.07% F-score 85.91% and MCC 76.78% on a PH^2 dataset.

Iqbal et al. [75] put forth a Deep CNN named as "CSLNet" for computerized detection of multiple classes of skin lesion using dermoscopic images. The proposed system classified eight different classes of skin lesion including NV, MEL, BCC, AK, DF, BKL, SCC and VASC. The DCNN model comprised of 68 convolutional layers. Softmax and LeakyRelu were implemented as activation function for output and convolutional layer respectively. The proposed CSLNet performed seemingly well with precision 91%, sensitivity 90%, accuracy 90% and specificity 98% on ISIC 2019 dataset. The proposed framework was also applied to ISIC 2017 and it showed ~1.6% less specificity score than an earlier proposed model. Also, DCNN couldn't classify NV very well as it had massive intra-class variations. However, specificity score of proposed model was similar to other best models that performed in ISIC 2018 challenge dataset.

Soenksen et al. [76] proposed a Suspicious Pigmented Lesion (SPL) detection method based on DCNNs. The detection of blob regions was done using SIFT and LoG (Laplacian of Gaussians) based blob identification approach. They used pre-trained ImageNet CNN model for classifying images into 6 various categories such as backgrounds, skin, SPLs, Non-SPLs (type and B) and skin edges. The proposed DCNN comprised of three groups of convolutional layers and max pooling layer. ReLU activation function was applied to a dense layer, followed by Softmax activation applied to determine the probability vector of final output class. Data augmentation was done that produced 300,000 images for training of classifier. Ugly duckling criteria was used to gain intra-patient lesion saliency using DCNN characteristics. The proposed method showed good score of sensitivity 90.3%, specificity (89.6–90.2%) in classifying SPLs vs. N-SPLS using 38,283 dermatological datasets gathered from 133 patients and other openly accessible images.

Safdar et al. [77] proposed an ensemble model combining two pretrained deep CNN architectures, DenseNet-201 and ResNet-50 for classification of melanoma disease from dermoscopic images. Dataset was collected from three publicly available datasets PH^2, Med-Node and DermIS. Advanced data purification techniques were applied for removal of unwanted artifacts and for contrast adjustment. Lesion segmentation was performed using K-mean clustering algorithm. Image augmentation techniques were also applied to balance the dataset. The proposed segmentation method achieved satisfactory results with 89% AUC, 91% precision, 95.7% specificity and 94% sensitivity. The ensemble model used majority voting technique for classification and achieved an accuracy of 95.2% with specificity, sensitivity and AUC of 96.7%, 92.8% and 98.5% respectively.

Safdar et al. [78] developed a deep learning framework for malignant melanoma detection using an ensemble approach with ResNet-50 and Inception-V3 models. The model used large datasets of ISIC-2020, Med-Node and PH^2 for testing and training. Image normalization and hair removal techniques were used in data preprocessing and semantic segmentation is performed using FCN-8. The segmentation achieved an accuracy of 94% with Dice coefficient and Jaccard similarity index of 88% and 89%. To tackle the dataset imbalancing, data augmentation was employed. For classification task, the ensemble of pre-trained ResNet-50 and Inception-V3 models achieved an accuracy of 93.4% with specificity of 96.0% and AUC of 98.8%.

An enhanced transformer network called SkinTrans was developed by Xin et al. [79] for the purpose of classification of skin cancer using dermoscopic images. In methodology, ViT was developed for patch embedding with overlapping and multi-scale sliding windows and to encode the data together, contrastive learning was used. Two datasets were used in this study. Ham10000 and a clinical dataset. In preprocessing, data augmentation and normalization was performed to enhance the dataset and balance it. SkinTrans achieved 94.1% accuracy on the clinical dataset whereas, it achieved 94.3% accuracy on HAM10000 dataset.

The study by Farooq et al. [80] presented a framework that consist of four phases for melanoma segmentation and classification. In first phase data is collected using publicly available dataset. In second phase of preprocessing, unwanted artifacts like hair and moles were removed from dermoscopic images using Dull Razor program which uses bilateral interpolation for hair removal and a median filter for image smoothing. In second phase of segmentation, U-Net architecture is used to capture detailed spatial information. The third phase involved classification using a modified AlexNet with six convolutional layers and four fully connected layers. The model is trained and validated using ISIC-2019 dataset with 25,441 images divided into 4522 melanoma and 20,802 non-melanoma images. Data augmentation is applied to address class imbalance. The proposed method achieved an accuracy of 98.82% with specificity of 98.5%, precision of 97.4% and sensitivity of 96.34%.

A ViT-based architecture was presented by Cirrincione et al. [81] to distinguish melanoma from non-cancerous skin images. Dermoscopic images from the ISIC-2017 challenge dataset were used in the study. Multi-head attention mechanism is used in the analysis and classification of skin lesions. Experimentation included the use of Adam optimizer, optimizing the learning rate and investigating the effects of network depth through analysis. Experiments were performed using Google Colab. 94.8% accuracy was achieved by the proposed model with sensitivity of 92.8%, specificity of 96.7% and 94.8% AUROC.

In 2023, Khan and Khan [82] presented an approach to classify skin malignancies, including melanoma and nonmelanoma. The authors proposed an architecture called SkinVit, which contains A transformer block, an MLP head and a vision outlook attention mechanism, to capture both local and global aspects and to overcome the CNN's inability in capturing the global contextual information. This study used three datasets. The first dataset used was the ISIC-2019, the second dataset included images that were collected from several databases and the third dataset was a blend of the two. The proposed model achieved an accuracy of 91.09% on the ISIC-2019 dataset, which was followed closely by third dataset with 89.11% and the second dataset achieved 86.11% accuracy.

For detecting melanoma in the early stage, Moturi et al. [83] customized CNN and performed experimentation on MobileNetV2 architecture. In the first step, the data was collected and then preprocessing was applied on it to enhance the quality of the data. After preprocessing, feature extraction was performed along with data partitioning and model training. Dataset that was used in this study was HAM1000 dataset which consist of 10,015 images. CNN was customized through modifications in convolution and pooling layers. Activation functions and regularization techniques were used for fine-tuning the model. Both models MobileNetV2 and CNN were trained on the dataset. MobileNetV2 achieved a validation accuracy of 85%, whereas, customized CNN gave an accuracy of 95%.

In 2024, the author Rashad et al. [84] classified melanoma using deep learning models of CNN, specifically GoogleNet and VGG-19. The dataset used in this study was ISIC-2019 with 4522 images of melanoma and 2624 other images, in total 7146 images. Data balancing was performed on the dataset, images were resized and then normalized. VGG-19 was implemented using python and TensorFlow with 16 convolution layers and

GoogleNet was implemented using the same with 21 layers. To distinguish between melanoma and non-melanoma images, GoogleNet achieved 80.07% accuracy on the dataset, whereas, VGG-19 achieved 85.57% accuracy.

Senthil Sivakumar et al. [85] used deep learning approach to create a model for diagnostic of malignant melanoma cancer. The methodology involved included developing a CNN model from scratch using ResNet50. The dataset used in this study was ISIC dataset. This dataset included 3000 various images of melanoma and non-melanoma for the purpose of training and testing of model. Data preprocessing was applied on dataset along with segmentation and feature extraction. Using ResNet50, the model achieved a 93.9% F1-score and 94% accuracy.

The SkinSwimViT model was introduced by Tang et al. [86] for the classification of skin lesions. It combined a global attention mechanism with a Swin Transformer design allowing for lightweight transformer design. ISIC 2018 dataset was used in this study which included 10,015 samples. To balance the dataset data augmentation techniques such as rotation of data, flipping images, and translation were used. Swim Transformer block for local feature extraction along with global attention block was added in SkinSwimViT. By using transfer learning the model was fine-tuned. With recall of 97.55%, precision of 97.83%, specificity of 99.36%, and 97.79% F1-score, the proposed model achieved an accuracy of 97.88%.

Using dermoscopic images, Aksoy et al. [87] used advance deep learning models to classify between benign and malignant melanoma. By using ViT, ConvNeXt architecture and various transformer models, the study aimed to increase accuracy for classification task. A dataset from Kaggle was used in this study which consisted 13,900 preprocessed photos of size 224 × 224. Predefined weights and transformations were used throughout the training of models of ConvNext, ViT and SwinTransformerV2. With an accuracy of 91.5%, a precision of 90.45% for benign cases and 92.61% for malignant cases and a recall of 92.8% for benign cases and 90.2% for malignant cases, the ConvNeXt model outperformed other models.

Reis and Turk [88] presented a novel deep learning architecture with lightweight ViT, a hybrid framework and an ensemble learning model for early stage detection of melanoma using skin lesion images. The proposed model, named MABSCNET, utilized both CNN and transformer elements for improved feature extraction. The dataset used in this study is ISIC-2018, ISIC-2020 and another dataset from Kaggle. Preprocessing steps included image enhancement techniques of CLAHE and the hypercolumn method. MSBSCNET model achieved an accuracy of 78.63% on ISIC-2020 dataset and 100% accuracy on ISIC-2018 dataset outperforming ViT and Hybrid model which achieved 76.50% and 92.74% accuracy respectively on ISIC-2020 dataset.

Ramkumar et al. [89] created a hybrid deep learning approach for skin cancer segmentation and classification using Swin Transformer and Fast Neural Networks (FNN) to manage the limitation of non-uniformly distributed data, thus enhancing the accuracy. The proposed model used Swim Transformers with U-Net for segmentation and Swim Transformers with FNN in the classification layer. The first step in this model included data collection. Three datasets were utilized in this study which are ISIC-2018, PH-2 and HAM10000 dataset with 10,015 dermatoscopic images. In the second step, data preprocessing was performed to eliminate noise and enhance image quality by using image histogram algorithms and HR-IQE. Data augmentation is performed in the third step to prevent the issue of overfitting due to lack of sufficient labeled data by applying the transformation, translation, resizing image and rotation. In forth step segmentation is performed by incorporating Swin Transformers in U-Net architecture, utilizing Swin Transformers encoder and decoder blocks in U-shaped architecture, along with up sampling and down sampling blocks interconnected with skip connections to extract high-resolution features. In the final step, the model was trained with FNNs for classification. The model was implemented using Keras with TensorFlow on an Intel i7 processor with GeForce GTX. In segmentation, the model achieved a DICE score of 98.6% on the PH^2 dataset which is higher than 98.5% on ISIC-2018 and 98% on HAM-10000. The average model-building time for the proposed model was 9.98 s for training on all three datasets. For the classification task, on all three datasets, the model achieved an accuracy of 98.6% with precision, recall and F1-score of 97.6%, 97.8% and 98% respectively exhibiting superior classification.

Wang et al. [90] proposed a deep learning two-stage framework for melanoma detection using StyleGAN2-ADA for image synthesis and Vision Transformer for classification. The proposed model used progressive growing techniques and adaptive discriminator for data augmentation and conditional specifications to balance the dataset. The generator consisted of coarse and fine components to refine the image. Melanoma BatchFormer vision transformer used a dual-branch training strategy to address data scarcity. It used self-attention mechanisms to process image patches and enhance feature extraction to classify melanoma malignancy. The dataset used in this study is ISIC-2020 with 33,126 dermoscopic images. The training process involved augmenting the dataset with synthetic images generated by StyleGAN2-ADA. Various techniques like noise addition and random flipping were used to increase the quality of image. The method was implemented using PyTorch on RTX 3090 GPU. During image generation, 5000 synthetic melanoma images were generated, which closely resembled real ISIC2020 images, with Inception Score of 3.91 and Frechet Inception Distance of 0.79. The proposed model achieved an accuracy of 98.43% with AUC of 98.63, Sensitivity of 99.01% and F1-score of 96.30%. A comparative analysis of DL approaches is given in Table 3.

Table 3 Comparative analysis of deep learning approaches in automated melanoma detection.

Sr. no.	Author/s	Year	Technique/approach	Dataset	Results
1	Sheha et al. [46]	2012	Traditional & Automatic MLP classifier Feature extraction → (GLCM)	Dermoscopy images dataset	Traditional MLP outperformed 92% and 100% accuracy in training & testing
2	Codella et al. [47]	2015	Ensemble, Caffe CNN, sparse coding, hybrid Non-linear SVM feature normalization	ISIC dataset	Hybrid Accuracy 93.1% & 73.9% Sensitivity 94.9% & 73.8% Specificity 92.8% & 74.3% in both tasks
3	Premaladha and Ravichandran [48]	2016	Classification → Hybrid-Ada Boost and DLNN algorithms Contrast enhanced images → CLAHE preprocessing Image segmentation → NOS technique Feature extraction → GLCM and Geometrical based feature	PH^2 and Med-Node databases (992-dermoscopy image)	Accuracy 92.89% (DLNN), 91.73% (Hybrid-Ada Boost), Sensitivity 94.83% (DLNN), 94.08% (Hybrid-Ada Boost), Specificity 90.45% (DLNN), 88.71% (Hybrid-Ada Boost, Kappa value 0.856 (DLNN), 0.831 (Hybrid-Ada Boost)
4	Nasr-Esfahani et al. [49]	2016	CNN ((5 × 5 kernel) convolving) Segmentation → K-mean classifier	UMGC digital images archive	Sensitivity 81% Specificity 80% PPV 75% NPV 86% Accuracy 81.70%
5	Jafari et al. [50]	2016	CNN Guided filter → edge preserving	DermQuest database (126 digital images)	Segmentation performance analysis Accuracy 98.5% Sensitivity 95% Specificity 98.9%

Continued

#	Author	Year	Methodology	Dataset	Results
6	Sabouri and Gholam Hosseini [51]	2016	CNN Training → ReLU activation function	Online sources (52 most challenging images)	Accuracy 86.67%
7	Shoieb et al. [52]	2016	SVM classifier Image smoothing → Median filter Segmentation → K-means & Gabor filter Feature extraction → CNN	DermQuest and DermIS dataset (337 skin images)	Specificity 88% Sensitivity 100% Accuracy 94%
8	Sabbaghi et al. [53]	2016	Stacked sparse-auto-encoder framework Deep BOF (Bag of features) Clustering → K-means	National Institute of Health United States (8-bit RGB images)	Specificity 94.9% Sensitivity 95.4% Best ROC
9	Yang et al. [54]	2017	GoogleNet and U-Net architecture Multi-task DCNN	ISIC 2017 challenge datasets	AUC value 92.6% Jaccard index 72.4%
10	Zhang [55]	2017	Enhanced deep neural network framework (Softmax layer) Non-linear activation function (ELU and ReLU) Adam algorithms	2017 ISBI dataset	Accuracy 91% Specificity 94% Sensitivity 95%
11	Ali and Al-Marzouqi [56]	2017	LightNet CNN architecture Softmax layer (classification) Reduced output error → Gradient descent algorithm	ISIC dataset (1279 dermoscopic images)	Accuracy 81.6% Sensitivity 1.49% Specificity 98%

Table 3 Comparative analysis of deep learning approaches in automated melanoma detection—cont'd

Sr. no.	Author/s	Year	Technique/approach	Dataset	Results
12	Kaymak et al. [57]	2018	Alex Net	ISIC challenge data set (2018) 10,015 dermoscopy images	Accuracy 84% Sensitivity 84.7% Specificity 83.8%
13	Maia et al. [58]	2018	Feature extraction → ResNet50, InceptionV3, VGG16 and VGG19 architectures 10-Fold cross validation Classification → Logistic regression coupled with VGG19 architecture	PH^2 database (200 dermoscopic images)	Accuracy 92.5% Precision 85.71% Recall 75% Specificity 96.88% F1-score 80%
14	Salido and Ruiz [59]	2018	AlexNet Network Training → SGD (stochastic gradient descent) algorithm	PH^2 dataset	Specificity 96% Accuracy 96% Sensitivity 100% Combine classification performance Accuracy 90%
15	Jadhav et al. [60]	2018	Feature extraction & segmentation → 3-layer CNN Sub-region classification → Cubic Support Vector machine (SVM)	Local dataset (12 most tough and challenging images)	Precision 97.8% Accuracy 98.7% Jaccard index 95.8%
16	Hosny et al. [61]	2018	Pre-trained DCNN (Deep CNN) "AlexNet"	PH^2 dataset	Accuracy 98.61% Sensitivity 98.33% Specificity 98.93% Precision 97.73%
17	Ünlü and Çınar [62]	2018	"AlexNet" with ReLU activation function SoftMax (last layer)	ISIC dataset (2000 images)	Accuracy 86% Average accuracy 81%

#	Author	Year	Method	Dataset	Results
18	Jaworek-Korjakowska [63]	2018	U-Net encoder-decoder architecture Rectified linear (ReLu) activation function Softmax classifier	PH^2 database Interactive atlas of Dermoscopy	Sensitivity 85% Specificity 81% DSC 84%
19	Dorj et al. [64]	2018	ECOC SVM classifier Feature extraction → Alex Net	N/A	Accuracy Squamous cell carcinoma 95.1% Actinic keratosis 92.3% Basal cell carcinoma 91.8% Melanoma 94.2%
20	Devassy et al. [65]	2018	Deep feature extraction → ResNet Hand crafted feature extraction → SIFT algorithm SVM and RUSBoot classifier	ISIC 2017 challenge dataset	**SVM (with deep features) outperformed** Accuracy 83.24% Specificity 95.68% Sensitivity 30.63% Recall 30.63% Precision 60.22%
21	Nie et al. [66]	2019	YOLO Network (YoloV1, YoloV2 and YoloV3) Darknet-19, Darknet53 and 24 convolutional layer	ISIC dataset (200 skin images)	**YOLO-V2 outperformed** (using Daknet-19 features extraction) mAP 83%
22	Adegun and Viriri [67]	2019	Pixel wise classification → Softmax classifier Strong feature extraction → CNN architecture Enhanced supervised Encoder Decoder network	ISIC 2017 dataset PH^2 dataset	**ISIC** Accuracy 95% Dice score 92% Sensitivity 97% Specificity 96% **PH^2** Accuracy 95% Dice score 93% Sensitivity 93% Specificity 95%

Continued

Table 3 Comparative analysis of deep learning approaches in automated melanoma detection—cont'd

Sr. no.	Author/s	Year	Technique/approach	Dataset	Results
23	Hagerty et al. [68]	2019	Ensemble of ResNet 50 network, Logistic regression (LR)	NIH dermoscopy studies	Accuracy 94%
24	Gavrilov et al. [69]	2019	Inception V3 architecture (Ensemble of 5 networks)	ImageNet dataset	Specificity 92% Sensitivity 85% Accuracy 91% (AUC–ROC) 96%
25	Hekler et al. [70]	2019	ResNet50 Convolutional Neural Network	Largest regional institute of dermatohistopathologic institute	Misclassification rate of the CNN 20% (nevi) 18% (melanoma) 19% (all set of images)
26	Hirano et al. [71]	2020	"GoogleNet" and Hyperspectral data (HSD) 5-Folds cross validation	Augmented dataset	Accuracy 77.2% Sensitivity 72.3% Specificity 81.2%
27	Razzak et al. [72]	2020	FcResNet architecture	ISIC 2018–19 dataset challenge	F-score 96.1% Precision 96% Specificity 96% Sensitivity 96.4% Accuracy 96.1% AUC 98.4%
28	Polat and Koc [73]	2020	CNN and OVA (one versus All) approach	HAM10000 dataset (10,015 dermoscopic images)	CNN and OVA Improved accuracy from 77% to 92.90%
29	Bama et al. [74]	2020	GMM (Gaussian mixture model) for segmentation	PH^2 dataset	Accuracy 86.83% Sensitivity 86.45% Specificity 97.32% Precision 97.07% F-score 85.91% MCC 76.78%
30	Iqbal et al. [75]	2021	DCNN model → multi-class classification Softmax, LeakyRelu activation function	ISIC (2017, 18, 19)	**Out-performed on ISIC 2019** Precision 91% Sensitivity 90% Accuracy 90% Specificity 98%

#	Reference	Year	Technique	Dataset	Results
31	Soenksen et al. [76]	2021	ImageNet CNN → (SPL) detection Blob region detection → SIFT and LoG (Laplacian of Gaussians) Ugly duckling criteria → intra-patient lesion saliency	38,283 dermatological datasets (openly accessible images)	Sensitivity 90.3% Specificity (89.6–90.2%)
32	Safdar et al. [77]	2021	DenseNet-201 and ResNet-50	PH^2, Med-Node and DermIS	Accuracy 95.2% Specificity 96.7% Sensitivity 92.8% AUC 98.5%
33	Safdar et al. [78]	2021	ResNet-50 and Inception-V3	ISIC-2020, Med-Node and PH^2	Accuracy 93.4% Specificity 96.00% AUC 98.8%
34	Xin et al. [79]	2022	SkinTrans: A Transformer Network	Ham10000 and clinical dataset	Accuracy 94.1% on clinical dataset 94.3% Accuracy on HAM10000 dataset
35	Farooq et al. [80]	2023	U-Net and AlexNet	ISIC-2019	Accuracy 98.82% Specificity of 98.5% Precision of 97.4% Sensitivity of 96.34%
36	Cirrincione et al. [81]	2023	ViT-based architecture	ISIC-2017	Sensitivity 92.8% Specificity of 96.7% AUROC 94.8%
37	Khan and Khan [82]	2023	Vision Transformer based SkinVit	ISIC-2019, other collection from public dataset, private dataset, HAM1000 dataset	Accuracy 91.09% on ISIC-2019 dataset Accuracy 86.11% on second dataset Accuracy 89.11% on third dataset
38	Moturi et al. [83]	2024	Customized CNN and MobileNetV2	HAM10000 dataset	Accuracy 85% on MobileNetV2 Accuracy 95% on customized CNN
39	Rashad et al. [84]	2024	GoogleNet and VGG-19	ISIC-2019	Accuracy 80.07% on GoogleNet Accuracy 85.57% on VGG-19

Continued

Table 3 Comparative analysis of deep learning approaches in automated melanoma detection—cont'd

Sr. no.	Author/s	Year	Technique/approach	Dataset	Results
40	Senthil Sivakumar, et al. [85]	2024	Modified CNN with ResNet50	ISIC dataset	Accuracy 94% F1-score 93.9%
41	Tang et al. [86]	2024	SkinSwimViT Architecture	ISIC 2018	Accuracy of 97.88% Recall of 97.55% Precision of 97.83% Specificity of 99.36% F1-score 97.79%
42	Aksoy et al. [87]	2024	ViT and ConvNeXt architecture	Kaggle Dataset (13,900 images)	Accuracy 91.5% Precision 90.45% (benign cases) Precision 92.61% (malignant cases) Recall of 92.8% (benign cases) Recall 90.2% (malignant cases)
43	Reis and Turk [88]	2024	MABSCNET	ISIC–2018, ISIC–2020	Accuracy 78.63% on ISIC–2020 dataset Accuracy 100% on ISIC–2018 dataset
44	Ramkumar et al. [89]	2024	Swin Transformer with U-Net and Fast Neural Network	ISIC–2018, PH–2 and HAM10000	Accuracy 98.6% Precision 97.6% Recall 97.8% F1-score 98%
45	Wang et al. [90]	2024	StyleGAN2-ADA and Vision Transformer	ISIC–2020	Accuracy 98.43% AUC 98.63 Sensitivity 99.01% F1-score 96.30%

3. Discussion

The detection of melanoma using machine and deep learning approaches has significantly improved over the years. Nevertheless, there exists some challenges as well as some key aspects that should be catered well. This paper yields a comprehensive review of various machine and deep learning approaches toward automated melanoma detection. Generally, the ML techniques are based on pre-processing and post-processing phases for classifying skin lesion to malignant or benign. Image processing (noise reduction, illumination correction, artifact removal, contrast enhancement, smoothing), lesion segmentation, feature extraction, feature reduction and image classification are the key stages in melanoma diagnosis.

The hand-crafted features are normalized and fed to the classifiers for testing. Image pre-processing, hand-crafted features extraction and segmentation serves as an additional over-head and makes the processing of ML complex and slow. The study shows that many researchers perform successful feature extraction using ABCD rules of dermatology. In addition to it, some of them also perform smoothing and adjusting of irregular borders of skin lesion images. Deep learning, on the other hand, omits the need of exhaustive and complex pre-processing. It mostly by-passes the extensive pre-processing, feature extraction and segmentation tasks and performs normalization of pixel values, cropping and re-sizing, etc.

In medical imaging, the most challenging task is acquiring labeled images. To counter this problem, the ISBI challenge (by ISIC archive) has put forth an effort to provide a large and diverse variety of images datasets. This provides a unique and publically available platform to academia and researchers for effectively evaluating their work in the medical field. A number of authors used ISIC 2016, ISIC 2017, and ISIC (2018–19) challenge datasets for their automated melanoma detection methods.

Several studies show that considering both local and deep (global) features provide a better accuracy score, however, increasing the accuracy rate of melanoma detection is still a laborious task. The subsequent goal is to achieve maximum score for sensitivity and optimize specificity which will increase the overall accuracy of the model. One of the researchers put forth a modern melanoma detection approach with a unique combination of deep (global) features and hand–crafted (local) features. Their proposed model showed good accuracy score overall.

The size of training dataset is of great significance. The future trends aim to increase the volume as well as variety of training datasets. The disproportion of class distribution (positive & negative ratio) of skin lesion images result into biased classification. A well-balanced dataset can be achieved using several data augmentation approaches and external sources of data. Various augmentation methods such as random brightness, vertical and horizontal flip, random contrast have been implemented by researchers to lessen the problem of over-fitting.

Furthermore, the need to use clinical data in melanoma detection is mostly overlooked. The clinical images represent different details and highlight key aspects of the lesion images. Thus, only adopting the dermoscopic images for model training is not practicable. An effort toward building a clinical image library is put forth by ISIC which has provided large number of clinical images of skin lesions. Moreover, it is observed that some of the authors have used raw images from open source databases for classification of malignant melanoma. Also, only few researchers have attained the ground truth images for their classifier's training and testing. Therefore, ground truth images should also be produced for improved classifier training and evaluation.

In general, a classifier's job is to choose a disease label with the maximum probability. Solely classifier's prediction doesn't provide a full support to the target practitioners, patients or medical students. Including metadata (patient's demographics) may enable CAD systems to provide better visual description of skin lesions to the users. Therefore, it is also helpful to acquire patient's metadata and carry out pattern analysis for better representation of results.

One important aspect this survey highlights is that most of the researchers carried out their melanoma detection technique on people with light skin. For instance, the images presented by ISIC database are primarily gathered from Europe, Australia and United States. This bounds the CNN models to train and perform well on light skinned images only. CNN should be trained and evaluated on various skin colors including the dark skinned images too. This practice will make the deep learning models diverse and serviceable worldwide.

One of the key benefits of applying deep learning is its ability to cut off the over-head of feature engineering, ultimately making the model less complicated and less laborious. Furthermore, it is easier to ensemble or amend the deep learning models as per the new problems. Deep learning also holds some drawbacks such as a need to train on a large volume of dataset for outstanding performance. Also, the training phase of deep learning models is comparatively more complex, expensive and time consuming. Summarily, this survey indicates that sparse representations, deep learning based models or their ensemble can largely improve the success rate of computerized melanoma detection methods.

4. Conclusion

Melanoma is an extremely fatal type of skin cancer. To assist the dermatologists in critical inspection of melanoma, various automated melanoma detection tools are developed across the globe. These dermoscopic tools are based on machine and deep learning models which are trained and evaluated on skin lesion image datasets. The classification approach lies on image pre-processing, feature extraction, segmentation and lastly classification. This review paper intends to deliver a full-scale overview of existing work on malignant melanoma diagnosis. Several first-class models were deeply analyzed.

Generally, neural network based models such as ResNet, Inception V3 along with an ensemble of other advanced models performed seemingly well. However, these models are costly and require excessive training time. It is evident that melanoma classification results are significantly improved when the deep models are applied to well segmented and pre-processed images. Study shows that applying deep learning based lesion segmentation methods have a greater significance as they produce fine grained lesion images. Furthermore, this paper highlights that the skin image archives should also focus on gathering images with varied skin colors, not solely light skinned images. This enables the classifier to be more diverse and practical. Future advancements in deep learning approaches and the ensemble of existing ones, along with a diverse and large collection of supporting dataset can help in achieving advance and early detection of malignant melanoma.

Conflict of interest

No conflict among authors.

Acknowledgment

This study has been carried out under Riphah Artificial Intelligence Research (RAIR), Lab.

References

[1] Mayo Clinic, Melanoma, 2020, March 10, Available from: https://www.mayoclinic.org/diseases-conditions/melanoma/symptoms-causes/syc-20374884.
[2] World Health Organization, Radiation: Ultraviolet (UV) Radiation and Skin Cancer, 2017, October 16, Available from: https://www.who.int/news-room/q-a-detail/radiation-ultraviolet-(uv)-radiation-and-skin-cancer.
[3] kaggle, Melanoma, 2025, January 12, Available from: https://www.kaggle.com/datasets/drscarlat/melanoma.
[4] American Cancer Society, Key Statistics for Melanoma Skin Cancer, 2021, January 12, Available from: https://www.cancer.org/cancer/melanoma-skin-cancer/about/key-statistics.html.
[5] S. Akbar, M.U. Akram, M. Sharif, A. Tariq, U.U. Yasin, Decision support system for detection of papilledema through fundus retinal images, J. Med. Syst. 41 (2017) 66.
[6] A. Jamal, M. Hazim Alkawaz, A. Rehman, T. Saba, Retinal imaging analysis based on vessel detection, Microsc. Res. Tech. 80 (7) (2017) 799–811.
[7] S. Akbar, M.U. Akram, M. Sharif, A. Tariq, S.A. Khan, Decision support system for detection of hypertensive retinopathy using arteriovenous ratio, Artif. Intell. Med. 90 (2018) 15–24.
[8] S. Akbar, M. Sharif, M.U. Akram, T. Saba, T. Mahmood, M. Kolivand, Automated techniques for blood vessels segmentation through fundus retinal images: a review, Microsc. Res. Tech. 82 (2019) 153–170.
[9] M.A. Khan, M.I. Sharif, M. Raza, A. Anjum, T. Saba, S.A. Shad, Skin lesion segmentation and classification: a unified framework of deep neural network features fusion and selection, Expert. Syst. 39 (7) (2022) e12497.
[10] B. Mughal, N. Muhammad, M. Sharif, A. Rehman, T. Saba, Removal of pectoral muscle based on topographic map and shape-shifting silhouette, BMC Cancer 18 (1) (2018) 1–14.

[11] H.A. Miot, M.P. Paixão, F.M. Paschoal, Basics of digital photography in dermatology, An. Bras. Dermatol. 81 (2006) 174–180.
[12] Kaggle, Melanoma, 2019, February 27, Available from: https://www.kaggle.com/drscarlat/melanoma.
[13] B. Mughal, M. Sharif, N. Muhammad, et al., A novel classification scheme to decline the mortality rate among women due to breast tumor, Microsc. Res. Tech. 81 (2) (2018) 171–180.
[14] A. Adegun, S. Viriri, Deep learning techniques for skin lesion analysis and melanoma cancer detection: a survey of state-of-the-art, Artif. Intell. Rev. 54 (2021) 811–841.
[15] K. Meethongjan, M. Dzulkifli, A. Rehman, A. Altameem, et al., An intelligent fused approach for face recognition, J. Intell. Syst. 22 (2) (2013) 197–212.
[16] H.M. Fahad, M.U. Ghani Khan, et al., Microscopic abnormality classification of cardiac murmurs using ANFIS and HMM, Microsc. Res. Tech. 81 (5) (2018) 449–457.
[17] A. Husham, M. Hazim Alkawaz, T. Saba, A. Rehman, J. Saleh Alghamdi, Automated nuclei segmentation of malignant using level sets, Microsc. Res. Tech. 79 (10) (2016) 993–997.
[18] A. Rehman, Brain stroke prediction through deep learning techniques with ADASYN strategy, in: 2023 16th International Conference on Developments in eSystems Engineering (DeSE), Istanbul, Turkiye, 2023, pp. 679–684.
[19] K. Yousaf, Z. Mehmood, T. Saba, A. Rehman, A.M. Munshi, R. Alharbey, M. Rashid, Mobile-health applications for the efficient delivery of health care facility to people with dementia (PwD) and support to their carers: a survey, Biomed Res. Int. 2019 (2019) 1–26.
[20] Kaggle, Melanoma Detection Dataset, 2020, May 12, Available from: https://www.kaggle.com/wanderdust/skin-lesion-analysis-toward-melanoma-detection.
[21] T. Mendonça, P.M. Ferreira, J.S. Marques, A.R. Marcal, J. Rozeira, PH 2-A dermoscopic image database for research and benchmarking, in: 2013 35th Annual International Conference of the IEEE Engineering in Medicine and Biology Society (EMBC), 2013, pp. 5437–5440.
[22] Dermis, Malignant Melanoma, Metastatic, 2010, March 21, Available from: https://www.dermis.net/dermisroot/en/18195/diagnose.htm.
[23] N.Z. Dermnet, Melanoma In Situ Images, 2019, January 1, Available from: https://dermnetnz.org/topics/melanoma-in-situ-images.
[24] I. Giotis, N. Molders, S. Land, M. Biehl, M.F. Jonkman, N. Petkov, MED-NODE: a computer-assisted melanoma diagnosis system using non-dermoscopic images, Expert Syst. Appl. 42 (2015) 6578–6585.
[25] Kaggle, SIIM-ISIC Melanoma Classification, 2020, August 31, Available from: https://www.kaggle.com/c/siim-isic-melanoma-classification.
[26] IEEE International Symposium on Biomedical Imaging, Challenges, 2017, April 21, Available from: https://biomedicalimaging.org/2017/challenges/.
[27] D. Gutman, N.C. Codella, E. Celebi, B. Helba, M. Marchetti, N. Mishra, et al., Skin lesion analysis toward melanoma detection: a challenge at the international symposium on biomedical imaging (ISBI) 2016, hosted by the international skin imaging collaboration (ISIC), arXiv preprint arXiv:1605.01397, 2016.
[28] P. Tschandl, C. Rosendahl, H. Kittler, The HAM10000 dataset, a large collection of multi-source dermatoscopic images of common pigmented skin lesions, Sci. Data 5 (2018) 1–9.
[29] N.N. Sultana, N.B. Puhan, Recent deep learning methods for melanoma detection: a review, in: International Conference on Mathematics and Computing, 2018, pp. 118–132.
[30] M. Mohammed, M.B. Khan, E.B.M. Bashier, Machine Learning: Algorithms and Applications, CRC Press, 2016.
[31] S. Lloyd, M. Mohseni, P. Rebentrost, Quantum algorithms for supervised and unsupervised machine learning, arXiv preprint arXiv:1307.0411, 2013.
[32] Java Point, Unsupervised Machine Learning, 2018, December 31, Available from: https://www.javatpoint.com/unsupervised-machine-learning.
[33] L. Li, Q. Zhang, Y. Ding, H. Jiang, B.H. Thiers, J.Z. Wang, Automatic diagnosis of melanoma using machine learning methods on a spectroscopic system, BMC Med. Imaging 14 (2014) 1–12.

[34] M.H. Jafari, S. Samavi, N. Karimi, S.M.R. Soroushmehr, K. Ward, K. Najarian, Automatic detection of melanoma using broad extraction of features from digital images, in: 2016 38th Annual International Conference of the IEEE Engineering in Medicine and Biology Society (EMBC), 2016, pp. 1357–1360.

[35] R. Moussa, F. Gerges, C. Salem, R. Akiki, O. Falou, D. Azar, Computer-aided detection of melanoma using geometric features, in: 2016 3rd Middle East Conference on Biomedical Engineering (MECBME), 2016, pp. 125–128.

[36] Z. Waheed, A. Waheed, M. Zafar, F. Riaz, An efficient machine learning approach for the detection of melanoma using dermoscopic images, in: 2017 International Conference on Communication, Computing and Digital Systems (C-CODE), 2017, pp. 316–319.

[37] T.T.K. Munia, M.N. Alam, J. Neubert, R. Fazel-Rezai, Automatic diagnosis of melanoma using linear and nonlinear features from digital image, in: 2017 39th Annual International Conference of the IEEE Engineering in Medicine and Biology Society (EMBC), 2017, pp. 4281–4284.

[38] I.A. Ozkan, M. Koklu, Skin lesion classification using machine learning algorithms, Int. J. Intell. Syst. Appl. Eng. 5 (2017) 285–289.

[39] S.E. Roslin, Classification of melanoma from dermoscopic data using machine learning techniques, Multimed. Tools Appl. 79 (2020) 3713–3728.

[40] N. Hameed, A. Shabut, M.A. Hossain, A computer-aided diagnosis system for classifying prominent skin lesions using machine learning, in: 2018 10th Computer Science and Electronic Engineering (CEEC), 2018, pp. 186–191.

[41] M. Ramezani, A. Karimian, P. Moallem, Automatic detection of malignant melanoma using macroscopic images, J. Med. Signals Sens. 4 (2014) 281.

[42] A.P. Chakkaravarthy, A. Chandrasekar, Automatic detection and segmentation of melanoma using fuzzy c-means, in: 2019 Fifth International Conference on Science Technology Engineering and Mathematics (ICONSTEM), 2019, pp. 132–136.

[43] E. Vocaturo, E. Zumpano, Automatic Detection of Dysplastic Nevi: A Multiple Instance, Learning Solution, 2020.

[44] J.-G. Lee, S. Jun, Y.-W. Cho, H. Lee, G.B. Kim, J.B. Seo, et al., Deep learning in medical imaging: general overview, Korean J. Radiol. 18 (2017) 570–584.

[45] M. Coşkun, Ö. Yildirim, U. Ayşegül, Y. Demir, An overview of popular deep learning methods, Eur. J. Tech. 7 (2017) 165–176.

[46] M.A. Sheha, M.S. Mabrouk, A. Sharawy, Automatic detection of melanoma skin cancer using texture analysis, J. Comput. Appl. 42 (2012) 22–26.

[47] N. Codella, J. Cai, M. Abedini, R. Garnavi, A. Halpern, J.R. Smith, Deep learning, sparse coding, and SVM for melanoma recognition in dermoscopy images, in: International Workshop on Machine Learning in Medical Imaging, 2015, pp. 118–126.

[48] J. Premaladha, K. Ravichandran, Novel approaches for diagnosing melanoma skin lesions through supervised and deep learning algorithms, J. Med. Syst. 40 (2016) 1–12.

[49] E. Nasr-Esfahani, S. Samavi, N. Karimi, S.M.R. Soroushmehr, M.H. Jafari, K. Ward, et al., Melanoma detection by analysis of clinical images using convolutional neural network, in: 2016 38th Annual International Conference of the IEEE Engineering in Medicine and Biology Society (EMBC), 2016, pp. 1373–1376.

[50] M.H. Jafari, N. Karimi, E. Nasr-Esfahani, S. Samavi, S.M.R. Soroushmehr, K. Ward, et al., Skin lesion segmentation in clinical images using deep learning, in: 2016 23rd International Conference on Pattern Recognition (ICPR), 2016, pp. 337–342.

[51] P. Sabouri, H. Gholam Hosseini, Lesion border detection using deep learning, in: 2016 IEEE Congress on Evolutionary Computation (CEC), 2016, pp. 1416–1421.

[52] D.A. Shoieb, S.M. Youssef, W.M. Aly, Computer-aided model for skin diagnosis using deep learning, J. Image Graphics 4 (2016) 122–129.

[53] S. Sabbaghi, M. Aldeen, R. Garnavi, A deep bag-of-features model for the classification of melanomas in dermoscopy images, in: 2016 38th Annual International Conference of the IEEE Engineering in Medicine and Biology Society (EMBC), 2016, pp. 1369–1372.

[54] X. Yang, Z. Zeng, S.Y. Yeo, C. Tan, H.L. Tey, Y. Su, A novel multi-task deep learning model for skin lesion segmentation and classification, arXiv preprint arXiv:1703.01025, 2017.

[55] X. Zhang, Melanoma segmentation based on deep learning, Comput. Assist. Surg. 22 (2017) 267–277.

[56] A.A. Ali, H. Al-Marzouqi, Melanoma detection using regular convolutional neural networks, in: 2017 International Conference on Electrical and Computing Technologies and Applications (ICECTA), 2017, pp. 1–5.

[57] S. Kaymak, P. Esmaili, A. Serener, Deep learning for two-step classification of malignant pigmented skin lesions, in: 2018 14th Symposium on Neural Networks and Applications (NEUREL), 2018, pp. 1–6.

[58] L.B. Maia, A. Lima, R.M.P. Pereira, G.B. Junior, J.D.S. de Almeida, A.C. de Paiva, Evaluation of melanoma diagnosis using deep features, in: 2018 25th International Conference on Systems, Signals and Image Processing (IWSSIP), 2018, pp. 1–4.

[59] J.A.A. Salido, C. Ruiz, Using deep learning to detect melanoma in dermoscopy images, Int. J. Mach. Learn. Comput 8 (2018) 61–68.

[60] A.R. Jadhav, A.G. Ghontale, V.K. Shrivastava, Segmentation and border detection of melanoma lesions using convolutional neural network and SVM, in: Computational Intelligence: Theories, Applications and Future Directions-Volume I, Springer, 2019, pp. 97–108.

[61] K.M. Hosny, M.A. Kassem, M.M. Foaud, Skin cancer classification using deep learning and transfer learning, in: 2018 9th Cairo International Biomedical Engineering Conference (CIBEC), 2018, pp. 90–93.

[62] E.I. Ünlü, A. Çınar, Classification of skin images with respect to melanoma and nonmelanoma using the deep neural network, IOSR J. Eng. 8 (12) (2018) 35–40.

[63] J. Jaworek-Korjakowska, A deep learning approach to vascular structure segmentation in dermoscopy colour images, Biomed. Res. Int. 2018 (2018) 1–8.

[64] U.-O. Dorj, K.-K. Lee, J.-Y. Choi, M. Lee, The skin cancer classification using deep convolutional neural network, Multimed. Tools Appl. 77 (2018) 9909–9924.

[65] B.M. Devassy, S. Yildirim-Yayilgan, J.Y. Hardeberg, The impact of replacing complex hand-crafted features with standard features for melanoma classification using both hand-crafted and deep features, in: Proceedings of SAI Intelligent Systems Conference, 2018, pp. 150–159.

[66] Y. Nie, P. Sommella, M. O'Nils, C. Liguori, J. Lundgren, Automatic detection of melanoma with yolo deep convolutional neural networks, in: E-Health and Bioengineering Conference (EHB), 2019, 2019, pp. 1–4.

[67] A.A. Adegun, S. Viriri, Deep learning-based system for automatic melanoma detection, IEEE Access 8 (2019) 7160–7172.

[68] J.R. Hagerty, R.J. Stanley, H.A. Almubarak, N. Lama, R. Kasmi, P. Guo, et al., Deep learning and handcrafted method fusion: higher diagnostic accuracy for melanoma dermoscopy images, IEEE J. Biomed. Health Inform. 23 (2019) 1385–1391.

[69] D. Gavrilov, A. Melerzanov, N. Shchelkunov, E. Zakirov, Use of neural network-based deep learning techniques for the diagnostics of skin diseases, Biomed. Eng. 52 (2019) 348–352.

[70] A. Hekler, J.S. Utikal, A.H. Enk, C. Berking, J. Klode, D. Schadendorf, et al., Pathologist-level classification of histopathological melanoma images with deep neural networks, Eur. J. Cancer 115 (2019) 79–83.

[71] G. Hirano, M. Nemoto, Y. Kimura, Y. Kiyohara, H. Koga, N. Yamazaki, et al., Automatic diagnosis of melanoma using hyperspectral data and GoogLeNet, Skin Res. Technol. 26 (2020) 891–897.

[72] I. Razzak, G. Shoukat, S. Naz, T.M. Khan, Skin lesion analysis toward accurate detection of melanoma using multistage fully connected residual network, in: 2020 International Joint Conference on Neural Networks (IJCNN), 2020, pp. 1–8.

[73] K. Polat, K.O. Koc, Detection of skin diseases from dermoscopy image using the combination of convolutional neural network and one-versus-all, J. Artif. Intell. Syst. 2 (2020) 80–97.

[74] S. Bama, R. Velumani, N. Prakash, G. Hemalakshmi, A. Mohanarathinam, Automatic segmentation of melanoma using superpixel region growing technique, Mater. Today Proc. 45 (2020) 1726–1732.

[75] I. Iqbal, M. Younus, K. Walayat, M.U. Kakar, J. Ma, Automated multi-class classification of skin lesions through deep convolutional neural network with dermoscopic images, Comput. Med. Imaging Graph. 88 (2021) 101843.

[76] L.R. Soenksen, T. Kassis, S.T. Conover, B. Marti-Fuster, J.S. Birkenfeld, J. Tucker-Schwartz, et al., Using deep learning for dermatologist-level detection of suspicious pigmented skin lesions from wide-field images, Sci. Transl. Med. 13 (2021).

[77] K. Safdar, S. Akbar, A. Shoukat, A majority voting based ensemble approach of deep learning classifiers for automated melanoma detection, in: 2021 International Conference on Innovative Computing (ICIC), 2021.

[78] K. Safdar, S. Akbar, S. Gull, An automated deep learning based ensemble approach for malignant melanoma detection using dermoscopy images, IEEE Xplore (2021).

[79] C. Xin, et al., An improved transformer network for skin cancer classification, Comput. Biol. Med. 149 (2022) 105939.

[80] A.B. Farooq, S. Akbar, R. Arif, S.A. Hassan, S. Gull, Melanoma classification through deep learning using dermoscopic images, in: 2023 International Conference on IT and Industrial Technologies (ICIT), IEEE, 2023, pp. 1–6.

[81] G. Cirrincione, et al., Transformer-based approach to melanoma detection, Sensors 23 (12) (2023) 5677.

[82] S. Khan, A. Khan, SkinViT: a transformer based method for melanoma and nonmelanoma classification, PLoS One 18 (12) (2023) e0295151.

[83] D. Moturi, R.K. Surapaneni, V. Sai, Developing an efficient method for melanoma detection using CNN techniques, J. Egypt. Natl. Canc. Inst. 36 (1) (2024).

[84] M. Rashad, M. Mansour, M. Taha, An efficient approach for automatic melanoma detection based on data balance and deep neural network, J. Comput. Commun. 3 (1) (2024) 22–32.

[85] M. Senthil Sivakumar, L. Megalan Leo, T. Gurumekala, V. Sindhu, A. Saraswathi Priyadharshini, Deep learning in skin lesion analysis for malignant melanoma cancer identification, Multimed. Tools Appl. 83 (6) (2024) 17833–17853.

[86] K. Tang, J. Su, R. Chen, R. Huang, M. Dai, Y. Li, SkinSwinViT: a lightweight transformer-based method for multiclass skin lesion classification with enhanced generalization capabilities, Appl. Sci. 14 (10) (2024) 4005.

[87] S. Aksoy, P. Demircioglu, I. Bogrekci, Enhancing melanoma diagnosis with advanced deep learning models focusing on vision transformer, swin transformer, and ConvNeXt, Dermatopathology 11 (3) (2024) 239–252.

[88] H.C. Reis, V. Turk, Fusion of transformer attention and CNN features for skin cancer detection, Appl. Soft Comput. (2024) 112013.

[89] K. Ramkumar, E.P. Medeiros, A. Dong, V.H.C. de Albuquerque, M.R. Hassan, M.M. Hassan, A novel deep learning framework based swin transformer for dermal cancer cell classification, Eng. Appl. Artif. Intell. 133 (2024) 108097.

[90] R. Wang, et al., A novel approach for melanoma detection utilizing GAN synthesis and vision transformer, Comput. Biol. Med. 176 (2024) 108572.

CHAPTER 17

Enhancing multi-omics cancer subtype classification using explainable convolutional neural networks

Nabaa Abd Mohammed[a] and Mazin Abed Mohammed[b]

[a]Department of Computer Science, College of Computer Science and Information Technology, University of Anbar, Anbar, Iraq
[b]Department of Artificial Intelligence, College of Computer Science and Information Technology, University of Anbar, Anbar, Iraq

1. Introduction

Because high-throughput sequencing technologies have rapidly developed in recent years, the multi-omics study has quickened the development of precision medicine [1]. Therefore, the use of precision medicine in tumors' invasion prevention, diagnosis, and therapy continues to be suboptimal. Multi-omics methods tend to combine different omics profiles from the same patient and as a result, over time more studies have attempted to define the molecular and clinical profiles of cancers utilizing multi-omics perspectives. Integrated multi-omics techniques can be classified into two groups are Euclidean and non-Euclidean integration. The first one takes expression matrix as input and build machine learning models for both clustering and classification. For instance, Chaudhary et al. initially used a deep autoencoder (AE) or the class of deep-belief neural networks [2] for predicting the survival of hepatocellular carcinoma patients [3]. Deep Type, a cancer-relevant data representation and learning framework that incorporates supervised classification, unsupervised clustering, and dimensionality reduction While these methods provide excellent outcome in managing big datasets, they involve a lot of steps to explain how some of the features influence the predetermined results. On the other hand, the Non-Euclidean data integration is based on the fact that the training models are based on the topology data of the network. These methods can distinguish between cancer subtypes according to similarities from different omics datasets where such approaches include; SNF, Grassmann Cluster, and HOPES. Compared to network-based methods, the clinically more intuitive approach is currently at the foreground of methodological development in multi-omics data integration largely concentrating on the unsupervised setup [4].

Tumor classification on different subtypes is relevant while treating cancer and determining the outcome. Specifically, it as a multi-class classification task and presents certain

Explainable AI in Healthcare Imaging for Medical Diagnoses
https://doi.org/10.1016/B978-0-443-23979-3.00017-8

433

difficulties when analyzed with machine learning when integrating multi-omics data [5]. To the best of my knowledge, the existing theories do not accurately reflect the complex reality of a multi-class network with both cancer subtype discriminations and biomarker detection requirements. There are new directions in the usage of graph structures in deep learning, which are convolutional neural networks (CNNs). Thus, CNNs categories all the multiple nodes which are initially unlabeled, by using both structural features of the nodes and their feature representations, making the model inherently interpretable. Cisco has shown that mechanized CNNs can fundamentally be utilized in complex genome-disease and drug-disease affiliation forecasts, which reflects their capability of handling the issues of multi-omics cancer subtype classification.

Thus, in the current study, we have proposed a multi-omics integration model using convolutional neural network for the identification of cancer subtypes. The approach proposed in this paper develops a novel diagnostic model, which enforces the pipeline of "multiple omics data integration, and then classification." This method allows for a more accurate and detailed cancer subtype partition because it integrates multi-omics data optimally into a deep learning structure.

- The study present a comprehensive data preprocessing pipeline that guarantees the provision of high quality input for multi-omics cancer subtype classification. This step becomes crucial for managing large datasets, determine data variability and remove noise to enhance model performance.
- The classifying feature of K-Means clustering for feature extraction is relatively new in multi-omics analysis. It enables capture of significant data patterns while aggregating features that are similar to one another, which lowers the amount of data that is being used in cancer subtype classification although the important features that are necessary for classification are preserved.
- SelectKBest is the technique used in the study to select the most appropriate features from the multi-omics data. This makes the process deliver the best performing models by improving the specificity of features being used, the computation effort and classification of the models.
- The use of preprocessing, feature extraction process, feature selection step, and the classification using CNN leads to overall higher accuracy in cancer subtype classification. Also, the big advantage of the model reported in the study is that it provides better interpretability through allowing the embedding of additional steps like K-Means and SelectKBest enabling insight into the most significant features influencing the classification.

The rest organization of this study are: Section 2 presented the related works on Multi-Omics Cancer classification with AI methods. The proposed methodology for Enhancing Multi-Omics Cancer Subtype Classification Using CNN in Section 3. Section 4 presented the results and discussion with features selected of multi omics data. Finally, The conclusion and future works presented in Section 5.

2. Related work

For understanding the multifaceted aspects of the diseases including Alzheimer's, Parkinson's and cancer, the integration of multi-omics data is a critical part of study [6–9]. Nevertheless, this task is not trivial at all and demands the application of advanced computational tools. New methods of data analysis and integration become problematic and require the development of new mathematical models to extract biologically significant information from multi omics data and integrate the results into a unified picture of the pathogenesis of the disease. However, due to the nature of high dimensionality and structural complexity involved in such data, multi-omics integration has the potential to greatly enhance our understanding of cancer biology. According to Zhang et al. [10] proposed a novel deep learning method based on multi-omics and attention mechanism to effectively identify cancer subtypes called multiGATAE. multiGATAE first constructs a similarity graph by integrating multi-omics data, and then feeds the similarity graph and omics data into a graph autoencoder network consisting of a graph attention network and an omics-level attention mechanism to obtain the embedding representation. Once the embedding representation is obtained, the K-means clustering method is applied to it to identify the subtypes. multiGATAE is compared with eight public cancer datasets. The C-index, which evaluates the predictive performance of the model, showed that Multi-GATAE outperformed other methods on the datasets where MultiGATAE achieved a C-index of 0.644, outperforming others. The study used eight cancer datasets from TCGA, but the sample sizes for each dataset were relatively small. A single log-rank test was also used to assess differences between subtypes [10].

Also, García-Díaz et al. [11] proposed a gene clustering algorithm (GGA) to solve the maximum diversity clustering problem. It was applied to classify an imbalanced database of 801 samples of RNA-Seq data for gene expression in 5 cancer types. The samples consist of 20,531 genes. GGA extracts several sets of genes that achieve high accuracy in multiple classification. The accuracy was evaluated by the Extreme Learning Machine algorithm and was found to be slightly higher in balanced databases than in imbalanced databases. The final classification decision was made by a weighted majority voting system among feature sets. The proposed algorithm finally selects 49 genes to classify the samples with an average accuracy of 98.81% and a standard deviation of 0.0174. The study identifies a major limitation of the gene clustering algorithm (GGA) related to exploring the solution space. Specifically, it is noted that the huge number of features (20,531 genes) is a significant challenge, as selecting several dozen genes from such a large set represents a huge search space. Thus, adequate exploration of this space is not guaranteed, which may lead to incomplete solutions [11]. Another study by Ma [12] proposes a novel method with deep multi-network embedding, called DeepMNE, to discover potential associations between long noncoding RNAs and diseases, especially for novel diseases and long noncoding RNAs. DeepMNE extracts multi-omics data to describe diseases and long noncoding RNAs, and proposes a deep learning-based network fusion method to integrate multi-source information. Moreover, DeepMNE complements the sparse association network and uses kernel neighbor similarity to construct disease similarity and long noncoding RNA similarity networks. Moreover, a graph embedding method is adopted to predict potential associations. Experimental results show that DeepMNE has higher predictive performance on novel associations, novel lncRNAs, and novel diseases. In addition, DeepMNE also achieves significant predictive performance on disordered datasets with F-score = 87%, AUC = 94%. The accuracy of predicting lncRNA-disease associations often depends on the availability of large, high-quality datasets. However, public datasets on lncRNA-disease associations are incomplete or imbalanced, limiting the model's ability to generalize well to new data [12]. New study by Cristovao et al. [13] propose to use machine learning capabilities, especially deep learning, to identify breast cancer subtypes. Given the scarcity of publicly available data, both cancer and non-cancer data are used to design semi-supervised setups. Multi-gene data, including microRNA expression and copy number alterations, are utilized. Accuracy results obtained show that simpler models perform at least as well as deep semi-supervised methods on gene expression data. When multi-gene data types are combined together, the performance of deep models shows a slight improvement in accuracy, reaching 94%, suggesting the need for further analysis on larger multi-gene datasets. Due to the lack of publicly available large datasets, further analysis of larger datasets, especially those containing multigene data, is needed to fully evaluate the potential of deep learning models. Also, while exploring the combination of multiple types of multigene data (e.g., gene expression, microRNA, and copy number alterations), deep models showed only modest improvements in accuracy. This suggests the need for more comprehensive multigene datasets for further investigation [13]. Another study by Gao et al. [14] proposed a supervised cancer classification framework, Deep Cancer Subtype Classification (DeepCC), based on deep learning of functional spectra that identify biological pathway activities. In two case studies on colorectal cancer and breast cancer classification, both DeepCC classifiers and DeepCC single-sample predictors achieved overall higher sensitivity, specificity, and accuracy compared to other widely used classification methods such as random forests (RF), support vector machine (SVM), gradient boosting machine (GBM), and multinomial logistic regression algorithms. DeepCC showed higher sensitivity, specificity, and accuracy on the validation datasets with P-values indicating strong statistical significance (e.g., $P = 8.47 \times 10^{-20}$ for sensitivity, $P = 1.61 \times 10^{-10}$ for specificity, and $P = 3.07 \times 10^{-21}$ for overall accuracy) when compared to other classifiers like RF, SVM, GBM, and multinomial logistic regression Simulation analysis based on random sampling of genes demonstrated the power of DeepCC in dealing with missing data. One limitation of reliance on gene expression signature-based classification methods is their poor robustness to batch effects and platform variability, leading to inconsistent results [14].

The study by Li and Nabavi [15] proposed a multi-gene end-to-end graph neural network (GNN)-based framework for accurate classification of cancer subtypes. The proposed model uses multi-gene data in the form of heterogeneous multi-class graphs, which combines inter-gene and intra-gene connections from established biological knowledge. The proposed model incorporates learned graph features and global genome features for accurate classification. The proposed model is tested on The Cancer Genome Atlas (TCGA) cancer dataset and Invasive Breast Cancer (BRCA) dataset to classify molecular subtype and cancer subtype, respectively. The proposed model shows superior performance compared to four state-of-the-art baseline models in terms of accuracy, F1 score, precision, and recall. The proposed model with GAT achieved accuracy 83.9% and F1 score: 84% while the proposed model with GCN achieved accuracy: 81.2% and F1 score: 81% while FC–NN achieved accuracy: 78.4% and F1 score: 75%. One of the limitations of this study is that two well-established and widely adopted GNN models were used. Due to the rapid development of GNN models, new models have emerged with the recent boom in GNN models. Also, as the graph size grows or more genomes are added, GAT-based models become more sensitive to parameters and take longer to train [15]. Also, Ma et al. [16] proposed a classification model by leveraging the power of extreme gradient boosting (XGBoost) and using increasingly complex multi-genome data to separate early-stage and late-stage cancers. The XGBoost model was applied to four types of cancer data (HNSC, LUSC, KIRP, and KIRC) downloaded from TCGA and its performance was compared with other popular machine learning methods. Experimental results showed that the proposed method obtained statistically significantly better or comparable predictive performance, achieving an accuracy of 80.8% on the KIRC dataset, 85.2% on the KIRP dataset, 79.4% on the HNSC dataset, and 83.4% on the LUSC dataset. Moreover, the integration of multi-genome data by autoencoder can enhance the accuracy of cancer stage classification. Finally, bioinformatics analysis was conducted to evaluate the medical utility of the important genes ranked by their importance using the XGBoost algorithm. Large-scale comparative experiments have shown that the XGBoost method has excellent performance in predicting cancer stage using multi-genome data. Moreover, identifying new candidate genes associated with cancer stage would contribute to further elucidating the disease progression mechanism and developing new treatments. A general limitation of this study is that the sample size for each cancer type in The Cancer Genome Atlas (TCGA) is relatively small, which may affect the generalizability of the results [16].

However, identifying correct cancer subtypes is essential for individualized treatments and using complex data of multiple layers is a promising approach for understanding disease mechanisms including cancer. The first limitation of the existing machine learning techniques is that they are incapable of capturing the complex interactions existing across diverse omics datasets. To overcome CNN based approach has been developed to detect latent features and distinct graph structures of different omics data to improve the cancer prediction and offer greater distinguishability of patients for better disease control in medical study.

3. The proposed CNN Omics approach

The methodology of enriching the multi–omics cancer subtype identification is based on the explainable deep learning model, CNN, and includes critical steps for better functionality. First, the genomic, transcriptomics and proteomics data for comparison are normalized using Standard Scaler in order to apply the same model training. Second, the k-mean clustering analysis is used to extract features where similar data points are clustered to minimize noise levels. Feature selection is done using SelectKBest to retain the first k features with the highest ANOVA F-value meaning that only the most statistically important features are used for input data organization for enhanced efficiency and precision. The filtered features are then passed into an Explainable CNN, which generates the class specific high level features and provides interpretability using Grad CAM to visualize the regions that make the classification possible. Lastly, the model is trained using cross-entropy loss and the Adam optimizer, and quantified the reliability and effectiveness of our cancer subtype prediction using accuracy, F1-score, and AUC analysis. The proposed Explainable CNN Omics approach presented in Fig. 1.

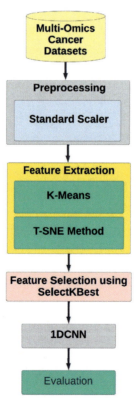

Fig. 1 The proposed Explainable CNN Omics approach.

3.1 Omics dataset

The dataset is unrestricted for academic use and can be found on the Internet. The dataset we have in combined form has huge amount of data on different kinds of cancer. In the context of the dataset, each raw can be considered as a distinct patient sample, while each column of the matrix offers important insight of the molecular nature of cancer [17]. An integrated view of many types of cancers has been provided such as hepatocellular carcinoma (HCC), brain cancer (GBM), pancreatic cancer (PaCa), breast cancer (BrCa), colon cancer (CoCa), gastric cancer (GaCa) and lung cancer. A comprehensive picture of these disorders is now emerging at a molecular level, and special attention is being paid to the cells that make up cancer. The current data includes 20,244 samples and 69 features Total number of features for dataset of multi-omics of cancer subtypes are described in Table 1.

Table 1 The details of omics dataset used.

No.	Cancer type/healthy	Number of features
1	Healthy	8
2	Lung cancer	15
3	Hepatocellular carcinoma	10
4	Pancreatic cancer	7
5	Breast cancer	4
6	Colon cancer	4
7	Gastric cancer	5
8	Brain cancer	4
9	Hepatitis B (HBV)	7
10	Blood	4
	Total	68

3.2 The preprocessing using Standard Scaler

In this step the multi-omics data is scaled to a standard deviation of zero mean and a variance of one. This is important in getting a good result from the machine learning algorithm. The Standard Scaler works by standardizing the input data so that it's mean is zero and variance is one. Namely, it standardizes each feature, that is, it moves every value of a given feature to around the same mean and scales the spread of values to the same standard deviation. This makes it possible that all features are of scale during learning and do not in their estimate influence the learning since possibly one feature could be more influential than the other. Standardization is especially pertinent for multi-omics data analysis since genomic, transcriptomic, as well as proteomic data vary significantly in their range of values. The use of the Standard Scaler prepares the data for analysis in a more comparable form across dimensions to enhance the stability of machine learning algorithms like K-means clustering and the Explainable CNN used in classifying cancer subtypes in our model.

3.3 Omics feature extraction using K-Means

During this phase, the K–Means clustering algorithm is used to extract the relevant feature about the multi-omics datasets. The K–Means algorithms working on data is split into K cluster based on the resemble in features. In cancer subtype classification, K–Means is useful in discovering internal structures in the multiple omics data through the patients or data point that has close molecular similarity [18]. As a result of clustering the data, K–Means simplifies the high-dimensional omics landscape while identifying essential groupings for subclassification of cancer. Such extracted cluster-based feature is further applied to improve the interpretability and accuracy for the classification work, and form a more standard input stream for the data digesting model—Explainable CNN. This step prevents overloading the subsequent analysis with many unnecessary and less noticeable patterns, which were detected in the multi-omics data. Steps for Feature Extraction using K-Means:

- **Initialization**: K is given, which is the number of clusters, that can be directly translated as the number of subtypes or key groups in the multi-omics data.
- **Cluster assignment**: There is a formation of K groups of the dataset where each data point is associated with the cluster with the shortest Euclidean distance. It means that the commonly used random centroid is applied as the starting centers for clustering.
- **Update centroids**: Centroids (mean points of clusters) are updated, by taking the average of data points in the particular cluster.
- **Iterative reassignment**: Assigning data points to the closest centroid and re-estimating centroid values is repeated repeatedly until all clusters do not change their categories or until some convergence criterion is achieved.
- **Cluster representation**: The last clusters stand for patients or data points that have high molecular similarity and can be grouped into separate clusters. These clusters are what is referred to as the feature extraction dataset.
- **Feature transformation**: The cluster assignments (labels) are feature mapped, in other words, they are transformed into features that represent the Multi-omics data more structured and feature-based.
- **Output**: Cluster-based representation extracted is taken to next step of feature selection and classification through Explainable CNNs.

3.4 Feature selection using SelectKBest

Feature selection is a part of the data preprocessing phase that identifies and preserves the most informative features of the database while removing uninformative and redundant features [19]. In this work, we utilize the SelectKBest method to improve the explanatory outcomes of our Explainable CNN since it helps in filtering the important features known for improving the model's accuracy. Steps for Feature Selection using SelectKBest:

- **Define the target variable**: Determine which variable we would like to predict that refers to the cancer subtypes.
- **Select the scoring function**: However, it depends on the type of features and the problem at hand, and the scoring function to be used is as follows. Commonly used functions include:
  - ➢ For continuous data: f_regression (for regression type problems).
  - ➢ For categorical data: chi2 or mutual_info_classif when the case is for classification.
- **Fit the SelectKBest model**: Filter the multi-omics dataset (features) and the target variable using the SelectKBest in order to select only the most relevant features. This model quantifies all the features using the selected scoring function.
- **Select the top K features**: Suggest the number of features that must be retained say K. The SelectKBest model sorts the features according to the score they have obtained and includes only the K numbers of features that show potential to be most relevant to the target variable.
- **Transform the dataset**: New-feature set = Select only K features of the original-feature set. Such a transformation helps decrease dimensionality and bring the model's attention to the set of features that can have the most impact on the classification of the sample.
- **Output**: The resulting set of selected features is then formatted to feed into the Explainable CNN, to enhance the learning process and increase the model interpretability.

3.5 Classification based on Explainable CNN

The analysis of the classification using Explainable CNN aims at probing how the identified features from the multi-omics are useful in predicting different cancer subtypes with an emphasis on model interpretability. The methodology emphasizes not only achieving high accuracy in classification but also providing an explanation of the machine's decision-making process. Steps for Classification using Explainable CNN:

1. **Input preparation**: These features selected in the previous step (SelectKBest) will then be used as the input to the Explainable CNN. These inputs represent the most informative part of the multi-omics data, focusing on the substituent variability.
2. **Network architecture design**: Design the layers of the Explainable CNN; most of the Explainable CNNs contain two or more convolutional layers accompanied by pooling layers. To achieve this, the above architecture provides a way of obtaining hierarchical feature representation of the input data by design. The architecture may also incorporate:
   (a) **Dropout layers**: In order to reduce the overfitting of this type, a given percentage of neurons are randomly set to zero during the training process.

(b) **Batch normalization**: For purposes of making training more stable and faster, the output of one layer is normalized with the purpose of feeding it into the subsequent layer.

3. **Activation functions**: Employ non-symmetric activation functions such as ReLU (Rectified Linear Unit) activation function after the convolutional layers since it adds non-linearity that enables the model to learn non-linear dependencies on the training data.

4. **Output layer**: The last layer depends on the problem type but for the multi-class classification their output should be probabilities of the classes, if they are cancer subtypes.

5. **Loss function and optimizer**: A loss function is specified, which for multi-class classification is usually categorical cross-entropy. Select an optimizer (Adam or SGD) that works on updating the parameters within the model during the training process.

6. **Training the model**: Therefore Explainable CNN is trained using the prepared dataset which involves:
   • Passing the input features through the network used.
   • Calculating the loss depending on the estimated outcomes and the real target values.
   • Propagating the error in order to update the weights until it reaches the correct solution.

7. **Model evaluation**: XAI follows the training/validation/emphasis paradigm, after training, the model should be tested for its accuracy, precision, recall, and F1 score etc., on a validation dataset. This step determines how well the model performs when tested with new data which is never used in training.

4. Results and discussions

In this study, we integrated an Explainable CNN into the multi-omics data for cancer classification on the subtype of cancer. The overall objective was to improve the predictive models that define cancer subtypes as such diagnostic methods are severely lacked in modern oncology. The Explainable CNN model builds on convolutional neural networks advantage while implementing the features to make the computation process more transparent. Our approach started with the preprocessing of Multi-omics data in which the data undergoes a scaling and normalization in order maximize their format for analysis. We fine-tuned the K-Means clustering for feature extraction that familiarize the system with patterns present in multi-omics data. After this, in order to select the optimal features important to classification use the SelectKBest and Chi-square scores were used to provide the best features that made the most contribution to the classification process.

The results of the Explainable CNN model proposed in this study were assessed according to conventional evaluation measures including accuracy, precision, recall, and F1-score. Moreover, the focus was made on interpretability and visualization of feature importance alongside the decision–jumping paths by using feature importance and decision plots, necessary for understanding the molecular mechanisms related to different cancer subtypes. In summary, the results proved that our Explainable CNN model could automatically classify the images effectively, and the understandability of the features improved the situation awareness for clinicians to make decisions according to the patient's conditions.

In this section, a CNN is demonstrated on a multi-omics cancer dataset. The proposed deep learning algorithm for CNN will be fed with features. After the features are obtained, they will be passed to the three convolutional layers, flat and dense layers that constitute the deep learning CNN algorithms proposed in this thesis. The input data is subjected to 32 filters by the first convolutional layer, each of which performs a convolutional operation that scans the input with a kernel size of 3. Low-level features can be extracted using this procedure. Max pooling is used to reduce the computational burden and spatial complexity while retaining the most important features after the initial convolution. A subsequent max pooling layer further reduces the dimensionality, ensuring that the most salient features are preserved, while a second convolutional layer analyzes the pooled feature maps using 64 filters, capturing more complex patterns. In order to capture more complex and abstract features, 128 filters are applied by the third convolutional layer. The 2D feature maps are then converted to a 1D vector by the regularization layer, making them suitable for use in dense layers that perform classification. To add nonlinearity, a first dense layer with 128 ReLU activation units is applied. Then, an output layer consisting of two Softmax activation units is used to generate a probability distribution of multi-omics cancer classes. For 100 epochs, the model is trained using the Adam optimizer with the sparse_categorical_crossentropy loss function. After testing with 16, 64, and 128 iterations, this value is calculated using experimental epochs with a batch size of 32. The results of the experiments show that 32 iterations per epoch is the optimal batch size, with memory requirements and iterations per epoch increasing by 64 and 128 iterations per epoch, respectively. Validation is the part of training that helps monitor how well the model handles new data. After a number of encounters across different time periods. To avoid overfitting and obtain better experimental results, it is important to choose the right number of epochs. The aim of determining these values empirically is to avoid overfitting. The performance of the CNN classifier is shown in Table 2 on the multi-omics cancer dataset.

Table 2 Evaluation metric on multi-omics cancer dataset with CNN.

Class	Precision	Recall	F1-measure	Accuracy
Cluster 0	94%	92%	93%	93.9%
Cluster 1	77%	67%	71%	
Cluster 2	87%	93%	90%	
Cluster 3	100%	98%	99%	
Cluster 4	67%	100%	80%	
Cluster 5	97%	90%	94%	
Cluster 6	91%	97%	94%	
Cluster 7	88%	92%	90%	
Cluster 8	98%	95%	96%	
Cluster 9	86%	92%	89%	

The accuracy, precision, recall, and F1 score values for each class are shown in Table 2 and the outcomes are as follows: accuracy 93.9%, precision 94.1%, recall 93.9%, and F1 score 93.9% on the multi-omics cancer dataset, as shown in Fig. 2.

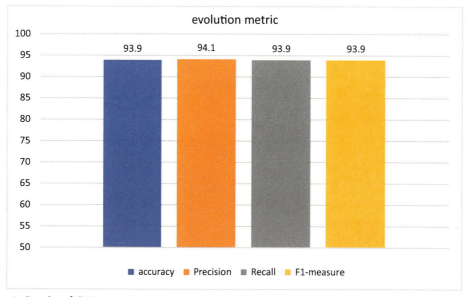

Fig. 2 Results of CNN metrics.

A confusion matrix is a graph that evaluates the performance of a classification method. A confusion matrix is created to thoroughly evaluate the model's performance. The efficacy of the CNN approach is assessed by computing its confusion matrix, seen in Fig. 3, after the implementation of CNN using multi-omics cancer dataset and attaining notable accuracy.

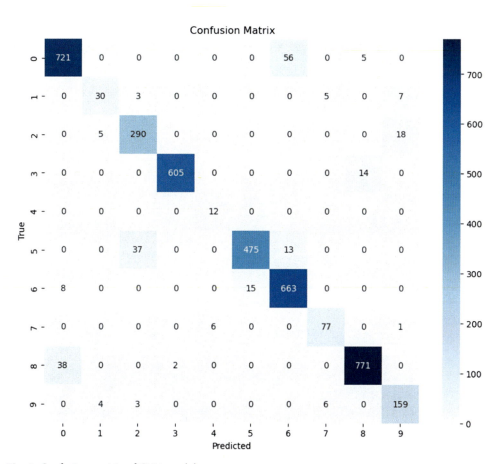

Fig. 3 Confusion matrix of CNN model.

After training is completed, the model is evaluated on the test set to determine its accuracy level. Using the CNN model which is used in deep learning for multi-omics cancer data, an accuracy of up to 93.9% is achieved. The training and evaluation accuracy is predicted using an equal number of epochs. The loss function is optimized via the Adam optimizer. Training and validation accuracy of the proposed model. The structure of the CNN algorithm is shown in Fig. 4, while the training and validation loss is represented in Fig. 5.

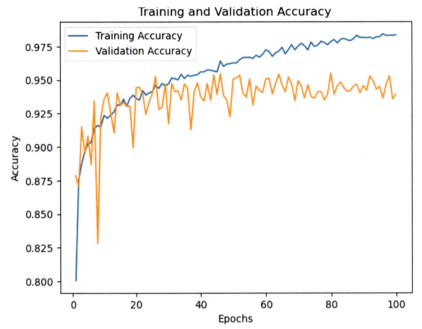

Fig. 4 The training and validation accuracy for CNN.

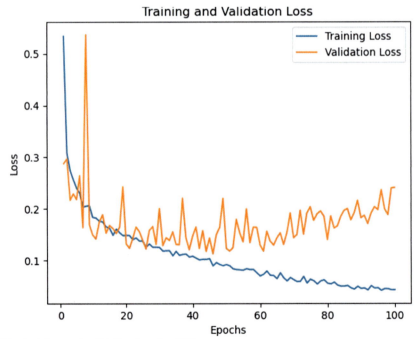

Fig. 5 The training and validation loss for CNN.

4.1 Discussion

These outcomes show the CNN classification model's performance metrics for each of the 10 groups or classes. The model shows balanced performance in both recognizing genuine positives and avoiding false positives in Group 0, as seen by its good precision, recall, and F1 score when classifying cases in Group 0. With precision at 94%, recall at 92%, and F1 score at 93%, there are very few misclassifications to be indicated by the minor decline in precision. Group 1's F1 score is lower at 71% due to weaker recall (67%), but comparatively good precision (77%). This shows that the model has poor recall because, while it is adept at properly detecting Group 1 cases, it misses some of them. Low recall suggests that some real benefits go unnoticed. Classifying examples from Group 2 yields similar results for the model: 87% precision, 93% recall, and 90% F1 score. This shows that the model for this set is well-balanced, efficiently detecting real positives while averting false positives. All of the anticipated occurrences for Set 3 are true because precision is 100% flawless. Recall, however, is little lower at 98%, suggesting that a tiny percentage of actual positives are being overlooked. Even at 98%, the F1 score is still quite high, indicating a solid overall performance. For Set 4, recall is excellent at 100%, while precision is at 67%, suggesting that there are erroneous positives. This implies that there aren't many true positives being missed by the model, which would affect the 80% F1 score.

The discrepancy between recall and precision points to an imbalance, which may be brought about by the model's more cautious forecasts for this collection. With a high recall of 90% and a precision of 97%, the model performs well and has a solid F1 score of 94%. This suggests that the model is thorough in recognizing the real instances of Cluster 5 and accurate in predicting them. The majority of the real positives in Cluster 6 were accurately detected, as shown by the high recall of 97%. Although there are fewer false positives, precision is somewhat greater at 97%, indicating good overall performance as seen by the high F1 score of 94%. With an F1 score of 90%, the model accurately recognizes the majority of Cluster 7 examples with a decent recall of 88%. However, precision is somewhat higher at 92%, indicating that the model maintains a reasonable balance between precision and F1 score. Cluster 8 has extremely high recall and precision (98% precision, 95% recall), and the model does very well in identifying this cluster. As a consequence, the model for this cluster has an excellent F1 score of 96%, suggesting that it is balanced and effective. While recall in Cluster 9 is somewhat higher at 92%, showing that some instances are missed, precision in Cluster 9 is strong at 86%, indicating that the majority of anticipated positives are right. Even with the F1 score of 89%, this cluster is doing well overall. Excellent results on the majority of metrics across the majority of clusters point to a well-rounded model that can successfully manage the classification process.

5. Conclusion

In this study, we presented the Explainable CNN for the classification of cancer subtypes based on multi-omics data. This way of working was proven successful in integrating state of the art machine learning issues with a focus on interpretability which is crucial in clinical applications. Thus, approaching the problem in a systematic way including data cleaning, feature extraction by clustering and KBest feature selection we were able to provide the model with the most important biological markers for cancer differentiation. Moreover, Explainable CNN-based has increased the accuracy 93.9%, precision 94.1%, recall 93.9%, and F1 score 93.9% effectively than the traditional cancer subtypes. Experiments conducted for our proposed Explainable CNN show significant results, further asserting our research and advance work for precise cancer subtypes classification. The integration of explainability features ensure that increased understanding was achieved on outcomes of relative classification by incorporating aspects of transparency of the models. This is especially the case in oncology, as appreciation of the underpinning biology can hugely inform management plans. In summary, the application of Explainable CNNs can be promisingly promising in the multisite field and provides an opportunity to improve cancer treatment. For these, future studies can expand the present study by utilizing a larger dataset and improving the architecture of the models used for the investigation to improve both accuracy and comprehensibility of the predictions. Moreover, future studies should attempt to integrate other layers of omics layers including metabolomics and epigenomics to Explainable CNN model. This integration could improve the predictive certainty since molecular profiling offers a greater amount of detail regarding cancer subtypes and the way they communicate. Also, could use Explainable CNNs, as well as, capture multi-omics data at different time points to perform longitudinal studies. This approach would aid the study of cancer subtype evolution and effects of treatment in order to better understand the progression of cancer and provide more tailored treatment.

References

[1] A.M. Ali, M.A. Mohammed, A comprehensive review of artificial intelligence approaches in omics data processing: evaluating progress and challenges, Int. J. Math. Stat. Comput. Sci. 2 (2024) 114–167.

[2] G.E. Hinton, R.R. Salakhutdinov, Reducing the dimensionality of data with neural networks, Science 313 (5786) (2006) 504–507, https://doi.org/10.1126/science.1127647.

[3] K. Chaudhary, O.B. Poirion, L. Lu, L.X. Garmire, Deep learning-based multi-omics integration robustly predicts survival in liver cancer, Clin. Cancer Res. 24 (6) (2018) 1248–1259, https://doi.org/10.1158/1078-0432.CCR-17-0853.

[4] M.A. Mohammed, K.H. Abdulkareem, A.M. Dinar, B.G. Zapirain, Rise of deep learning clinical applications and challenges in omics data: a systematic review, Diagnostics 13 (4) (2023) 664.

[5] M.A. Mohammed, A. Lakhan, K.H. Abdulkareem, B. Garcia-Zapirain, Federated auto-encoder and XGBoost schemes for multi-omics cancer detection in distributed fog computing paradigm, Chemom. Intell. Lab. Syst. 241 (2023) 104932.

[6] M. Mohammed, Enhanced cancer subclassification using multi-omics clustering and quantum cat swarm optimization, Iraqi J. Comput. Sci. Math. 5 (3) (2024) 552–582.

[7] M.A. Mohammed, A. Lakhan, K.H. Abdulkareem, B. Garcia-Zapirain, A hybrid cancer prediction based on multi-omics data and reinforcement learning state action reward state action (SARSA), Comput. Biol. Med. 154 (2023) 106617.

[8] M.S. Karthika, H. Rajaguru, A.R. Nair, Performance enhancement of classifiers through bio inspired feature selection methods for early detection of lung cancer from microarray genes, Heliyon 10 (16) (2024) e36419.

[9] Z. Momeni, E. Hassanzadeh, M. Saniee Abadeh, R. Bellazzi, A survey on single and multi omics data mining methods in cancer data classification, J. Biomed. Inform. 107 (December 2019) (2020) 103466.

[10] G. Zhang, Z. Peng, C. Yan, J. Wang, J. Luo, H. Luo, MultiGATAE: a novel cancer subtype identification method based on multi-omics and attention mechanism, Front. Genet. 13 (March) (2022) 1–10.

[11] P. García-Díaz, I. Sánchez-Berriel, J.A. Martínez-Rojas, A.M. Diez-Pascual, Unsupervised feature selection algorithm for multiclass cancer classification of gene expression RNA-Seq data, Genomics 112 (2) (2020) 1916–1925.

[12] Y. Ma, DeepMNE: deep multi-network embedding for lncRNA-disease association prediction, IEEE J. Biomed. Health Inform. 26 (7) (2022) 3539–3549.

[13] F. Cristovao, S. Cascianelli, A. Canakoglu, et al., Investigating deep learning based breast cancer subtyping using pan-cancer and multi-omic data, IEEE/ACM Trans. Comput. Biol. Bioinform. 5963(c) (2020) 1–14.

[14] F. Gao, W. Wang, M. Tan, et al., DeepCC: a novel deep learning-based framework for cancer molecular subtype classification, Oncogenesis 8 (9) (2019).

[15] B. Li, S. Nabavi, A multimodal graph neural network framework for cancer molecular subtype classification, BMC Bioinformatics 25 (1) (2024) 1–19.

[16] B. Ma, F. Meng, G. Yan, H. Yan, B. Chai, F. Song, Diagnostic classification of cancers using extreme gradient boosting algorithm and multi-omics data, Comput. Biol. Med. 121 (2020) 103761.

[17] CfOmics, A Cell-Free MultiOmics Database for Diseases, 2024. https://cfomics.ncrnalab.org.

[18] L.Y. Guo, A.H. Wu, Y.X. Wang, L.P. Zhang, H. Chai, X.F. Liang, Deep learning-based ovarian cancer subtypes identification using multi-omics data, BioData Min. 13 (2020) 1–12.

[19] Y. Huang, P. Zeng, C. Zhong, Classifying breast cancer subtypes on multi-omics data via sparse canonical correlation analysis and deep learning, BMC Bioinformatics 25 (1) (2024) 132.

CHAPTER 18

Explainable convolutional neural network for Parkinson's disease detection

Aiesha Mahmoud Ibrahim[a] and Mazin Abed Mohammed[b]
[a]Department of Computer Science, College of Computer Science and Information Technology, University of Anbar, Anbar, Iraq
[b]Department of Artificial Intelligence, College of Computer Science and Information Technology, University of Anbar, Anbar, Iraq

1. Introduction

Parkinson's disease (PD) is a neurological disorder that develops progressively and attacks millions of people around the world [1]. It is marked by tremors, rigidity, and poor balance. Early diagnosis of PD is crucial for relieving symptoms and bettering the patient's life quality. However, the current diagnostic methods are generally subjective and are heavily dependent on the clinical assessment, often resulting in not only later detection but also misdiagnosis, especially, in the early stages of the disease [2]. This underlines the necessity for more objective, data-based endeavors to enhance diagnostic precision.

With developments in the area of artificial intelligence and deep learning, there has been a growing interest in the application of these technologies in the medical field [3]. Convolutional Neural Networks (CNNs), a high-quality deep learning model, have shown the significantly remarkable success in the fields of image recognition and classification tasks [4]. Several recent studies have delved into the feasibility of CNNs to thoroughly analyze the complex biomedical data including voice records, motion signals, and brain scans with the intention of detecting PD-causing patterns [5]. The aforementioned algorithms have the ability to disrupt the conventional ways of diagnosing diseases through automation and higher precision diagnostic. Many challenges are faced as a result of the latest techniques used for Parkinson's disease detection. One of the main problems is the lack of data caused both by the scarcity of data on Parkinson's disease and the lack of diversity, leading to potential overfitting and reduced generalizability of the model. Besides, the variation in what input data the CNN uses, like MRI scans, gait analysis, or speech recordings, will make the system fail in detecting Parkinson's accurately [6]. The next critical issue is the automatic feature learning process in CNNs, which can lead to the inability to detect the subtle and early-stage

Explainable AI in Healthcare Imaging for Medical Diagnoses
https://doi.org/10.1016/B978-0-443-23979-3.00018-X

features; to ensure that this fault is not detected, some issues such as pre-processing the data or domain-specific feature engineering can be done [7]. Moreover, CNN architectures are quite often compared to "black boxes," which make it difficult even for the experts to understand what the models have done/learned. Despite their profitable application in medicine, clinicians are still in need of transparent and explainable models [8]. The lack of clear insights into the decision-making process can limit the clinical adoption of CNN-based methods. Class imbalance is another challenge as Parkinson's disease datasets are often unbalanced. They have significantly fewer patients than healthy individuals, which leads to biased predictions toward the majority class and a decrease in sensitivity to PD cases. Lastly, these CNN models are non-robust to the real-world setting [9]. Not only the data is noisier and more variable but the models cannot retain performance across different clinical environments with various data acquisition techniques, patient demographics, and environmental conditions, either [10].

In this study, we aim to develop and evaluate a CNN-based model for Parkinson's disease detection. By leveraging a dataset that includes various biological markers, we train and test the model to classify patients as either PD-positive or healthy controls. Our approach not only focuses on optimizing the model's accuracy but also on providing a robust framework that could eventually be implemented in clinical settings to assist healthcare professionals in diagnosing PD at an earlier stage. The results show promising improvements in detection accuracy, offering a valuable tool for PD diagnosis. The main contributions of this study are:

- A well-designed preprocessing approach was implemented, combining various techniques to enhance data quality and representation. These improvements allowed the CNN to detect subtle patterns, even without explicit examples of well-established models for Parkinson's disease.
- By utilizing Generative Adversarial Networks (GANs), the dataset was expanded from 195 to 1000 samples, effectively addressing the issue of limited data. This enhancement improved the model's training process, resulting in better generalization and reduced overfitting.
- The Explainable Convolutional Neural Network (CNN) was carefully optimized, incorporating crucial steps such as feature extraction and selection. The model demonstrated high accuracy and robustness in Parkinson's disease classification, outperforming traditional methods and offering broader applications.

The rest organization of this study are: Section 2 presented the related works on Parkinson's disease classification with AI methods. The proposed methodology for optimized CNN for Parkinson's disease detection in Section 3. Section 4 presented the results and discussion with features selected of Parkinson's disease data. Finally, The conclusion and future works presented in Section 5.

2. Related works

This part of the study discusses machine learning (ML) applications used for PD diagnosis. The prior study about CNNs for the medical image analysis, feature extraction, and classification are examined in the paper. The authors used these models to detect neurodegenerative diseases and they were found to be better in the case of the AD diagnosis and in general there was an improvement in the accuracy of the cancer cell model. The section also reviews the hurdles of the early-stage detection, data rarity, and the utility of data augmentation methods such as GANs for enhancing classification accuracy. According to Sharanyaa et al. [11] developed a method for diagnosing Parkinson's disease using audio signals and advanced preprocessing to remove noise. Key features, including the novel delta-amplitude modulated signal (delta-AMS), were extracted and selected using a custom Squirrel Water Search Algorithm (SSWA). The selected features were then input into an attention-based LSTM model, which was trained using SSWA. The model achieved 92.5% accuracy, 95.4% sensitivity, and 91.4% specificity in detecting Parkinson's disease [11]. Also, Wodzinski et al. [12] proposed a method to detect Parkinson's disease using continuous-vowel sounds and a ResNet architecture. Spectrograms of audio recordings were used as image inputs to a ResNet pre-trained on ImageNet and SVD databases. The dataset was boosted in the time domain to prevent overfitting, and the Parkinson's dataset included 100 patients (50 healthy/50 diagnosed). The model achieved over 90% accuracy on the validation set, demonstrating the effective transfer of image-based features to audio spectrograms for Parkinson's detection [12]. Another study by Rahman et al. [13] applied deep learning and machine learning methods to predict and classify early-stage Parkinson's patients using speech signals from the UCI Parkinson's Disease Repository. Multiple classifiers, including XGBoost, Random Forest, SVM, and deep neural networks (DNN), were evaluated. XGBoost achieved the highest accuracy of 92.18% among machine learning models, while the DNN2 model achieved the best accuracy of 95.41% among deep learning techniques. The study concluded that deep neural networks outperformed traditional machine learning methods for Parkinson's classification [13].

Gunduz [14] proposed two CNN-based frameworks for Parkinson's disease classification using speech features from 252 individuals (188 PD patients, 64 healthy). The first framework combines feature sets before feeding them into a 9-layer CNN, while the second framework processes feature sets through parallel input layers connected to convolution layers. The models were trained using the UCI Machine Learning Repository dataset and validated with Leave-One-Out Cross-Validation (LOPO CV). The second framework showed better performance, achieving 86.9% accuracy compared to 84.7% for the first. A key limitation was the small dataset size [14]. Also, Mian [15] conducted an experimental study on the open-source Kaggle PD speech dataset, using novel techniques to identify Parkinson's disease. An unsupervised autoencoder was proposed for feature selection, with compressed features passed to machine learning algorithms like SVM, logistic regression, random forest, and naive Bayes, alongside a 1D CNN. The classifiers were evaluated using accuracy, precision, recall, and F1-score. The 1D CNN achieved the highest F1-score of 92.7%, with logistic regression close behind at 92.2%. The study's key contribution is the novel use of unsupervised neural network feature selection for Parkinson's disease detection [15].

According to the surveys mentioned, the research gap can be found in several important spots. The first key problem of many is that the existing studies are challenged by the small and also imbalanced datasets which result in models that are unable to generalize over different patient populations. The second problem with CNNs is that they are not focused enough on the integration of multi-modal data sources such as clinical data, speech, gait analysis to get the most out of the detection accuracy. However, the CNN model does not lend itself to easy interpretability of the key structures of these models; thus, clinicians need to have clear insights into this decision-making process in order to trust and use these models in practice. Doing so can improve the diagnostic accuracy, and the robustness and clinical applicability as well.

3. The proposed methodology

This paper describes how we shall make use of a procedure with a multi-step approach to getting Parkinson's disease data using the Explainable CNN. The process starts with the data preprocessing use of different methods to clean, normalize, and improve raw data. This guarantees a better representation of the CNN model's input data. A solved issue of data scarcity is confronted by incorporating a Generative Adversarial Network (GAN) to the process of data augmentation thus, we increase the number of samples from 195 to 1000. The effect of this data augmentation results in the enhancement of the model's ability to generalize, which in turn, is achieved through the reduction of overfitting. Furthermore, feature extraction and selection techniques are applied to the preprocessed data to reveal the most significant patterns. These features are then put into an optimized CNN architecture, particularly designed for the classification of Parkinson's disease. It should be clarified here that the CNN consists of a specific number of convolutional layers to automatically identify features and the followed pooling and fully-connected layers to classification. Eventually, the parameters such as accuracy, precision, Recall, and F1 score are used performance metrics to evaluate the CNN's engines' efficacy of Parkinson's disease classification. The results of applying the respective metrics of the CNN to the practical classification problems are compared to the basic/traditional ones for the highlighting of the model's improvements. The proposed CNN for Parkinson's disease detection multi steps presented in Fig. 1.

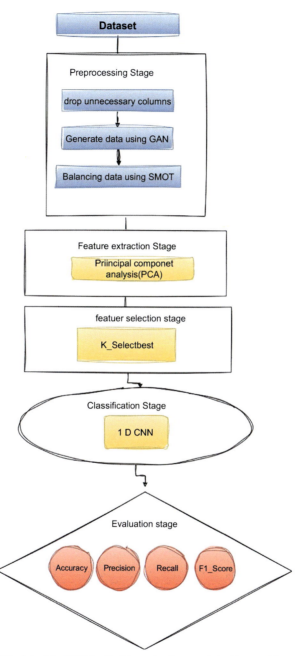

Fig. 1 The proposed Explainable CNN for Parkinson's disease detection multi steps.

3.1 PD dataset

The PD dataset that was taken from the Kaggle is an extensive compilation of clinical data targeted at people who are PD diagnosed [16]. Created by Max Little, the University of Oxford, this data primarily records a speech signal from a diverse group of PD patients. The dataset's main goal is to increase the PD classification accuracy. The dataset is made up of 195 samples and 23 features in both male and female sets, giving equilibrium to the representation of both the sexes. Those documented features are fundamental in that they contain the basic characteristics of patients which are necessary to improve disease classifications. A detailed list of these features that also show their relevance to PD detection can be found in Table 1.

Table 1 PD dataset features.

Feature name	Description
Name	ASCII subject name and recording number
MDVP:Fo(Hz)	Average vocal fundamental frequency
MDVP:Fhi(Hz)	Maximum vocal fundamental frequency
MDVP:Flo(Hz)	Minimum vocal fundamental frequency
MDVP:Jitter(%), MDVP:Jitter(Abs), MDVP:RAP, MDVP:PPQ, Jitter:DDP	Several measures of variation in fundamental frequency
MDVP:Shimmer, MDVP:Shimmer(dB), Shimmer:APQ3, Shimmer:APQ5, MDVP:APQ, Shimmer:DDA	Several measures of variation in amplitude
NHR, HNR	Two measures of ratio of noise to tonal components in the voice
Status	Health status of the subject (one)—Parkinson's (zero)—healthy
RPDE, D2	Two nonlinear dynamical complexity measures
DFA	Signal fractal scaling exponent
spread, spread2, PPE	Three nonlinear measures of fundamental frequency variation

3.2 Preprocessing stage based on GANs

In the GANs-based preprocessing phase, we have introduced the PD dataset subsequently that it becomes more efficient by the study. Due to the restricted initial sample size of 195, GANs take on the capability to produce artificial data that improves the training set, and thus they alleviate the problem of data scarcity and the model's performance advantages. The naming process will start by training the GAN the main dataset, which in this case comprises clinical features and the patient voice record. The GAN architecture is based on two components: the generator, which is the child of the synthetic data, and the discriminator that checks the lyrics of the fake data. The GAN learns the generation of perfect samples by constantly updating the parameters of both the generator and the discriminator, so that the model is able to produce new samples similar to the initial dataset in the terms of the distribution of data [17].

After the training is completed, the GAN will be able to produce extra data which contributes to the dataset with quantity up to 1000 examples. This solution not only dramatically increases the amount of training data but it also injects variability that helps the model learn the general features of the data, and thus become a good classifier even with unrelated examples. Furthermore, the newly created data objects are then blended with the original dataset in such a way that they conserve the overall ethnic distribution of the patient population. Eventually, the augmented dataset is further prepared by standardization and feature scaling sequences for feeding to the CNN. This whole preprocessing flow, including the presence of GANs, clearly demonstrates the superiority of solving PD with the quality and quantity of the data being increased greatly, which consequently leads to the PD categorization. Training a GAN on a Parkinson's dataset will result in the creation of new synthetic data that is done by sampling the output of the generator and thus the maximum number of additional data points is generated. These synthetic data points are subsequently included in the Parkinson's dataset that is used to train the CNN model. The use of GANs on the Parkinson's data involved the generation of a new dataset similar to the first one, with 600 sample size, and it was named dataset (A). The next time, the dataset with 1000 samples has been prepared to be called (B). Thus, the first dataset represents the real Parkinson's data, and the other two are the data produced by GAN.

3.2.1 Data Balancing stage

For Data Balancing stage we deal with the difficulty which comes naturally from the data and which can be seen as the reward for our efforts related to PD treatment. This class imbalance can gull the accuracy of the model where the classifier may be overly sensitive to the majority class, thereby losing its ability to do that which is the main thing in the detection of Parkinson's disease.

There are two strategies we use for data balance, which are:
- Synthetic Minority Over-sampling Technique (SMOTE): We employ the use of SMOTE to create synthetic samples for the minority class (patients with Parkinson's disease). By this method, it is possible to generate new, synthetic instances which are based on the feature space of the existing minority class samples, thereby enriching the representation of this class without simply duplicating current data.
- Under-sampling of Majority Class: Besides SMOTE, the possibility of under-sampling the majority class (normal persons) is presented to balance the dataset. This

procedure implies the random selection of a subset of the samples of the majority class to decrease their homogeneity. While this method can cause some degree of information loss, it helps in creating a more equitable training environment.

3.2.2 Feature Standardization stage

Feature Standardization Phase encompasses the process of the dataset's features being balanced and shifted to improve the CNN performance. The primary importance of the standardization process becomes obvious when multiple features are usually represented in disparate units and ranges. This firstly includes the following steps:

- **Identifying features**: We start by identifying the numerical features available in the dataset, which may consist of clinical measurements, signal amplitudes, and other relevant attributes related to Parkinson's disease.
- **Calculating mean and standard deviation**: Users first calculate the mean and standard deviation for each feature with the training set of the dataset. These statistics are the pivot for transforming the features to a common scale.
- **Applying standardization**: To standardize them, the following formula is used:

$$Z = \frac{(X - \mu)}{\sigma}$$

where ZZZ is the standardized value, XXX is the original feature value, μ\mu$\mu$ is the mean of the feature, and σ\sigma$\sigma$ is the standard deviation. The transformation, in this case, is a zero-mean and a unit variance conversion that enables the model to learn more effectively.

- **Handling outliers**: Moreover, in the process of standing, we are also looking at the outliers that can affect the mean and standard deviation. If we face such cases, we apply the outlier detection methods before the transformation.
- **Validation**: At the end of the standardization process, we also confirm that the transformed features keep their statistical distributions and relationships. This is a step to ascertain that the conversion did not result in the data patterns being distorted.

3.3 PD Feature Extraction phase using principal component analysis (PCA)

In the Process of Feature Extraction, we use PCA (Principal Component Analysis) to decrease the size of the data set to those dimensions necessary for the most critical information to be preserved [18]. In this phase, a model execution time is reduced, a set of the most important features that the model takes into account are selected, and the problem of overfitting is mitigated.

- **Features standardization**: Before doing PCA we must bring the mean of all the features to zero, and their standard deviation to one. We also need to do this because

PCA is sensitive to different variances of the features. We rescale every feature so as to have a mean of 0 and variance of 1.

- **Covariance matrix calculation**: We calculate the covariance matrix of the standardized features and this matrix shows the associations between different aspects, thus we can see how they move together.
- **Eigenvalue and Eigenvector calculation**: Subsequently, we determine the eigenvalues and eigenvectors of the covariance matrix. The eigenvalues are the representation of the degree of variance each principal component generates, and the eigenvectors tell you the direction each component is in the function space.
- **Principal components selection**: We call the eigenvalues in the ranking to know the main principal components that absorb the most variability in the dataset. A portion of these are the ones we choose by setting a certain threshold which is usually determined by the cumulative explained variance ratio. Usually, elements comprising 95% of the variance explain it.
- **Transforming**: First, the prime dataset is depicted onto the selected principal components, and thus, a new feature space with lower dimensionality is created. This way the core and important details of the data are preserved and the noisy and redundant features are thrown away.
- **Feature validation**: At the end, we verify the extracted features through their direct impact on the classification task. This can be accomplished using visualization techniques like scatter plots to show the relationship between the reduced features and class separation (e.g. Parkinson's disease vs healthy controls).

3.4 Feature Selection phase using the Select KBest algorithm

In the Feature Selection process, we utilize The Select KBest algorithm that aids the identification and preservation of the most important features for PD determining [19]. This stage is the most important concerning the fact it helps overfitting, increases model performance and brings out interpretability. The following steps compose the process:

- **Feature ranking**: The Select KBest algorithm is responsible for ranking all features based on their individual performance in the prediction of the target variable such as PD classification. It inspects each feature using a statistical scoring function, which can be a choice made depending on data type like mutual information, ANOVA F-statistic, and chi-squared.
- **Select the features number (K)**: We find the appropriate number of features to be taken (K) on the basis of the domain knowledge and validation methods. It includes the tray of exploitations of various values for K and also checking the circuit of the models through cross-validation to allow them to turn in only one without adding redundancy.

- **Selecting best features**: In case the processing of the setting of K is soaked, the algorithm gets the top K factors that links the most with target variable. The rest of the relevant features are put together for model training, while the unimportant features are thrown out.
- **Validation of selected features**: Once the features are picked, then we check their relevance by testing the mode's operation with such issues versus the full feature set. In this way, the model can be trained, and the measures accuracy, precision, recall, and F1-score among others can be assessed to make sure that the chosen feature supports the task of the classification.
- **Finally, we examine individual features** that are known to be the cause of Parkinson's ailment, so as to understand their significance more clearly in the context of Parkinson's disease.

3.5 Dataset Splitting phase

The Dataset Splitting process holds in its hands the handled and augmented set among them is separated such that a part of it is for the CNN testing. The split adopts a widely used method in the domain of machine learning, where the whole data are divided into 80% for training and 20% for testing. This process makes sure that we have a fair model where the bias is controlled while we use a different set to evaluate the model on.

- **Training set (80%)**: The 80% version of the whole dataset takes form of the training set hat is subsequently used to train the CNN. In this case, the model is provided with the training sample to comprehend the actual background of the dataset. The model will adjust its parameters to minimize the loss function's feedback. The training process consists of the different times when the model learns by the information it gets from the loss function.
- **Testing set (20%)**: Conversely, 20% of the data is devoted to executing the battery of tests on the trained model and then comparing the models' performance. Once we finish the training phase, we use the testing set independently to evaluate the model. These set Izabela analyzes the extent to distinguish among the imbibed data. The testing set will serve us to see how precise the model is on the data set and other parameters like accuracy, sensitivity, specificity, and so on.
- **Randomized splitting**: The dataset will be more representative of the entire population if we randomly split it, this is why the scientist uses a random splitting protocol. The scientific community highly recommends the randomization of the dataset split as it eliminates the possibility of influencing the data by either one side, thereby escaping the aforementioned situation.

3.6 The CNN model for PD classification

In this work, one-dimensional convolutional neural networks (1DCNN) were designed because dealing with one-dimensional dataset [1]. To process and analyze the Parkinson's dataset, the CNN takes the input data, processes it, and classifies it into two categories, either affected or unaffected [20]. The input layer accepts data in the sampled form. After preprocessing, the data is resampled into (samples, features, 1), where each sample is a sequence of features. The Conv1D layer moves a set of filters (also called kernels) over the input data and the filter size is 3. Each filter is a small matrix of weights learned during training. The filter performs element-by-element multiplication with the input data and sums the results to produce a single value. After convolution, an activation function (ReLU) is applied to introduce nonlinearity to the data generated by the filter.

The MaxPooling1D layer reduces the dimensionality (length) of the input data set. This is done by taking the maximum value in each input patch where applying a MaxPooling1D layer with pool size 2 to the output of the previous Conv1D layer will reduce its shape. The data then passes through the flattening layer, where this layer converts the output of the two-dimensional matrix from the previous layer into a one-dimensional vector. This is necessary before data is entered into the dense (fully connected) layers. When the data is fed into the dense layer, it performs a weighted sum of the inputs, adds a bias term, and applies the ReLU activation function. The data then passes through the output layer, which is the last dense layer that contains a single neuron with a sigmoid activation function, which produces a probability value between 0 and 1, which represents either With or without Parkinson's disease. Fig. 2 shows the proposed 1DCNN architecture.

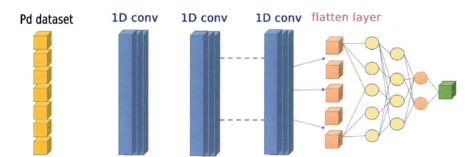

Fig. 2 The proposed 1DCNN architecture.

Activation function (ReLU), dense layer SoftMax activation function, flatten layer, 1D convolution layer, MaxPooling layer, and one-dimensional convolutional neural network (CNN) layers are obtained and their parameters in this study. Algorithm 1 explains the working steps of the proposed 1DCNN algorithm.

Algorithm 1: CNN Steps
Input: X_train: Training data, X_test: Testing data
Output: Status 0 or 1
Start
Step 1: assign each parameter a random value.
epoch = 100, batch's size = 16, Pool size = 2.
Step2: Contact CNN to implement network layers.
• Conv1D (with filter= 32, kernel_size=3, activation='relu', MaxPooling1D(pool_size=2), input_shape = (num_ features, 1))
• Conv1D (with filter= 64, kernel_size=3, activation='relu', MaxPooling1D(pool_size=2)
• Conv1D (with filter= 128, kernel_size=3, activation='relu', MaxPooling1D(pool_size=2)
• Flatten output of Conv1D
• Fully connected layer with one-layer activation function = sigmoid
• Categorize the result for the provided dataset.
• Update weight
Step 3: Go back to step 2
When epoch equals 100
Then stop
Step 4: Evaluate the model
Step 5: END

In the CNN model defined previously, there are a total of 12 layers:

1. **Input layer**: This is not counted as a trainable layer but represents the input data's shape.
2. **Reshape layer**: Reshapes the input tensor to have the appropriate shape for convolutional operations.
3. **Convolutional layer (Conv1D)**: First convolutional layer with 32 filters and a kernel size of 3.
4. **MaxPooling1D layer**: Performs max pooling operation to reduce the spatial dimensions of the feature map.
5. **Convolutional layer (Conv1D)**: Second convolutional layer with 64 filters and a kernel size of 3.
6. **MaxPooling1D layer**: Performs max pooling operation.
7. **Convolutional layer (Conv1D)**: Third convolutional layer with 128 filters and a kernel size of 3.
8. **MaxPooling1D layer**: Performs max pooling operation.
9. **Flatten layer**: Flattens the output of the convolutional layers into a 1D tensor.
10. **Dense layer**: Fully connected layer with 256 units and ReLU activation function.
11. **Dropout layer**: Applies dropout regularization to prevent overfitting.
12. **Output layer**: Dense layer with a single unit and sigmoid activation function for binary classification.

Therefore, there are 9 trainable layers (Conv1D, Dense, and Dropout layers), and 3 non-trainable layers (Input, Reshape, and MaxPooling1D layers) in the model. Our CNN model assigns starting values to all parameters required in the CNN, which are summarized in Table 2.

Table 2 The proposed CNN initialized parameters.

Gradient descent optimizer	ReLU, Softmax and Sigmoid
Activation function	Adam
Batch-size	16
Epoch	100

3.7 Evaluation phase

This section presents the metrics used to assess the performance of the DL models. In the case of classification algorithms, the assessment relies on the following metrics: accuracy, recall, precision, and f1_score. The computation of these metrics relies on the confusion matrices. A confusion matrix, sometimes referred to as an error matrix, is a specific table-like structure that enables the evaluation of algorithmic performance. The matrix displays actual instances in each row and predicted cases in each column. Occasionally, the depiction is inverted. Fig. 9 depicts a simple binary classification confusion matrix as an example. For instance, the examples here may be categorized as either positive or negative. Class 1 corresponds to positive situations, while class 2 corresponds to negative cases. These cases are further divided into four predictions as shown in Fig. 3.

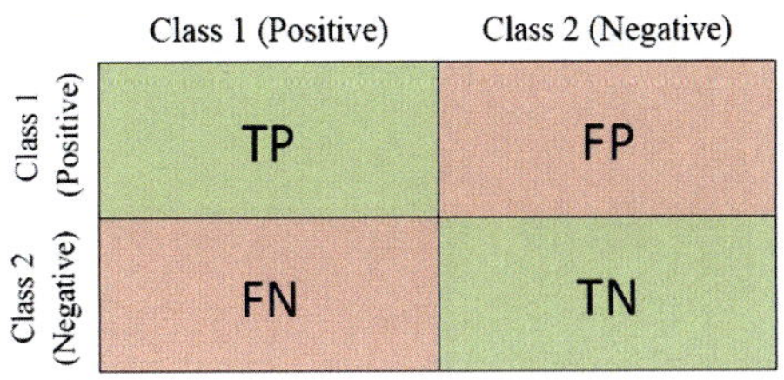

Fig. 3 Confusion matrix.

True positive (TP) denotes that the actual and the predicted values are positive.
False negative (FN) refers to an observation that is positive, yet a negative prediction is made.
True negative (TN) denotes a positive actual value with a negative prediction.
False positive (FP) refers to an actual value that is negative but is expected to be positive.

The following measures are described as follows:

Accuracy: The proportion of properly categorized samples among the tested samples is known as the accuracy rate. In other words, accurate categorizations of negative and positive specimens are identified. Eq. (1) provides a formal definition.

$$Accuracy = \frac{TP + TN}{TP + TN + FP + FN} \tag{1}$$

Precision: Precision is determined by dividing the number of properly anticipated positive instances by the sum of correctly predicted positive cases and mistakenly predicted positive cases. Precision is a valuable metric when it is more important to minimize false positives rather than false negatives. The precision metric is determined using Eq. (2).

$$Precision = \frac{TP}{TP + FP} \tag{2}$$

Recall: Recall is calculated by dividing the total number of properly anticipated positive instances by the sum of the correctly predicted negative cases and the number of correctly predicted positive cases. The usefulness of recall as a metric becomes apparent when the risk of false negatives outweighs that of false positives. Recall is computed using Eq. (3).

$$Recall = \frac{TP}{TP + FN} \tag{3}$$

The F1 score combines precision and recall using their harmonic mean, and maximizing the F1 score implies simultaneously maximizing precision and recall. In cases of an uneven distribution of classes, F1 score tends to be more valuable than accuracy. The F1 score is computed using Eq. (4)

$$f1\_score\ Rate = \frac{2*(Precision)*(Recall)}{(Recall + Precision)} * 100\% \tag{4}$$

4. CNN results of classification stage

Feature extraction from data related to Parkinson's disease is used in classification which is the final step of the proposed system. The proposed CNN algorithm is used to train, test and compare features, making classification the most important stage in the system. In this section, the performance of CNN will be presented on the Parkinson's dataset, which is the original dataset A. Moreover, the performance will be compared with data augmentation techniques, which are the dataset B and C.

4.1 Classification CNN results on dataset A

The CNN model proposed in this thesis consists of three convolutional layers followed by pooling layers, flattening layers, and dense layers. The first convolutional layer applies

64 filters to the input data, each of which performs a convolutional operation that scans the input with a kernel size of 2. This operation helps in extracting low-level features. After the first convolution, max-pooling is applied to reduce the spatial dimensionality and computational load while retaining the most important features. The second convolutional layer processes pooled feature maps using 128 filters, capturing more complex patterns as the subsequent max pooling layer again reduces dimensionality, ensuring that the most prominent features are retained. The third convolutional layer applies 256 filters to capture more complex and abstract features. The flattening layer then converts the 2D feature maps into a 1D vector, making it suitable for dense layers that perform the classification task. The first dense layer, which contains 64 ReLU activation units, is applied to introduce nonlinearity. After that, the output layer consisting of 2 softmax activation units is applied to produce Probability distribution for the two categories: presence and absence of Parkinson's disease. Fig. 4 shows detailed information about each layer in the proposed CNN algorithm.

Layer (type)	Output Shape	Param #
convid (Conv1D)	(None, 19, 64)	192
max_poolingid (MaxPooling1D)	(None, 9, 64)	0
convid (Conv1D)	(None, 8, 128)	16,512
max_poolingid (MaxPooling1D)	(None, 4, 128)	0
convid (Conv1D)	(None, 3, 256)	65,792
flatten (Flatten)	(None, 768)	0
dense (Dense)	(None, 64)	49,216
dense (Dense)	(None, 2)	130

Total params: 131,842 (515.01 KB)

Trainable params: 131,842 (515.01 KB)

Non-trainable params: 0 (0.00 B)

Fig. 4 The detailed information about the proposed CNN algorithm.

The model is trained using the Adam optimizer and categorical entropy loss for 100 epochs. This value was determined through experimental epochs with a batch size of 32 after testing 16, 64, and 128. Based on experimental results, the ideal batch size is 32 iterations per epoch, while larger batches of 64 and 128 increase the number of iterations per epoch and memory needs, respectively. The training process includes validation to monitor the model's performance on unseen data. After several Experiences in different eras. Using the optimal number of epochs is crucial to prevent overfitting and improve experimental results. In order to prevent overfitting, these values are determined experimentally. Table 3 shows the performance of the classifier on the Parkinson's A dataset.

Table 3 The values of evolution metric on the Parkinson's A dataset.

Class	Precision	Recall	F1-measure	Accuracy
No Parkinson's	71%	71%	71%	89.7%
Parkinson's	94%	94%	94%	

The results in Table 3 show the values of accuracy, precision, recall, and F1 score for each class and as shown in Fig. 5, the results obtained are: accuracy 89.7%, Precision 93%, recall 93%, F1 score 93%, on the data set.

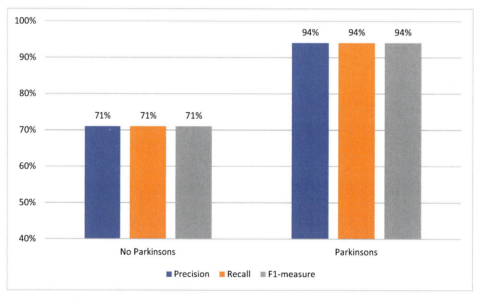

Fig. 5 Results of CNN metrics on dataset A.

After training, the model is evaluated on the test set to measure its accuracy. The CNN algorithm, used for deep learning on dataset A, results in an accuracy of up to 89%. The same number of epochs is used to predict the accuracy of training and evaluation. The loss function is optimized using the "Adam" optimizer. The training and validation accuracy of the proposed CNN is clearly illustrated in Fig. 6, while the training and validation loss is visualized in Fig. 7.

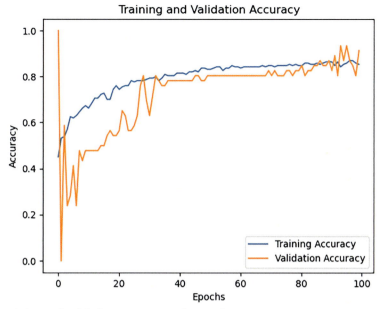

Fig. 6 The training and validation accuracy on dataset A.

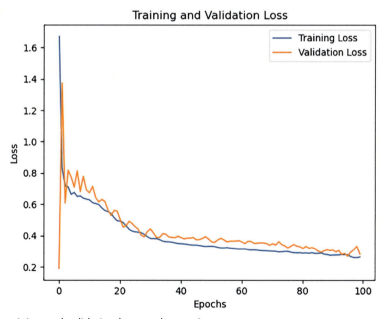

Fig. 7 The training and validation loss on dataset A.

A confusion matrix is generated to comprehensively evaluate the performance of the model. The effectiveness of the CNN approach is evaluated by calculating its confusion matrix, as shown in Fig. 8, after implementing the CNN with the Parkinson A dataset and achieving significant accuracy.

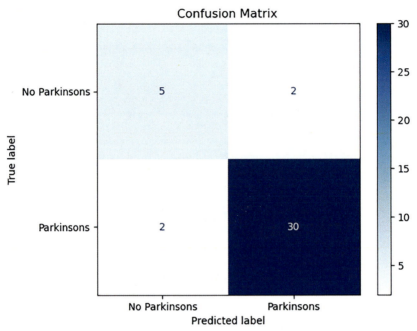

Fig. 8 Confusion matrix for CNN on dataset A.

4.2 Classification CNN results on dataset B

This section will show how the CNN performs at this point using the 600-samples Parkinson B dataset, which has been enhanced using data augmentation techniques. After the features are acquired, they will be passed into the three convolutional layers, flattening, and dense layers that make up the deep learning CNN algorithms suggested in this thesis. The input data is subjected to 64 filters by the first convolutional layer, each of which does a convolutional operation that scans the input using a kernel size of 2. Low-level characteristics may be extracted with the use of this procedure. Max-pooling is used to decrease the computational burden and spatial complexity while keeping the most significant features after the initial convolution. The succeeding max pooling layer further decreases dimensionality, ensuring that the most prominent features are kept, while the second convolutional layer analyzes pooled feature maps employing 128 filters, capturing more complicated patterns. In order to capture more intricate and abstract characteristics, 256 filters are applied by the third convolutional layer. The 2D feature maps are then transformed into a 1D vector by the flattening layer, which qualifies them for use in dense layers that carry out the classification operation. To add nonlinearity, the first dense layer which has 64 ReLU activation units is applied. Next, the output layer which consists of two softmax activation units is used to generate a probability distribution for the two categories of Parkinson's disease presence and absence.

For 100 epochs, the model is trained using the Adam optimizer with class entropy loss. After examining 16, 64 and 128 samples, this value was chosen using experimental epochs with a batch size of 32 which is the best batch size, according to the experimental data, which is 32 iterations per epoch; Larger batches, 64 and 128 respectively, increase the number of iterations per epoch and memory requirements. Validation is a step in the training process that tracks a model's performance on unobserved data. After several experiments at different times. It is necessary to choose the optimal number of epochs in order to reduce overfitting and enhance experimental results. Empirical determination of these values is necessary to avoid overfitting. The performance of the classifier on the Parkinson A dataset is shown in Table 4.

Table 4 The evolution metric on the Parkinson's B dataset.

Class	Precision	Recall	F1-measure	Accuracy
No Parkinson's	95%	74%	83%	93%
Parkinson's	93%	99%	96%	

The results in Table 4 show the values of accuracy, precision, recall, and F1 score for each class and as shown in Fig. 9, the results obtained are: accuracy 93.3%, Precision 92.9%, recall 98.9%, F1 score 95.8%, on the data set.

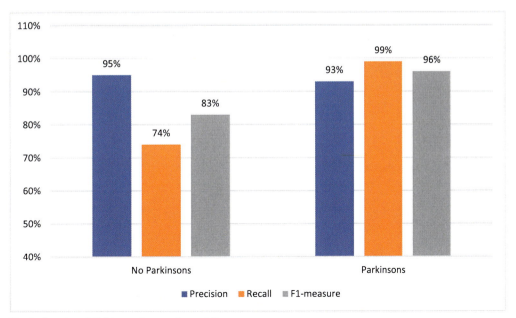

Fig. 9 Results of CNN metrics on dataset B.

After training, the model is evaluated on the test set to measure its accuracy. The CNN algorithm, used for deep learning on dataset A, results in an accuracy of up to 93%. The same number of epochs is used to predict the accuracy of training and evaluation. The loss function is optimized using the "Adam" optimizer. The training and validation accuracy of the proposed CNN is clearly illustrated in Fig. 10, while the training and validation loss is visualized in Fig. 11.

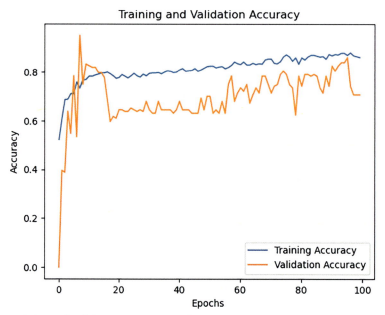

Fig. 10 The training and validation accuracy on dataset B.

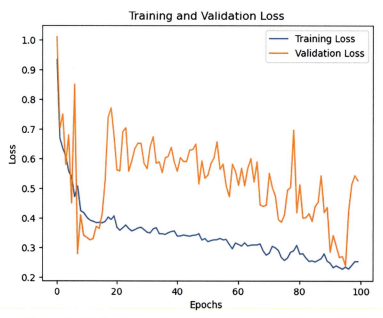

Fig. 11 The training and validation loss on dataset B.

A confusion matrix is created to thoroughly assess the effectiveness of the model. The effectiveness of the CNN technique is assessed by computing its confusion matrix, as shown in Fig. 12, subsequent to the implementation of the CNN using the Parkinson's disease B dataset and attaining good accuracy.

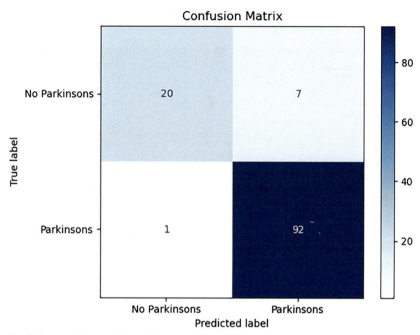

Fig. 12 Confusion matrix on dataset B.

4.3 Classification CNN results on dataset C

In this section, a CNN will be demonstrated using the 1000-sample Parkinson C dataset that has been improved using data augmentation techniques. The proposed deep learning algorithms for QCNN will be fed with features. After the features are obtained, they will be passed to the three convolutional layers, the flat and dense layers that make up the deep learning CNN algorithms proposed in this thesis. The input data is subjected to 64 filters by the first convolutional layer, each of which performs a convolutional operation that scans the input using a kernel size of 2. Low-level features can be extracted using this procedure. Maximum pooling is used to reduce the computational burden and spatial complexity while retaining the most important features after the initial convolution. A subsequent max-pooling layer further reduces dimensions, ensuring that the most prominent features are preserved, while a second convolutional layer analyzes pooled feature maps using 128 filters, capturing more complex patterns. In order to capture more complex and abstract features, 256 filters were applied by the third convolutional layer.

The 2D feature maps are then converted into a 1D vector by the flattening layer, making them suitable for use in dense layers that perform classification. To add nonlinearity, the first dense layer containing 64 ReLU activation units is applied. Next, an output layer consisting of two softmax activation units is used to generate a probability distribution for the presence and absence categories of Parkinson's disease. For 100 epochs, the model is trained using the Adam optimizer with category entropy loss. After testing with 16, 64, and 128 iterations, this value was calculated using experimental epochs with a batch size of 32. Results from experiments show that 32 iterations per epoch is the sweet spot for batch size, with 64 and 128 iterations per epoch increasing memory requirements and iterations per epoch, respectively. Validation is a part of training that helps keep an eye on how well the model does on new data. After a number of encounters across various time periods. To avoid overfitting and get better experimental findings, it's important to choose the right number of epochs. The goal of determining these values empirically is to avoid overfitting. On the Parkinson C dataset, the classifier's performance is shown in Table 5.

Table 5 The evolution metric on the Parkinson's C dataset.

Class	Precision	Recall	F1-measure	Accuracy
No Parkinson's	95%	94%	94.4%	95.5%
Parkinson's	96%	95%	96%	

The results in Table 5 show the values of accuracy, precision, recall, and F1 score for each class and as shown in Fig. 13, the results obtained are: accuracy 95.5%, Precision 96%, recall 94%, F1 score 95%, on the data set.

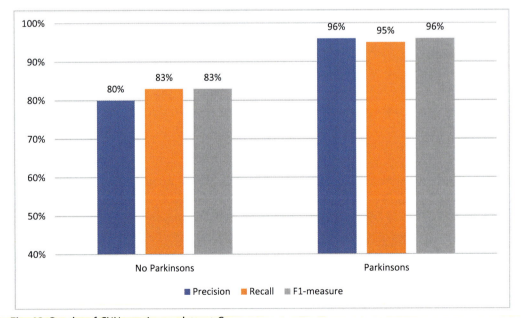

Fig. 13 Results of CNN metrics on dataset C.

After training, the model is evaluated on the test set to measure its accuracy. The CNN algorithm, used for deep learning on dataset A, results in an accuracy of up to 93%. The same number of epochs is used to predict the accuracy of training and evaluation. The loss function is optimized using the "Adam" optimizer. The training and validation accuracy of the proposed CNN is clearly illustrated in Fig. 14, while the training and validation loss is visualized in Fig. 15.

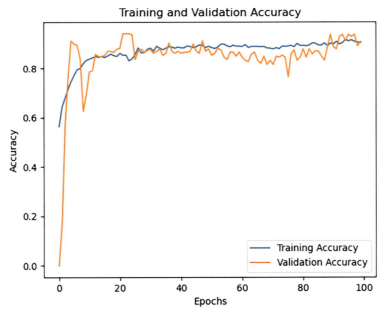

Fig. 14 The training and validation accuracy on dataset C.

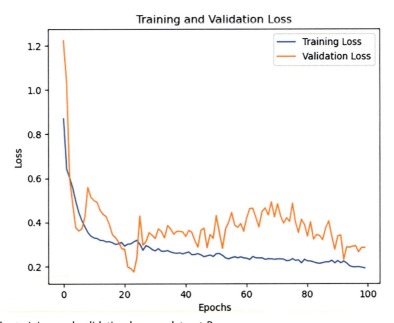

Fig. 15 The training and validation loss on dataset B.

A confusion matrix is generated to comprehensively evaluate the performance of the model. The effectiveness of the CNN approach is evaluated by calculating its confusion matrix, as shown in Fig. 16, after implementing the CNN with the Parkinson C dataset and achieving significant accuracy.

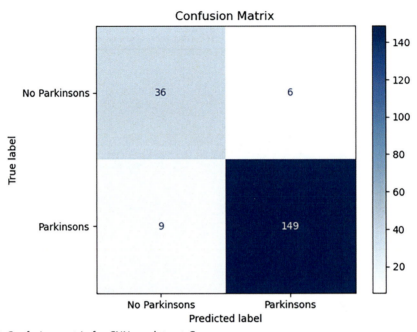

Fig. 16 Confusion matrix for CNN on dataset C.

5. Conclusion

This research discusses a use of CNN approach for enhancing the PD classification. We not only focused on addressing data imbalance, feature scaling, and reduction of dimensionality but also solved definition issues by training. The extension of the database from 195 to 1000 samples with the help of the GANs beat those challenges and moved model learning higher and the generalization further. Feature extraction through PCA and using Select KBest for feature selection made it possible to keep only the most suitable ones and thus, to enhance the learning power of the model. Splitting the dataset with an 80:20 ratio for training and testing allowed us to evaluate the models in a robust way, to prevent biased performance. Parkinson's disease could be identified by the CNN model, which was developed and enhanced through this multi-stage process, with high accuracy and robustness. The study has given meaning to those things whose contribution is not only the detection of symptoms but also the classification and scale of diseases related to the neurological system. The experimental results presented a significant increase in precision of detection compared to the traditional methods, with a 95.5% accuracy and 95% precision, a 94% recall, and a 94.4% F1 score. In the future, its recommended use swarm optimizations algorithms such CSO. Practical swarm and other methods to choose the best features.

Acknowledgments

We would like to confirm that all the figures were originally created by us.

Funding information

The authors received no specific funding for this study.

Author contributions

Aiesha Mahmoud Ibrahim: Conceptualization, Methodology, Formal analysis, Investigation, Writing—original draft, Visualization, Writing—review & editing. **Mazin Abed Mohammed**: Conceptualization, Methodology, Writing—review & editing, Project administration, Supervision.

Conflict of interest

The authors declare no conflict of interest.

Data availability statement

The data used in this study is publicly available and can be accessed from the following link: Oxford, U. of Parkinson's Disease Dataset. https://www.kaggle.com/datasets/gargmanas/parkinsonsdataset.

References

[1] A. Mahmoud Ibrahim, M. Abed Mohammed, A comprehensive review on advancements in artificial intelligence approaches and future perspectives for early diagnosis of Parkinson's disease, Int. J. Math. Stat. Comput. Sci. 2 (2024) 173–182, https://doi.org/10.59543/ijmscs.v2i.8915.

[2] V. Rajinikanth, S. Yassine, S.A. Bukhari, Hand-Sketchs based Parkinson's disease screening using lightweight deep-learning with two-fold training and fused optimal features, Int. J. Math. Stat. Comput. Sci. 2 (2024) 9–18, https://doi.org/10.59543/ijmscs.v2i.7821.

[3] S.J. Priya, A.J. Rani, M.S.P. Subathra, M.A. Mohammed, R. Damaševičius, N. Ubendran, Local pattern transformation based feature extraction for recognition of Parkinson's disease based on gait signals, Diagnostics 11 (8) (2021) 1395.

[4] M.A. Mohammed, M. Elhoseny, K.H. Abdulkareem, S.A. Mostafa, M.S. Maashi, A multi-agent feature selection and hybrid classification model for Parkinson's disease diagnosis, ACM Trans. Multimedia Comput. Commun. Appl. 17 (2s) (2021) 1–22.

[5] S.A. Mostafa, A. Mustapha, M.A. Mohammed, R.I. Hamed, N. Arunkumar, M.K. Abd Ghani, M.M. Jaber, S.H. Khaleefah, Examining multiple feature evaluation and classification methods for improving the diagnosis of Parkinson's disease, Cogn. Syst. Res. 54 (2019) 90–99.

[6] S.A. Mostafa, A. Mustapha, S.H. Khaleefah, M.S. Ahmad, M.A. Mohammed, Evaluating the performance of three classification methods in diagnosis of Parkinson's disease, in: Recent Advances on Soft Computing and Data Mining: Proceedings of the Third International Conference on Soft Computing and Data Mining (SCDM 2018), Johor, Malaysia, February 06-07, 2018, Springer International Publishing, 2018, pp. 43–52.

[7] S. Saravanan, K. Ramkumar, K. Adalarasu, V. Sivanandam, S.R. Kumar, S. Stalin, R. Amirtharajan, A systematic review of artificial intelligence (AI) based approaches for the diagnosis of Parkinson's disease, Arch. Comput. Methods Eng. 29 (6) (2022) 3639–3653.

[8] Z.K. Senturk, Early diagnosis of Parkinson's disease using machine learning algorithms, Med. Hypotheses 138 (2020) 109603.

[9] B. Palumbo, F. Bianconi, S. Nuvoli, A. Spanu, M.L. Fravolini, Artificial intelligence techniques support nuclear medicine modalities to improve the diagnosis of Parkinson's disease and Parkinsonian syndromes, Clin. Translat. Imaging 9 (2021) 19–35.

[10] S. Dixit, K. Bohre, Y. Singh, Y. Himeur, W. Mansoor, S. Atalla, K. Srinivasan, A comprehensive review on AI-enabled models for Parkinson's disease diagnosis, Electronics 12 (4) (2023) 783.

[11] S. Sharanyaa, M. Sambath, P.N. Renjith, Optimized deep learning for the classification of Parkinson's disease based on voice features, Crit. Rev. Biomed. Eng. 50 (5) (2022) 1–28.

[12] M. Wodzinski, A. Skalski, D. Hemmerling, J.R. Orozco-Arroyave, E. Nöth, Deep learning approach to Parkinson's disease detection using voice recordings and convolutional neural network dedicated to image classification, in: 2019 41st Annual International Conference of the IEEE Engineering in Medicine and Biology Society (EMBC), IEEE, 2019, July, pp. 717–720.

[13] S. Rahman, M. Hasan, A.K. Sarkar, F. Khan, Classification of Parkinson's disease using speech signal with machine learning and deep learning approaches, Eur. J. Electr. Eng. Comput. Sci. 7 (2) (2023) 20–27.

[14] H. Gunduz, Deep learning-based Parkinson's disease classification using vocal feature sets, IEEE Access 7 (2019) 115540–115551.

[15] T.S. Mian, An unsupervised neural network feature selection and 1D convolution neural network classification for screening of parkinsonism, Diagnostics 12 (8) (2022) 1796.

[16] Oxford, U. of. Parkinson's Disease Dataset. https://www.kaggle.com/datasets/gargmanas/parkinsonsdataset.

[17] N. Peppes, P. Tsakanikas, E. Daskalakis, T. Alexakis, E. Adamopoulou, K. Demestichas, FoGGAN: generating realistic Parkinson's disease freezing of gait data using GANs, Sensors 23 (19) (2023) 8158.

[18] D.V. Rao, Y. Sucharitha, D. Venkatesh, K. Mahamthy, S.M. Yasin, Diagnosis of Parkinson's disease using principal component analysis and machine learning algorithms with vocal features, in: 2022 International Conference on Sustainable Computing and Data Communication Systems (ICSCDS), IEEE, 2022, April, pp. 200–206.

[19] S. Srinivasan, P. Ramadass, S.K. Mathivanan, K. Panneer Selvam, B.D. Shivahare, M.A. Shah, Detection of Parkinson disease using multiclass machine learning approach, Sci. Rep. 14 (1) (2024) 13813.

[20] S. Kaur, H. Aggarwal, R. Rani, Diagnosis of Parkinson's disease using deep CNN with transfer learning and data augmentation, Multimed. Tools Appl. 80 (7) (2021) 10113–10139.

CHAPTER 19

Data analytics and cognitive computing for digital health: A generic approach and a review of emerging technologies, challenges, and research directions

K. Aditya Shastry[a] and D.N. Disha[b]
[a]Department of Information Science and Engineering, Nitte Meenakshi Institute of Technology, Bengaluru, India
[b]Department of Artificial Intelligence and Data Science, NMAMIT, Udupi, India

1. Introduction

The medical domain is undergoing a data revolution with the rapid adoption of digital health technologies. These technologies generate enormous quantities of information, ranging from "electronic health records" (EHRs) to records collected through wearable devices and mobile applications. This wealth of data presents a significant opportunity to leverage advanced data analytics and cognitive computing techniques to extract insights and knowledge from health data and improve healthcare outcomes [1]. Data analytics and cognitive computing are essential tools for processing and analyzing complex data and have become increasingly relevant in the healthcare sector. Data analytics, which comprises descriptive, predictive, and prescriptive analytics, can be used to reveal patterns, associations, and correlations in large datasets. It can help healthcare providers gain insights into patient care and outcomes, identify factors that contribute to diseases, and improve the accuracy of clinical decision-making [2]. Cognitive computing, on the other hand, is an advanced computing technology that enables machines to understand, analyze, and resolve issues comparable to a person. This technology is critical to enhancing clinical decision-making and patient care by processing and interpreting vast amounts of complex health data [3]. Despite the benefits that data analytics and cognitive computing can offer in healthcare, their implementation faces several challenges. These challenges include data privacy and security concerns, limited access to high-quality data, and a shortage of skilled professionals with expertise in these technologies. Therefore, it is essential to identify research directions and opportunities that can help overcome these challenges and enable the full potential of data analytics and cognitive computing in healthcare [4].

Explainable AI in Healthcare Imaging for Medical Diagnoses
https://doi.org/10.1016/B978-0-443-23979-3.00019-1

477

This research paper proposes a novel generic approach that utilizes deep learning (DL) and natural language processing (NLP) for continuous remote monitoring in digital health. The approach involves using DL algorithms to analyze data collected from wearable devices and remote monitoring tools, while NLP is used to analyze patient feedback and electronic medical records. Various DL algorithms like convolutional neural networks (CNNs) and recurrent neural networks (RNNs) are used to analyze medical images and time-series data, and unsupervised learning techniques such as autoencoders are used for feature extraction and anomaly detection. Stanford CoreNLP and Apache OpenNLP are used for analyzing patient feedback and medical records to identify important phrases and topics. The combination of DL and NLP can enable continuous remote monitoring of patient health and provide healthcare providers with real-time insights for early intervention and more personalized treatment plans, ultimately leading to improved patient outcomes, reduced healthcare costs, and enhanced quality of care. The proposed approach can enable health professionals to remotely monitor patients' health conditions in real time and predict the likelihood of adverse health events. This research paper also offers a comprehensive review of the prevailing works in the domain of data analytics and cognitive computing in digital health. The paper aims to examine the applications of these technologies in various areas of healthcare, identify the challenges and prospects of implementing data analytics and cognitive computing in healthcare, and propose research directions that can help accelerate the adoption of these technologies. By providing insights into current trends, challenges, and research directions, this paper could aid inform future research and development in the field of digital health.

This research paper makes the following contributions:

- Proposes a generic approach for continuous remote monitoring in digital health using DL and NLP.
- Comprehensive comparison of the proposed approach with previous related works with research observations.
- Discusses the development of data analytics and cognitive computing in digital health and their contemporary applications.
- Examines the applications of data analytics and cognitive computing in various areas of healthcare, including disease diagnosis, treatment planning, and patient monitoring.
- Identifies the concerns and possibilities of implementing data analytics and cognitive computing in healthcare, such as data quality issues, privacy concerns, and the need for skilled professionals.
- Identifies key research directions that can help accelerate the adoption of data analytics and cognitive computing in healthcare, such as the development of personalized healthcare solutions using ML and big data analytics, the integration of cognitive computing with EHRs for clinical decision support, and the use of NLP for patient engagement and communication.

The remaining parts of this work are structured as follows. Section 2 discusses the related work in data analytics and cognitive computing for digital health. Section 3 discusses the proposed generic approach for using DL and NLP for continuous remote monitoring in digital health. Sections 4 and 5 describe the emerging technologies and the applications of data analytics and cognitive computing, respectively. Section 6 elaborates on the challenges and opportunities in implementing data analytics and cognitive computing in healthcare. The future research directions in data analytics and cognitive computing in healthcare are discussed in Section 7. This chapter ends with a conclusion and future scope in Section 8.

2. Related work

In this section, we discuss some of the related work where DL and NLP are being used for continuous remote monitoring in digital health. We provide a comparison between our proposed approach and previous related works. We also identify and discuss the limitations of existing works and explain how our approach overcomes these limitations. There has been a significant amount of research in the area of using DL and NLP for continuous remote monitoring in digital health. Here are some notable examples.

The study [5] reviews remote patient monitoring (RPM) systems, covering innovative tools, artificial intelligence (AI) impact, tasks, and developments. The study emphasizes patient-centric RPM using wearable devices, sensors, and technology in the cloud, fog, edge, and blockchain. AI plays a vital role in monitoring physical activity, chronic diseases, and vital signs. AI-enabled RPM detects early health deterioration, personalizes monitoring, and learns human behavior. The study discusses challenges, trends, and future directions for AI in RPM applications. The review [6] focuses on time series forecasting using AI/ML methods and the recent work on wearable technology in heart failure. Limitations include challenges, security, and data aggregations in applying AI/ML in the medical domain. The review [7] explores the possibilities, difficulties, and consequences of NLP approaches, as well as the relevant literature regarding the utilization of EHRs for medical trials in these domains. The benefits of AI in precision medicine, as well as the potential of big data analytics and nanotechnology, are discussed in Sahu et al. [8] in depth. The article also discusses the benefits and drawbacks of using AI in targeted therapy.

A systematic literature review was conducted in Behera et al. [9] to examine earlier study linked to cognitive computing in the health domain. The review covered techniques, procedures, applications, findings, advantages, and limitations. The review highlights the abilities and effect of cognitive computing on clinical outcomes and proposes future research directions. The study concludes with managerial implications, limitations, and areas for future work. The review [10] discusses the progress and challenges related to the integration of cutting-edge technologies like blockchain, big data, telehealth, and smart gadgets in healthcare and training for medical professionals. The paper highlights the importance of digital health technologies in improving healthcare accessibility and availability, but also the need for proper regulation and validation of these technologies. The literature search showed that the recent COVID-19 pandemic has increased the need for telehealth and e-health technologies. The study concludes that scientific research is necessary to ensure the reliability, safety, and ethical aspects of digital health technologies before they are implemented in the healthcare sector.

Navaz et al. [11] discuss how "smart and connected health" (SCH) can help address the increasing prevalence of cardiovascular diseases, which are the primary source of mortality worldwide. SCH is described as a solution that can make healthcare more planned, precautionary, and customized, with value-added services. The article proposes an architectural model to identify SCH attributes and detect skill-linked tasks in its implementation. A COVID-19 case study is presented to illustrate how different countries have leveraged the power of various SCH technologies to prevent the epidemic. The article also highlights the issues with SCH acceptance and possible research guidelines for improved healthcare focused on patients. This work [12] discusses how big data are being used in healthcare to extract relevant information and develop smart applications to solve real-life problems. The article notes that the digital transformation of healthcare systems has posed challenges to researchers and caretakers, including storage and minimizing treatment costs. The article reviews existing literature from 2011 to 2021 and concludes that to lower healthcare expenditure, shorten computational cost, and boost quality care, adopting modern hybrid technologies that utilize ML and cloud computing is essential. The purpose of this comprehensive analysis is to improve healthcare providers', academics', and regulators' ability to draw on the results of the investigation in their own work. Table 1 shows the comparison between the proposed work and the related works.

Overall, the proposed work overcomes the limitations of these previous works by proposing a comprehensive approach that utilizes DL and NLP algorithms for continuous remote monitoring in digital health, providing a more holistic approach to patient monitoring. Additionally, the proposed work offers an exhaustive analysis of the current state of data analytics and cognitive computing in digital health, identifying the challenges and opportunities of implementing these technologies in healthcare, and proposing research directions that can help accelerate their adoption.

3. A generic approach using DL and NLP for continuous remote monitoring in digital health

In this section, we suggest a generic approach using DL and NLP for continuous remote monitoring in digital health. In this approach, DL algorithms are used to analyze data collected from wearable devices and other remote monitoring tools, while NLP is

Table 1 Comparison of proposed work with existing works.

Existing works	Overcoming limitations in proposed work
Shaik et al. [5]	• The proposed work goes beyond a comprehensive review of RPM systems by presenting a generic approach using DL and NLP for continuous remote monitoring in digital health. The proposed work suggests various DL algorithms such as CNNs and RNNs to analyze medical images and time-series data, while unsupervised learning techniques such as autoencoders are recommended for feature extraction. • In contrast, Shaik et al. [5] only explore the categorization of physical exercise, tracking of degenerative illnesses, and tracking of physiological parameters in crisis situations are all areas where AI plays a significant part in RPM devices.
Gautam et al. [6]	• Both proposed work and Gautam et al. [6] acknowledge the potential of distant monitoring through smart devices to revolutionize healthcare, but the proposed work offers a more comprehensive approach using DL and NLP to analyze the data collected from these devices. • Additionally, the proposed work suggests unsupervised learning techniques for feature extraction, while Gautam et al. [6] mainly focus on the fundamentals of AI/ML methods, time-series prediction, and the prevailing scenario of AI/ML in the domain of smart devices. • The proposed work offers a broader perspective by reviewing the current state of data analytics and cognitive computing in digital health, examining the applications of these technologies in various areas of healthcare, identifying the challenges and opportunities of implementing them in healthcare, and proposing research directions that can help accelerate their adoption.
Juhn et al. [7]	• The proposed work complements [7] by leveraging NLP to analyze patient feedback and electronic medical records in addition to data collected from wearable devices and remote monitoring tools. • Additionally, the proposed work suggests unsupervised learning techniques for feature extraction, while Gautam et al. [6] mainly focus on the fundamentals of AI/ML methods, time-series prediction, and the prevailing scenario of AI/ML in the domain of smart devices. • By doing so, the proposed work aims to enable health professionals to remotely monitor patients' health conditions in real time and predict the likelihood of adverse health events, leading to improved patient outcomes, reduced healthcare costs, and enhanced quality of care.
Sahu et al. [8]	• While both proposed work and Sahu et al. [8] discuss the integration of AI in healthcare, the proposed work suggests a more generic approach using DL and NLP for continuous remote monitoring in digital health, whereas Sahu et al. [8] mainly focus on the combination of AI in high-precision healthcare.

Continued

Table 1 Comparison of proposed work with existing works—cont'd

Existing works	Overcoming limitations in proposed work
Behera et al. [9]	• The proposed work goes beyond discussing the part of AI in healthcare by presenting a comprehensive evaluation of the current state of data analytics and cognitive computing in digital health, identifying the challenges and opportunities of implementing them in healthcare, and proposing research directions that can help accelerate their adoption. • The proposed work and Behera et al. [9] are similar in that they discuss the possibility of cognitive computing in healthcare, but the proposed work suggests a more specific approach using DL and NLP for continuous remote monitoring in digital health. • In contrast, Behera et al. [9] provide a comprehensive literature review of previous study associated with cognitive computing in healthcare, including methods, algorithms, applications, and results. • The proposed work offers a more focused approach by suggesting various DL algorithms, such as CNNs and RNNs, to analyze medical images and time-series data, and unsupervised learning techniques, such as autoencoders, for feature extraction. • Additionally, the proposed work reviews the current state of data analytics and cognitive computing in digital health, examining the applications of these technologies in various areas of healthcare, identifying the challenges and opportunities of implementing them in healthcare, and proposing research directions that can help accelerate their adoption.
Senbekov et al. [10]	• In Senbekov et al. [10], the authors provide a limited review of digital health technologies and their reliability, safety, testing, and ethical aspects. • Proposed work conducts a comprehensive analysis of the existing literature to detect gaps in healthcare big data analytics and suggest enhanced solutions to address them.
Navaz et al. [11]	• There is narrow focus on smart and connected health (SCH) technologies and their challenges in Navaz et al. [11]. • Proposed work expands on this focus by thoroughly analyzing the literature to identify essential characteristics and best practices for data administration.
Khan et al. [12]	• Limited exploration of recent medical big data analytics approaches in the cloud using ML in Khan et al. [12]. • Proposed work suggests the adaptation of these newer models to lessen the financial burden of therapy, speed up modeling times, and boost therapeutic effectiveness.

utilized to assess the information collected from patient feedback and electronic medical records. The DL methods used may include CNNs for analyzing medical images, such as X-rays and MRIs, and RNNs for analyzing time-series data, such as heart rate and blood pressure. Additionally, unsupervised learning techniques, such as autoencoders, may be used for feature extraction and anomaly detection.

For NLP, algorithms such as the Stanford CoreNLP and Apache OpenNLP can be used to analyze patient feedback and electronic medical records. These algorithms can identify important phrases and topics related to patient symptoms and concerns, allowing healthcare providers to quickly address patient needs. The combination of DL and NLP can enable continuous remote monitoring of patient health, providing healthcare providers with real-time insights into patient health status and allowing for early intervention and more personalized treatment plans. This strategy may lead to better clinical satisfaction, lower medical expenditures, and higher care standards all around.

Fig. 1 shows the proposed generic approach using DL and NLP for continuous remote monitoring in digital health.

Fig. 1 shows that the output from both modules is combined to provide a comprehensive view of patient health status, allowing for continuous remote monitoring. The output can be shared with healthcare providers for real-time insights into patient health status, allowing for early intervention and personalized treatment plans. The results for patients may improve as a result of this, reduced healthcare costs, and enhanced overall

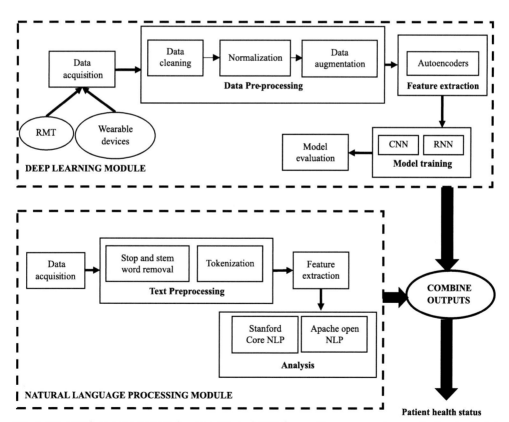

Fig. 1 Proposed generic approach using DL and NLP for continuous remote monitoring in digital health.

quality of care. As shown in Fig. 1, the proposed generic approach includes the following two main modules.

3.1 Deep learning module

The deep learning module is responsible for processing and analyzing the physiological data collected from wearable devices and remote monitoring tools. It can be broken down into several submodules.

3.1.1 Data acquisition

The data acquisition submodule is the first step in the proposed generic approach for continuous remote monitoring in digital health. It is accountable for accumulating physiological information from wearable devices and remote monitoring tools. The data collected can include measurements of the body's "vitals," including pulse, hypertension, respiration, and other health-related information. Wearable devices can include smartwatches, fitness trackers, biosensors, and other monitoring tools that can be worn or attached to the body. Remote monitoring tools can include medical devices such as electrocardiograms, blood glucose meters, and spirometers, which may be employed to examine the patient's health condition remotely. The information collected from these devices can be in various formats and may differ in terms of sampling rate, resolution, and accuracy. Therefore, the data acquisition submodule should be designed to handle different data sources and formats. For example, some devices may provide data in real time, while others may provide data at regular intervals. Additionally, some devices may provide continuous data streams, while others may provide discrete data.

The data acquisition submodule can involve both hardware and software components. The hardware component involves selecting and setting up the wearable devices and remote monitoring tools, ensuring they are properly calibrated and functioning correctly. The software component involves developing the interfaces and protocols to communicate with the devices and retrieve the information. The data acquisition submodule is a critical step in the proposed approach, as the correctness and consistency of the information gathered can directly impact the quality of the analysis and the effectiveness of the predictive model. Therefore, the data acquisition submodule should be designed to ensure data quality, accuracy, and completeness, while also being flexible enough to handle a variety of data sources and formats. Algorithm 1 shows the steps involved in the data-acquisition module.

Algorithm 1 DataAcquisition().

```
1:        ▷ Select and set up the wearable devices and remote monitoring tools
2:            ▷ Ensure devices are properly calibrated and functioning correctly
3:                        ▷ Connect to the devices and retrieve data
```

```
 4:  while True do
 5:            ▷ Retrieve data from wearable devices and remote monitoring tools
 6:      data = get_data()
 7:                                    ▷ Process and store the collected data
 8:      store_data(data)
 9:                                    ▷ Wait for the next data sample
10:        sleep(sampling_interval)
11:                                    ▷ Handle errors and exceptions
12:        continue
13: end while
```

In Algorithm 1, the data-acquisition() function is responsible for collecting physiological data from wearable devices and remote monitoring tools. The function uses a loop to continuously retrieve data from the devices and store it for further analysis. The get-data() function is used to retrieve data from the devices, while the store_data() function is used to store the collected data. The time.sleep() function is used to set the sampling interval for the data acquisition. If an error or exception occurs during data collection, the loop continues without interrupting the data acquisition process. The sampling-interval variable can be adjusted to control the rate at which data are collected from the devices.

3.1.2 Preprocessing

This submodule performs cleaning, normalizing, and augmenting the collected data. This can involve removing any noise or outliers, normalizing the data to a standard scale, and generating additional data using techniques such as data augmentation. The preprocessing submodule is a crucial step in the proposed generic approach for continuous remote monitoring in digital health. It is responsible for cleaning, normalizing, and augmenting the collected data from wearable devices and remote monitoring tools. The quality of the input data significantly impacts the reliance on the DL model's precision and accuracy. Therefore, the preprocessing submodule must be designed to ensure the data are clean, consistent, and complete. The preprocessing submodule can involve several steps, including data cleaning, normalization, and augmentation. Data cleaning involves removing any noise or outliers in the data. This can include identifying and removing anomalous data points or filtering out any irrelevant data. The objective of information cleaning is to guarantee that the information is accurate and consistent across all patients and devices.

Normalization is another important step in the preprocessing submodule. It involves scaling the data to a standard range to facilitate comparisons across different data sources. For instance, heart rate data collected from different devices may have different units of measurement and sampling rates. Normalization helps to convert the data to a common scale, enabling meaningful comparisons. Data augmentation is a technique used to

generate additional data to enhance the performance of the DL model. It involves applying various transformations to the existing data, such as rotations, scaling, or translation.
These transformations can create new data points that are similar to the original data, but
not identical, helping the model to learn more effectively. Data augmentation is particularly useful when the labeled dataset is small, and the DL model has limited exposure to
the data. By cleaning, normalizing, and augmenting the collected data, the preprocessing
submodule can improve the performance of the DL model and enhance the accuracy of
the predictions. Algorithm 2 shows the steps in this module.

Algorithm 2 Preprocessing.

1:   **function** REMOVE_OUTLIERS(data, threshold=3)
2:       Avg ← mean(data)
3:       SD ← std(data)
4:       z_scores ← (data − Avg)/SD
5:       filtered_data ← data[abs(z_scores) < threshold]
6:       **return** filtered_data
7:   **end function**
8:   **function** NORMALIZE_DATA(data)
9:       normalized_data ← (data − min(data))/(max(data) − min(data))
10:      **return** normalized_data
11: **end function**
12: **function** AUGMENT_DATA(data)
13:              ▷ Perform data augmentation using techniques such as flipping,
        rotation, etc.
14:      **return** augmented_data
15: **end function**
16:                              ▷ Example usage of the preprocessing functions
17: raw_data ← ...   ▷ Raw physiological data from wearable devices and remote
        monitoring tools
18: clean_data ← REMOVE_OUTLIERS(raw_data)
19: normalized_data ← NORMALIZE_DATA(clean_data)
20: augmented_data ← AUGMENT_DATA(normalized_data)

The modules in Algorithm 2 are explained here:

- *Feature extraction*: This submodule is responsible for identifying relevant features from
  the preprocessed data. This can involve utilizing unsupervised learning methods like
  autoencoders to detect features that are most pertinent to predicting the patient's
  health status.
- *Model training*: This submodule is responsible for training the DL model on the
  extracted features. This can involve utilizing DL algorithms like CNNs and RNNs
  to predict the patient's health status. CNNs are commonly used in image and signal

processing tasks where the input has a longitudinal or temporal structure. CNNs are particularly effective at identifying local patterns in the data, such as edges or corners that can be used to classify the input. In healthcare, CNNs have been used to analyze medical images and identify abnormalities, such as tumors or lesions. RNNs are commonly used in sequential data analysis tasks, such as time-series data or NLP. RNNs are particularly effective at modeling long-term dependencies in the data, allowing them to make predictions based on previous inputs. In healthcare, RNNs have been used to predict patient outcomes based on time-series physiological data or electronic medical records. The training process involves using a labeled dataset to train the model to predict a patient's health status based on the input data. Patient's medical condition is indicated by tags in the labeled dataset, which consists of input information. The model is adjusted based on the error between its predictions and the actual output labels in the dataset. To avoid overfitting, in which the model is specialized to the training information and functions inadequately on new information, regularization techniques such as dropout and weight decay can be used. Cross-validation techniques can also be used to evaluate the model's performance on new data. Overall, the model training submodule is an essential step in the proposed generic approach for continuous remote monitoring in digital health. By training a DL model on a large dataset of patient data, healthcare providers can make more accurate predictions about the patient's health status and provide personalized treatment plans.

- *Evaluation*: This submodule is responsible for evaluating the accuracy and performance of the trained model. This can involve testing the model on a validation dataset and comparing its performance against other models.

3.2 Natural language processing (NLP) module

The NLP module is responsible for processing and analyzing text-based data collected from patient feedback and electronic medical records. It can be broken down into several submodules:

- *Data acquisition*: This submodule is responsible for collecting patient feedback and electronic medical records. This can include data such as patient-reported symptoms, medical history, and medication information.
- *Text preprocessing*: This submodule cleans and prepares the text data for analysis. This can involve removing stop words, stemming words, and tokenizing the text into individual words.
- *Feature extraction*: This submodule is responsible for identifying important phrases and topics related to patient symptoms and concerns. This can involve using techniques such as named entity recognition and topic modeling to extract relevant features from the text data.
- *Analysis*: This submodule is responsible for analyzing the extracted features to derive intuitions into the patient's health status. This can involve using algorithms such as

Stanford CoreNLP and Apache OpenNLP to perform sentiment analysis and identify patterns in the text data. To perform sentiment analysis, algorithms such as Stanford CoreNLP and Apache OpenNLP can be used. These algorithms use machine learning (ML) techniques to classify text based on its sentiment. They can identify positive, bad, or unbiased sentiment in a piece of content and assign a numerical score to represent the intensity of the sentiment.

Stanford CoreNLP is a suite of NLP tools that can perform a variety of NLP jobs, including entity identification, sentiment analysis, and more. In the proposed approach, the sentiment analysis component of CoreNLP is used to classify the sentiment of patient feedback and electronic medical records [13]. To use Stanford CoreNLP for sentiment analysis, the text data first need to be preprocessed and tokenized. Preprocessing involves removing unnecessary text and symbols, such as punctuation marks and numbers. Tokenization involves breaking the text data down into individual words or tokens. Once the text data have been preprocessed and tokenized, the sentiment analysis component of CoreNLP can be applied. This component uses an ML algorithm to classify the sentiment of each token in the text data as positive, negative, or neutral. The sentiment analysis algorithm in CoreNLP uses a lexicon-based approach, where each word in the text data is assigned a score based on its sentiment. Words with a positive sentiment, such as "good" or "happy," are assigned a positive score, while words with a negative sentiment, such as "bad" or "sad," are assigned a negative score. The scores for each word are then combined to produce an overall sentiment score for the text data.

The sentiment score produced by CoreNLP can be used to gain insights into the patient's health status. For example, if the sentiment score is predominantly negative, it could indicate that the patient is experiencing a decline in their health status. Healthcare providers can use this information to intervene early and provide personalized treatment plans. Overall, by using the sentiment analysis component of Stanford CoreNLP in the proposed approach, healthcare providers can gain insights into the patient's health state and improve the quality of care.

Apache OpenNLP is an NLP toolkit that can perform a many tasks, including sentence detection, and more [14]. In the proposed approach, the named entity recognition component of OpenNLP is used to identify important phrases and topics related to patient symptoms and concerns. To use Apache OpenNLP for named entity recognition, the text data first need to be preprocessed and tokenized. Preprocessing involves removing unnecessary text and symbols, such as punctuation marks and numbers. Tokenization involves breaking the text data down into individual words or tokens. Once the text data have been preprocessed and tokenized, the named entity recognition component of OpenNLP can be applied. This component uses ML algorithms to identify named entities in the text data, such as people, locations, organizations, and medical concepts. The named entities identified by OpenNLP can be used to gain insights into the patient's symptoms and concerns. For example, if the NER component identifies a named entity such as "diabetes," it could indicate that the patient has a history of diabetes and may require special treatment or monitoring. Similarly, if the NER component identifies a named entity such as "hospital" or "emergency room," it could indicate that the patient has recently been admitted to a hospital or had an emergency, requiring prompt attention from healthcare providers. Overall, by using the named entity recognition component of Apache OpenNLP in the proposed approach, healthcare providers can gain insights into the patient's symptoms and concerns and improve the quality of care.

In addition to sentiment analysis, the Analysis submodule can also use other algorithms to identify patterns in the text data. For example, topic modeling algorithms can identify important topics and themes in the patient feedback and electronic medical records. These topics can be used to gain insights into the patient's concerns and symptoms, and to identify areas where additional treatment or support may be needed. Overall, the Analysis submodule plays a crucial role in the proposed generic approach for continuous remote monitoring in digital health. By analyzing the extracted features from the NLP module, healthcare providers can gain a more comprehensive view of the patient's health status and make more informed treatment decisions.

4. Emerging technologies in data analytics and cognitive computing in healthcare

This section discusses some of the emerging technologies as shown here [15]:

- *NLP*: NLP is a technology that can analyze and understand human language. In healthcare, NLP can be used to analyze medical records and patient feedback to understand patient needs and concerns [16].
- *Predictive analytics*: Predictive analytics uses statistical algorithms to identify patterns and predict future outcomes. In healthcare, predictive analytics may be utilized to detect high risk patients, predict disease progression, and develop personalized treatment plans [17].
- *ML*: ML is a kind of AI that uses algorithms to understand patterns from data. In healthcare, ML can be used to analyze patient information and detect patterns which are not visible to human experts. This may result in more precise disease diagnosis and more effective treatment planning [18].
- *DL*: DL is a type of ML that utilizes ANNs to learn patterns from data. In healthcare, DL can be used to analyze medical images, identify tumors, and develop predictive models to identify high-risk patients [19].
- *Computer vision*: Computer vision is a technology that allows machines to interpret and analyze visual information. In healthcare, computer vision can be used to analyze medical images and identify abnormalities, such as tumors or lesions [20].
- *Blockchain*: Blockchain is a secure and decentralized way of storing and sharing data. In healthcare, it is possible to safely preserve and exchange health information using blockchain solutions [21].

- *Internet of Things (IoT)*: IoT is a network of devices that can collect and share data. Many IoT sensors are capable of helping track physiological parameters of individuals in medical settings, track medication adherence, and identify potential health issues before they become serious [22].
- *Augmented reality (AR) and virtual reality (VR)*: AR and VR are technologies that allow users to interact with a virtual environment. In healthcare, AR and VR can be used to visualize medical data and simulate medical procedures [23].
- *Robotic process automation (RPA)*: Automating routine processes with intelligent automation is the goal of RPA. Managerial chores in the healthcare industry, including as entering data and invoicing, could be automated by RPA so that doctors can devote more time to direct treatment for patients [24].
- *Chatbots*: Chatbots are computer programs that can simulate human conversation. In healthcare, chatbots can be used to triage patient concerns and answer common questions, reducing the workload on healthcare providers [25].

Table 2 summarizes these emerging technologies.

These evolving tools have the ability to further enhance the capabilities of data analytics and cognitive computing in healthcare, improving patient outcomes, and transforming the healthcare industry.

Table 2 Emerging technologies in data analytics and cognitive computing for healthcare.

Technology	Description	Applications	Benefits	Examples
Natural language processing (NLP)	Analyzes and understands human language	Analyzing medical records and patient feedback to understand patient needs and concerns	Improved patient care, reduced administrative workload	Analyzing patient feedback to improve healthcare service quality
Predictive analytics	Uses statistical algorithms to identify patterns and predict future outcomes	Identifying high-risk patients, predicting disease progression, and developing personalized treatment plans	Improved patient outcomes, reduced healthcare costs	Predicting patient readmissions and proactively intervening
Machine learning (ML)	Uses algorithms to learn patterns from data	Analyzing patient data to detect patterns which might be invisible to humans	Improved disease diagnosis, more effective treatment planning	Identifying at-risk patients and improving care outcomes

Table 2 Emerging technologies in data analytics and cognitive computing for healthcare—cont'd

Technology	Description	Applications	Benefits	Examples
Deep learning (DL)	Uses artificial neural networks to learn patterns from data	Analyzing medical images, identifying tumors, and developing predictive models to identify high-risk patients	Improved disease diagnosis, more effective treatment planning	Analyzing MRIs to identify Alzheimer's disease
Computer vision	Allows machines to interpret and analyze visual information	Analyzing medical images and identifying abnormalities, such as tumors or lesions	Improved disease diagnosis, more effective treatment planning	Identifying skin lesions and detecting skin cancer
Cognitive computing	Simulates human thought processes, such as reasoning and problem-solving	Analyzing patient data and developing personalized treatment plans based on individual patient needs	Improved patient care, more effective treatment planning	Analyzing patient data to identify individualized care plans
Blockchain	Secure and decentralized way of storing and sharing data	Securely storing and sharing patient data, tracking medical devices, and pharmaceutical products	Increased data security and privacy, improved patient outcomes	Sharing electronic health records between healthcare providers
Internet of Things (IoT)	Network of devices that can collect and share data	Monitoring patient vital signs, tracking medication adherence, and identifying potential health issues before they become serious	Improved patient care, more effective treatment planning	Remote patient monitoring and identifying changes in patient health status
Augmented reality (AR) and virtual reality (VR)	Tools which permit customers to communicate with a virtual environment	Visualizing medical data and simulating medical procedures	Improved medical training, enhanced patient education	Simulating surgeries to improve training for medical students

Continued

Table 2 Emerging technologies in data analytics and cognitive computing for healthcare—cont'd

Technology	Description	Applications	Benefits	Examples
Robotic process automation (RPA)	makes use of computer programs called robots to handle mundane jobs	Automating administrative tasks, such as data entry and billing	Reduced administrative workload, increased efficiency	Automating claims processing and medical coding
Chatbots	Computer programs that can simulate human conversation	Triaging patient concerns and answering common questions	Improved patient engagement, reduced workload on healthcare providers	Answering common healthcare questions and providing basic medical advice

5. Real world case studies of data analytics and cognitive computing in healthcare

Data analytics and cognitive computing have the potential to transform healthcare by providing insights into large and complex healthcare data sets. In this research paper, we examine the applications of data analytics and cognitive computing in various areas of healthcare, including disease diagnosis, treatment planning, and patient monitoring.

5.1 Disease diagnosis

Data analytics and cognitive computing could be employed to enhance disease diagnosis by analyzing large sets of patient data. By analyzing patterns and identifying correlations, these technologies can assist healthcare providers in making accurate diagnoses and provide personalized treatment options. For instance, in oncology, data analytics and cognitive computing can help identify genetic markers of cancer and classify different types of tumors, which can lead to more precise and personalized treatment options [24]. There are many real-world case studies that demonstrate how data analytics and cognitive computing can improve disease diagnosis, especially in the field of oncology. Here are a few examples:

- *IBM Watson Genomic Analysis*: IBM Watson Genomic Evaluation is a platform that utilizes cognitive computing to examine genomic information and provide personalized cancer treatment recommendations. The platform has been used to help diagnose and treat patients with rare and complex cancers. In one case, a patient with a rare form of leukemia was diagnosed using Watson Genomic Evaluation, and the platform recommended a personalized treatment plan that led to remission [25].
- *Memorial Sloan Kettering (MSK) Cancer Center*: The MSK has developed a platform called Oncology Expert Advisor that uses data analytics to help clinicians make

treatment decisions. The platform analyzes data from EHRs, pathology reports, and radiology reports to provide clinicians with personalized treatment recommendations. In one study, the platform was used to identify treatment options for patients with advanced pancreatic cancer, resulting in improved patient outcomes [26].

- *Stanford University*: Stanford University researchers used data analytics to identify genetic mutations in patients with rare pediatric cancers. The researchers analyzed genomic data from 91 pediatric cancer patients and identified 51 genetic mutations that could be targeted by existing drugs. By using data analytics to identify these mutations, the researchers were able to provide personalized treatment options for patients with rare and difficult-to-treat cancers [27].

These case studies demonstrate the potential of data analytics and cognitive computing to improve disease diagnosis and provide personalized treatment options, especially in the field of oncology. By analyzing large sets of patient data, these technologies can assist healthcare providers in making accurate diagnoses and identifying targeted treatment options that can lead to improved patient outcomes.

5.2 Treatment planning

Data analytics and cognitive computing can also be used to enhance treatment planning. By analyzing patient data, including medical histories, genetic information, and medication histories, these technologies can help predict how a patient will respond to a particular treatment. This knowledge may be employed to optimize therapy strategies, reduce side effects, and increase treatment efficacy. For example, data analytics can be used to analyze large clinical trial datasets to detect subgroups of patients that may help the majority from a particular treatment [28]. There are several real-world case studies that demonstrate how data analytics and cognitive computing can enhance treatment planning in healthcare:

- In [29], the researchers data analytics and cognitive computing to improve treatment planning for patients with multiple myeloma. The researchers used ML algorithms to analyze clinical and genomic data from patients and identify the most effective combination of treatments for each patient. By using this approach, the researchers were able to increase the number of patients who achieved complete remission and improve patient outcomes [29].
- In [30], the authors used data analytics and cognitive computing to predict the efficacy of immunotherapy treatments for patients with advanced melanoma. The researchers developed an ML model that analyzed patient data, including tumor genomic data and immune cell data, to predict which patients would respond best to a particular treatment. By using this approach, the researchers were able to optimize treatment plans and increase treatment efficacy [30].
- The authors [31] used data analytics to optimize treatment plans for patients with traumatic brain injuries (TBI). The researchers developed an ML model that analyzed

patient data, including CT scans and medical records, to predict which patients were at high risk of poor outcomes after TBI. By using this approach, the researchers were able to optimize treatment plans for patients and improve patient outcomes [31].

These case studies demonstrate the potential of data analytics and cognitive computing to enhance treatment planning in healthcare. By analyzing patient data and predicting treatment efficacy, these technologies can help healthcare providers optimize treatment plans and increase treatment efficacy while minimizing side effects.

5.3 Patient monitoring

Data analytics and cognitive computing can also help monitor patients and identify early signs of potential health issues. Wearable devices and mobile applications can collect data on patient vital signs and other health-related behaviors, which can then be analyzed to identify patterns and predict health issues. For example, by analyzing patient data, data analytics and cognitive computing can help detect early signs of heart disease and other chronic conditions, allowing healthcare providers intervene before the disease progresses [32]. There are several real-world case studies that demonstrate how data analytics and cognitive computing can be used to monitor patients and identify early signs of potential health issues:

- The work [33] used data analytics and cognitive computing to monitor patients with heart failure. The researchers used a combination of wearable devices and mobile applications to collect data on patient vital signs, physical activity, and other health-related behaviors. By analyzing this data, the researchers were able to detect early signs of heart failure exacerbation and intervene before the condition progressed. As a result, the researchers were able to reduce hospital readmissions and improve patient outcomes.

- In [34], the researchers developed an ML model that can calculate the probability of hospital readmissions for patients with chronic obstructive pulmonary disease. The model analyzed patient data, including vital signs, medication adherence, and other health-related behaviors, to classify patients who were at high risk of hospital readmission. By using this approach, healthcare providers can intervene before the condition progresses and reduce the risk of hospital readmissions.

- In [35], the authors developed a remote monitoring system for patients with diabetes. The system collects data from patients' continuous glucose monitoring devices and delivers the information to a cloud-based platform. The data are then analyzed using ML algorithms to predict hypoglycemic and hyperglycemic events. Healthcare providers can then use this information to adjust treatment plans and reduce the risk of diabetes-related complications.

These case studies demonstrate the potential of data analytics and cognitive computing to monitor patients and identify early signs of potential health issues. By analyzing patient data, these technologies can help healthcare providers intervene before the condition progresses and improve patient outcomes.

6. Challenges and opportunities of implementing data analytics and cognitive computing in healthcare

Data analytics and cognitive computing have tremendous potential to transform the healthcare industry by enabling more accurate diagnosis, personalized treatment, and improved patient outcomes. However, there are also significant obstacles that need to be overcome before these benefits [36]. Fig. 2 shows the challenges and opportunities faced during the implementation of data analytics and cognitive computing in healthcare.

6.1 Challenges

Following are certain challenges that may be faced while implementing data analytics and cognitive computing in healthcare [37]:

- *Data quality and interoperability*: One of the biggest challenges is the quality and interoperability of healthcare data. Healthcare data are often fragmented, inconsistent, and siloed in different systems, making it difficult to integrate and analyze. Additionally, data privacy regulations and concerns further complicate data sharing and integration.
- *Data governance*: The use of data analytics and cognitive computing raises important questions around data governance, including who owns the data, how it can be used, and who has access to it. Regulations and ethical considerations must be taken into account to ensure that patient data are protected and used appropriately.
- *Infrastructure and resource constraints*: Implementing data analytics and cognitive computing requires significant infrastructure and resources, including advanced hardware and software, skilled staff, and specialized training. These requirements can be a significant barrier for smaller healthcare organizations or those in low-resource settings.
- *Algorithm bias and accuracy*: The accuracy and bias of algorithms used in data analytics and cognitive computing can be a major challenge, particularly when dealing with complex or rare diseases. Biases can occur due to factors such as inadequate data representation, inadequate data cleaning, or insufficiently diverse datasets, leading to inaccurate diagnoses and treatments.
- *Data privacy*: Healthcare data are highly sensitive, and protecting patient privacy is of utmost importance. However, this can be challenging in the context of data analytics, as data need to be shared and analyzed across different organizations and systems.
- *Data security*: Similar to data privacy, ensuring the security of healthcare data is critical. Healthcare data are often targeted by cyber criminals, and breaches can result in serious harm to patients.
- *Lack of standards*: Healthcare data are frequently captured in diverse systems and structures that cause difficulty in combining and analyzing data from different sources. Standardization is necessary to enable interoperability and facilitate data sharing.

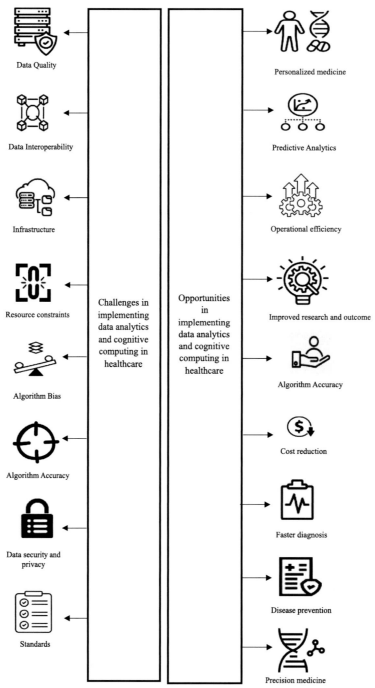

Fig. 2 Challenges and opportunities of implementing data analytics and cognitive computing in healthcare.

6.2 Opportunities

Following are certain opportunities present in data analytics and cognitive computing in healthcare [38]:

- *Personalized medicine*: Data analytics and cognitive computing can enable personalized medicine by analyzing individual patient data and providing tailored treatment recommendations based on genetic, lifestyle, and environmental factors.
- *Predictive analytics*: Individuals at high danger of contracting certain disorders could be identified using predictive modeling, enabling for timely treatment and prevention. It can also help healthcare providers predict patient outcomes and optimize treatment plans.
- *Operational efficiency*: By automating routine tasks and optimizing workflows, data analytics and cognitive computing can improve operational efficiency, reduce costs, and improve patient flow and experience.
- *Improved research and development*: Data analytics and cognitive computing can accelerate research and development in the healthcare industry by enabling large-scale data analysis and insights that were previously not possible.
- *Better benefits for patients*: Using predictive analytics, doctors could better pinpoint patterns and trends that can lead to improved patient outcomes, such as reducing hospital readmissions and improving medication adherence.
- *Cost reduction*: By analyzing healthcare data, organizations can identify areas where they can reduce costs, such as by reducing waste, optimizing resource utilization, and improving supply chain management.
- *Faster diagnosis*: Cognitive computing can help healthcare providers analyze large volumes of patient data to arrive at a diagnosis more quickly, potentially saving lives in critical situations.
- *Disease prevention*: By analyzing population health data, healthcare providers can identify trends and patterns that can help them develop effective disease prevention strategies.
- *Precision medicine*: The term "precision medicine" describes the practice of customizing a patient's healthcare services to their unique set of characteristics, including their genes, surroundings, and daily routine. With the use of data analytics, doctors could determine whether therapies are more than likely to enhance a patient's condition.

Data analytics and cognitive computing have the possibility to transform the medical domain by facilitating precise analysis, individualized treatment, and superior patient outcomes. However, significant challenges must be addressed, including data quality and interoperability, data governance, resource constraints, and algorithm bias and accuracy. Addressing these challenges will require collaboration between healthcare providers, technology vendors, and regulatory bodies to ensure that the benefits of these technologies are realized while protecting patient privacy and safety.

7. Future research directions in data analytics and cognitive computing in healthcare

Data analytics and cognitive computing have a promising future in healthcare, and there are several research directions that can be explored to further improve the applications of these technologies. Some of the future research directions are highlighted here [39]:

- *Personalized medicine*: The use of data analytics and cognitive computing can lead to personalized medicine, where treatments are tailored to individual patients based on their medical history, genetics, and lifestyle. This approach could be more effective and efficient treatments and can significantly improve patient outcomes.
- *Predictive analytics*: Predictive analytics can be used to forecast health outcomes and identify patients at risk of developing certain diseases. This can help healthcare providers to intervene early and prevent the onset of diseases or complications.
- *Integration of healthcare data*: There are vast healthcare records that are being produced from several resources like the "EHRs," digital gadgets, and additional healthcare gadgets. Integrating this data can provide a more comprehensive view of a patient's health and can help identify patterns and trends that can be used to improve patient outcomes.
- *Explainable AI*: As AI and ML become more prevalent in healthcare, there is a growing need for explainable AI models that can provide insights into how decisions are being made. This can help improve transparency, build trust, and enable healthcare providers to make informed decisions.
- *Real-time monitoring*: Real-time monitoring of patient data can provide healthcare providers with immediate feedback on a patient's condition and allow for early intervention when necessary. This can be particularly useful in critical care settings and can help prevent adverse events.
- *Data security and privacy*: As healthcare data become more valuable, there is a growing need for improved data security and privacy measures. Research is needed to develop secure and privacy-preserving methods for data analytics and cognitive computing in healthcare.
- *Integration of cognitive computing with clinical decision support systems*: The integration of cognitive computing with mechanisms that enhance professional judgment can help healthcare providers make more informed decisions by providing them with appropriate data at the time of support. This enhances the quality of care and decrease errors.

Additional studies are required to discover the various applications of these tools and address the challenges associated with their implementation in healthcare. The future of healthcare is exciting, and data analytics and cognitive computing will play a significant role in shaping it.

8. Conclusion and future scope

In conclusion, this research work proposes a generic methodology that combines DL and NLP for continuous remote monitoring in digital health. The approach has been explained in detail with various DL algorithms, such as CNNs and RNNs, being suggested to analyze medical images and time-series data, and unsupervised learning techniques such as autoencoders being recommended for feature extraction. By applying this approach, health professionals can remotely monitor patients' health conditions in real time, leading to improved patient outcomes, reduced healthcare costs, and enhanced quality of care. This chapter also offers and exhaustive review of the current state of data analytics and cognitive computing in digital health, identifying the challenges and opportunities of implementing them in healthcare, and proposing research directions that can help accelerate their adoption. By analyzing the current trends, limitations of existing works, and research gaps, this chapter offers valuable insights for future research and development in the field of digital health. The proposed approach has several future scopes such as expanding its applicability to other parts of medical domain, like the psychological well-being and chronic disease management, developing a more comprehensive and robust model by incorporating more features and data sources, and integrating it with other technologies such as blockchain for secure and decentralized data sharing. Overall, this research work offers a promising approach for leveraging advanced data analytics and cognitive computing to transform healthcare and improve patient outcomes.

References

[1] S. Vyas, D. Bhargava, Big data analytics and cognitive computing in smart health systems, in: Smart Health Systems, Springer, Singapore, 2021, https://doi.org/10.1007/978-981-16-4201-2-8.

[2] R.K. Behera, P.K. Bala, A. Dhir, The emerging role of cognitive computing in healthcare: a systematic literature review, Int. J. Med. Inform. 129 (2019) 154–166, https://doi.org/10.1016/j.ijmedinf.2019.04.024.

[3] M. Supriya, V.K. Chattu, A review of artificial intelligence, big data, and blockchain technology applications in medicine and global health, Big Data Cogn. Comput. 5 (3) (2021) 41, https://doi.org/10.3390/bdcc5030041.

[4] B.J. Awrahman, C.A. Fatah, M.Y. Hamaamin, A review of the role and challenges of big data in healthcare informatics and analytics, Comput. Intell. Neurosci. 2022 (2022) 5317760, https://doi.org/10.1155/2022/5317760.

[5] T. Shaik, X. Tao, N. Higgins, L. Li, R. Gururajan, X. Zhou, U.R. Acharya, Remote patient monitoring using artificial intelligence: current state, applications, and challenges, Adv. Rev. 13 (2) (2023) e1485.

[6] N. Gautam, S.N. Ghanta, J. Mueller, M. Mansour, Z. Chen, C. Puente, Y.M. Ha, T. Tarun, G. Dhar, K. Sivakumar, Y. Zhang, A.A. Halimeh, U. Nakarmi, S. Al-Kindi, D. DeMazumder, S.J. Al'Aref, Artificial intelligence, wearables and remote monitoring for heart failure: current and future applications, Diagnostics 12 (12) (2022) 2964, https://doi.org/10.3390/diagnostics12122964.

[7] Y. Juhn, H. Liu, Artificial intelligence approaches using natural language processing to advance EHR-based clinical research, J. Allergy Clin. Immunol. 145 (2) (2020) 463–469, https://doi.org/10.1016/j.jaci.2019.12.897.

[8] M. Sahu, R. Gupta, R.K. Ambasta, P. Kumar, Artificial intelligence and machine learning in precision medicine: a paradigm shift in big data analysis, Prog. Mol. Biol. Transl. Sci. 190 (1) (2022) 57–100, https://doi.org/10.1016/bs.pmbts.2022.03.002.

[9] R.K. Behera, P.K. Bala, A. Dhir, The emerging role of cognitive computing in healthcare: a systematic literature review, Int. J. Med. Inform. 129 (2019) 154–166, https://doi.org/10.1016/j.ijmedinf.2019.04.024.

[10] M. Senbekov, T. Saliev, Z. Bukeyeva, A. Almabayeva, M. Zhanaliyeva, N. Aitenova, Y. Toishibekov, I. Fakhradiyev, The recent progress and applications of digital technologies in healthcare: a review, Int. J. Telemed. Appl. 2020 (2020) 8830200, https://doi.org/10.1155/2020/8830200. PMID: 33343657. PMCID: PMC7732404.

[11] A.N. Navaz, M.A. Serhani, H.T. El Kassabi, N. Al-Qirim, I.H. Trends, Technologies, and key challenges in smart and connected healthcare, IEEE Access 9 (2021) 74044–74067, https://doi.org/10.1109/ACCESS.2021.3079217. PMID: 34812394. PMCID: PMC8545204.

[12] S. Khan, H.U. Khan, S. Nazir, Systematic analysis of healthcare big data analytics for efficient care and disease diagnosing, Sci. Rep. 12 (1) (2022) 22377, https://doi.org/10.1038/s41598-022-26090-5. PMID: 36572709. PMCID: PMC9792582.

[13] C. Manning, M. Surdeanu, J. Bauer, J. Finkel, S. Bethard, D. McClosky, The Stanford CoreNLP Natural Language Processing Toolkit, in: Proceedings of 52nd Annual Meeting of the Association for Computational Linguistics: System Demonstrations, Association for Computational Linguistics, Baltimore, MD, 2014, pp. 55–60.

[14] M.M. van Buchem, O.M. Neve, I.M.J. Kant, Analyzing patient experiences using natural language processing: development and validation of the artificial intelligence patient reported experience measure (AI-PREM), BMC Med. Inform. Decis. Mak. 22 (2022) 183, https://doi.org/10.1186/s12911-022-01923-5.

[15] A.G. Sreedevi, T.N. Harshitha, V. Sugumaran, P. Shankar, Application of cognitive computing in healthcare, cybersecurity, big data and IoT: a literature review, Inf. Process. Manag. 59 (2) (2022), https://doi.org/10.1016/j.ipm.2022.102888.

[16] S. Locke, A. Bashall, S. Al-Adely, J. Moore, A. Wilson, G. Kitchen, Natural language processing in medicine: a review, Trends Anaesth. Crit. Care 38 (2021) 4–9.

[17] Z.E. Rasjid, Predictive analytics in healthcare: the use of machine learning for diagnoses, in: International Conference on Electrical, Computer and Energy Technologies (ICECET), Cape Town, South Africa, 2021, pp. 1–6.

[18] M. Javaid, A. Haleem, R.P. Singh, R. Suman, S. Rab, Significance of machine learning in healthcare: features, pillars and applications, Int. J. Intell. Netw. 3 (2022) 58–73, https://doi.org/10.1016/j.ijin.2022.05.002.

[19] T. Igbe, Z. Nie, Y. Alhandarish, L. Wang, Y. Liu, A. Kandwal, Deep learning intervention for health care challenges: some biomedical domain considerations, JMIR mHealth uHealth 7 (8) (2019) e11966.

[20] A. Esteva, K. Chou, S. Yeung, N. Naik, A. Madani, A. Mottaghi, Y. Liu, E. Topol, J. Dean, R. Socher, Deep learning-enabled medical computer vision, npj Digit. Med. 4 (5) (2021), https://doi.org/10.1038/s41746-020-00376-2.

[21] A. Haleem, M. Javaid, R.P. Singh, R. Suman, S. Rab, Blockchain technology applications in healthcare: an overview, Int. J. Intell. Netw. 2 (2021) 130–139, https://doi.org/10.1016/j.ijin.2021.09.005.

[22] M. Alam, I.R. Khan, M.A. Alam, F. Siddiqui, S. Tanweer, IoT framework for healthcare: a review, in: IEEE World Conference on Applied Intelligence and Computing (AIC), Sonbhadra, India, vol., 2022, pp. 925–934.

[23] J. Orji, A. Hernandez, B. Selema, R. Orji, Virtual and augmented reality applications for promoting safety and security: a systematic review, in: IEEE 10th International Conference on Serious Games and Applications for Health (SeGAH), Sydney, Australia, 2022, pp. 1–7.

[24] Y. Kumar, A. Koul, R. Singla, Artificial intelligence in disease diagnosis: a systematic literature review, synthesizing framework and future research agenda, J. Ambient Intell. Humaniz. Comput. 14 (2022) 8459–8486.

[25] K. Itahashi, S. Kondo, T. Kubo, Y. Fujiwara, M. Kato, H. Ichikawa, T. Koyama, R. Tokumasu, J. Xu, C.S. Huettner, V.V. Michelini, L. Parida, T. Kohno, N. Yamamoto, Evaluating clinical genome sequence analysis by Watson for genomics, Front. Med. 5 (2018), https://doi.org/10.3389/fmed.2018.00305.

[26] G. Simon, C.D. DiNardo, K. Takahashi, T. Cascone, C. Powers, R. Stevens, J. Allen, M.B. Antonoff, D. Gomez, P. Keane, S.F. Suarez, Q. Nguyen, E. Roarty, S. Pierce, J. Zhang, B.E. Hardeman, K. Lakhani, K. Shaw, B. Smith, S. Swisher, R. High, P.A. Futreal, J. Heymach, L. Chin, Applying artificial intelligence to address the knowledge gaps in cancer care, Oncologist 24 (6) (2019) 772–782, https://doi.org/10.1634/theoncologist.2018-0257. epub 2018 Nov 16 PMID: 30446581. PMCID: PMC6656515.

[27] S. Marwaha, J.W. Knowles, E.A. Ashley, A guide for the diagnosis of rare and undiagnosed disease: beyond the exome, Genome Med. 14 (2022) 23, https://doi.org/10.1186/s13073-022-01026-w.

[28] D. Nguyen, M.-H. Lin, D. Sher, W. Lu, X. Jia, S. Jiang, Advances in automated treatment planning, Semin. Radiat. Oncol. 32 (4) (2022) 343–350, https://doi.org/10.1016/j.semradonc.2022.06.004.

[29] L. Jana, R.J. Chen, B. Chen, M.Y. Lu, M. Barbieri, D. Shao, A.J. Vaidya, C. Chen, L. Zhuang, D.F.K. Williamson, M. Shaban, T.Y. Chen, F. Mahmood, Artificial intelligence for multimodal data integration in oncology, Cancer Cell 40 (10) (2022) 1095–1110, https://doi.org/10.1016/j.ccell.2022.09.012.

[30] H. Du, Y. He, W. Lu, Y. Han, Q. Wan, Machine learning analysis of immune cells for diagnosis and prognosis of cutaneous melanoma, J. Oncol. 2022 (2022) 7357637, https://doi.org/10.1155/2022/7357637. PMID: 35126517. PMCID: PMC8813285.

[31] R.A. Stocker, Intensive care in traumatic brain injury including multi-modal monitoring and neuro-protection, Med. Sci. 7 (3) (2019) 37, https://doi.org/10.3390/medsci7030037.

[32] N.M. Al-Zidi, M. Tawfik, A.M. Al-Hejri, I. Fathail, T.A. Aldhaheri, Q. Al-Tashi, Smart system for real-time remote patient monitoring based on internet of things, in: 2nd International Conference on Computational Methods in Science & Technology (ICCMST), Mohali, India, 2021, pp. 1–6.

[33] S.M.A. Iqbal, I. Mahgoub, E. Du, Advances in healthcare wearable devices, npj Flex Electron. 5 (2021) 9.

[34] X. Min, B. Yu, F. Wang, Predictive modeling of the hospital readmission risk from patients' claims data using machine learning: a case study on COPD, Sci. Rep. 9 (1) (2019) 2362.

[35] P. Dupenloup, R.L. Pei, A. Chang, M.Z. Gao, P. Prahalad, R. Johari, K. Schulman, A. Addala, D.P. Zaharieva, D.M. Maahs, D. Scheinker, A model to design financially sustainable algorithm-enabled remote patient monitoring for pediatric type 1 diabetes care, Front. Endocrinol. (Lausanne) 13 (2022) 1021982, https://doi.org/10.3389/fendo.2022.1021982. PMID: 36440201. PMCID: PMC9691757.

[36] A. Haleem, M. Javaid, R.P. Singh, R. Suman, Medical 4.0 technologies for healthcare: features, capabilities, and applications, Internet Things Cyber-Phys. Syst. 2 (2022) 12–30, https://doi.org/10.1016/j.iotcps.2022.04.001.

[37] T. Taipalus, V. Isomöttönen, H. Erkkilä, Data analytics in healthcare: a tertiary study, SN Comput. Sci. 4 (2023) 87, https://doi.org/10.1007/s42979-022-01507-0.

[38] B.K.S.P. Kumar Raju Allur, Research challenges and future directions in applying cognitive computing in the healthcare domain, in: Cognitive Intelligence and Big Data in Healthcare, Wiley, 2022.

[39] R. Sabharwal, S.J. Miah, S. Fosso Wamba, Extending artificial intelligence research in the clinical domain: a theoretical perspective, Ann. Oper. Res. (2022), https://doi.org/10.1007/s10479-022-05035-1.

CHAPTER 20

New challenges and opportunities to explainable artificial intelligence (XAI) in smart healthcare

Armin Shoughi[a], Mohammad Bagher Dowlatshahi[a], and Arefeh Amiri[b]
[a]Department of Computer Engineering, Faculty of Engineering, Lorestan University, Khorramabad, Lorestan, Iran
[b]Shahid Madani Hospital, Lorestan University of Medical Sciences, Khorramabad, Lorestan, Iran

1. Introduction

The rapid advancement of artificial intelligence (AI) and machine learning (ML) has opened up a plethora of opportunities to revolutionize healthcare. AI-powered systems are being developed to diagnose diseases and personalize treatment plans, with the potential to improve patient outcomes and reduce healthcare costs. However, the widespread adoption of AI in healthcare has also raised concerns about transparency and explainability. In the realm of healthcare applications, two predominant AI technologies are prominently employed: ML and deep learning (DL). ML algorithms, adept at discerning precise relationships within data, necessitate feature engineering due to the pervasive challenges posed by the "curse of dimensionality." Conversely, DL, a subset of ML, manifests as models featuring a multitude of hidden layers, mandating expansive datasets and high-performance computational infrastructure. Despite the commendable attributes of DL—namely, its versatility, high performance, and capacity for generalization—both ML and DL are commonly perceived as inscrutable "black boxes." Without a clear understanding of how AI systems make decisions, it is difficult to assess their validity, identify potential biases, and ensure that they align with ethical principles and patient preferences. This lack of transparency poses a significant barrier to the responsible and effective implementation of AI in healthcare. A critical objective of AI in medicine is to personalize medical decisions, health practices, and medications for unique people. Nevertheless, the current state of AI in medicine has been described as "strong on promise but rather lacking in evidence and demonstration." To make a precise diagnosis in precision medicine, doctors need substantially more information than a simple binary prediction [1]. Specifically, researchers have noted that the inability to perform a tailored assessment by doctors is a significant limitation of AI in dermatology [2]. There must be evidence that a high-performance DL algorithm truly recognizes the correct region of the image and does not overemphasize insignificant details [3]. In recent years, new methods

Explainable AI in Healthcare Imaging for Medical Diagnoses
https://doi.org/10.1016/B978-0-443-23979-3.00020-8

for visualizing AI models have emerged, including attention maps [4], class activation maps [5], salience maps [6], and occlusion maps [7]. These methods provide insights into the decision-making process of AI models and enable users to understand how specific features contribute to the AI's output. Localization and segmentation techniques are relatively straightforward to grasp due to their visual nature. In contrast, delving into the intricacies of deep neural network (DNN)-based models trained on nonimaging medical data remains a formidable undertaking [8, 9]. These models often deal with complex data structures and relationships, making it difficult to provide clear and concise explanations. To address this exigency, the imperative of elucidating the decision-making processes of AI models becomes apparent. This exigency is met by the conceptual framework of XAI, which delineates a set of attributes expressly designed to explicate the intricate process by which AI models formulate predictions. The integration of XAI not only serves to demystify the predictive mechanisms inherent to ML and DL but also augurs well for enhancing comprehension and engendering a more favorable reception among healthcare practitioners, thereby fostering the judicious integration of these systems within various clinical applications. Also, it is worth mentioning that IAI stands as a complementary field to explainable artificial intelligence (XAI). While XAI primarily focuses on providing explanations for individual AI decisions, IAI takes a broader approach, aiming to make AI systems more comprehensible in general. This comprehensive understanding encompasses the AI system's internal mechanisms, the development of inherently interpretable models, and the seamless integration of XAI techniques into the AI development process. Generally, XAI and interpretable artificial intelligence (IAI) are two emerging fields that aim to make AI systems more understandable and transparent. While both share the common goal of enhancing the interpretability of AI, they differ in their scope and focus. Let us delve into these concepts in more detail. XAI is focused on providing explanations for AI decisions in a way that is understandable to both humans and other AI systems. This can be done by using a variety of techniques, such as feature importance, which identifies the most important features that contributed to a particular decision, or sensitivity analysis, which measures the impact of changing individual features on the decision. IAI is a broader approach that seeks to make AI systems more understandable in general, not just when they are making decisions. This can be done by using techniques such as visualization to represent the internal workings of the AI system or debugging to identify and fix errors in the AI system. XAI, fundamentally, elucidates the rationale behind a decision without delving into the procedural mechanisms leading to that decision. Conversely, IAI is concerned with elucidating how a decision was arrived at, without necessarily explicating the reasonableness of the criteria employed [10]. XAI, therefore, facilitates the explication of learning models, concentrating on the elucidation of why a particular decision was made, and delving into the analytical paradigms that underpin such decisions. Conversely, the interpretability of ML aims at empowering users to grasp the outcomes of learning models by unveiling the rationale governing their decisions [11]. In accordance with extant literature, explainability implies the ability to articulate the findings of an ML model in human-understandable terms. Adadi and Berrada [12] posit that interpretable systems achieve explainability when their processes are intelligible to humans, thereby establishing a close correlation between the concepts of explainability and interpretability.

This chapter delves into the burgeoning landscape of XAI in smart healthcare, exploring its emerging challenges and opportunities. It commences by defining XAI and its significance in the context of healthcare, highlighting its potential to enhance trust, identify biases, and optimize decision-making, then unveils the complexities and intricacies of XAI implementation in healthcare, examining the diverse range of AI models, data sources, and ethical considerations that shape its application. Key challenges are addressed, including the need for standardized evaluation metrics, the development of human-interpretable explanations, and the management of sensitive patient data. In light of these challenges, the chapter identifies promising opportunities for XAI in smart healthcare, including personalized medicine, risk stratification, and clinical trial design. By embracing XAI, healthcare systems can harness the power of AI while maintaining transparency, accountability, and ethical integrity. The remaining sections are organized as follows: Section 2 elaborates on the challenges posed by black-box AI models in healthcare and the importance of XAI in addressing them. It also illustrates the potential applications of XAI in various healthcare domains. Section 3 identifies and discusses emerging challenges in applying XAI effectively in the healthcare context. It addresses the challenges of selecting appropriate XAI methods and techniques for different healthcare applications, the complexities of generating meaningful explanations for complex AI models, and the ethical considerations and potential biases associated with XAI in healthcare. Section 4 highlights the transformative potential of XAI in enhancing the quality and accessibility of healthcare services. It discusses the opportunities to improve patient engagement and empower individuals to make informed healthcare decisions, the potential to address healthcare disparities and promote equitable access to care, and the opportunities for personalized medicine and precision diagnostics enabled by XAI. Section 5 presents real-world case studies showcasing the successful implementation of XAI in various healthcare applications. It demonstrates how XAI has improved the transparency, trust, and accuracy of AI-based healthcare solutions and has had a positive impact on patient care, decision-making, and medical research. Section 6 outlines promising research directions in XAI for smart healthcare. It focuses on novel explainability methods, integration with clinical workflows, and human-AI collaboration. It emphasizes the need for continuous evaluation and validation of XAI techniques in real-world healthcare settings and proposes collaborative efforts among researchers, healthcare professionals, and policymakers to advance XAI in a responsible and ethical manner. Section 7 summarizes the key findings and conclusions of the chapter. It reaffirms the importance of XAI in addressing the challenges and unlocking the opportunities of AI in smart healthcare and emphasizes the need for continued research, collaboration, and ethical considerations to ensure the responsible and beneficial development of XAI in healthcare.

2. The role of XAI in smart healthcare

AI-based algorithms are gaining a major role in healthcare. However, the opaque nature of certain AI models, particularly DNNs, presents a significant hurdle in the adoption of these technologies. The lack of transparency inherent in black-box models can erode trust, hinder explainability, and pose challenges for accountability, particularly in high-stakes healthcare settings. Among the challenges of black-box AI models in healthcare [13, 14]:

- *Limited explainability*: Without access to the internal workings of the model, it is difficult to understand how it arrives at its conclusions. This can hinder trust and adoption among healthcare professionals and patients.
- *Bias and fairness*: Black-box models can inherit biases from the training data, leading to unfair or discriminatory outcomes. This can undermine the ethical principles of healthcare and disproportionately impact vulnerable populations.
- *Debugging and error identification*: Identifying and fixing errors in black-box models is challenging due to the lack of transparency. This can hinder the development of robust and reliable healthcare AI systems.
- *Accountability and liability*: In the event of adverse outcomes, it can be difficult to attribute responsibility to the AI model or its developers. This can create challenges for liability and legal recourse.

Despite the challenges posed by black-box AI models, their potential to improve healthcare outcomes is undeniable. By addressing the explainable, fair, and accountable aspects of black-box AI, researchers, healthcare professionals, and policymakers can harness the power of AI while maintaining trust, equity, and patient safety in the healthcare system. XAI plays a pivotal role in tackling the challenges raised and fostering trust in AI-driven healthcare solutions. Among the most important roles are [15, 16]:

- *Transparency or unveiling the black box*: The opacity of AI models, often referred to as "black boxes," hinders comprehension of how they arrive at their decisions. This lack of transparency breeds skepticism among healthcare providers and patients, leading to reluctance to adopt AI-driven solutions. XAI sheds light on these black boxes, providing insights into the factors that influence AI decisions. By demystifying the inner workings of AI models, XAI fosters trust and enhances the acceptability of AI in healthcare.
- *Bias mitigation or ensuring equitable care*: AI models, trained on vast datasets, may inadvertently inherit biases embedded within the data. These biases can translate into discriminatory outcomes, potentially exacerbating existing societal inequities in healthcare.

XAI, with its ability to identify and analyze potential biases, empowers healthcare professionals to mitigate these biases, ensuring that AI is used to promote equitable and unbiased care.

- *Improved decision-making or informed clinical judgment*: XAI offers a wealth of information that can enrich clinical decision-making. By providing insights into the rationale behind AI predictions and recommendations, XAI can guide clinicians in making informed and nuanced assessments, leading to more personalized and effective patient care. This collaborative approach, where AI serves as an advisor rather than a replacement for human expertise, empowers healthcare providers to leverage AI's capabilities while maintaining their autonomy and judgment.

The realm of XAI is undergoing a remarkable transformation, propelled by the development of innovative tools and frameworks that aim to simplify the selection and application of XAI methodologies. This burgeoning interest is evident in the rising frequency of XAI mentions in scientific literature, as illustrated in Fig. 1. XAI's maiden appearance in a PubMed title, abstract, or keywords occurred in 2018, embedded within a study examining ML applications in neuroscience [17, 18]. Since then, XAI mentions have witnessed an exponential ascent, amassing a total of 488 citations, with over 63% (311) originating from 2022 and the opening quarters of 2023 [19]. This trajectory underscores the burgeoning significance of XAI in both biomedical and healthcare domains. Among the notable XAI tools, DALEX [20] stands out as an open-source software platform that streamlines the process of employing XAI methods. Its user-friendly interface and comprehensive suite of tools empower users to effectively interpret ML models. DALEX's modular design facilitates seamless integration with various modeling frameworks, enabling researchers to readily explore diverse XAI techniques. SHAP (SHapley Additive exPlanations) [21] presents a game-theoretic approach to decomposing the contribution of each feature to a model's prediction. This granular understanding enables users to meticulously examine how individual inputs influence the model's output. SHAP's versatility makes it a valuable tool for various XAI applications. Local interpretable model-agnostic explanations (LIME) [22] stands as another prominent XAI technique that generates local explanations by approximating the model with a simpler surrogate model. This approach preserves model privacy while maintaining interpretability, making it suitable for sensitive data settings. XAIFlow [23] emerges as an end-to-end XAI platform that bridges the gap between model training and XAI. It enables researchers to seamlessly integrate XAI methods into their ML workflows, streamlining the process of gaining insights from ML models. These notable examples, along with numerous other XAI tools, collectively demonstrate the diverse approaches and tools shaping the rapidly evolving landscape of XAI. As the field continues to mature, XAI holds immense promise for enhancing the transparency, reliability, and trust in ML applications across various domains, including healthcare, finance, and environmental protection.

Despite this surge in XAI discourse, the number of manuscripts dedicated to thoroughly examining the concepts and challenges of XAI in healthcare remains relatively limited. This underrepresentation presents an opportunity for researchers and academics to delve into this critical area, addressing the complexities and ethical considerations surrounding the application of XAI in healthcare settings. By fostering a deeper understanding of XAI, we can ensure its responsible and effective integration into clinical practice, maximizing its potential to improve patient outcomes while upholding the principles of transparency and accountability.

3. Emerging challenges of XAI in smart healthcare

As mentioned earlier, XAI is a rapidly growing field that aims to make AI systems more transparent, understandable, and accountable. This is particularly important in healthcare, where AI is being increasingly used to make critical decisions about patient care. However, there are several emerging challenges in applying XAI effectively in the healthcare context. Here are some of the most pressing issues.

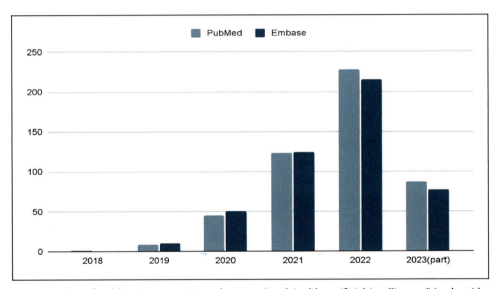

Fig. 1 Number of publications containing the term "explainable artificial intelligence" in the titles, abstracts, and keywords of the PubMed and Embase databases per year. Queries performed on March 26, 2023.

3.1 Data heterogeneity

Data heterogeneity poses a substantial barrier to the effective application of XAI in healthcare settings. The complexity, noise, and heterogeneity of clinical data render it difficult to extract meaningful insights and develop accurate and transparent explanations for AI models. Diverse data formats, data structures, and data quality variations characterize heterogeneous healthcare data. This heterogeneity can impede XAI algorithms from effectively processing and analyzing data, leading to potentially inaccurate or misleading explanations. Several factors contribute to the inherent heterogeneity of healthcare data. First, healthcare data are often collected from a variety of sources, such as electronic health records (EHRs), patient registries, and research studies. Each source may adhere to distinct data formatting, structures, and quality standards, hindering the integration of data from disparate sources. Second, healthcare data are susceptible to noise, attributable to the subjective nature of medical data, the prevalence of errors, and the incompleteness of data sets. Patient-reported data may be inaccurate or incomplete, and laboratory results may be misinterpreted. This noise can obfuscate patterns and trends within data, challenging XAI algorithms to effectively identify meaningful insights. Third, healthcare data exhibit heterogeneity in the type of information recorded. Clinical data, including diagnoses, medications, and procedures, demographic data such as age, gender, and race, and lifestyle data such as smoking status and diet are all incorporated into healthcare datasets. This heterogeneity complicates the development of explanations relevant to specific patients or clinical scenarios by XAI algorithms [24]. In addition to technical challenges, data heterogeneity introduces ethical concerns in the context of XAI in healthcare, which highlights the potential for XAI methods to perpetuate or exacerbate existing healthcare disparities. It is crucial to ensure that XAI methods do not perpetuate or exacerbate these disparities. This necessitates careful assessment of the data sources employed for training and evaluating XAI models, as well as the potential for bias in the data. Despite the challenges, data heterogeneity also presents opportunities for XAI in healthcare. XAI methods can be employed to identify and interpret patterns within heterogeneous data that might not be readily apparent to clinicians. This can lead to the development of novel insights and the creation of more effective interventions.

3.2 Explainability of different AI models

The burgeoning field of AI has witnessed the emergence of a diverse array of AI models, each with its unique characteristics and decision-making processes. This diversity presents a formidable challenge in developing XAI methods capable of effectively explaining the outputs of these models. XAI methods need to be able to explain the outputs of a variety of AI models, including deep learning models, which are becoming increasingly popular in healthcare. Traditional XAI approaches, such as feature importance and sensitivity analysis, were primarily designed to explain the decision-making of simpler models, such as linear models and shallow neural networks. However, these methods are ill-suited for unraveling the intricate decision-making processes of DNNs, which are rapidly gaining traction in healthcare applications. DNNs, with their complex network structures and nonlinear activation functions, often operate as opaque "black boxes," making their decision-making processes difficult to interpret. This lack of transparency raises concerns about the fairness, accuracy, and trustworthiness of AI-powered healthcare systems. The inability to explain the reasoning behind AI-generated recommendations or decisions can erode patient trust and hinder the adoption of AI in healthcare settings. The generalizability and interpretability of XAI methods for diverse AI models are paramount for their successful application in healthcare. However, the challenges highlighted earlier underscore the need for sustained research and development in the area of XAI to ensure that AI-powered healthcare systems are transparent, trustworthy, and aligned with ethical principles. By addressing these challenges, XAI holds the potential to transform healthcare by improving patient outcomes and enhancing the efficiency of clinical care processes [25].

3.3 User-centered explanations

The burgeoning field of XAI holds immense promise for enhancing healthcare outcomes by bridging the gap between complex AI models and human understanding. However, the effectiveness of XAI hinges on its ability to provide user-centered explanations that are tailored to the diverse needs and backgrounds of healthcare professionals. User-centered explanations in XAI go beyond simply providing technical details of the AI model's decision-making process. Rather, they involve crafting explanations that are contextually relevant, engaging, and actionable for the intended audience. This requires a deep understanding of the user's knowledge, expertise, and goals to ensure that the explanations are not overly complex or irrelevant [26]. In the healthcare domain, user-centered explanations are particularly critical due to the diverse range of healthcare professionals who interact with AI-powered tools. Clinicians with varying levels of technical expertise and medical knowledge may require explanations tailored to their specific needs. For instance, a physician with extensive training in a particular medical specialty may require in-depth explanations that delve into the underlying medical concepts and decision-making processes. Conversely, a nurse with less technical expertise may benefit from simpler explanations that highlight the key factors influencing the AI's recommendation and its implications for patient care. so, user-centered explanations are not merely an optional feature but an essential element of trustworthy AI in healthcare [27]. By understanding and catering to the diverse needs of healthcare professionals, XAI can transform from a black box into a transparent and collaborative partner in the healthcare ecosystem. As XAI technology continues to evolve, user-centered explanations will remain a cornerstone of ensuring that AI-powered healthcare systems are aligned with patient-centered care and ethical principles.

3.4 Interpreting multiple causes

The advent of XAI has revolutionized the landscape of healthcare by bridging the gap between complex AI models and human understanding. However, the efficacy of XAI in healthcare settings hinges on its ability to elucidate the intricate interplay between multiple causal factors that contribute to a specific clinical outcome. Conventional XAI approaches often focus on isolating the impact of individual features or factors on the AI's decision-making process. While this approach can provide valuable insights into the influence of specific variables, it falls short of capturing the true complexity of clinical decision-making, which often involves a multitude of interconnected factors. This can lead to incomplete or misleading explanations that fail to reflect the holistic nature of healthcare decisions. To effectively interpret multiple causes in healthcare, XAI methods must go beyond mere feature isolation and delve into the intricate web of causal relationships that shape patient outcomes. This necessitates the development of XAI techniques that can identify and prioritize causal relationships and untangle causal complexity. The ability to effectively interpret multiple causes is a cornerstone of trustworthy and effective XAI in healthcare. By demystifying the intricate interplay between multiple causal factors, XAI can transform from a black box into a transparent and collaborative partner in the healthcare ecosystem. Empowered with insights into the causal network of patient outcomes, XAI can support healthcare professionals in making informed decisions that improve patient outcomes and enhance the quality of care. As XAI technology continues to evolve, the ability to interpret multiple causes will remain a critical factor in ensuring that AI-powered healthcare systems are aligned with patient-centered care and ethical principles.

3.5 Integrating XAI into existing healthcare systems

The integration of XAI into existing healthcare systems represents a critical challenge that must be addressed to facilitate the widespread adoption and successful implementation of XAI in healthcare. Healthcare systems, often characterized by their intricate legacy architecture and entrenched workflows, present a formidable obstacle to seamlessly integrating new technologies without disrupting existing operations. To effectively bridge the gap between XAI and existing healthcare systems, several crucial factors must be considered:

- *Data interoperability*: Ensuring seamless data exchange and integration is paramount, given the fragmented nature of healthcare data often scattered across disparate systems and formats. Standardized data formats and interoperable data exchange protocols are essential to enable XAI tools to access and comprehend the relevant data for generating meaningful explanations.
- *Workflow integration*: XAI tools must be integrated seamlessly into existing healthcare workflows without disrupting patient care or clinical processes. This may involve modifying existing systems or developing new interfaces that facilitate the integration of XAI explanations without compromising the efficiency of clinical operations.
- *User-centered design*: XAI tools must be designed with the specific needs and preferences of healthcare providers in mind. This includes providing clear, concise explanations tailored to the user's level of expertise and using intuitive interfaces that are easy to navigate. Visuals, interactive elements, and user-friendly explanations can enhance the accessibility and usability of XAI tools for diverse user groups.

- *Training and education*: Healthcare providers must be provided with adequate training and education to effectively utilize XAI tools and interpret the provided explanations. This may involve in-person training sessions, online tutorials, or interactive simulations that familiarize clinicians with the capabilities and nuances of XAI.
- *Regulatory compliance*: XAI tools must adhere to relevant healthcare regulations and data privacy standards to protect patient confidentiality and ensure responsible use of patient data. This includes complying with data protection frameworks, such as HIPAA and GDPR, and establishing robust data governance processes to safeguard sensitive information.
- *Ethical considerations*: The development and use of XAI tools must be guided by ethical principles, such as minimizing bias, ensuring fairness, and upholding patient autonomy. This involves incorporating ethical considerations into the design and implementation of XAI systems, ensuring that AI-generated decisions do not perpetuate discriminatory practices or infringe upon patient rights.
- *Continuous improvement*: XAI systems should undergo continuous evaluation and improvement based on user feedback and real-world usage data. This iterative process can ensure that XAI tools remain effective, user-friendly, and aligned with the evolving needs of the healthcare community. Healthcare organizations should foster an environment of open communication and feedback to gather insights from clinicians and continuously refine the XAI tools they adopt [3].

XAI has emerged as a critical component of responsible AI development, particularly in healthcare where trust and transparency are paramount. As XAI research continues to mature, we can expect to see XAI methods used to improve patient care, reduce healthcare costs, and address healthcare disparities. Despite the challenges mentioned earlier, XAI has the potential to revolutionize healthcare by making AI systems more transparent, understandable, and accountable. By addressing these challenges, healthcare organizations can effectively integrate XAI into existing systems, unlocking the potential of XAI to enhance healthcare decision-making, improve patient outcomes, and promote patient-centered care. As XAI technology continues to evolve, healthcare systems must adapt and integrate XAI seamlessly to maintain a competitive edge and deliver the best possible care to patients. In conclusion, integrating XAI into existing healthcare systems is not merely a technical challenge; it requires a multifaceted approach that encompasses technological advancements, user-centered design, ethical considerations, and continuous improvement. By bridging the gap between XAI and the complex ecosystem of healthcare, we can harness the power of XAI to transform healthcare delivery and empower clinicians to make informed decisions that enhance patient well-being. As XAI technology continues to mature, healthcare organizations must be proactive in adopting XAI solutions tailored to their specific needs and aligned with the ethical principles underpinning healthcare delivery.

While XAI holds immense promise for enhancing healthcare decision-making, improving patient outcomes, and promoting patient-centered care, selecting the most appropriate XAI methods for specific healthcare applications is a crucial challenge. Choosing the appropriate XAI method for a specific healthcare application requires careful consideration of several factors [23]:

- *Data type and complexity*: The nature and complexity of the data being analyzed significantly influence the suitability of XAI methods. For instance, image-based data, such as medical images, may necessitate XAI methods that can interpret visual patterns and features, while text-based data, such as patient records, may require methods that can analyze linguistic structures and semantic relationships.

- *Explainability goals*: The intended purpose of the explanations plays a pivotal role in guiding the selection of XAI methods. For example, if the goal is to identify biases or unfair treatment, methods that focus on model predictions and their underlying data patterns are more suitable. Conversely, if the goal is to understand the decision-making rationale behind a particular recommendation, methods that provide insights into individual features and their contribution to the overall prediction are more appropriate.

- *User group*: Tailoring explanations to the specific needs and expertise of the user group is crucial for effective communication and ensuring that explanations are meaningful and actionable. For clinicians, explanations should focus on clinical relevance and potential implications for patient care. For patients, explanations should be presented clearly and understandably, using nontechnical language and avoiding complex mathematical concepts. Generally, healthcare professionals, patients, or policymakers may have different levels of technical expertise and require different types of explanations.

- *Level of detail*: The level of detail in explanations provided by XAI methods should be tailored to the specific needs and expertise of the user group. Clinicians, who have a deeper understanding of medical concepts and data, may require more detailed explanations that delve into the underlying rationale of the AI model's predictions or recommendations. Conversely, patients may need simpler explanations that focus on the clinical implications of the AI's decision-making, avoiding overly technical jargon and complex mathematical concepts.

- *Communication goals*: The communication goals of XAI explanations should be clearly defined to ensure that explanations are effective and serve the intended purpose. For clinicians, explanation goals may include understanding the model's reasoning process, identifying potential biases or unfair treatment, and interpreting the clinical implications of the AI's recommendations. For patients, explanation goals may focus on understanding their diagnosis or treatment plan, making informed decisions about their care, and building trust in the AI-driven decision-making process.

- *Balancing level of detail and communication goals*: Balancing the level of detail and communication goals of XAI explanations is crucial for effective communication and ensuring that explanations are meaningful and actionable. Too much detail can overwhelm the user, making it difficult to grasp the key takeaways from the explanation. Conversely, too little detail may not provide sufficient information for the user to understand or make informed decisions. Achieving this balance requires a thoughtful approach to explanation design and tailoring the level of detail to the specific needs of the user group. For clinicians, explanations can be more detailed and technical, while for patients, explanations should be presented clearly and concisely, using nontechnical language and visual aids when appropriate.

- *Computational efficiency*: The computational cost of XAI methods can significantly impact their feasibility and deployment in real-world healthcare settings. XAI methods should ideally be computationally efficient to ensure they can be integrated seamlessly into existing AI systems without compromising performance or system responsiveness. This is particularly important for applications that require real-time or near real-time decision-making, such as medical imaging diagnosis or patient risk stratification.

- *Explainability quality*: The quality of explanations provided by XAI methods is crucial for their effectiveness and trustworthiness. Explanations should be accurate in reflecting the underlying decision-making process of the AI model. They should also be complete, addressing all relevant factors that influenced the model's prediction or recommendation. Additionally, explanations should be relevant to the specific needs and context of the user group. For clinicians, explanations should be clinically meaningful and provide insights into potential implications for patient care. For patients, explanations should be presented in a clear and understandable manner, avoiding complex technical jargon and mathematical concepts.

- *Interpretability*: Interpretability refers to the ease with which users can understand and make sense of the explanations provided by XAI methods. Explanations should be presented in a format that is accessible to the intended user group, using nontechnical language and visual aids when appropriate. Complex mathematical notation or overly technical jargon should be avoided, as these can hinder comprehension and hinder effective communication. Simplicity and clarity are paramount when conveying explanations in the context of healthcare, where decision-making often involves balancing technical expertise with patient understanding.

- *Fairness and bias assessment*: Fairness and bias assessment is essential to ensure that XAI methods do not perpetuate or amplify existing societal disparities in healthcare. XAI methods should be evaluated for potential biases in their treatment of different patient groups or demographics. This may involve analyzing the distribution of explanations across different groups and identifying any patterns that suggest unfair treatment. Additionally, XAI methods should be designed to avoid reinforcing existing societal biases that may be reflected in the training data.

- *Ethical adherence*: The development and deployment of XAI methods must adhere to established ethical principles in healthcare, such as patient confidentiality, data privacy, and informed consent. XAI methods should not compromise patient privacy by revealing sensitive information or by enabling unauthorized access to patient data. Data privacy safeguards should be in place to ensure that patient information is protected throughout the XAI process. Additionally, informed consent should be obtained from patients prior to the application of XAI methods, ensuring that they are aware of the potential implications and have a choice in whether or not to receive XAI-generated explanations.

The development of a comprehensive and well-defined framework for selecting appropriate XAI methods in healthcare is essential to ensure that AI-driven decision-making in healthcare is transparent, accountable, and aligned with ethical principles. This framework should consider the multifaceted nature of XAI, encompassing considerations related to data type, explainability goals, user group, computational efficiency, explainability quality, interpretability, fairness, bias assessment, and ethical adherence. By adopting a systematic and evidence-based approach to XAI method selection, healthcare organizations can harness the power of XAI to revolutionize healthcare delivery while maintaining patient trust and ensuring responsible AI development. Therefore, the selection of appropriate XAI methods and techniques is a critical aspect of ensuring the trustworthiness and effectiveness of AI-driven healthcare solutions and by carefully considering the abovementioned factors and leveraging emerging XAI tools and frameworks, healthcare professionals can make informed decisions about XAI implementation, ensuring that AI is used to improve patient care while upholding ethical and responsible AI practices.

4. Opportunities unveiled by XAI in smart healthcare

By making complex AI models more transparent and understandable, XAI empowers healthcare professionals, patients, and policymakers to make informed decisions, improve patient outcomes, and address healthcare disparities. XAI can be instrumental in improving diagnostic accuracy and risk assessment by providing insights into the reasoning behind AI-generated diagnoses and risk predictions. This allows healthcare professionals to better understand the underlying medical rationale, enhancing their confidence in AI-assisted decision-making. For instance, XAI can help radiologists interpret complex medical images, identifying subtle abnormalities that might otherwise go undetected [28]. Similarly, XAI can be used to analyze genomic data [29], enabling oncologists to make more personalized treatment decisions for cancer patients [30]. Furthermore, XAI facilitates the development of personalized treatment plans and drug recommendations by tailoring AI-generated insights to individual patient characteristics and preferences. This can optimize treatment outcomes and reduce the risk of adverse drug reactions [31]. For example, XAI can analyze patient data to identify genetic predispositions to certain diseases, enabling healthcare providers to prescribe preventive measures or personalized treatment regimens [32]. Additionally, XAI can be used to evaluate the effectiveness of different drug combinations, optimizing treatment regimens for individual patients [33]. In addition, XAI can foster better patient engagement and informed consent by providing patients with clear and understandable explanations of AI-generated recommendations. This can enhance patient trust and acceptance of AI-powered healthcare solutions [34]. For instance, XAI can be used to generate visualizations that explain the rationale behind AI-generated treatment plans [35]. Similarly, it can be used to create personalized decision-support tools that help patients understand the benefits and risks of different treatment options [36]. XAI also plays a pivotal role in addressing healthcare disparities by identifying and mitigating biases in AI models. This can ensure that AI-powered healthcare decisions are fair and equitable for all patients. For example, XAI can be used to monitor AI models for potential bias, detecting and correcting unfair or discriminatory patterns. Additionally, XAI can be used to identify and address sources of bias in the data used to train AI models, ensuring that AI-powered decisions are not influenced by factors such as race, ethnicity, or socioeconomic status. In the subsequent sections, a comprehensive analysis will be undertaken concerning the following thematic elements: precision diagnostics, the cultivation of patient engagement, the nuanced exploration of informed consent, monitoring for bias in AI models, identifying bias in training data, and fostering inclusive healthcare environments. The ensuing discussion aims to explicate and elucidate these facets with a focus on navigating the judicious and responsible implementation thereof.

4.1 Precision diagnostics

In the realm of personalized medicine, XAI plays a pivotal role in unlocking a new era of precision-driven treatment. By elucidating the factors that influence AI-generated recommendations, XAI empowers healthcare providers to tailor treatment plans to each patient's unique characteristics, medical history, and genetic predispositions. XAI's ability to analyze patient data, including genomic information, medical records, and lifestyle factors, enables healthcare professionals to identify potential drug interactions, optimize drug dosages, and predict treatment responses. This personalized approach leads to more effective and safer treatments, minimizing adverse effects and maximizing patient outcomes. For instance, XAI can analyze genetic information to tailor cancer treatment plans, ensuring that chemotherapy regimens are optimized for individual patient profiles. Additionally, XAI can monitor patient response to treatment and adjust medication dosages accordingly, ensuring that patients receive the most effective and personalized pharmacotherapy [37]. By carefully analyzing these images, XAI can detect subtle abnormalities and diagnoses that may be overlooked by human experts. This enhanced ability to detect early-stage diseases can enable timely intervention and improved patient outcomes. For example, XAI can analyze mammogram images to detect subtle lesions that may indicate the presence of breast cancer, enabling early diagnosis and treatment, which can significantly improve patient survival rates. Similarly, XAI can analyze retinal images to identify signs of diabetic retinopathy or glaucoma, enabling timely diagnosis and intervention to prevent vision loss [38]. Additionally, XAI can analyze MRI scans to identify patterns associated with neurodegenerative diseases, such as Alzheimer's disease or Parkinson's disease. This information can be used to diagnose these diseases earlier and develop more effective treatments.

4.2 Fostering patient engagement and informed consent

XAI's transparency also empowers patients to actively participate in their healthcare decisions. By providing clear and understandable explanations of AI-generated recommendations, XAI fosters patient engagement, informed consent, and trust in healthcare providers. This transparency helps patients make informed choices about their treatment and diagnostic options, ultimately leading to better patient outcomes. For instance, XAI can explain the rationale behind proposed treatment plans or diagnostic insights, allowing patients to understand the basis for the recommendations and make informed decisions about their healthcare. This transparency can also enhance informed consent processes, ensuring that patients are fully aware of the risks, benefits, and potential outcomes of proposed treatments or diagnostic procedures [39]. To effectively leverage XAI for patient engagement, healthcare providers should consider the following strategies: transparency, patients should be informed about the use of AI in their care and how it will affect their treatment. This includes disclosing the capabilities and limitations of AI models, as well as the potential benefits and risks associated with their use. Explainability, patients should be provided with explanations of how AI models make decisions about their care. This can be done through a variety of methods, such as providing patients with access to XAI tools or developing user-friendly interfaces that explain AI models in plain language. Active involvement, patients should be actively involved in the decision-making process when AI is used in their care. This includes providing patients with the opportunity to ask questions, raise concerns, and provide input on their treatment plans.

4.3 Monitoring for bias in AI models

XAI techniques can be employed to continuously monitor AI models for potential biases. This involves analyzing model predictions and identifying patterns that suggest unfair treatment based on factors such as race, ethnicity, or socioeconomic status [40]. By detecting these biases early on, healthcare providers can take corrective measures to ensure that AI models are not perpetuating discriminatory practices.

4.4 Identifying bias in training data

A significant source of bias in AI models can be the training data used to develop them. XAI techniques can help identify and address biases embedded in training data by examining the distribution of features and ensuring that the data accurately reflects the diverse patient population. This process involves removing or reweighting biased data points to minimize their impact on model predictions.

4.5 Fostering inclusive healthcare environments

By addressing healthcare disparities and promoting equitable treatment, XAI can contribute to the creation of more inclusive healthcare environments where all patients have equal access to quality care. It can help dismantle barriers to care and ensure that AI is used to enhance, rather than exacerbate, existing inequalities.

So, XAI's transformative potential for personalized medicine and precision diagnostics holds immense promise to revolutionize healthcare, offering a personalized and data-driven approach to treatment and diagnostics. By harnessing XAI's capabilities, healthcare professionals can tailor care to individual patient needs, leading to better patient outcomes, improved patient satisfaction, and reduced healthcare costs.

5. Case studies of XAI in smart healthcare

Real-world case studies have demonstrated the transformative potential of XAI in various healthcare applications. From skin lesion analysis to heart failure risk prediction and drug discovery, XAI techniques are enhancing the capabilities of AI models, empowering healthcare professionals to diagnose, treat, and develop new therapies with greater precision and effectiveness. This part delves into real-world case studies, showcasing how XAI is revolutionizing healthcare by bridging the gap between AI's predictive power and human understanding. We explore how XAI techniques are illuminating the decision-making processes of AI models, enabling healthcare professionals to gain deeper insights into the rationale behind AI recommendations. Through these examples, we will illustrate the profound impact of XAI on enhancing trust, collaboration, and decision-making in healthcare, paving the way for a future where AI seamlessly integrates into the healthcare ecosystem to improve patient outcomes.

5.1 Skin lesion analysis

Skin cancer is a common and often deadly form of cancer. Early detection and diagnosis of skin cancer can improve patient survival rates. AI models have been developed to automatically detect and classify skin lesions, with promising results. However, these models often lack interpretability, making it difficult for dermatologists to understand how they make decisions and trust their recommendations. To address this issue, researchers have developed XAI techniques to explain the decisions of skin lesion analysis models. In the field of dermatology, XAI is playing a crucial role in enhancing the accuracy and trustworthiness of AI-powered skin lesion analysis systems. For instance, Hauser et al. [41] demonstrated how XAI techniques, such as LIME and SHAP, were employed to explain the decision-making processes of an AI model for classifying skin cancer lesions. This approach enabled dermatologists to gain deeper insights into the model's reasoning, leading to improved trust and confidence in its recommendations. Metta et al. [42] explored the use of XAI to visualize the saliency of features in skin lesion images. This technique allowed dermatologists to identify the specific regions of an image that was most influential in the model's classification, providing valuable information for further diagnosis and treatment planning.

5.2 Heart failure risk prediction

Heart failure is a prevalent and debilitating condition that affects millions of individuals worldwide. Early detection and risk stratification are crucial for timely interventions and improved patient outcomes. In recent years, XAI techniques have emerged as promising tools for enhancing the interpretability and trustworthiness of AI-powered heart failure risk prediction models. Wrazen et al. [43] introduce a pipeline that explains black-box AI models. It uses the DeepSHAP algorithm to provide global and local explanations for the model's predictions. The pipeline is validated by consulting with medical experts and reviewing the literature. This approach enabled clinicians to gain insights into the factors most significantly contributing to an individual's risk assessment, guiding the development of personalized preventive measures and early intervention strategies. Moreno-Sanchez [44] evaluated the impact of XAI on the clinical adoption of an AI model for heart failure risk prediction. The study found that XAI-enhanced models were associated with higher clinician acceptance and usage, indicating the importance of interpretability in fostering trust in AI-based decision support tools. Similarly, Dave et al. [45] employed a combination of LIME and SHAP to explain another XGBoost model designed for identifying the presence of heart diseases. Jia et al. introduced counterfactual explanations to identify specific modifications to input data that would alter the model's prediction. Specifically, they focused on predicting readiness for weaning from mechanical support using a convolutional neural network developed on the MIMIC-III dataset [46]. However, these studies lacked validation by medical experts or integration with existing medical literature to establish their clinical relevance. In contrast, Pawar et al. proposed a theoretical approach to explain medical models in conjunction with clinical knowledge. Their framework aimed to enhance model transparency and enable the tracking of model performance [47]. However, they did not provide concrete examples or real-world applications of their proposed approach. Wang et al. presented a theory-driven framework for designing human-centered XAI, drawing upon philosophical and psychological principles. Their framework emphasized the importance of considering human cognition and understanding in the design of XAI tools [48]. This approach has the potential to address the limitations of previous XAI efforts by ensuring that the explanations generated are meaningful and actionable for healthcare professionals. The use of XAI in heart failure risk prediction holds immense potential for improving patient outcomes. By providing healthcare professionals with clear explanations for AI predictions, XAI can empower them to make informed decisions about patient management, ultimately leading to earlier diagnosis, more effective treatment, and reduced mortality.

5.3 Drug discovery

Drug discovery is a time-consuming and resource-intensive process. The application of AI models has emerged as a promising approach to expedite drug discovery by identifying potential drug targets and developing novel therapeutic agents. However, many AI models used in drug discovery are black-box models, hindering comprehension of their internal workings and interpretation of their outcomes. This opaqueness poses several challenges, including difficulty in understanding the rationale behind model recommendations, identifying the most influential factors driving predictions, and effectively debugging and enhancing the models. The development of AI-powered drug discovery tools is further complicated by domain-specific issues, such as the data format employed by these techniques. The choice of chemical "representation model" dictates the amount of chemical information preserved, including pharmacophores, physicochemical characteristics, and functional groups, ultimately influencing the extent of interpretability and performance of the resulting AI model [49]. Moreover, the limitations of our current understanding of molecular pathology restrict our ability to establish mathematical models of drug action and provide definitive explanations.

In this context, XAI presents a promising avenue for augmenting human creativity and capabilities in the design of novel bioactive compounds with desired properties [50]. XAI techniques enable mechanistic interpretation of drug mechanisms [51, 52], contribute to enhanced drug safety assessments, and assist in the planning of organic syntheses [53]. If fully realized, XAI will prove invaluable in processing and comprehending complex chemical data, fostering the generation of novel therapeutic hypotheses. In recent years, feature attribution approaches including (gradient-based, perturbation-based, and surrogate models) have become the most widely employed XAI family of tools for ligand- and structure-based drug development. McCloskey et al. [54] used gradient-based attribution to identify ligand pharmacophores necessary for affinity. Pope et al. [55] used gradient-based feature attribution to identify key functional groupings for adverse impact prediction [56]. SHAP was recently utilized to assess important aspects for the prediction of chemical strength and activities of combination therapy [57, 58].

The positive impact of XAI on patient care, decision-making, and medical research is undeniable. By providing healthcare professionals with clear explanations for AI predictions,

XAI is fostering trust, collaboration, and informed decision-making. This has led to improved patient outcomes, reduced healthcare costs, and accelerated the pace of medical innovation. The lessons learned from these case studies hold profound implications for future XAI development. First, it is of paramount importance to prioritize the development of XAI techniques that are not only transparent but also actionable and user-friendly. Second, the integration of XAI into healthcare systems must be carefully considered, ensuring that it aligns with existing clinical workflows and regulatory frameworks. As XAI continues to evolve, its transformative impact on healthcare is poised to grow even further. By bridging the chasm between predictive power and human understanding, XAI holds the key to unlocking the full potential of AI in healthcare, paving the way for a future where AI seamlessly integrates into the healthcare ecosystem to improve patient outcomes and revolutionize the way we approach medicine.

6. Future directions and research agenda

The emergence of XAI marks a pivotal moment in healthcare, ushering in a transformative solution to overcome the constraints posed by opaque AI models. This development not only addresses the longstanding challenge of comprehensibility but also establishes a foundation for trust and collaboration between human professionals and AI tools. As XAI evolves and matures, its far-reaching potential stands poised to revolutionize the entire healthcare landscape. Beyond merely elucidating the decision-making processes of AI models, XAI is positioned to reshape the very fabric of healthcare delivery. Its maturation holds the promise of not only demystifying the intricate workings of black-box algorithms but also fundamentally improving patient outcomes. This evolution in AI transparency has the potential to empower healthcare professionals with nuanced insights into diagnostics, treatment plans, and therapeutic strategies. Consequently, it sets the stage for a paradigm shift in healthcare, where the synergy between human expertise and AI capabilities leads to more precise, efficient, and personalized patient care. In this era of healthcare transformation, the maturation of XAI serves as a beacon of progress, with implications extending far beyond current limitations. The ongoing evolution of XAI not only promises a brighter future for healthcare but also invites exploration into novel frontiers, where the harmonious interplay between human intuition and machine intelligence redefines the possibilities of patient-centered care. The transformative potential of XAI in healthcare is undeniable. By bridging the trust gap between humans and AI, XAI has the power to revolutionize patient care, drug discovery, and healthcare operations. However, to fully realize the potential of XAI, substantial research and development are needed to address the challenges and opportunities that lie ahead. Several key research directions need to be explored.

6.1 Novel explainability methods

Current XAI techniques often provide superficial explanations that merely highlight the most influential features in a given prediction. However, such explanations fall short of illuminating the intricacies of decision-making process, particularly for complex tasks or when multiple models interact. To make XAI effective, it is essential to develop new and more effective XAI techniques that go beyond superficial explanations and provide deeper insights into the decision-making processes of AI models. This includes techniques that can explain the reasoning behind complex predictions, identify potential biases, and account for the impact of individual features [35]. In particular, there is a need for more research on developing explainability methods that are not only accurate but also understandable and trustable by humans. This includes exploring new visualization techniques, using natural language explanations, and considering user cognitive biases [8, 16]. These approaches can help humans better understand the rationale behind AI decisions, build trust in AI systems, and ultimately make more informed decisions.

6.2 Elevating explanatory techniques

To truly harness the potential of XAI, researchers must transcend mere feature importance metrics and delve into the underlying logic of AI models. This necessitates the development of advanced explainability techniques that can dissect the interplay between features, identify potential biases, and provide comprehensive insights into the model's decision-making process. One promising approach is to harness the power of interactive visualizations to elucidate the decision-making process of AI models clearly and intuitively. These visualizations can enable clinicians to grasp the intricate relationships between features and predictions, fostering a deeper understanding of the model's reasoning. In addition, researchers should prioritize the development of natural language explanations that provide insights into AI models in a way that is accessible and comprehensible to healthcare professionals without requiring deep technical expertise. These explanations can translate the intricate workings of AI models into a language easily comprehended by clinicians, fostering trust and adoption. Furthermore, researchers should focus on addressing the explainability of AI models that utilize multimodal data, such as images, videos, and text, which are commonplace in healthcare settings. This entails developing techniques that can effectively explain the decision-making process for these complex datasets, enabling clinicians to make informed decisions based on AI insights.

6.3 Seamless integration into clinical workflows

The adoption and effective utilization of XAI tools in healthcare depend on their seamless integration into existing clinical workflows. This integration is crucial for several reasons: first, clinicians must be able to easily access and use XAI tools without requiring extensive technical expertise. This necessitates user-friendly interfaces that are intuitive and adaptable to the varying levels of technical proficiency among healthcare professionals. Second, explanations generated by XAI tools must be tailored to the specific needs of clinicians, providing granular insights that align with their clinical decision-making processes. This contextual relevance ensures that explanations are meaningful and actionable for clinicians. Third, integrating XAI tools directly into EHRs and other healthcare systems streamline their adoption and ensure that clinicians have immediate access to AI-generated insights at their fingertips. This integration seamlessly integrates AI into the clinical workflow, fostering natural adoption. Some of the benefits of seamless integration are clinicians gain a deeper understanding of AI models and their decision-making processes, fostering trust and acceptance. Also, AI-generated insights can inform clinical decisions, leading to more accurate diagnoses, personalized treatment plans, and improved patient outcomes. XAI tools can automate repetitive tasks, freeing up clinicians' time to focus on patient care and collaboration. Also, XAI can uncover patterns and trends in healthcare data that may be overlooked by clinicians, leading to discoveries and advancements. XAI can empower patients to make informed decisions about their healthcare, fostering shared decision-making and improving patient engagement.

6.4 Cocreation of human-AI synergy

XAI is not about replacing human expertise with AI. Instead, it is about creating a powerful collaboration between humans and AI, where each complements the other's strengths. AI can provide vast amounts of data-driven insights and predictions, but it is ultimately up to human clinicians to interpret and apply these insights in the context of a patient's unique situation. Human clinicians bring years of experience, clinical judgment, and empathy to the table. They understand the nuances of patient care, the complexities of medical conditions, and the ethical considerations involved in decision-making. AI, on the other hand, can analyze vast amounts of data and identify patterns and correlations that may be overlooked by human clinicians. When human expertise and AI insights come together, the result is a powerful synergy that can lead to more informed and nuanced decisions. Clinicians can use AI-generated insights to validate their clinical judgment, identify areas for further investigation, and explore new treatment options. AI, in turn, can learn from human feedback and interactions, improving its ability to provide accurate and relevant predictions. XAI can also be used to foster shared decision-making between clinicians and patients. By providing patients with insights into AI-generated predictions, clinicians can empower them to make informed choices about their healthcare. This approach can lead to better patient engagement, adherence to treatment, and overall health outcomes.

The transformative potential of XAI extends far beyond the applications mentioned earlier. To ensure the reliability and impact of XAI techniques, rigorous evaluation, and validation in real-world healthcare settings are crucial. This involves conducting empirical studies to assess the accuracy, interpretability, and generalizability of XAI explanations across diverse patient populations and clinical scenarios. Additionally, gathering feedback from healthcare professionals on the usability and effectiveness of XAI tools is essential to refine their design and functionality. Advancing XAI responsibly and ethically requires collaboration among researchers, healthcare professionals, and policymakers. Researchers should prioritize the development of XAI techniques that are transparent, fair, and unbiased, ensuring that AI models do not perpetuate or amplify existing biases in healthcare. Additionally, safeguarding patient privacy and data security is paramount, and XAI tools should adhere to rigorous ethical guidelines.

By embracing these research directions and fostering a culture of collaboration, we can establish a strong foundation for XAI to become an integral part of the healthcare ecosystem. As XAI matures, it has the potential to drive innovation, improve patient outcomes, and enhance the delivery of healthcare worldwide, transforming the way we approach disease prevention, diagnosis, and treatment.

7. Conclusion

As we conclude this exploration of XAI in smart healthcare, a powerful message resonates: XAI transcends mere technology, acting as a transformative force poised to reshape the healthcare landscape. This chapter delved into the intricate dance between the burgeoning adoption of AI in healthcare and the critical need for explainability. We meticulously navigated the challenges that impede progress, from the opaque nature of black-box models to the complexities of crafting meaningful explanations for healthcare's unique data landscape. Yet, amidst these hurdles, we unearthed a treasure trove of opportunities XAI presents. The key findings echo a resounding truth: XAI is not an option, but an imperative. It serves as the cornerstone for unlocking the true potential of AI in healthcare, fostering trust and transparency where it matters most. By demystifying AI's decision-making processes, XAI empowers both clinicians and patients, enabling informed consent, shared understanding, and ultimately, better healthcare outcomes. Its applications weave across the healthcare tapestry, from diagnostics and treatment recommendations to drug discovery and patient engagement, each domain enriched by the illuminating power of XAI. However, the path forward is not without its thorns. Selecting the most suitable XAI methods for diverse healthcare scenarios demands meticulous consideration. Crafting explanations that resonate with the nuances of clinical practice necessitates innovative approaches tailored to the complexities of healthcare data. Ethical considerations loom large, necessitating careful attention to potential biases and the implications for patient privacy. Addressing these challenges effectively necessitates a concerted symphony of collaboration, a harmonious interplay between researchers, healthcare professionals, and policymakers. The future of XAI in smart healthcare pulsates with boundless potential. Novel explainability methods must be explored, and their ability to cater to the specific needs of healthcare data and complex AI models rigorously evaluated. Seamless integration with clinical workflows is paramount, ensuring that XAI's insights seamlessly flow into real-world practice. Fostering human-AI collaboration stands as a cornerstone, for it is in the harmonious interplay between human expertise and AI's analytical prowess that the true potential of XAI unfolds. As we move forward, relentless evaluation and validation must guide our steps. Real-world healthcare settings serve as the proving ground where the effectiveness and trustworthiness of XAI techniques are demonstrably validated. But the journey does not end there. Responsible and ethical development of XAI necessitates proactive collaboration that transcends disciplinary boundaries. Researchers must join hands with healthcare professionals, policymakers, and the broader public to address potential biases, navigate the complexities of regulatory frameworks, and ensure that AI's advancements serve the greater good. In conclusion, XAI stands at the precipice of a new era in healthcare, poised to usher in a future where data-driven insights illuminate clinical decision-making, empower patients, and ultimately, improve lives. By embracing the challenges, fostering collaboration, and prioritizing ethical considerations, we can collectively chart a course toward a future where XAI serves as a powerful force for good, transforming healthcare for the benefit of all. Remember, the journey has just begun, and each step we take, guided by the principles of explainability, collaboration, and ethical responsibility, paves the way for a healthier, more empowered future for generations to come.

References

[1] B. Mahbooba, M. Timilsina, R. Sahal, M. Serrano, Explainable artificial intelligence (XAI) to enhance trust management in intrusion detection systems using decision tree model, Complexity 2021 (2021) 1–11.
[2] A. Gomolin, E. Netchiporouk, R. Gniadecki, I.V. Litvinov, Artificial intelligence applications in dermatology: where do we stand? Front. Med. 7 (2020) 100.
[3] A. Shoughi, M.B. Dowlatshahi, A. Amiri, M. Kuchaki Rafsanjani, R.S. Batth, Automatic ECG classification using discrete wavelet transform and one-dimensional convolutional neural network, Computing 106 (2024) 1227–1248.
[4] D.C. Somers, S.L. Sheremata, Attention maps in the brain, Wiley Interdiscip. Rev. Cogn. Sci. 4 (4) (2013) 327–340.
[5] B.N. Patro, M. Lunayach, S. Patel, V.P. Namboodiri, U-CAM: visual explanation using uncertainty based class activation maps, in: Proceedings of the IEEE/CVF International Conference on Computer Vision, 2019, pp. 7444–7453.
[6] C. Roggeman, W. Fias, T. Verguts, Salience maps in parietal cortex: imaging and computational modeling, Neuroimage 52 (3) (2010) 1005–1014.
[7] M. Aminu, N.A. Ahmad, M.H.M. Noor, COVID-19 detection via deep neural network and occlusion sensitivity maps, Alex. Eng. J. 60 (5) (2021) 4829–4855.
[8] M.T. Ribeiro, S. Singh, C. Guestrin, "Why should I trust you?" Explaining the predictions of any classifier, in: Proceedings of the 22nd ACM SIGKDD International Conference on Knowledge Discovery and Data Mining, 2016, pp. 1135–1144.
[9] A. Shoughi, M.B. Dowlatshahi, A practical system based on CNN-BLSTM network for accurate classification of ECG heartbeats of MIT-BIH imbalanced dataset, in: 2021 26th International Computer Conference, Computer Society of Iran (CSICC), IEEE, 2021, pp. 1–6.
[10] V. Vishwarupe, P.M. Joshi, N. Mathias, S. Maheshwari, S. Mhaisalkar, V. Pawar, Explainable AI and interpretable machine learning: a case study in perspective, Procedia Comput. Sci. 204 (2022) 869–876.
[11] T. Miller, Explanation in artificial intelligence: insights from the social sciences, Artif. Intell. 267 (2019) 1–38.
[12] A. Adadi, M. Berrada, Peeking inside the black-box: a survey on explainable artificial intelligence (XAI), IEEE Access 6 (2018) 52138–52160.

[13] V. Hassija, V. Chamola, A. Mahapatra, A. Singal, D. Goel, K. Huang, S. Scardapane, I. Spinelli, M. Mahmud, A. Hussain, Interpreting black-box models: a review on explainable artificial intelligence, Cogn. Comput. 16 (1) (2024) 45–74.

[14] C. Manresa-Yee, M.F. Roig-Maimó, S. Ramis, R. Mas-Sansó, Advances in XAI: explanation interfaces in healthcare, in: Handbook of Artificial Intelligence in Healthcare: Vol. 2: Practicalities and Prospects, Springer, 2021, pp. 357–369.

[15] H.W. Loh, C.P. Ooi, S. Seoni, P.D. Barua, F. Molinari, U.R. Acharya, Application of explainable artificial intelligence for healthcare: a systematic review of the last decade (2011–2022), Comput. Methods Programs Biomed. 226 (2022) 107161.

[16] A. Rai, Explainable AI: from black box to glass box, J. Acad. Mark. Sci. 48 (2020) 137–141.

[17] M.-A.T. Vu, T. Adalı, D. Ba, G. Buzsáki, D. Carlson, K. Heller, C. Liston, C. Rudin, V.S. Sohal, A.S. Widge, A shared vision for machine learning in neuroscience, J. Neurosci. 38 (7) (2018) 1601–1607.

[18] T. Hulsen, Explainable artificial intelligence (XAI): concepts and challenges in healthcare. AI 4 (3), 652–666, https://doi.org/10.3390/ai4030034.

[19] S. Bharati, M.R.H. Mondal, P. Podder, A review on explainable artificial intelligence for healthcare: why, how, and when? IEEE Trans. Artif. Intell. 5 (4) (2024) 1429–1442, https://doi.org/10.1109/TAI.2023.3266418.

[20] H. Baniecki, W. Kretowicz, P. Piä, J. Wiĺ, Dalex: responsible machine learning with interactive explainability and fairness in Python, J. Mach. Learn. Res. 22 (214) (2021) 1–7.

[21] S.M. Lundberg, S.-I. Lee, A unified approach to interpreting model predictions, in: Advances in Neural Information Processing Systems, vol. 30, 2017.

[22] M.T. Ribeiro, S. Singh, C. Guestrin, "Why should I trust you?" Explaining the predictions of any classifier, in: Proceedings of the 22nd ACM SIGKDD International Conference on Knowledge Discovery and Data Mining, 2016, pp. 1135–1144.

[23] Z.C. Lipton, The mythos of model interpretability: in machine learning, the concept of interpretability is both important and slippery, Queue 16 (3) (2018) 31–57.

[24] D. Minh, H.X. Wang, Y.F. Li, T.N. Nguyen, Explainable artificial intelligence: a comprehensive review, Artif. Intell. Rev. 55 (5) (2024) 3503–3568.

[25] M.T. Ribeiro, S. Singh, C. Guestrin, "Why should I trust you?" Explaining the predictions of any classifier, in: Proceedings of the 22nd ACM SIGKDD International Conference on Knowledge Discovery and Data Mining, 2016, pp. 1135–1144.

[26] M. Ribera, À. Lapedriza García, Can we do better explanations? A proposal of user-centered explainable AI, in: CEUR Workshop Proceedings, 2019.

[27] D. Dave, H. Naik, S. Singhal, P. Patel, Explainable AI meets healthcare: a study on heart disease dataset, arXiv preprint arXiv:2011.03195 (2020).

[28] E. Neri, G. Aghakhanyan, M. Zerunian, N. Gandolfo, R. Grassi, V. Miele, A. Giovagnoni, A. Laghi, SIRM Expert Group on Artificial Intelligence, Explainable AI in radiology: a white paper of the Italian Society of Medical and Interventional Radiology, La Radiol. Med. 128 (2023) 755–764.

[29] J.B. Awotunde, E.A. Adeniyi, G.J. Ajamu, G.B. Balogun, F.A. Taofeek-Ibrahim, Explainable artificial intelligence in genomic sequence for healthcare systems prediction, in: Connected e-Health: Integrated IoT and Cloud Computing, Springer, 2022, pp. 417–437.

[30] M. Pocevičiūtė, G. Eilertsen, C. Lundström, Survey of XAI in digital pathology, in: Artificial Intelligence and Machine Learning for Digital Pathology: State-of-the-Art and Future Challenges, Springer, 2020, pp. 56–88.

[31] J.B. Awotunde, A.L. Imoize, A.E. Adeniyi, K.M. Abiodun, E.F. Ayo, K.V.N. Kavitha, G.J. Ajamu, R. O. Ogundokun, Explainable machine learning (XML) for multimedia-based healthcare systems: opportunities, challenges, ethical and future prospects, in: Explainable Machine Learning for Multimedia Based Healthcare Applications, Springer, 2023, pp. 21–46.

[32] D.S. Watson, Interpretable machine learning for genomics, Hum. Genet. 141 (9) (2022) 1499–1513.

[33] K.K. Kırboğa, S. Abbasi, E.U. Küçüksille, Explainability and white box in drug discovery, Chem. Biol. Drug Des. 102 (2023) 217–233.

[34] W. Saeed, C. Omlin, Explainable AI (XAI): a systematic meta-survey of current challenges and future opportunities, Knowl.-Based Syst. 263 (2023) 110273.

[35] G. Alicioglu, B. Sun, A survey of visual analytics for explainable artificial intelligence methods, Comput. Graph. 102 (2022) 502–520.

[36] A. Rehman, A. Farrakh, U.F. Mushtaq, Improving clinical decision support systems: explainable AI for enhanced disease prediction in healthcare, Int. J. Comput. Innov. Sci. 2 (2) (2023) 9–23.

[37] N.J. Schork, Artificial intelligence and personalized medicine, in: Precision Medicine in Cancer Therapy, Springer, 2019, pp. 265–283.

[38] R. Ghnemat, S. Alodibat, Q. Abu Al-Haija, Explainable artificial intelligence (XAI) for deep learning based medical imaging classification, J. Imaging 9 (9) (2023) 177.

[39] M. Dalvi-Esfahani, M. Mosharaf-Dehkordi, L.W. Leong, T. Ramayah, A.M.J. Kanaan-Jebna, Exploring the drivers of XAI-enhanced clinical decision support systems adoption: insights from a stimulus-organism-response perspective, Technol. Forecast. Soc. Change 195 (2023) 122768.

[40] M. Jeyaraman, S. Balaji, N. Jeyaraman, S. Yadav, Unraveling the ethical enigma: artificial intelligence in healthcare, Cureus 15 (8) (2023) 2168–8184.

[41] K. Hauser, A. Kurz, S. Haggenmueller, R.C. Maron, C. von Kalle, J.S. Utikal, F. Meier, S. Hobelsberger, F.F. Gellrich, M. Sergon, Explainable artificial intelligence in skin cancer recognition: a systematic review, Eur. J. Cancer 167 (2022) 54–69.

[42] C. Metta, A. Beretta, R. Guidotti, Y. Yin, P. Gallinari, S. Rinzivillo, F. Giannotti, Explainable deep image classifiers for skin lesion diagnosis, arXiv preprint arXiv:2111.11863 (2021).

[43] W. Wrazen, K. Gontarska, F. Grzelka, A. Polze, Explainable AI for medical event prediction for heart failure patients, in: J.M. Juarez, M. Marcos, G. Stiglic, A. Tucker (Eds.), Artificial Intelligence in Medicine, Springer, Cham, 2023, pp. 97–107, ISBN: 978-3-031-34344-5.

[44] P.A. Moreno-Sanchez, Improvement of a prediction model for heart failure survival through explainable artificial intelligence, arXiv preprint arXiv:2108.10717 (2021).

[45] D. Dave, H. Naik, S. Singhal, P. Patel, Explainable AI meets healthcare: a study on heart disease dataset, arXiv preprint arXiv:2011.03195 (2020).

[46] Y. Jia, J. McDermid, I. Habli, Enhancing the value of counterfactual explanations for deep learning, in: International Conference on Artificial Intelligence in Medicine, Springer, 2021, pp. 389–394.

[47] U. Pawar, D. O'Shea, S. Rea, R. O'Reilly, Explainable AI in healthcare, in: 2020 International Conference on Cyber Situational Awareness, Data Analytics and Assessment (CyberSA), 2020, pp. 1–2, https://doi.org/10.1109/CyberSA49311.2020.9139655.

[48] D. Wang, Q. Yang, A. Abdul, B.Y. Lim, Designing theory-driven user-centric explainable AI, in: Proceedings of the 2019 CHI Conference on Human Factors in Computing Systems, 2019, pp. 1–15.

[49] P.F. Bendassolli, Theory building in qualitative research: reconsidering the problem of induction, in: Forum Qualitative Sozialforschung Forum: Qualitative Social Research, vol. 14, 2013.

[50] P. Schneider, G. Schneider, De novo design at the edge of chaos: miniperspective, J. Med. Chem. 59 (9) (2016) 4077–4086.

[51] Y. Xu, J. Pei, L. Lai, Deep learning based regression and multiclass models for acute oral toxicity prediction with automatic chemical feature extraction, J. Chem. Inf. Model. 57 (11) (2017) 2672–2685.

[52] H.L. Ciallella, H. Zhu, Advancing computational toxicology in the big data era by artificial intelligence: data-driven and mechanism-driven modeling for chemical toxicity, Chem. Res. Toxicol. 32 (4) (2019) 536–547.

[53] S. Dey, H. Luo, A. Fokoue, J. Hu, P. Zhang, Predicting adverse drug reactions through interpretable deep learning framework, BMC Bioinform. 19 (21) (2018) 1–13.

[54] K. McCloskey, A. Taly, F. Monti, M.P. Brenner, L.J. Colwell, Using attribution to decode binding mechanism in neural network models for chemistry, Proc. Natl. Acad. Sci. 116 (24) (2019) 11624–11629.

[55] P.E. Pope, S. Kolouri, M. Rostami, C.E. Martin, H. Hoffmann, Explainability methods for graph convolutional neural networks, in: Proceedings of the IEEE/CVF Conference on Computer Vision and Pattern Recognition, 2019, pp. 10772–10781.

[56] R.R. Tice, C.P. Austin, R.J. Kavlock, J.R. Bucher, Improving the human hazard characterization of chemicals: a Tox21 update, Environ. Health Perspect. 121 (7) (2013) 756–765.

[57] S.M. Lundberg, S.-I. Lee, A unified approach to interpreting model predictions, Adv. Neural Inf. Process. Syst. 30 (2017).

[58] R. Rodríguez-Pérez, J. Bajorath, Interpretation of compound activity predictions from complex machine learning models using local approximations and Shapley values, J. Med. Chem. 63 (16) (2019) 8761–8777.

Index

Note: Page numbers followed by *f* indicate figures, *t* indicate tables, and *b* indicate boxes.